AF811803

Springer Tracts in Mechanical Engineering

Series Editors

Seung-Bok Choi, College of Engineering, Inha University, Incheon, Korea (Republic of)

Haibin Duan, Beijing University of Aeronautics and Astronautics, Beijing, China

Yili Fu, Harbin Institute of Technology, Harbin, China

Carlos Guardiola, CMT-Motores Termicos, Polytechnic University of Valencia, Valencia, Spain

Jian-Qiao Sun, University of California, Merced, CA, USA

Young W. Kwon, Naval Postgraduate School, Monterey, CA, USA

Francisco Cavas-Martínez, Departamento de Estructuras, Universidad Politécnica de Cartagena, Cartagena, Murcia, Spain

Fakher Chaari, National School of Engineers of Sfax, Sfax, Tunisia

Springer Tracts in Mechanical Engineering (STME) publishes the latest developments in Mechanical Engineering - quickly, informally and with high quality. The intent is to cover all the main branches of mechanical engineering, both theoretical and applied, including:

- Engineering Design
- Machinery and Machine Elements
- Mechanical Structures and Stress Analysis
- Automotive Engineering
- Engine Technology
- Aerospace Technology and Astronautics
- Nanotechnology and Microengineering
- Control, Robotics, Mechatronics
- MEMS
- Theoretical and Applied Mechanics
- Dynamical Systems, Control
- Fluids Mechanics
- Engineering Thermodynamics, Heat and Mass Transfer
- Manufacturing
- Precision Engineering, Instrumentation, Measurement
- Materials Engineering
- Tribology and Surface Technology

Within the scope of the series are monographs, professional books or graduate textbooks, edited volumes as well as outstanding PhD theses and books purposely devoted to support education in mechanical engineering at graduate and post-graduate levels.

Indexed by SCOPUS, zbMATH, SCImago.

Please check our Lecture Notes in Mechanical Engineering at http://www.springer.com/series/11236 if you are interested in conference proceedings.

To submit a proposal or for further inquiries, please contact the Springer Editor **in your region**:

Ms. Ella Zhang (China)
Email: ella.zhang@springernature.com
Priya Vyas (India)
Email: priya.vyas@springer.com
Dr. Leontina Di Cecco (All other countries)
Email: leontina.dicecco@springer.com

All books published in the series are submitted for consideration in Web of Science.

More information about this series at http://www.springer.com/series/11693

Esmaeal Ghavanloo · S. Ahmad Fazelzadeh ·
Francesco Marotti de Sciarra
Editors

Size-Dependent Continuum Mechanics Approaches

Theory and Applications

Springer

Editors
Esmaeal Ghavanloo
School of Mechanical Engineering
Shiraz University
Shiraz, Iran

S. Ahmad Fazelzadeh
School of Mechanical Engineering
Shiraz University
Shiraz, Iran

Francesco Marotti de Sciarra
Department of Structures for Engineering
and Architecture
University of Naples Federico II
Naples, Italy

ISSN 2195-9862 ISSN 2195-9870 (electronic)
Springer Tracts in Mechanical Engineering
ISBN 978-3-030-63049-2 ISBN 978-3-030-63050-8 (eBook)
https://doi.org/10.1007/978-3-030-63050-8

© The Editor(s) (if applicable) and The Author(s), under exclusive license to Springer Nature Switzerland AG 2021
This work is subject to copyright. All rights are solely and exclusively licensed by the Publisher, whether the whole or part of the material is concerned, specifically the rights of translation, reprinting, reuse of illustrations, recitation, broadcasting, reproduction on microfilms or in any other physical way, and transmission or information storage and retrieval, electronic adaptation, computer software, or by similar or dissimilar methodology now known or hereafter developed.
The use of general descriptive names, registered names, trademarks, service marks, etc. in this publication does not imply, even in the absence of a specific statement, that such names are exempt from the relevant protective laws and regulations and therefore free for general use.
The publisher, the authors and the editors are safe to assume that the advice and information in this book are believed to be true and accurate at the date of publication. Neither the publisher nor the authors or the editors give a warranty, expressed or implied, with respect to the material contained herein or for any errors or omissions that may have been made. The publisher remains neutral with regard to jurisdictional claims in published maps and institutional affiliations.

This Springer imprint is published by the registered company Springer Nature Switzerland AG
The registered company address is: Gewerbestrasse 11, 6330 Cham, Switzerland

To my wife, Nazanin, and my son, Alireza, whose valuable support and patience have enabled me to complete this project, and to my parents for their encouragement, and finally to my mentor Prof. Hashem Rafii-Tabar, who gave me the opportunity to learn from him and work alongside him.

Esmaeal Ghavanloo

To my wife, Zahra, and my children, Ali and Yeganeh, for their patience and unwavering support.

S. Ahmad Fazelzadeh

To Angela for her unwavering support, patience, and understanding

Francesco Marotti de Sciarra

Foreword

The present book "Size-Dependent Continuum Mechanics Approaches: Theory and Application" is a collection of papers edited by Esmaeal Ghavanloo, S. Ahmad Fazelzadeh and Francesco Marotti de Sciarra. It covers a lot of interesting information about continuum mechanics approaches taking into account size-dependent effects. Approaches present in the book are mostly based on nonlocal theories, but also on micromorphic, peridynamic and gradient theories. Instead of the classical continuum mechanics with taking into account the material point and its infinitesimal surrounding, the alternative theories include also far-distance (beyond the infinitesimal surrounding) information. The majority of presentations in this book are related to contributions in this field of Ahmed Cemal Eringen (see, for example, W.H. Müller, "Eringen, Ahmed Cemal", In H. Altenbach, A. Öchsner (eds), *Encyclopedia of Continuum Mechanics*, Springer, 2020, 860–862).

It should be noted that the non-classical continuum mechanics approaches are much more complicated in comparison with the classical ones. Many examples are presented in the book for beams and one can follow the discussion in a simple manner having only basic knowledge on the Euler-Bernoulli beam theory. In addition, several applications are given and one can see that non-classical approaches are helpful if the structural size is very small. In this case, size effects are obvious and classical theories failed. The advanced theories allow the solution of new problems, but they have also disadvantages: the number of constitutive parameters increases. The estimation of the constitutive parameters is not trivial and several suggestions are discussed in the literature.

The book contains 15 papers prepared by leading scientists in size-dependent theories. In the first chapter "Lattice-Based Nonlocal Elastic Structural Models" in the sense of Lagrange, Hencky and Eringen are discussed and compared. In the following chapter, "Eringen's Nonlocal Integral Elasticity and Applications for Structural Models" are presented. The focus of applications is on carbon nanotubes. The third chapter is devoted to "Nonlocal Mechanics in the Framework of the General Nonlocal Theory". Here the focus is on the general nonlocal theory and it is shown that the strain gradient theory can capture the same phenomena. In the next chapter "Displacement Based Nonlocal Models for Size Effect Simulation in Nanomechanics" are presented. The main results of such theory and the differences with other nonlocal models are described. In the fifth chapter "One-Dimensional Well-Posed Nonlocal Elasticity Models for Finite Domains" are introduced. It is shown that some paradoxical results disappear by using the suggested models. In chapter "Iterative Nonlocal Residual Elasticity", a new approach is presented allowing to avoid some complications of finding solutions of Eringen's nonlocal model. The seventh chapter deals with "Nonlocal Gradient Mechanics of Elastic Beams Under Torsion". The theory is applied to nano-electro-mechanical systems. The eighth chapter presents the "Reformulation of the Boundary Value Problems of Nonlocal Type Elasticity: Application to Beams". In the ninth chapter "Application of Combined Nonlocal and Surface Elasticity Theories to Vibration Response of a Graded Nanobeam" is discussed. It is well known that the correct modeling of small-size structures can be performed if the surface energy is taken into account. Special numerical techniques are necessary for nonlocal theories. In the tenth chapter, "Finite Element Nonlocal Integral Elasticity Approach" is introduced. In the focus of chapter "'Explicit' and 'Implicit' Non-local Continuum Descriptions: Plate with Circular Hole" is a model for materials with internal material organization when the internal and external length scales are of the same order. The twelfth chapter presents "Micromorphic Continuum Theory: Finite Element Analysis of 3D Elasticity with Applications in Beam- and Plate-Type Structures". A special 3D micromorphic element with 12 degrees of freedom (3 classical, 9 non-classical) is developed. "Peridynamic modeling of laminated composites" (which is in the focus of chapter "Peridynamic Modeling of Laminated Composites") presents a new approach to failure analysis. Chapter "Nonlocal Approaches to the Dynamics of Metamaterials" is devoted to a new class of materials with outstanding properties. In the last Chapter "Gradient Extension of Classical Material Models: From Nuclear & Condensed Matter Scales to Earth & Cosmological Scales" the extension of gradient theories to greater sizes is given.

All papers of the present book demonstrate that the modelling of small-size structures cannot be realized using the classical continuum approach. With respect to the scale effects, non-local theories can be applied, but this is not the only one possibility.

Magdeburg, Germany
September 2020

Holm Altenbach

Preface

Classical continuum mechanics has been widely utilized to solve fundamental problems in various fields of engineering and many aspects of physics. Despite its many successes, the classical continuum mechanics cannot always predict well experimentally-observed phenomena in both natural and man-made materials. For example, it fails to describe physical phenomena in which the long-range interactions play a major role. Furthermore, the mathematical modeling of matter via the classical continuum mechanics ignores the fact that the material is made of atoms and so it cannot be directly applied to study the discrete nature of the matter. In addition, the classical continuum mechanics is invariant with respect to length scale and cannot predict the size-dependency of mechanical properties of microstructures and nanostructures.

These limitations have motivated the development of various size-dependent continuum mechanics approaches including micromorphic, micropolar, nonlocal, and high-order strain/rotation gradient mechanics. The conception of these approaches was based on the query, *Is it possible to construct continuum-based approaches that can predict physical phenomena on the micro- or nano-scales?*

The starting point of the development of size-dependent continuum approaches was the monograph of the Cosserat brothers in 1909. Their work was forgotten over half a century since it was ahead of their time. After 1955, the Cosserat theory was extended by several research groups and many advances and criticisms were made since then. In two recent decades, these approaches have been recognized to be practical for mathematical representations of the physical world. In addition, several modifications and improvements of the size-dependent continuum mechanics approaches have been proposed, and their applications to describe the mechanics of various types of advanced materials and structures have been discussed.

The book presents a series of independent chapters written by scientists with worldwide expertise and international reputation in various fields of continuum and computational mechanics, as well as material science. In this book, recent advancements of size-dependent continuum mechanics approaches and their applications to describe the material behavior on different scales have been integrated. One main feature of this book is its in-depth discussions of the vast and rapidly expanding research works pertinent to the nonlocal continuum mechanics. By compiling different approaches into one book, a unique perspective is provided on the current state of the size-dependent continuum mechanics approaches and what the future holds. It is hoped that the reader will find this book a useful resource as he/she progresses in their study and research.

This book contains fifteen chapters by thirty-five researchers which are from Australia, China, France, Greece, India, Iran, Italy, Qatar, Tunisia, Turkey, UAE, and USA. Chapter "Lattice-Based Nonlocal Elastic Structural Models" describes the lattice-based nonlocal approach and presents some applications of the proposed approach for some lattice structural systems including axial lattices, beam lattices and plate lattices. The contemporary advances in Eringen's nonlocal elasticity theory with an emphasis on solving structural engineering problems are discussed in chapter "Eringen's Nonlocal Integral Elasticity and Applications for Structural Models". In chapter "Nonlocal Mechanics in the Framework of the General Nonlocal Theory", general nonlocal theory is introduced and it is shown that the theory can be reduced to the strain gradient theory and the couple stress theory. Chapter "Displacement Based Nonlocal Models for Size Effect Simulation in Nanomechanics" considers the displacement based nonlocal models which belong to the mechanically based nonlocality. A well-posed nonlocal differential model for finite domains is developed in chapter "One-Dimensional Well-Posed Nonlocal Elasticity Models for Finite Domains" and its applicability to predict the static behavior of nanorods and nanobeams is investigated. Motivated by the existing complications of finding solutions of Eringen's nonlocal model, iterative nonlocal residual elasticity is presented in chapter "Iterative Nonlocal Residual Elasticity". The nonlocal gradient elasticity theory of inflected nanobeams is extended in chapter "Nonlocal Gradient Mechanics of Elastic Beams Under Torsion" to the mechanics of elastic nanobeams under torsion. Chapter "Reformulation of the Boundary Value Problems of Nonlocal Type Elasticity: Application to Beams" is dedicated to the reformulation of the boundary value problems of nonlocal type elasticity. Application of the combined nonlocal and surface effects on the free and forced vibration response of a graded nanobeam is investigated in Chap. "Application of Combined Nonlocal and Surface Elasticity Theories to Vibration Response of a Graded Nanobeam".

In chapter "Finite Element Nonlocal Integral Elasticity Approach", a nonlocal finite element method is developed to study the bending, buckling, and vibration behavior of nanostructures. Chapter "'Explicit' and 'Implicit' Non-local Continuum Descriptions: Plate with Circular Hole" is focused on the correspondence between "implicit" type Cosserat (micropolar) and "explicit" type Eringen's two-phase local/nonlocal models, in terms of characteristic quantities. To investigate the mechanical behavior of small-scale structures, a new 12-DOF three-dimensional size-dependent micromorphic element is introduced in chapter "Micromorphic Continuum Theory: Finite Element Analysis of 3D Elasticity with Applications in Beam- and Plate-Type Structures". Chapter "Peridynamic Modeling of Laminated Composites" presents peridynamic modeling approaches namely bond-based, ordinary state-based, and peridynamic differential operator for predicting progressive damage in fiber-reinforced composite materials under general loading conditions. Chapter "Nonlocal Approaches to the Dynamics of Metamaterials" provides a concise review of nonlocal theories as applied to metamaterials, with special consideration given to vibrations and dynamics. Finally, in Chapter "Gradient Extension of Classical Material Models: From Nuclear & Condensed Matter Scales to Earth & Cosmological Scales", a concise review of gradient models (across scales, materials, and processes) is provided based on the internal length gradient approach.

The editors would like to thank all the contributing authors for their participation and cooperation, in spite of their busy work schedules during Covid-19 pandemic, which made this book possible. In addition, we wholeheartedly thank the anonymous reviewers for their carefully performed job and also the team of Springer, especially Dr. Leontina Di Cecco, for their excellent cooperation during the preparation of this edited book. Furthermore, we would like to thank Prof. Holm Altenbach for writing a foreword to the book. Finally, it should be noted that the completion of this book would not have been possible without the help, support and understanding of our families.

Shiraz, Iran — Esmaeal Ghavanloo
Shiraz, Iran — S. Ahmad Fazelzadeh
Naples, Italy — Francesco Marotti de Sciarra
September 2020

Contents

Editors and Contributors

About the Editors

Esmaeal Ghavanloo is currently Associate Professor of Mechanical Engineering at Shiraz University, Iran. He received his Ph.D. in Mechanical Engineering from Shiraz University, in 2013. His research interests include size-dependent continuum theories, nonlocal mechanics, mechanics of nanomaterials, nanomechanical systems, nanocomposites, and metamaterials. He has published more than 75 scientific articles in international peer-reviewed journals. Dr. Ghavanloo is an Associate Editor of "Micro & Nano Letters" and a member of the editorial boards of "Recent Patents on Nanotechnology", "Journal of Nanomaterials, Nanoengineering and Nanosystems" and "SN Applied Sciences". He is also an active reviewer of several scientific and international journals. His book "Computational Continuum Mechanics of Nanoscopic Structures: Nonlocal Elasticity Approaches" has been published by the Springer International Publishing in 2019.

S. Ahmad Fazelzadeh is currently the Distinguished Professor of Mechanical Engineering at Shiraz University, Iran. He was a Visiting Research Professor at the Zienkiewicz Centre for Computational Engineering, Swansea University, UK, and Department of Aerospace Science and Technology at Politecnico di Milano, Italy, in 2018 and 2020, respectively. He is an active researcher and dedicated teacher and has received numerous awards and recognitions for excellence in teaching and research. In the 50th anniversary celebration of Sharif University of Technology, "he was chosen as one of its fifty distinguished alumni in 2016." He is a member of the Iranian Society of Mechanical Engineering and was elected as a distinguished professor in 2018. His research and teaching span over 27 years and his expertise in the fields of Aeroelasticity, Electric Aircraft Design, Structural Dynamics and Stability, Nano Composites, Metamaterials, and Reliability Analysis. To his credit thus far he has two books, seven book chapters, over 120 technical papers in prestigious journals, 135 papers in the international conference proceedings, and one national Patent. He serves on the Editorial Boards of Applied Mathematical

Modelling (Elsevier), Journal of Mechanics of Advanced Composite Structures, Iranian Journal of Science and Technology, Transaction of Mechanical Engineering (Springer) and Sharif Journal of Mechanical Engineering. He also served as a reviewer for high prestige Journals.

Francesco Marotti de Sciarra is currently Full Professor of Solid and Structural Mechanics at the Department of Structures for Engineering and Architecture, University of Naples Federico II, Italy. His research and teaching have been going on for over 25 years and his publications cover several distinct areas of mechanics of nanostructures, functionally graded material and nonlocal elasticity. He is author of book chapters, four books, over 90 technical papers in prestigious journals, and 110 papers in the international conference proceedings. He is a member of the Italian Association of Theoretical and Applied Mechanics. He is a reviewer of several scientific and international journals.

Contributors

Elias C. Aifantis School of Engineering, Aristotle University of Thessaloniki, Thessaloniki, Greece;
Michigan Technological University, Houghton, MI, USA;
Mercator Fellow Friedrich-Alexander University, Erlangen-Nüremberg, Fürth, Germany

Gioacchino Alotta Department of Civil, Energy, Environment and Materials Engineering (DICEAM), Università Mediterranea di Reggio Calabria, Reggio Calabria, Italy

Reza Ansari University of Guilan, Rasht, Iran

Raffaele Barretta Department of Structures for Engineering and Architecture, University of Naples Federico II, Naples, Italy

Noël Challamel Université Bretagne Sud IRDL (CNRS UMR 6027), Centre de Recherche, Lorient Cedex, France

Constantinos Chr. Koutsoumaris Department of Civil Infrastructure and Environmental Engineering, Khalifa University of Science and Technology, Abu Dhabi, United Arab Emirates

Mehmet Dorduncu Department of Mechanical Engineering, Erciyes University, Kayseri, Turkey

Sami El-Borgi Mechanical Engineering Program, Texas A&M University at Qatar, Doha, Qatar

Isaac Elishakoff Department of Ocean and Mechanical Engineering, Florida Atlantic University, Boca Raton, FL, USA

Konstantinos G. Eptaimeros Institute of Applied and Computational Mathematics, Foundation for Research and Technology, Heraklion, Crete, Greece

S. Ali Faghidian Department of Mechanical Engineering, Science and Research Branch, Islamic Azad University, Tehran, Iran

Giuseppe Failla Department of Civil, Energy, Environmental and Materials Engineering, University of Reggio Calabria, Reggio Calabria, Italy

S. Ahmad Fazelzadeh School of Mechanical Engineering, Shiraz University, Shiraz, Iran;
Visiting Professor in Department of Aerospace Science and Technology Politecnico di Milano, Milan, Italy

Seyyed Amir Mahdi Ghannadpour New Technologies and Engineering Department, Shahid Beheshti University, Tehran, Iran

Esmaeal Ghavanloo School of Mechanical Engineering, Shiraz University, Shiraz, Iran

Mesut Kirca Istanbul Technical University, Istanbul, Turkey

Lorenzo Leonetti University of Calabria, Rende, Italy

Erdogan Madenci Department of Aerospace and Mechanical Engineering, University of Arizona, Tucson, AZ, USA

Mohammad Ali Maneshi School of Mechanical Engineering, Shiraz University, Shiraz, Iran

Francesco Marotti de Sciarra Department of Structures for Engineering and Architecture, University of Naples Federico II, Naples, Italy

Maysam Naghinejad Aerospace Engineering Department, Amirkabir University of Technology, Tehran, Iran

Amir Norouzzadeh University of Guilan, Rasht, Iran

Hamid Reza Ovesy Aerospace Engineering Department, Amirkabir University of Technology, Tehran, Iran

Francesco P. Pinnola Department of Structures for Engineering and Architecture, University of Naples Federico II, Naples, Italy

Prakash Rajendran Department of Mechanical Engineering, National Institute of Technology, Tiruchirappalli, Tamil Nadu, India

Hessam Rouhi University of Guilan, Rasht, Iran

Mohamed Shaat Mechanical Engineering Department, Abu Dhabi University, Al Ain, United Arab Emirates

Mohsen Taghizadeh Mechanical Engineering Department, Hakim Sabzevari University, Sabzevar, Iran

Mohamed Trabelssi Applied Mechanics and Systems Research Laboratory, Tunisia Polytechnic School, University of Carthage, La Marsa, Tunisia;
Department of Mechanical Engineering, Tunis Higher National Engineering School, University of Tunis, Tunis, Tunisia

Patrizia Trovalusci Sapienza University of Rome, Rome, Italy

Meral Tuna Yasar University, Izmir, Turkey

Marzia S. Vaccaro Department of Structures for Engineering and Architecture, University of Naples Federico II, Naples, Italy

Chien Ming Wang School of Civil Engineering, The University of Queensland, Brisbane, Australia

Xiao-Jian Xu School of Highway Chang'an University, Xi'an, People's Republic of China

Hong Zhang School of Mechatronical Engineering, Beijing Institute of Technology, Beijing, China

Lattice-Based Nonlocal Elastic Structural Models

Noël Challamel, Chien Ming Wang, Hong Zhang, and Isaac Elishakoff

Abstract This chapter is devoted to lattice-based nonlocal approaches in relation with elastic microstructured elements. Nonlocal continuous approaches are shown to be relevant for capturing length scale effects in discrete structural mechanics models. The chapter contains three complementary parts. In the first part, axial lattices, as already studied by Lagrange during the XVIIIth century are investigated, both for statics and dynamics problems. This discrete model is also called the Born-Kármán lattice model with direct neighbouring interactions. Exact solutions are presented for general boundary conditions. A nonlocal elastic rod model is then constructed from the lattice difference equations. The nonlocal model is similar to the nonlocal model proposed by Eringen in 1983 that is based on a stress gradient approach, although the small length scale of the nonlocal model may differ from statics to dynamics applications. This part is closed with a discussion on generalized lattices with direct and indirect neighbouring interactions and their possible nonlocal modelling. The second part of this study deals with lattice beam elements called Hencky-Bar-Chain models, due to the fact that the discrete beam model was introduced by Hencky in 1920. Exact solutions are presented for general boundary conditions in both statics and dynamics settings. A nonlocal elastic Euler-Bernoulli beam model is then developed from the lattice difference equations. The nonlocal model is similar to a stress

N. Challamel (✉)
Université Bretagne Sud IRDL (CNRS UMR 6027), Centre de Recherche,
Rue de Saint Maudé, BP92116, 56321 Lorient Cedex, France
e-mail: noel.challamel@univ-ubs.fr

C. M. Wang
School of Civil Engineering, The University of Queensland, Brisbane, Australia
e-mail: cm.wang@uq.edu.au

H. Zhang
School of Mechatronical Engineering, Beijing Institute of Technology, Beijing, China
e-mail: hong.zhang@bit.edu.cn

I. Elishakoff
Department of Ocean and Mechanical Engineering, Florida Atlantic University,
Boca Raton, FL 33431-0991, USA
e-mail: elishako@fau.edu

© The Author(s), under exclusive license to Springer Nature Switzerland AG 2021
E. Ghavanloo et al. (eds.), *Size-Dependent Continuum Mechanics Approaches*,
Springer Tracts in Mechanical Engineering,
https://doi.org/10.1007/978-3-030-63050-8_1

gradient nonlocal Euler-Bernoulli beam, where the nonlocality is of the Eringen type, although the length scale of the nonlocal model may also differ for statics or dynamics applications. The last part is devoted to lattice plates as introduced by Wifi et al. in 1988, and El Naschie in 1990, in connection with the finite difference formulation of Kirchhoff-Love plate models. Exact solutions for lattice plate statics and dynamics problems are presented for the Navier-type boundary conditions. A nonlocal elastic Kirchhoff-Love plate model is then derived based on the difference equations of the lattice plate. The microstructure-based nonlocal model slightly differs from an Eringen stress gradient Kirchhoff-Love plate model. The methodology followed from fundamental lattice microstructures shows that new nonlocal structural elements may be built from physical discrete structural approaches. The nonlocal models derived herein are closely related to the lattice microstructure assumed at the discrete level. It is expected that alternative nonlocal models may be achieved for some other microstructures.

1 Introduction

In this chapter, the behaviour of discrete structural elements (also labelled as microstructured elements or lattice models) is studied both in statics and in dynamics settings. The structural elements considered are typically discrete rods, discrete beams and discrete plates in the elastic range. The discrete models are composed of a finite number of elements connected by some elastic interactions, mainly axial and rotational springs. It is shown that these elastic structural elements possess some scale effects, as compared to their continuous analogues for an asymptotic large number of elements. Exact solutions of lattice structural mechanics problems (including discrete beams and plates) are available in the recent book of Wang et al. [1]. Scale effects in structural mechanics may be also captured within nonlocal mechanics, as extensively developed by Eringen and Kim [2], Eringen [3] (see also the seminal book of Eringen [4]). Owing to some common scale effects, Eringen [3] calibrated a stress gradient model (also called differential nonlocal model) from axial lattice dynamics results. He showed the possibility of approximating some wave dispersive properties of an axial lattice with a nonlocal axial model. This result opens some new directions in the connection of lattice mechanics with nonlocal mechanics. This chapter presents theoretical results in the same direction, for introducing nonlocal rod, beam and plate models from lattice structural mechanics. It is worth mentioning that exact solutions of nonlocal rod, beam and plate models are available in monographs such as the one of Elishakoff et al. [5], Gopalakrishnan and Narendar [6], Karlic et al. [7] or Ghavanloo et al. [8]. However, the connection of these nonlocal models with some discrete lattice formulation has not been studied in details in these books.

Therefore, we will connect the behaviour of lattice structural elements with nonlocal structural mechanics, using phenomenological or lattice-based nonlocal models.

The chapter consists of three complementary parts. In the first part, axial lattices (as already studied by Lagrange) are investigated for some dynamics problems including axial elastic supports. Historically, the theoretical investigation of one-dimensional lattices goes back to the XVIIIth century with the pioneering works of Lagrange. Lagrange [9, 10] calculated the exact eigenfrequencies of finite strings with concentrated masses, which can be viewed as a lattice string. Such a system composed of a finite number of degrees-of-freedom is governed by difference equations, as opposed to continuous systems governed by differential or partial differential equations. The mathematical problem of this difference eigenvalue problem valid for the finite string is in fact equivalent to the vibration problem of a finite microstructured rod in the axial direction or finite shaft in the torsional direction. The discrete axial model composed of masses connected by elastic springs, is also referred to as the Born-Kármán lattice model with direct neighbouring interactions. In this chapter, this Born-Kármán lattice model (or Lagrange model) will be investigated in presence of axial external supports, thus generalizing the results of Lagrange for general boundary conditions and elastic supports. A nonlocal elastic rod model is then developed from the lattice difference equations. The nonlocal model is similar to a stress gradient nonlocal model of Eringen [3], although the length scale of the nonlocal model may differ from statics to dynamics applications. This part ends with a discussion on generalized lattices with direct and indirect neighbouring interactions and their possible nonlocal modelling. The second part deals with lattice beam elements called Hencky-Bar-Chain models [11]. In 1920, Hencky developed a discrete model composed of a finite number of rigid elements connected by rotational springs [11]. Among various discrete structural problems (in-plane buckling of discrete columns, out-of-plane buckling of discrete beams, in-plane buckling of discrete arches and etc.), he solved the buckling problem of this Hencky-Bar-Chain model for some finite number of elements (typically for two, three and four elements). Exact solutions of this buckling problem, whatever the number of elements, have been derived later by Wang [12, 13], from the exact resolution of a linear difference equation. In the chapter, exact solutions of the buckling and vibration of Hencky-Bar-Chain model (labelled as Hencky beam model) are presented for general boundary conditions. A nonlocal elastic Euler-Bernoulli beam model is then built from the lattice difference equations of Hencky-Bar-Chain formulation. The nonlocal model is similar to a stress gradient nonlocal Euler-Bernoulli beam, where the nonlocality is of the Eringen type [3], although the length scale of the nonlocal model may also differ for statics or dynamics applications. The last part is devoted to lattice plates (or microstructured plates) as introduced by Wifi et al. [14] and El Naschie [15], in connection with the finite difference formulation of Kirchhoff-Love plate models. Exact solutions are presented for the Navier-type boundary conditions, any number of elements, both in statics and in dynamics settings. A nonlocal elastic Kirchhoff-Love plate model is then constructed from the lattice difference equations of the plate lattices. The microstructure-based nonlocal model slightly differs from an Eringen stress gradient Kirchhoff-Love plate model. More generally, exact solutions of lattice structural mechanics problems are available in the recent book of Wang et al. [1], including straight, curved Hencky beam models and lattice plate models. Lerbet et al. [16] also investigated non-conservative discrete structural mechanics problems with circulatory loading. In this chapter, nonlocal rod, beam and plate models are elaborated for each lattice structural model. The methodology followed from fundamental lattice microstructures shows that new nonlocal structural elements may be built from physical discrete structural approaches.

2 Discrete and Nonlocal Rods

2.1 Axial Lattices

This chapter studies the vibration behaviour of axial lattices composed of concentrated masses connected by elastic springs (lattice with direct elastic neighbouring interaction). Exact solutions of this discrete rod problem can be found for general boundary conditions, from the resolution of some linear difference equations. Historically, Lagrange [9, 10] was apparently the first to have derived the exact solutions of fixed-fixed finite lattices. He determined the eigenfrequencies of a lattice string (string with concentrated masses) with fixed-fixed boundary conditions. This problem has been recently revisited by Zhang et al. [17] within the theory of nonlocal string mechanics. From a mathematical point of view, the discrete string problem (or lattice string) is mathematically analogous to the one of a discrete rod (or lattice rod), as considered herein (see also Lagrange [9, 10]). Such a lattice with direct neighbouring interactions may be referred to as a Lagrange lattice or a Born-Kármán lattice [18]. This lattice problem is actually to be solved via a set of linear difference equations with corresponding discrete boundary conditions. The calculation of the eigenfrequencies of such finite lattices is available in many textbooks, for various boundary conditions (see for instance [19–21]). For instance, the vibration frequencies of such a general one-dimensional lattice system were initially calculated by Lagrange [9, 10] for fixed-fixed case (and later by [19], Tong et al. [22], Thomson and Dahleh [20], Blevins [21] or more recently by Challamel et al. [23]). The modelling of the free end boundary condition may be achieved by considering only half of the lumped mass at the lattice border. With such an assumption for the free end condition, the eigenfrequencies of a clamped-free axial lattice has been calculated by Thomson and Dahleh [20] and more recently by Challamel et al. [23]. As detailed in Challamel et al. [24], the lattice string is mathematically equivalent to the axial lattice, but it is also analogous to a torsional lattice, or a shear lattice at some extent. A one-dimensional shear lattice model could capture the shear properties of a multistorey building (see Thomson and Dahleh [20] or Luongo and Zulli [25] who solved a shear lattice problem with clamped-free ends and assuming a full lumped mass at the border). In this chapter, we will mainly focus on the vibrational behaviour of axial lattices. It is worth mentioning that exact results are also available for the static problem of axial lattices under concentrated load and distributed axial load. For instance, Triantafyllidis and Bardenhagen [26] studied the static behaviours of a nonlinear axial lattice loaded by a tension at the ends. Hérisson et al. [27] used Hurwitz Zeta functions to derive exact solutions for nonlinear lattices with elastic quadratic interactions and subjected to distributed axial load. Gazis and Wallis [28] derived the exact solutions for an axial lattice of semi-infinite length with direct and indirect neighbouring interactions . The derivation considered linear interaction inside the lattice while except for the lattice boundaries. Charlotte and Truskinovsky [29] studied a four end forces loaded lattice with direct and indirect neighbouring interactions.

Lagrange [9, 10] already in 1759 discussed the link between finite lattice string and the continuous string. The finite string will asymptotically converge to a continuum string when the number of elements increases. The vibration of a continuous string is ruled by a "local" continuous wave equation. The introduction of length scale effects representing the discreteness of the lattice structure in a corrected (or enriched) continuous model is more recent. The development of higher-order continuous models from the lattice formulation may be achieved by expanding the pseudo-differential operators with high-order continuous differential operators. This methodology, also called continualization technique, has been initiated in the 1960s especially for applications in the field of discrete wave equations, or wave in nonlinear axial lattices (see for instance Kruskal and Zabusky [30]). Various approximations of the pseudo-differential operators may be used, using power series (Taylor expansion) or rational series (Padé approximants). Enriched continua, also called quasicontinua by Collins [31] can be viewed as a continuous approximation of the exact (or reference) lattice problem. They mathematically differ from the lattice problem, due to the asymptotic expansion of the pseudo-differential operators at a given order. For instance, Kruskal and Zabusky [30] expanded the pseudo-differential operator involved in a nonlinear axial lattice in power series, by considering a fourth-order Taylor-based asymptotic expansion. It can be shown that the additional term responsible of small length scale effects due to this expansion modifies the potential energy functional, which is then no more positive definite. Some alternative expansions of the pseudodiffential operators have been used, for deriving alternative quasicontinua. For instance, Benjamin et al. [32] transformed Korteweg-de Vries wave equation by replacing higher-order spatial derivatives with coupled spatio-temporal derivatives. Collins [31] and Rosenau [33] used the same methodology to avoid higher-order uncoupled spatial derivatives, by inverting the spatial pseudo-differential operator. For linear elastic interactions, Jaberolanssar and Peddieson [34] obtained a nonlocal wave equation with coupled spatio-temporal derivatives (without higher-order spatial derivatives). The result has been generalized by Rosenau [33] who expanded the pseudo-differential operator of the nonlinear lattice, with a Padé approximant, for deriving a consistent nonlocal (and nonlinear) wave equation without higher-order spatial derivatives. Jaberolanssar and Peddieson [34] and Rosenau [33] obtained a nonlocal wave equation built from the difference equations of the axial elastic lattice. Eringen [3] postulated a nonlocal elastic model from a phenomenological point of view, in a differential format [3, 4]. The stress gradient of Eringen [3] relates the stress and the strain in an implicit differential form. The obtained nonlocal wave equation is mathematically similar to the one issued of a continualization process applied to the difference equations of the axial lattice, as followed by Jaberolanssar and Peddieson [34] and Rosenau [33]. Therefore, there is a strong relation between the discrete lattice mechanics theory and the continuous nonlocal mechanics theory, where the nonlocal terms of the approximated continuous model can be calibrated with respect to the lattice spacing. Enriched continuous formulations of axial lattices have been already derived in dynamics [3, 33] as well as in statics applications. For instance, Triantafyllidis and Bardenhagen [26] used a Taylor expansion of the pseudo-differential operators (continualization technique) to derive a gradient elasticity theory applied to the static response of an axial lattice. Hérisson et al. [27] used Padé approximant for characterizing nonlinear lattices with a nonlocal nonlinear continuous bar model. Gul et al. [35] also used a Taylor expansion of the pseudo-differential operator to approximate the vibration behavior of finite axial lattices with a gradient elasticity model (with no definite positive energy functional).

It has been already commented that the continualization technique applied to the vibration of the lattice problem could be reformulated in terms of nonlocal wave equation. The continuous nonlocal phenomenological model of Eringen [3] (stress gradient model) has been identified based on the wave dispersive properties of the axial lattice. Aydogdu [36] calculated the analytical vibration frequencies of finite rods with clamped-clamped and clamped-free ends based on Eringen's nonlocal theory. A hybrid gradient/nonlocal model (which includes the strain gradient elasticity model and the stress gradient model of Eringen) has been developed by Challamel et al. [37], in order to better calibrate the dispersive parameters of the linear axial lattice. Aydogdu [38] generalized his earlier results [36] by considering a nonlocal elastic rod embedded in elastic medium (with linear elastic interaction with the substrate). The nonlocal longitudinal wave propagation problem for multi-walled carbon nanotubes with van der Waals interactions between each nanotube walls has been studied by Aydogdu [39].

This chapter is mainly focused on the axial lattice problem. Axial lattice with fixed-fixed and fixed-free boundary conditions in the presence of elastic neighbouring interactions are analyzed in statics and dynamics. It should be noted that elastic interactions on an elastic substrate are also included in Rosenau [33] as we shall study in this chapter. Also, a nonlocal model will be continualized from the discrete equations to fit the behavior of an axial lattice on elastic support (see also Challamel et al. [40]).

2.2 Lattice Formulation: Governing Equations

In the first part of this chapter, we study a one-dimensional uniform lattice composed of $n + 1$ masses connected by n springs with identical stiffness as shown in Fig. 1.

The mass m_i of each particle are identical, except for the border mass which is $m_i/2$. L is the total length of the lattice system and $L = n \times a$, where a is the nodal spacing in the lattice. Fixed-fixed and fixed-free boundary conditions are specifically studied. For fixed-fixed boundary conditions, there are $n + 1$ particles with two of

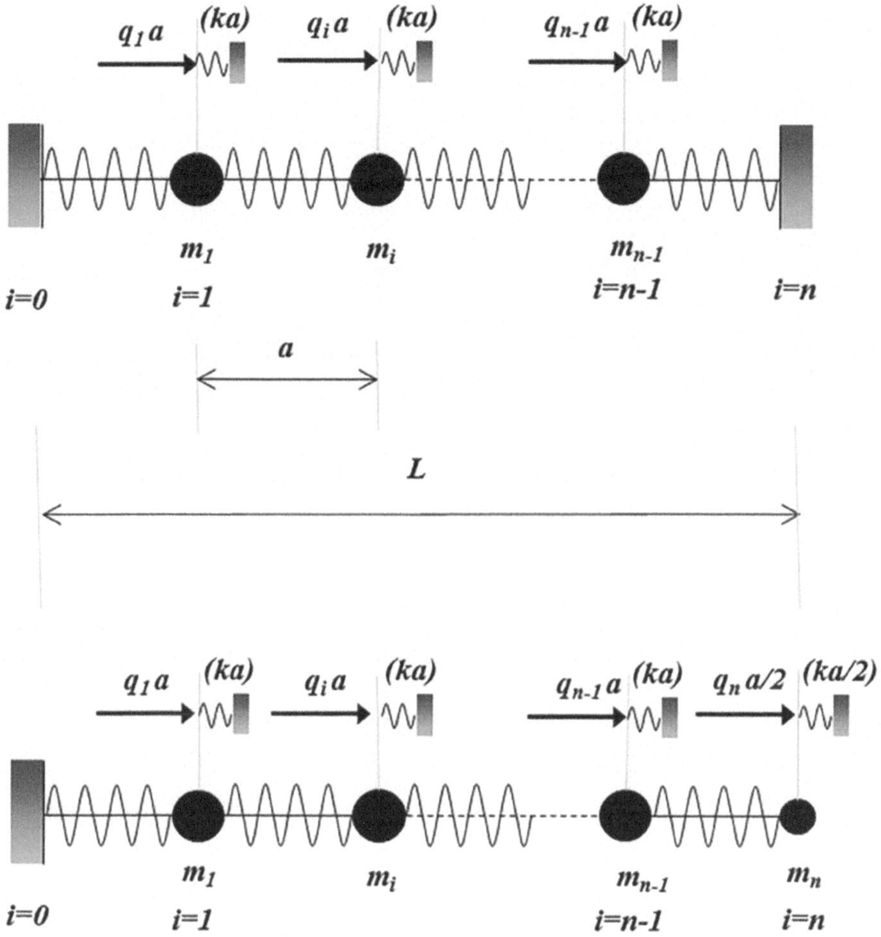

Fig. 1 One dimensional lattice with fixed-fixed and fixed-free boundary conditions under distributed load

them attached at the border, i.e. $n - 1$ free particles, whereas for the fixed-free chain, there are $n + 1$ particles, one of them attached, and the other free i.e. n free particles. The balance equations of the axial lattice with only direct neighbouring interactions including the presence of distributed load can be obtained from:

$$\frac{N_{i+1/2} - N_{i-1/2}}{a} = \rho A \frac{d^2 u_i}{dt^2} + k u_i - q_i \tag{1}$$

where k is the equivalent stiffness of the elastic substrate, ρA denotes the mass density per unit length, q_i is the distributed axial load applied at node i, u_i is the displacement of node i and N_i denotes the normal force in the ith spring (with half a shift for the considered element). The mass distribution is uniform $m_i = \rho Aa$ except at the border where $m_0 = m_n = \rho Aa/2$. Equations are derived for a one-dimensional axial lattice, but the developments could also be used for string lattices, shear lattices and torsional lattices (see Ref. [23] or Ref. [24] for a discussion on the analogies between these different kinds of lattices). We shall mainly focus on the calculation of eigenfrequencies of this lattice. The distributed axial force q_i is assumed to vanish, i.e. $q_i = 0$ (see Challamel et al. [40] for the general calculations valid in case of parabolic axial forces).

The normal force in the spring $i + 1/2$ (defined in the element) is relative to the axial displacement between two adjacent nodes:

$$N_{i+1/2} = EA\frac{u_{i+1} - u_i}{a} \tag{2}$$

where EA/a is the stiffness of the axial spring. The governing mixed functional differential equation (or mixed differential-difference equation – see Myshkis [41]) is obtained by substituting Eq. (2) into Eq. (1):

$$EA\frac{u_{i+1} - 2u_i + u_{i-1}}{a^2} - \rho A\frac{d^2u_i}{dt^2} - ku_i = -q_i \tag{3}$$

In this chapter, we will focus on the dynamic analysis and will not take into account the distributed axial forces so that $q_i = 0$. Distributed forces and elastic medium are absent in the usual Born-von Kármán lattice equations, i.e. for $q_i = 0$ and $k = 0$ so that the governing Eq. (3) becomes

$$q_i = k = 0 \quad \Rightarrow \quad EA\frac{u_{i+1} - 2u_i + u_{i-1}}{a^2} - \rho A\frac{d^2u_i}{dt^2} = 0 \tag{4}$$

The dynamic behaviours of this lattice will be studied for two archetypal boundary conditions, namely for fixed-fixed and fixed-free boundary restraints. The boundary conditions for fixed-fixed ends (as shown in Fig. 1) are given by:

$$u_0 = 0 \quad \text{and} \quad u_n = 0 \tag{5}$$

whereas the boundary conditions for fixed-free ends should be written as:

$$u_0 = 0 \quad \text{and} \quad \frac{N_{n+1/2} - N_{n-1/2}}{a} = \frac{\rho A}{2}\frac{d^2u_n}{dt^2} + \frac{k}{2}u_n \quad \text{with} \quad N_{n+1/2} = 0 \tag{6}$$

which may be equivalently formulated as

$$u_0 = 0 \quad \text{and} \quad EA\frac{u_n - u_{n-1}}{a} = -a\frac{\rho A}{2}\frac{d^2u_n}{dt^2} - a\frac{k}{2}u_n \tag{7}$$

Note that half mass is attached at the last spring of half elastic interaction. As notably mentioned by Maugin [42] (see more recently Challamel et al. [24] for structural mechanics applications), the difference equations valid for lattice elasticity exactly corresponds to the finite difference approximation for the continuous medium (elastic medium). Hence, the theoretical foundation of lattice mechanics is firmly related to the numerical analysis of continuum elasticity as solved by the finite difference method. The free boundary condition considered is related to the central finite difference discretization of the continuous equations. The last mass $m_n = m_i/2$ naturally appears in this scheme, as shown by LeVeque [43] who treated Neumann type continuous boundary conditions by using a central finite difference approximation. Challamel et al. [23] also considered this kind of discrete boundary condition for formulating the Neumann boundary condition. Of course, it would have been possible to get an exact solution of the lattice problem with a mass m_n equal to the other values mi for instance (or a border mass not equal to $m_i/2$), but the resulting discrete solution would be slightly different from the solution presented herein. In addition, Kivshar et al. [44] set last particle mass m_n different from half of the internal particle mass and then the associated continuous boundary conditions are in a mixed-type in presence of both displacement and its derivative.

It is possible, equivalently, to derive the governing difference Eq. (3) from energy arguments, based on the potential energy W given by

$$W = \sum_{i=0}^{n-1} \frac{EA}{2} a \left[\left(\frac{u_{i+1} - u_i}{a} \right)^2 \right] + \sum_{i=1}^{n-1} \frac{k}{2} a u_i^2 + \frac{k}{4} a u_0^2 + \frac{k}{4} a u_n^2 \tag{8}$$

With half of the mass at the ends, the kinetic energy T could be written as

$$T = \sum_{i=1}^{n-1} \frac{\rho A}{2} a \left(\frac{du_i}{dt}\right)^2 + \frac{\rho A}{4} a \left(\frac{du_0}{dt}\right)^2 + \frac{\rho A}{4} a \left(\frac{du_n}{dt}\right)^2 \tag{9}$$

By using Hamilton's principle, the mixed differential-difference governing equation is found again. Considering a harmonic motion $u_i(t) = u_i e^{j\omega t}$ where $j = \sqrt{-1}$ and ω is the angular frequency of vibration, the linear mixed differential-difference Eq. (3) may reduce in a linear second-order difference equation in space:

$$EA \frac{u_{i+1} - 2u_i + u_{i-1}}{a^2} + \left(\rho A \omega^2 - k\right) u_i = 0 \tag{10}$$

2.3 Lattice Formulation: Resolution

By introducing the following dimensionless parameters

$$q^* = \frac{qL^2}{EA}, \; k^* = \frac{kL^2}{EA}, \; \beta = \rho \frac{\omega^2 L^2}{E} \tag{11}$$

the linear second-order difference equation can be expressed as:

$$u_{i+1} - \left(2 + \frac{k^* - \beta}{n^2}\right) u_i + u_{i-1} = 0 \tag{12}$$

This linear difference equation can be solved by assuming a displacement solution in a form of power function (Goldberg [45]):

$$u_i = u_0 \, \lambda^i \tag{13}$$

The substitution of Eq. (13) into the difference Eq. (12) furnishes the following auxiliary equation:

$$\lambda^2 - \left(2 + \frac{k^* - \beta}{n^2}\right) \lambda + 1 = 0 \tag{14}$$

Note that this equation is palindromic, implying that if λ is its solution, so is its reciprocal $1/\lambda$.

The auxiliary equation admits two complex conjugate solutions

$$\lambda = 1 - \frac{\beta - k^*}{2n^2} \pm \sqrt{\left(1 - \frac{\beta - k^*}{2n^2}\right)^2 - 1} = \cos\phi \pm j \sin\phi \quad \text{with} \quad \cos\phi = 1 - \frac{\beta - k^*}{2n^2} \tag{15}$$

The general solution for the linear difference Eq. (12) can finally be expressed in trigonometric functions:

$$u_i = A\cos(\phi\, i) + B\sin(\phi\, i) \tag{16}$$

For fixed-fixed boundary conditions, using one boundary condition $u_0 = 0$ necessarily shows that the eigenmode is reduced to a simple sinusoidal function:

$$u_i = B\sin(\phi\, i) \tag{17}$$

The eigenfrequencies are obtained by using the second boundary condition $u_n = 0$ and consequently the frequency equation is given by

$$\sin(n\phi) = 0 \quad \text{with} \quad \phi = \arccos\left[1 - \frac{\beta - k^*}{2n^2}\right] \tag{18}$$

The exact eigenfrequencies of the lattice system with fixed-fixed boundary conditions are obtained for the mth mode from Eq. (18):

$$\beta_{m,n} = 4n^2 \sin^2\left(\frac{m\pi}{2n}\right) + k^* \quad \text{for} \quad m \in \{1, 2, ..., n-1, n\} \tag{19}$$

For fixed-free ends, the two boundary conditions are given by the kinematic constraint and the free end boundary condition $u_n - u_{n-1} = (\beta - k^*)\ u_n/2n^2$. The consideration of the essential boundary condition $u_0 = 0$ leads again to the trigonometric solution (17). One obtains from the second boundary condition:

$$\cos(n\phi) = 0 \quad \text{with} \quad \phi = \arccos\left[1 - \frac{\beta - k^*}{2n^2}\right] \tag{20}$$

The natural frequencies of the mth mode, valid for fixed-free boundary conditions are finally obtained from Eq. (20):

$$\beta_{m,n} = 4n^2 \sin^2\left[\frac{(2m-1)\,\pi}{4n}\right] + k^* \quad \text{for} \quad m \in \{1, 2, ..., n-1, n\} \tag{21}$$

Both formula valid for fixed-fixed and fixed-free boundary conditions coincide with the ones presented in Challamel et al. [23] without considering the external medium, i.e. for $k^* = 0$.

2.4 Nonlocal Continualized Model and Eringen's Model

Following the methodology introduced by Kruskal and Zabusky [30] for nonlinear lattices, the mixed differential-difference Eq. (3) is continualized to:

$$EA\frac{u(x+a,t) - 2u(x,t) + u(x-a,t)}{a^2} - \rho A\frac{\partial^2 u(x,t)}{\partial t^2} - ku(x,t) = -q(x) \tag{22}$$

It is possible to expand the spatial difference operator in Taylor series for the spatial difference operator, for sufficiently smooth displacement fields (see Salvadori [46], Kruskal and Zabusky [30], Rosenau [33], Andrianov and Awrejcewicz [47, 48] and Andrianov et al. [49]):

$$u(x \pm a, t) = \sum_{k=0}^{\infty} \frac{(\pm a)^k}{k!}\frac{\partial^k}{\partial x^k} u(x,t) = \left[e^{a\partial/\partial x}\right] u(x,t) \tag{23}$$

so that Eq. (22) could also be rewritten by using a pseudo-differential operator:

$$\frac{4EA}{a^2}\left[\sinh^2\left(\frac{a}{2}\frac{\partial}{\partial x}\right)\right] u(x,t) - \rho A\frac{\partial^2 u(x,t)}{\partial t^2} - ku(x,t) = -q(x) \tag{24}$$

The pseudo-differential operator can be truncated at order 4, by using a Taylor-based expansion [30]:

$$\frac{4EA}{a^2}\left[\sinh^2\left(\frac{a}{2}\frac{\partial}{\partial x}\right)\right]u\,(x,t) = EA\frac{\partial^2}{\partial x^2}\left[1+\frac{a^2}{12}\frac{\partial^2}{\partial x^2}\right]u\,(x,t)+O\left(a^4\right) \quad (25)$$

Rosenau [33], following the idea of Benjamin et al. [32] or Collins [31], replace the higher-order spatial derivatives using Padé approximant of order [2, 2] to expand the pseudo-differential operator (see Baker and Graves-Morris [50] for an extensive analysis of Padé approximants):

$$\frac{4EA}{a^2}\left[\sinh^2\left(\frac{a}{2}\frac{\partial}{\partial x}\right)\right]u\,(x,t) = EA\frac{\frac{\partial^2}{\partial x^2}}{1-\frac{a^2}{12}\frac{\partial^2}{\partial x^2}}u\,(x,t)+\ldots \quad (26)$$

The pseudo-differential equation would then be written with coupled spatio-temporal derivatives based on Padé approximant:

$$EA\frac{\partial^2 u\,(x,t)}{\partial x^2}-\rho A\left[1-\frac{a^2}{12}\frac{\partial^2}{\partial x^2}\right]\frac{\partial^2 u\,(x,t)}{\partial t^2}$$
$$-k\left[1-\frac{a^2}{12}\frac{\partial^2}{\partial x^2}\right]u\,(x,t) = -\left[1-\frac{a^2}{12}\frac{\partial^2}{\partial x^2}\right]q\,(x) \quad (27)$$

Jaberolanssar and Peddieson [34] obtained Eq. (27) in the context of dynamics and without considering elastic medium interaction, namely for $q(x)=0$ and $k=0$. Rosenau [33] derived a similar equation in a dynamic context, i.e. for $q(x)=0$, in the presence of elastic external interaction.

It is possible to derive equivalently the governing partial differential Eq. (27) from the following total potential energy

$$W=\int_0^L\left(\frac{1}{2}EA\left(\frac{\partial u}{\partial x}\right)^2+\frac{1}{2}ku^2+\frac{1}{2}l_c^2k\left(\frac{\partial u}{\partial x}\right)^2-qu-l_c^2\left(\frac{dq}{dx}\right)\left(\frac{\partial u}{\partial x}\right)\right)dx$$
$$\text{with}\quad l_c^2=\frac{a^2}{12} \quad (28)$$

coupled with the modified kinetic energy:

$$T=\int_0^L\left(\frac{1}{2}\rho A\left(\frac{\partial u}{\partial t}\right)^2+\frac{1}{2}l_c^2\rho A\left(\frac{\partial^2 u}{\partial x\partial t}\right)^2\right)dx \quad (29)$$

This modified kinetic energy which includes additional small length scale terms, has been obtained equivalently by Rosenau [51] from continualizing the lattice kinetic energy. Rosenau [51] also derived the modified energy potential from the continualization of the lattice potential energy including the elastic medium interaction while without distributed forces.

Furthermore, it is possible to re-express Eq. (27) from the following nonlocal (differential) constitutive law, coupled to local balance equations:

$$\left[1-\frac{a^2}{12}\frac{\partial^2}{\partial x^2}\right]N(x,t)=EA\frac{\partial u}{\partial x}(x,t)$$
$$\text{and}\quad \frac{\partial N(x,t)}{\partial x}=\rho A\frac{\partial^2 u(x,t)}{\partial t^2}+ku(x,t)-q(x) \tag{30}$$

One easily recognizes the stress gradient model of Eringen [3] in Eq. (30) where the length scale l_c of the nonlocal Eringen's model, is calibrated with respect to the lattice spacing a:

$$N-l_c^2\frac{\partial^2 N}{\partial x^2}=EA\frac{\partial u}{\partial x}\quad\text{with}\quad l_c^2=(e_0a)^2=\frac{a^2}{12}\quad\text{and}\quad e_0=\frac{1}{2\sqrt{3}} \tag{31}$$

It is shown from Eq. (31) that the small length scale coefficient of the nonlocal model can be calibrated from the lattice problem and is found to be $e_0=1/2\sqrt{3}$. The nonlocal model derived from the lattice equations exactly coincides with the nonlocal Eringen's stress gradient model, where the small length scale parameter of the nonlocal model is calculated regarding the nodal spacing in lattice. Eringen's nonlocal model [3] and the lattice-based nonlocal model (as obtained by Jaberolanssar and Peddieson [34] or Rosenau [33]) are both formulated in a modified wave equation where the additional small length scale term uses coupled second-order spatial and time derivatives. Similar models with additional spatio-temporal coupling terms have been proposed for other applications of one-dimensional continuum system by Love [52] for axial vibration problems in the presence of lateral inertia effects, Mindlin [53] and Polyzos and Fotiadis [54] for axial vibration problems, and by Rayleigh [55] for the bending problem of beam including the effect of rotary inertia.

It is possible to extract the normal force (of Eringen's stress gradient model) from Eqs. (30) and (31), thus leading to:

$$N=\left(EA+kl_c^2\right)\frac{\partial u}{\partial x}+l_c^2\rho A\frac{\partial^3 u}{\partial x\partial t^2}-l_c^2\frac{dq}{dx} \tag{32}$$

This expression of normal force corresponds to the natural boundary condition derived from application of Hamilton's principle. The nonlocal boundary condition raised from potential energy Eq. (28) and kinetic energy formula Eq. (29) is given by

$$\left[\left(EA+kl_c^2\right)\frac{\partial u}{\partial x}+l_c^2\rho A\frac{\partial^3 u}{\partial x\partial t^2}-l_c^2\frac{dq}{dx}\right]\delta u\,\Big|_0^L=0 \tag{33}$$

It is shown from Eq. (33), that no additional boundary condition is required for such a nonlocal model, as opposed to gradient elasticity models based on higher-order differential equations with additional extra boundary conditions.

The normal force expression Eq. (32) derived from Eringen's differential nonlocal model is consistent with the normal force formulae issued of the energy approach. This formula is obtained from continualization of the mixed differential-difference equation in displacement, which is a second-order difference equation in space. The normal force identification would have been different if one would have continualized directly the normal force in the difference equations of the centred lattice problem (first-order difference equation in space). The lattice difference equations (Eqs. (1) and (2)) are formulated with a central finite difference scheme. A similar centered difference scheme was already used by Jaberolanssar and Peddieson [34]. Such a scheme implicitly assumes that the normal force is an element quantity (defined at the middle of the spring element), as opposed to the displacement exactly defined at the nodal points. Now continualizing each first-order difference equation based on the Padé approximant of the pseudo-differential operator leads to the nonlocal balance equation and the nonlocal constitutive law:

$$\frac{\partial N}{\partial x} = \left(1 - \frac{a^2}{24}\frac{\partial^2}{\partial x^2}\right)\left(\rho A \frac{\partial^2 u}{\partial t^2} + ku - q\right)$$
$$\text{and} \quad \left(1 - \frac{a^2}{24}\frac{\partial^2}{\partial x^2}\right) N = EA\frac{\partial u}{\partial x} \tag{34}$$

A nonlocal coupling of both the balance equation and the constitutive law was also reported by Challamel et al. [56] in the continualization process of lattice beam equations (in bending). Going back to the axial problem investigated in this part, the normal force (of the continualized nonlocal model) can be obtained from Eq. (34), which after neglecting higher-order terms in a^4, gives:

$$N = \left(EA + k\frac{l_c^2}{2}\right)\frac{\partial u}{\partial x} + \frac{l_c^2}{2}\rho A \frac{\partial^3 u}{\partial x \partial t^2} - \frac{l_c^2}{2}\frac{dq}{dx} \quad \text{with} \quad l_c^2 = \frac{a^2}{12} \tag{35}$$

This continualization process is equivalent in term of displacement Eq. (27), to the one directly applied to the mixed differential-difference displacement equation. However, the boundary conditions in terms of normal force calculated by both methods differ, as highlighted by the formulae issued of Eringen's nonlocal model (see Eq. (32)) and the one issued of the continualized nonlocal model (see Eq. (35)).

2.5 Nonlocal Solutions

To summarize, the governing equation of the nonlocal rod embedded in an elastic medium is given by a linear wave equation corrected by small length scale terms:

$$\left(EA + kl_c^2\right)\frac{\partial^2 u}{\partial x^2} - \rho A \frac{\partial^2 u}{\partial t^2} + \rho A l_c^2 \frac{\partial^4 u}{\partial x^2 \partial t^2} - ku = -q + l_c^2 \frac{d^2 q}{dx^2} \tag{36}$$

where $l_c = e_0 a$. Here a is the characteristic intrinsic length of the elastic medium and e_0 is a parameter used to approximate lattice dynamic results. This equation is valid for Eringen's nonlocal model and for the continualized nonlocal model of lattice equations. $e_0 = 1/2\sqrt{3}$ has been found previously for the calibration of the small length scale parameter. It is worth mentioning that Eringen [3] obtained by equating the nonlocal wave dispersion frequency and lattice dynamic results at the end of the first Brillouin zone. We shall discuss such results later in the chapter.

The dimensionless parameters are introduced:

$$q^* = \frac{qL^2}{EA}, \quad k^* = \frac{kL^2}{EA}, \quad \beta = \rho\frac{\omega^2 L^2}{E}, \quad \overline{x} = \frac{x}{L}, \quad \overline{u} = \frac{u}{L}, \quad \overline{l_c}^2 = \frac{l_c^2}{L^2} \tag{37}$$

The length scale ratio $\overline{l_c}$ depends on the number of elements in the lattice considered. More specifically, we have $\overline{l_c}^2 = 1/(12n^2)$.

The nonlocal rod problem is governed by a linear second-order differential equation, which includes small length scale terms and interaction with elastic medium:

$$\left(1 + k^*\overline{l_c}^2 - \beta\overline{l_c}^2\right)\frac{d^2\overline{u}}{dx^2} + (\beta - k^*)\,\overline{u} = 0 \tag{38}$$

This differential equation can be easily integrated in term of trigonometric functions. For fixed-fixed boundary condition, the eigenmode is expressed in sinusoidal form:

$$\overline{u}\,(\overline{x}) = \overline{u}^0 \sin\,(m\pi\overline{x}) \tag{39}$$

The dimensionless frequency can then be obtained by inserting the eigenmode expression in the second-order differential Eq. (38):

$$\beta_m = k^* + \frac{(m\pi)^2}{1 + \overline{l_c}^2 (m\pi)^2} \tag{40}$$

The nonlocal continuous approximation Eq. (40) is a good approximation of the exact lattice solution given by Eq. (19) (Fig. 2). The same result was also reported by Aydogdu [38], who also studied the free vibration behaviour of a nonlocal rod embedded in elastic medium.

For fixed-free ends, the vibration mode shape can be also expressed in sinusoidal form:

$$\overline{u}\,(\overline{x}) = \overline{u}^0 \sin\left[(2m - 1)\,\pi\frac{\overline{x}}{2}\right] \tag{41}$$

The substitution of this solution into the governing differential Eq. (38) gives the dimensionless frequency Eq. (42), which was also reported by Aydogdu [38]

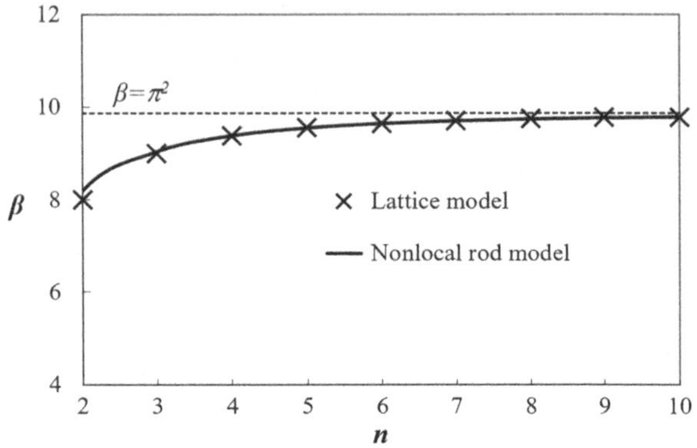

Fig. 2 Comparison of the axial lattice and the nonlocal rod models-Frequency versus the number n of elements; Fixed-fixed boundary conditions

for the clamped-free nonlocal rod. This vibration mode satisfies the displacement and the normal force boundary conditions of both Eringen's nonlocal model and the continualized nonlocal models.

$$\beta_m = k^* + \frac{\left[(2m-1)\frac{\pi}{2}\right]^2}{1+\overline{l_c}^2\left[(2m-1)\frac{\pi}{2}\right]^2} \tag{42}$$

The same results could have been derived from application of the stationarity of the Rayleigh quotient based on the augmented (nonlocal) potential energy and kinetic energy, as defined by Eqs. (28) and (29).

$$\omega_m^2 = \frac{\int_0^L \left(EA(\frac{du}{dx})^2 + ku^2 + l_c^2 k(\frac{du}{dx})^2\right)dx}{\int_0^L \left(\rho A u^2 + l_c^2 \rho A(\frac{du}{dx})^2\right)dx} \tag{43}$$

In conjunction with the trigonometric solution assumed for the vibration mode in this Rayleigh quotient (for both boundary conditions), one obtains the frequency solution for each boundary condition (Eqs. (40) and (42)).

The lattice solution can be compared to the nonlocal one, via expanding the lower frequency solutions. For instance, for fixed-fixed boundary conditions, the asymptotic expansion of the exact solution for a sufficiently large number n of elements gives:

$$\beta_{m,n} = 4n^2\sin^2\left(\frac{m\pi}{2n}\right) + k^* = k^* + (m\pi)^2\left[1-\frac{(m\pi)^2}{12n^2}\right] + o\left(\frac{1}{n^2}\right) \tag{44}$$

whereas the corresponding nonlocal continuous solution has been calculated before and may be expanded in the same way:

$$\beta_{m,n} = k^* + \frac{(m\pi)^2}{1 + \left(\frac{e_0 a}{L}\right)^2 (m\pi)^2} = k^* + (m\pi)^2 \left[1 - (m\pi)^2 \left(\frac{e_0 a}{L}\right)^2\right] + o\left(\frac{1}{n^4}\right) \tag{45}$$

The small length scale coefficient can be easily identified for low frequencies and a sufficient number of elements n, by comparing Eqs. (44) and (45):

$$e_0 = \frac{1}{2\sqrt{3}} \approx 0.289 \tag{46}$$

For the highest frequency, one obtains for $m = n$:

$$\beta_{m,n} = k^* + 4n^2 = k^* + \frac{(n\pi)^2}{1 + \pi^2 e_0^2} \quad \Rightarrow \quad e_0 = \sqrt{\frac{1}{\pi^2}\left(\frac{\pi^2}{4} - 1\right)} \approx 0.386 \tag{47}$$

The value $e_0 = 0.39$ is reported by Eringen [3] who equated the nonlocal wave dispersion frequency and the corresponding lattice dynamic results at the end of the first Brillouin zone (in the absence of elastic medium interaction-see also Challamel et al. [37]). It is worth mentioning that the stiffness k of the elastic medium does not affect this length scale calibration in this case.

2.6 Lattice with Direct and Indirect Interactions

The one-dimensional lattice studied in the previous part is the lattice with direct interaction, where each particle only interacts with its neighbours, without long range interactions. This direct interaction has been investigated for instance by Born and von von Kármán [18] for one-dimensional or three-dimensional lattices. As already commented, for one-dimensional system, the one-dimensional axial lattice is equivalent to a lattice string, that was first studied by Lagrange [9, 10], who discovered the exact vibration frequencies for fixed-fixed boundary conditions. This initial approach based on the simplest interaction can be generalized for considering direct and indirect neighbouring interactions, with a so-called general N-neighbour interaction problem (see Fig. 3). Such generalized lattices with N-neighbour interactions may be encountered in the field of molecular mechanics, when both short- and long-range interactions are included. These generalized interactions may be taken into account, to derive material properties at different scales, from nano to micro or to macro scales (see also Challamel et al. [57] or Challamel et al. [58]). Exact solutions of such generalized lattices for various boundary conditions are available in the literature. Pipes [59], Chen [60], Chen [61], Eaton and Peddieson [62] calculated the eigenfrequencies of a generalized lattice (lattice with short and long range interaction) of finite size for specific long range dependence. Brillouin [63], Eaton and Peddieson [62] and Roseneau [64] investigated more specifically

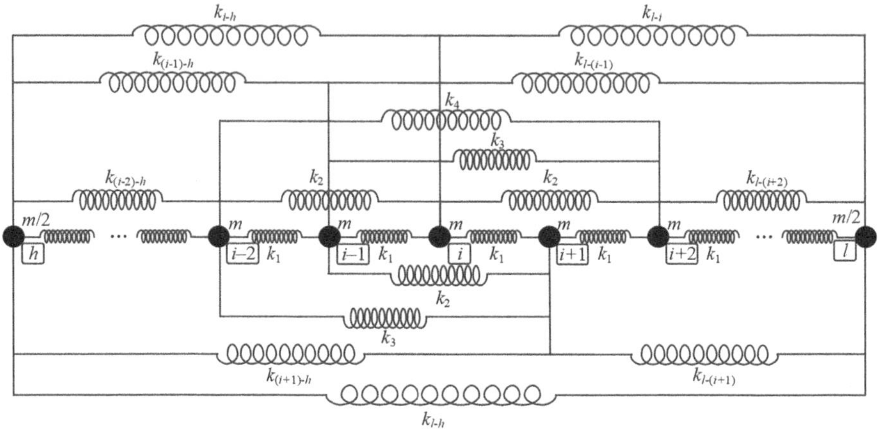

Fig. 3 Axial lattice composed of concentrated masses with p-neighbour elastic interaction

the wave dispersive properties of an infinite generalized lattice (lattice with short and long range interaction). Rosenau [64] studied the specific problem of a generalized lattice with an interaction that includes two closest neighbours (Rosenau [64] also derived the results for more neighbours). Pipes [59], Chen [60] and Chen [61] studied a specific generalized lattice with N-neighbour interaction, characterized by equal stiffness for each interaction. The exact eigenfrequencies of this specific generalized lattice have been obtained. The problem studied by Pipes [59], Chen [60] and Chen [61] is very specific in the sense that short- and long-range interactions are equal. For many physical systems, the effect of long range interaction decreases with respect to the distance so that a differentiation between short- and long-range interactions is necessary. When considering the N-neighbour interaction problem with generalized interactions, Eaton and Peddieson [62], calculated the exact eigenfrequencies of a generalized lattice with N-interaction for fixed-free, fixed-fixed and free-free boundary conditions. They also used a continualization procedure based on a Taylor-based asymptotic expansion of the pseudo-differential operator involved in the difference formulation of the generalized lattice (associated with a difference operator of order 2N) (this methodology was also followed by Rosenau [64]). Both Eaton and Peddieson [62] and Rosenau [64] derived with such a continualization procedure a higher-order gradient approach associated with the generalized lattice. The wave dispersion relation of this generalized lattice, was studied by Brillouin [63], Eaton and Peddieson [62] and Rosenau [64]. Rosenau [64] followed the same methodology used for the direct interaction problem, and presented a lower order spatial differential wave equation (with spatio-temporal derivatives), for capturing the inherent scale effects in the generalized lattice. We shall follow this approach by considering a nonlocal equivalent rod model associated with the finite generalized lattice. Triantafyllidis and Bardenhagen [26] also investigated the static behaviour of a generalized lattice and formulated an equivalent gradient elasticity approach at the macroscopic scale. More generally, generalized and complex interactions in lattice mechanics are still important to be studied, for a better understanding of the macroscopic properties of nonlocal and nonlinear generalized wave propagation properties (see Maugin [42]).

Nonlocal medium and generalized lattices with generalized interactions have been already related in the literature. Eringen and Kim [2] studied an N-neighbour interaction model as a discrete formulation of a strain-based nonlocal continuum model. The N-neighbour interaction lattice model can be also considered as the discrete formulation of a peridynamic model (nonlocal relative displacement-based model) [65]. Carcaterra et al. [66] revisited this problem with generalized interactions, and built a microscopic generalized lattice model (with direct and indirect neighbour interactions) based on a macroscopic higher-order gradient continuum media. Tarasov [67] studied the generalized elastic lattice with two closest neighbours. Tarasov [67] also continualized the higher-order difference operators to derive a higher-order constitutive law which is similar to the one derived by Eaton and Peddieson [62] or Rosenau [64]. Tarasov [67] also commented the possible loss of definite positiveness of the macroscopic elastic energy (issued of the continualization procedure) from the structure of the generalized interactions at the lattice level. The definite positivity of the generalized interaction is not necessarily associated with the definite positivity of the equivalent gradient elasticity approach. Michelitsch et al. [68] investigated the properties of generalized lattices with both short range and long-range interactions governed by a power law. They showed the link between such generalized lattice and some equivalent nonlocal media based on fractional nonlocal mechanics. Ghavanloo et al. [69] studied the wave propagation of diatomic lattices, with both a discrete and a nonlocal lattice-based approach formulated from a continualization procedure. Ghavanloo and Fazelzadeh [70] generalized the lattice problem with long-range interactions with additional coupling with internal mass called metamaterial.

The generalized axial lattice comprises concentrated masses m_i connected by an elastic network as shown in Fig. 3 ($m_i = \rho A a$ except at the border). Fixed-fixed boundary conditions will be assumed for the generalized lattice. The generalized lattice is characterized by a distribution of elastic stiffnesses k_j, where k_j is the axial stiffness of the j-th spring associated with its neighbouring interaction, for $j \in \{1, 2, ..., p\}$. This stiffness can be rewritten via a scaling law as:

$$k_j = \alpha_j \frac{EA}{j^2 a} \tag{48}$$

where a is the lattice spacing (identical to the lattice problem with direct interactions), α_j is a dimensionless influence function which characterizes the contribution of the long range interaction forces. EA is the equivalent axial rigidity of the continuum rod (local continuum rod). The higher-order mixed differential-difference equation of the generalized lattice is given by (see for instance Challamel et al. [57])

$$EA \sum_{j=1}^{p} \alpha_j \frac{u_{i+j} - 2u_i + u_{i-j}}{(ja)^2} - \rho A \frac{d^2 u_i}{dt^2} = 0 \tag{49}$$

with $\alpha_j \geq 0$ is a positive parameter (generally a decreasing function with the interaction distance) and the new normalization requirement $\sum_{j=1}^{p} \alpha_j = 1$.

By assuming a harmonic motion, the eigenvalue problem is governed by difference equation of order $2p$

$$EA \sum_{j=1}^{p} \alpha_j \frac{u_{i+j} - 2u_i + u_{i-j}}{(ja)^2} + \rho A \omega^2 u_i = 0 \tag{50}$$

This difference equation can be equivalently written as:

$$\sum_{j=1}^{p} \alpha_j \frac{u_{i+j} - 2u_i + u_{i-j}}{j^2} + \frac{\beta}{n^2} u_i = 0 \tag{51}$$

We consider fixed-fixed boundary conditions with the higher-order anti-periodic boundary conditions:

$$u_{-(k-1)} = -u_{k-1}, \quad u_{n-(k-1)} = -u_{n+k-1} \quad \text{for } k \in \{1, 2, ..., p\} \tag{52}$$

For this generalized lattice with such a generalized boundary condition, the exact vibration mode can be assumed in a sinusoidal form as

$$u_i = u^0 \sin\left(\pi \frac{i}{n}\right) \tag{53}$$

The fundamental frequency is obtained from the substitution of Eq. (53) into Eq. (52):

$$\beta = \sum_{j=1}^{p} \alpha_j \left(\frac{2n}{j}\right)^2 \sin^2\left(\frac{\pi j}{2n}\right) \tag{54}$$

Eaton and Peddieson [62] also obtained a similar formulae by using dimensional quantities without scaling factors.

It is possible to develop a nonlocal approach associated with the lattice problem, by using a continualization procedure. The higher-order difference Eq. (50) can be continualized with the introduction of a pseudo-differential operator:

$$EA \left[\sum_{j=1}^{p} \frac{4\alpha_j}{(ja)^2} \sinh^2\left(\frac{ja}{2}\frac{d}{dx}\right) \right] u + \rho A \omega^2 u = 0 \tag{55}$$

The pseudo-differential operator is expanded with a Taylor-based asymptotic expansion truncated at order 2:

$$\left[\sum_{j=1}^{p}\frac{4\alpha_j}{(ja)^2}\sinh^2\left(\frac{ja}{2}\frac{d}{dx}\right)\right]u(x)=\frac{d^2}{dx^2}\left[1+\frac{a^2}{12}\sum_{j=1}^{p}j^2\alpha_j\frac{d^2}{dx^2}\right]u(x)+O\left(a^4\right) \tag{56}$$

The Padé approximant of order [2,2] for the pseudo-differential operator can be alternatively used, as proposed by Rosenau [64]:

$$\left[\sum_{j=1}^{p}\frac{4\alpha_j}{(ja)^2}\sinh^2\left(\frac{ja}{2}\frac{d}{dx}\right)\right]u(x)=\frac{\frac{d^2}{dx^2}}{1+\frac{a^2}{12}\sum_{j=1}^{p}j^2\alpha_j\frac{d^2}{dx^2}}u(x)+\ldots \tag{57}$$

The pseudo-differential equation is then approximated by a linear second-order differential equation which is Eringen's type nonlocal model defined by

$$\left(1-\beta\overline{l_c}^2\right)\frac{d^2\overline{u}}{dx^2}+\beta\overline{u}=0 \tag{58}$$

where the nonlocal length scale ratio calibrated from:

$$l_c^2=a^2\sum_{j=1}^{p}j^2\frac{\alpha_j}{12} \tag{59}$$

For direct neighbor interaction, this general equation simplifies to:

$$\alpha_j=0 \text{ if } j\geq 2 \quad\Rightarrow\quad l_c^2=\frac{a^2}{12} \tag{60}$$

For fixed-fixed boundary conditions, the dimensionless frequency of the approximated nonlocal rod (of Eringen's type) is then given by

$$\beta=\frac{\pi^2}{1+\frac{\pi^2 l_c^2}{L^2}}=\frac{\pi^2}{1+\frac{\pi^2}{12n^2}\sum_{j=1}^{p}j^2\alpha_j} \tag{61}$$

Equation (61) is an approximation of Eq. (54) which is the exact eigenfrequency of the generalized lattice problem.

3 Discrete and Nonlocal Beams

3.1 Hencky-Bar-Chain Model

The microstructured beam model considered in this part (or lattice beam model) comprises rigid periodic segments linked by concentrated elastic rotational springs, sometimes referred to as Hencky's chain model (Hencky [11]-see also the discussion in Silverman [71] or El Naschie [15]). Hencky [11] introduced this discrete elastic model for capturing the behaviour of continuous Euler-Bernoulli beam elements for an infinite number of elements. Among various discrete structural problems (in-plane buckling of columns, lateral-torsional buckling of beams, buckling of curved elements, buckling of frames, etc.), Hencky [11] solved the buckling problem of this Hencky-Bar-Chain model for some finite number of elements (typically for two, three and four elements). Exact solutions of this buckling problem whatever the number of elements has been found later by Wang [12, 13] from the exact resolution of the second-order linear difference equation. In this chapter, exact solutions of the buckling and vibration of the Hencky-Bar-Chain model (labelled as Hencky beam model) are presented for general boundary conditions. Hencky's Bar Chain model has been widely used in the literature (see for instance Zaslavsky [72], El Naschie [15] or more recently Challamel et al. [56] or Wang et al. [1]). A nonlocal elastic Euler-Bernoulli beam model is then built from the lattice difference equations of the Hencky-Bar-Chain formulation. The nonlocal model is similar to a stress gradient nonlocal Euler-Bernoulli beam, where the nonlocality is of Eringen type [3], although the length scale of the nonlocal model may also differ for statics or dynamics applications [56]. Recently, Challamel et al. [73] compared equation and results of the continualized nonlocal beam models to those of the lattice models and highlighted the special role of boundary conditions. Wang et al. [74] derived a set of lattice-based continualized boundary conditions (static and kinematic boundary conditions) by using the displacement field at the adjacent nodes of the boundary nodes.

In this chapter, the buckling, vibration and bending of a Hencky-Bar Chain structural system are studied. Hencky-Bar Chain is represented in Fig. 4 with simply supported boundary conditions. Hencky-Bar Chain is a microstructured system (or discrete beam) that comprises n cells or n rigid elements connected by some elastic rotational springs of stiffness C. The length of each rigid segment is denoted by a while the length of the whole structure is L and $L = n \times a$, where n is the number of rigid bars. $C = EI/a$ is the spring stiffness (the scaling law relates the rotational spring stiffness C to the equivalent beam stiffness EI of the continuous Euler-Bernoulli beam achieved for an infinite number n of elements). The beam is loaded by an axial compressive force P, and by the uniformly distributed vertical force q at the nodes. The masses m_i are concentrated at each node i. The scaling law $m_i = \mu a$ can be used for the correspondence between the lattice and the continuous beam and μ is the mass per unit length of the beam (of the equivalent continuum model achieved for an infinite number n of elements).

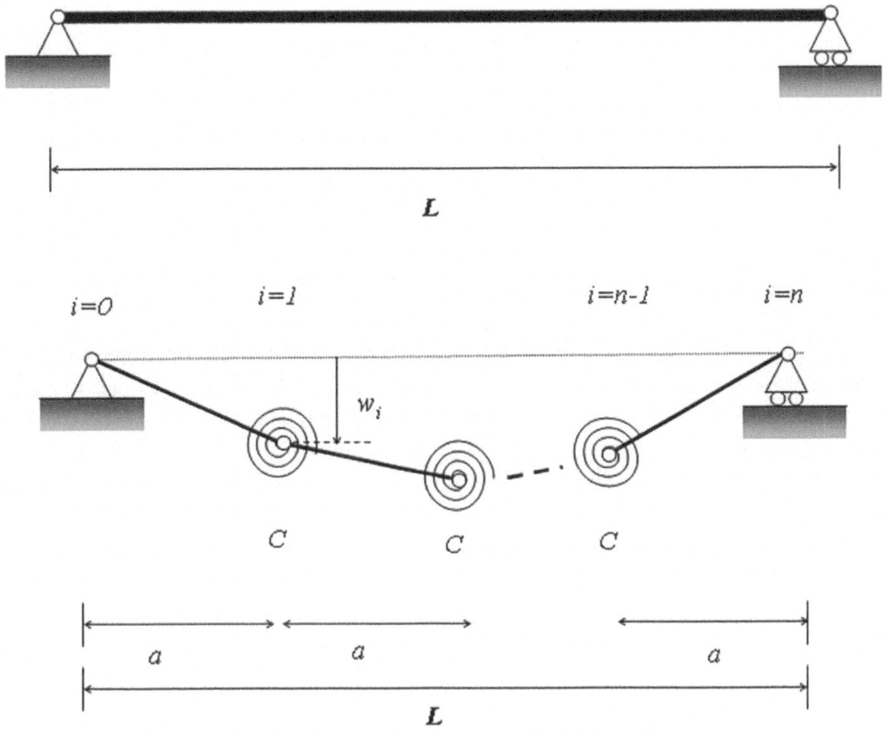

Fig. 4 Microstructured beam model (Hencky-Bar-Chain model); Simply supported boundary conditions $w_0 = w_n = 0$

The elastic potential energy U due to the deformation of elastic rotational springs in the Hencky-Bar-Chain system is written as

$$U = \frac{C}{2} \sum_{i=1}^{n-1} \left(\frac{w_{i+1} - 2w_i + w_{i-1}}{a} \right)^2 \tag{62}$$

where $w_i = w\,(x = x_i = ia)$. The work done W_e by the transverse distributed load and the axial load is obtained from

$$W_e = \sum_{i=1}^{n-1} qaw_i + \frac{P}{2} a \sum_{i=0}^{n-1} \left(\frac{w_{i+1} - w_i}{a} \right)^2 \tag{63}$$

Hencky-Bar-Chain model is composed of concentrated masses of equal value $m_i = \mu a$, of which the motion leads to the following kinetic energy T:

$$T = \sum_{i=1}^{n-1} \frac{1}{2} m_i (\frac{dw_i}{dt})^2 \tag{64}$$

The equation of motion is obtained from application of Hamilton's principle $\int_{t_1}^{t_2} (\delta T - \delta U + \delta W_e)\, dt \ = \ 0$ for the discrete structural system:

$$EI \frac{w_{i+2} - 4w_{i+1} + 6w_i - 4w_{i-1} + w_{i-2}}{a^4} + P \frac{w_{i+1} - 2w_i + w_{i-1}}{a^2} + \mu \frac{d^2 w_i}{dt^2} = q \tag{65}$$

One immediately recognizes that Eq. (65) is a mixed differential-difference equation which is actually the finite difference approximation for a continuous Euler-Bernoulli beam problem. Therefore, the Hencky-Bar Chain system can be treated as the physical support for the finite difference method applied to the continuous Euler-Bernoulli beam (see the remark of Silverman [71], and the physical interest of Hencky's chain versus the "abstract" numerical-based finite difference method).

The above governing equation in space can also be directly obtained from the discrete constitutive law and the second-order equilibrium equations for the bending problem

$$M_i \ = EI \frac{w_{i-1} - 2w_i + w_{i+1}}{a^2} \tag{66}$$

$$\frac{M_{i-1} - 2M_i + M_{i+1}}{a^2} + P \frac{w_{i-1} - 2w_i + w_{i+1}}{a^2} + \mu \frac{d^2 w_i}{dt^2} = q \tag{67}$$

3.2 Continualised Nonlocal Beam Model

As followed for the axial lattice, the discrete equations of the beam lattice (Hencky-Bar-Chain model) are continualized to derive an equivalent nonlocal beam continuum. The relation between the discrete and continuous displacements $w_i = w\,(x = x_i = ia)$ also requires a smooth function as:

$$w\,(x + a, t) = \sum_{k=0}^{\infty} \frac{a^k}{k!} \frac{\partial^k w(x,t)}{\partial x^k} = [e^{a\partial/\partial x}]w(x,t) \tag{68}$$

The following pseudo-differential operators can be expressed as:

$$w_{i-1} + w_{i+1} - 2w_i = \left[e^{a\partial/\partial x} + e^{-a\partial/\partial x} - 2\right] w\,(x,t) = 4\sinh^2\left(\frac{a}{2}\frac{\partial}{\partial x}\right) w\,(x,t) \tag{69}$$

and

$$
\begin{aligned}
& w_{i+2} - 4w_{i+1} + 6w_i - 4w_{i-1} + w_{i-2} = \\
& \left[e^{2a\partial/\partial x} - 4e^{a\partial/\partial x} + 6 - 4e^{-a\partial/\partial x} + e^{-2a\partial/\partial x}\right] w(x,t) = \\
& 16\sinh^4\left(\frac{a}{2}\frac{\partial}{\partial x}\right) w(x,t)
\end{aligned} \tag{70}
$$

The generalized governing equation for bending problem is then expressed as follows

$$
\mu\frac{\partial^2 w}{\partial t^2} + 16\frac{EI}{a^4}\sinh^4\left(\frac{a}{2}\frac{\partial}{\partial x}\right) w + \frac{4P}{a^2}\sinh^2\left(\frac{a}{2}\frac{\partial}{\partial x}\right) w = q \tag{71}
$$

The pseudo-differential operator can be efficiently approximated by Padé's approximant (Rosenau [33]; Wattis [75] or Andrianov et al. [49]):

$$
\frac{4}{a^2}\sinh^2\left(\frac{a}{2}\frac{\partial}{\partial x}\right) = \frac{\frac{\partial^2}{\partial x^2}}{1 - l_c^2\frac{\partial^2}{\partial x^2}} + \ldots \quad \text{with } l_c^2 = \frac{a^2}{12} \tag{72}
$$

The Padé approximant is applied to the lattice beam equations. Hencky-Bar-Chain model can then be approximated by a nonlocal continuous formulation given by

$$
\begin{aligned}
& \mu\left(1 - 2l_c^2\frac{\partial^2}{\partial x^2}\right)\frac{\partial^2 w}{\partial t^2} + EI\frac{\partial^4 w}{\partial x^4} + P\left(1 - l_c^2\frac{\partial^2}{\partial x^2}\right)\frac{\partial^2 w}{\partial x^2} \\
& = \left(1 - 2l_c^2\frac{d^2}{dx^2}\right) q
\end{aligned} \tag{73}
$$

which can be considered as a nonlocal continuous approximation of the lattice beam model. For the pure buckling problem ($\mu = 0$ and $q = 0$), Eq. (73) simplifies in

$$
EI\frac{d^4 w}{dx^4} + P\left(1 - l_c^2\frac{d^2}{dx^2}\right)\frac{d^2 w}{dx^2} = 0 \tag{74}
$$

Equation (74) can be equivalently obtained from a nonlocal Euler-Bernoulli beam where the nonlocality is of Eringen's type (in the sense that the bending moment-curvature differential law is a stress gradient model, first introduced by Eringen [3] for one-dimensional or three-dimensional nonlocal elasticity):

$$
M - l_c^2\frac{d^2 M}{dx^2} = EI\frac{d^2 w}{dx^2} \quad \text{and} \quad \frac{d^2 M}{dx^2} = -P\frac{d^2 w}{dx^2} \tag{75}
$$

A review of various nonlocal beam formulations (based on Eringen's nonlocality) applied to Euler-Bernoulli, Bresse-Timoshenko or higher-order kinematics is available in Elishakoff et al. [5] or Challamel [76]. Equation (75) shows that the

Hencky-Bar-Chain system can be captured by an Eringen's type nonlocal elastic model [3] applied at the beam scale (stress gradient model), with a scaling factor $l_c = a/\left(2\sqrt{3}\right)$ of the nonlocal model calibrated from the length of the rigid elements. For the uncoupled bending problem ($\mu = 0$ and $P = 0$), Eq. (73) reduces to

$$EI\frac{d^4w}{dx^4} = q - 2l_c^2\frac{d^2q}{dx^2} \tag{76}$$

This above differential equation can be equivalently derived from the following second-order differential equations:

$$M - 2l_c^2\frac{d^2M}{dx^2} = EI\frac{d^2w}{dx^2} \quad \text{and} \quad \frac{d^2M}{dx^2} = q \tag{77}$$

One recognizes an Eringen's based nonlocal beam model, but with a scaling factor $l_c = a/\sqrt{6}$, which is different from the buckling problem. Finally, the pure vibration problem ($q = 0$ and $P = 0$), is governed by

$$\mu\left(1 - 2l_c^2\frac{\partial^2}{\partial x^2}\right)\frac{\partial^2 w}{\partial t^2} + EI\frac{\partial^4 w}{\partial x^4} = 0 \tag{78}$$

This Rayleigh-type equation (in the sense that the nonlocal bending wave equation is corrected by some nonlocal rotary effects similar to the rotary effect introduced by Bresse (1859) and Rayleigh [55] for correcting the Euler-Bernoulli beam model) is equivalent to that considered in an Eringen's based nonlocal model applied at the beam scale (see Challamel [76]; Zhang et al. [77]):

$$M - 2l_c^2\frac{d^2M}{dx^2} = EI\frac{d^2w}{dx^2} \quad \text{and} \quad \frac{d^2M}{dx^2} = -\mu\frac{\partial^2 w}{\partial t^2} \tag{79}$$

The small length scale parameter $l_c = a/\sqrt{6}$ appears again in the vibration analysis (Challamel et al. [78]; Wang et al. [79]). A surprising finding is that the calibration of Eringen's length scale parameter is apparently dependent on the type of analysis, namely bending, buckling or vibration (see the discussion in Challamel et al. [78]; Wang et al. [79]). This would suggest that the Hencky-Bar-Chain for statics/dynamics problems is not strictly speaking captured by a stress gradient of Eringen's type with a constant length scale. Regarding this length-scale dependence on the type of analysis (buckling, vibration, static bending), Wang et al. [79] calibrated the nonlocal length scale with respect to the axial load intensity.

It is also possible to change the point of view, and to consider local constitutive laws but additional nonlocal inertia contributions.

$$M = EI\frac{\partial^2 w}{\partial x^2} \quad \text{and} \quad \frac{\partial^2 M}{\partial x^2} = -\mu\frac{\partial^2 w}{\partial t^2} + 2\mu l_c^2\frac{\partial^4 w}{\partial x^2\partial t^2} \tag{80}$$

This set of equation can be obtained from a "local" elastic potential energy formula and a revised form (or "nonlocal") of kinetic energy:

$$U[w] = \frac{1}{2}\int_0^L EI(\frac{\partial^2 w}{\partial x^2})^2 dx$$

$$\text{and} \quad T[w] = \frac{1}{2}\int_0^L \left(\mu(\frac{\partial w}{\partial t})^2 + 2\mu {l_c}^2(\frac{\partial^2 w}{\partial x \partial t})^2\right) \mathrm{d}x \tag{81}$$

This modification of the kinetic energy in the vibration analysis of microstructures is followed by Mindlin [53] and by Polyzos and Fotiadis [54]. The enriched kinetic energy Eq. (81) clearly contains an additional nonlocal rotary inertia.

An alternative model is derived based on the nonlocal corrections of the moment-curvature relationship and the balance equation with a nonlocal rotary term:

$$M - l_c^2 \frac{\partial^2 M}{\partial x^2} = EI\frac{\partial^2 w}{\partial x^2} \quad \text{and} \quad \frac{\partial^2 M}{\partial x^2} = -\mu\frac{\partial^2 w}{\partial t^2} + \mu l_c^2 \frac{\partial^4 w}{\partial x^2 \partial t^2} \tag{82}$$

The coupling of both equations in Eq. (82) again gives a bending wave equation very close to Eq. (78), when the higher-order term in l_c^4 is neglected:

$$\mu\left(1 - 2l_c^2\frac{\partial^2}{\partial x^2}\right)\frac{\partial^2 w}{\partial t^2} + \left(EI + \mu l_c^4 \frac{\partial^2}{\partial t^2}\right)\frac{\partial^4 w}{\partial x^4} = 0 \tag{83}$$

In order to resolve the apparent paradox of the dependence of Erigen's length scale on the type of problem considered, we develop herein another point of view which considers that the Hencky-Bar-Chain system behaves like a nonlocal beam with just one small length scale but with modified equilibrium equations.

This nonlocal model can be elaborated from the discrete equilibrium equation and the discrete formulation of the bending constitutive law Eq. (66), continualized as follows:

$$M(x) = \frac{4EI}{a^2}\sinh^2\left(\frac{a}{2}\frac{\partial}{\partial x}\right) w(x) \tag{84}$$

When applying the Padé approximant to the pseudo-differential operator in Eq. (84), Eringen's nonlocal elastic constitutive law with only one length scale is clearly recognized from this last equation:

$$M - l_c^2\frac{\partial^2 M}{\partial x^2} = EI\frac{\partial^2 w}{\partial x^2} \quad \text{and} \quad l_c = \frac{a}{2\sqrt{3}} \tag{85}$$

The governing equation can also be built from the continualization of Eq. (67), which again is approximated using a Padé approximant, as:

$$\frac{\partial^2 M}{\partial x^2} + P\frac{\partial^2 w}{\partial x^2} + \mu\left(1 - l_c^2\frac{\partial^2}{\partial x^2}\right)\frac{\partial^2 w}{\partial t^2} = \left(1 - l_c^2\frac{d^2}{dx^2}\right)q \tag{86}$$

The governing equation is no more local, and has been enriched with some nonlocal rotary inertia and an additional external load term. It is then possible to solve the paradox of apparent dependence of the length scale, with a nonlocal model coupled to modified nonlocal equilibrium equation. The nonlocal generalization of Newton's second law has already been formulated in a different context by Milton and Willis [80] or Charlotte and Truskinovsky [81]. For the beam lattice model considered in this chapter (Hencky-Bar-Chain model), the nonlocal moment-curvature relationship and the governing equation are written by

$$M = EI\left\langle\frac{\partial^2 w}{\partial x^2}\right\rangle \quad \text{and} \quad \left\langle\frac{\partial^2 M}{\partial x^2}\right\rangle + P\left\langle\frac{\partial^2 w}{\partial x^2}\right\rangle + \mu\frac{\partial^2 w}{\partial t^2} = q \tag{87}$$

where the angle bracket $\langle\rangle$ represents the nonlocal average operator associated with the pseudo-differential operator, given by Eq. (84) for the moment-curvature law, or Eq. (85) for its Padé approximant.

This correction Eq. (86) in the equilibrium equation may resolve the paradox of the dependence of length scale on the analysis type (buckling or vibrations analyses). When correcting both the balance equation and the constitutive law by nonlocal terms, the nonlocal length scale calibrated from the beam lattice is no more dependent on the configuration.

3.3 Buckling and Vibrations Analyses of Discretized Beam

In this section, the buckling and vibration solutions for the Hencky beam model with simply supported ends are presented. It should be noted that the post-buckling behavior of Hencky-bar-chain (also called as the discrete Euler beam) has been studied by Domokos [82], and more recently Challamel et al. [83]. We shall restrict the presentation to linear behavior.

We first solve the buckling problem of the Hencky-Bar-Chain system, formulated by a second-order linear difference equation:

$$Pw_i + \frac{EI}{a^2}\left(w_{i+1} - 2w_i + w_{i-1}\right) = 0 \quad \text{with} \quad w_0 = w_n = 0 \tag{88}$$

The exact analytical solution of this problem has been given by Wang [12, 13] or Seide [84]. By introducing dimensionless parameters, the second-order difference equation can be transformed to

$$w_{i+1} + \left(\frac{\bar{\beta}}{n^2} - 2\right) w_i + w_{i-1} = 0 \quad \text{with} \quad \bar{\beta} = \frac{PL^2}{EI} \tag{89}$$

The characteristic equation is obtained by replacing the displacement expressed as a power function $w_i = W\lambda^i$ in Eq. (89) which leads to

$$\lambda + \frac{1}{\lambda} = 2 - \frac{\bar{\beta}}{n^2} \tag{90}$$

which again is a palindromic equation. This last equation admits the following two solutions:

$$\lambda_{1,2} = 1 - \frac{\bar{\beta}}{2n^2} \pm j\sqrt{1 - \left(1 - \frac{\bar{\beta}}{2n^2}\right)^2} \quad \text{with} \quad j^2 = -1$$

$$\text{and for} \quad \left|1 - \frac{\bar{\beta}}{2n^2}\right| < 1 \tag{91}$$

It can be shown that the shape of solution w_i changes for the other cases if $\left|1 - \frac{\bar{\beta}}{2n^2}\right| \geq 1$ (in term of hyperbolic functions). In this particular case, the solution is identically vanishing with the considered boundary conditions. The solution of the characteristic equation is therefore given, with the conditions Eq. (91), by

$$\lambda_{1,2} = \cos\theta \pm j\sin\theta \quad \text{with} \quad \theta = \arccos\left(1 - \frac{\bar{\beta}}{2n^2}\right) \tag{92}$$

The general solution for the linear difference equation can be expressed in a trigonometric function:

$$w_i = A\cos(\theta i) + B\sin(\theta i) \quad \text{with} \quad \theta = \arccos\left(1 - \frac{\bar{\beta}}{2n^2}\right) \tag{93}$$

The two boundary conditions require a fundamental buckling mode $w_i = B\sin(\theta i)$ and then the formula for buckling load is given by:

$$\sin(\theta n) = 0 \quad \Rightarrow \quad \theta n = \pi \quad \Rightarrow \quad \cos\frac{\pi}{n} = 1 - \frac{\bar{\beta}}{2n^2}$$
$$\Rightarrow \quad \bar{\beta} = 4n^2 \sin^2\left(\frac{\pi}{2n}\right) \tag{94}$$

Equation (94) is consistent with the buckling load reported by Wang [12, 13]. Challamel et al. [85] obtained the same result with an alternative method based on recursive formula and Chebyschev polynomials. More generally, the higher eigenvalues are obtained from $\theta n = k\pi$ with the exact k-th buckling load:

$$\bar{\beta}_{k,n} = 4n^2 \sin^2\left(\frac{k\pi}{2n}\right) \tag{95}$$

We are mainly interested in the calculation of the critical buckling load, i.e. $k = 1$. However, we could also compare the exact formulae associated with the higher modes, by approximating the Eringen's nonlocal buckling solution via:

$$\bar{\beta}_{k,n} = \frac{(k\pi)^2}{1 + \frac{(k\pi)^2}{n^2}\left(\frac{l_c}{a}\right)^2} \tag{96}$$

The comparison between Eqs. (95) and (96) shows that the best matching for the critical buckling load ($k = 1$) is different from the highest buckling load (i.e., $k = n$):

$$\begin{aligned} & k = 1 \quad \Rightarrow \quad e_0^2 = \left(\frac{l_c}{a}\right)^2 = \frac{1}{12} \\ \text{and} \quad & k = n \quad \Rightarrow \quad e_0^2 = \left(\frac{l_c}{a}\right)^2 = \frac{1}{4} - \frac{1}{\pi^2} \end{aligned} \tag{97}$$

Andrianov et al. [49] obtained both values of e_0 reported in Eq. (97) via the matching of low frequency and high frequency of an axial discrete chain. As highlighted from the differential equations of the continualised nonlocal beam model, the value $e_0 = 1/\left(2\sqrt{3}\right) \approx 0.289$ appears again for the buckling problem when n is large. The last value $e_0 = \sqrt{1/4 - 1/\pi^2} \approx 0.386$ that is valid for the highest buckling loads (same as axial dynamics) corresponds to that in Eringen [3] in which the nonlocal model is compared with the dispersive wave equations of the Born-Kármán model (Eringen [3]; see also Challamel et al. [37] for a two-length-scale nonlocal model).

For practical applications, the exact buckling load formulae restricted to the critical bucking case obtained for $k = 1$ can be asymptotically expanded as follows,

$$\bar{\beta} = 4n^2 \sin^2\left(\frac{\pi}{2n}\right) = \pi^2\left(1 - \frac{\pi^2}{12n^2}\right) + o\left(\frac{1}{n^4}\right) \tag{98}$$

This formulae can be re-expressed in a more general equation. As mentioned by Seide [84], the dimensionless buckling load for the clamped-free, clamped-clamped or hinged-hinged boundary conditions can be formulated in a general form, which covers the previous structural case:

$$\frac{P_{\text{discrete}}}{P_{\text{E}}} = 1 - \frac{P_{\text{E}} L^2}{12EI}\frac{1}{n^2} + o\left(\frac{1}{n^4}\right) \tag{99}$$

where P_E is the buckling load of the Euler-Bernoulli column. A very similar length scale dependence would have been obtained from the buckling formulae of a hinge-hinge nonlocal Euler-Bernoulli beam model based on Eringen's nonlocal theory [56]:

$$\frac{P_{\text{nonlocal}}}{P_{\text{E}}} = \frac{1}{1+\pi^2\left(\frac{l_c}{L}\right)^2} = 1 - \frac{\pi^2}{12n^2} + O\left(\frac{1}{n^4}\right)$$
$$\text{with} \quad \frac{l_c}{a} = \frac{1}{2\sqrt{3}} \quad \text{and} \quad P_E = \frac{\pi^2 EI}{L^2} \tag{100}$$

More generally, we find that for other classical boundary conditions such as clamped-free, clamped-clamped, hinged-hinged or clamped-hinged ends, the buckling load of the nonlocal column can be calculated from the following formulae (see Wang et al. [86], Reddy and Pang [87], Challamel et al. [78]):

$$\frac{P_{\text{nonlocal}}}{P_{\text{E}}} = \frac{1}{1+\frac{P_{\text{E}} l_c^2}{EI}} = 1 - \frac{P_{\text{E}} L^2}{12EI}\frac{1}{n^2} + O\left(\frac{1}{n^4}\right) = 1 - \frac{P_{\text{E}} a^2}{12EI} + O\left(\frac{a^4}{L^4}\right) \tag{101}$$

The length scale dependence is of order 2 which is similar to the asymptotic formulae of the Hencky-Bar- Chain model. The lattice column presents a lower buckling load than that of the "local" Euler column. From Eq. (101), and in agreement with Salvadori [46] and Wang [13], the convergence rate of finite difference method shows that the error in the buckling load is of the a^2-type ($a = L/n$) in the present homogeneous difference eigenvalue problem. The Finite Difference approximation (or its equivalent lattice formulation) for linear buckling problem provides a lower bound buckling load to its corresponding continuous local problem (the comparison between local and nonlocal continuum formulations yield the same tendency).

The eigenfrequencies of the Hencky-Chain-Bar can be exactly calculated in the same manner, as detailed for instance by Leckie and Lindberg [88] or more recently Santoro and Elishakoff [89] (for the equivalent finite difference formulation), from the following linear fourth-order difference Eq. (102):

$$w_{i+2} - 4w_{i+1} + 6w_i - 4w_{i-1} + w_{i-2} - \frac{\Omega^2}{n^4} w_i = 0 \quad \text{with} \quad \Omega^2 = \omega^2 \frac{\mu L^4}{EI} \tag{102}$$

Similar to the buckling problem, the vibration mode can also assume a power function $w_i = W\lambda^i$, and then by substituting the function into Eq. (102), one obtains the following characteristic equation:

$$\left(\frac{1}{\lambda} + \lambda\right)^2 - 4\left(\frac{1}{\lambda} + \lambda\right) + 4 - \frac{\Omega^2}{n^4} = 0 \tag{103}$$

This quartic equation generates the following four solutions [89]:

$$\lambda_{1,2} = \cos\theta \pm j\sin\theta \quad \text{and} \quad \lambda_{3,4} = 2-\cos\theta \pm \sqrt{(2-\cos\theta)^2 - 1}$$
$$\text{with } \theta = \arccos\left(1 - \frac{\Omega}{2n^2}\right) \tag{104}$$

For simply supported boundary conditions, the eigenmodes are obtained from the trigonometric shape function $w_i = B\sin(\theta i)$. The fundamental eigenfrequency is then obtained by injecting the trigonometric solution in the fourth-order difference Eq. (102), as obtained by Leckie and Lindberg [88]:

$$\Omega = 4n^2\sin^2\left(\frac{\pi}{2n}\right) = \pi^2\left(1 - \frac{\pi^2}{12n^2}\right) + O\left(\frac{1}{n^4}\right) \tag{105}$$

This formulae is very similar in its form to the non-dimensional buckling formulae. From Eq. (105), we then have the square of the fundamental frequency as

$$\frac{\omega_{discrete}^2}{\omega_E^2} = 1 - \frac{\pi^2}{6n^2} + O\left(\frac{1}{n^4}\right) \quad \text{with } \omega_E^2 = \frac{EI}{\mu}\left(\frac{\pi}{L}\right)^4 \tag{106}$$

As for the buckling problem, the lattice beam also presents lower natural frequencies than those of the continuous "local" beams.

$$\frac{\omega_{nonlocal}^2}{\omega_E^2} = \frac{1}{1+\left(\frac{\pi l_c}{L}\right)^2} = 1 - \frac{\pi^2}{6n^2} + O\left(\frac{1}{n^4}\right) \quad \text{with } \frac{l_c}{a} = \frac{1}{\sqrt{6}} \tag{107}$$

Eringen's nonlocal model is an efficient continuous theory to get an approximation of the fundamental frequency of the Hencky-Bar-Chain model while the Finite Difference approximation gives a lower bound of the asymptotic Euler-Bernoulli beam.

The capability of the nonlocal beam theory to predict the eigenfrequencies of the Hencky-Bar-Chain model is also valid for higher frequencies. The frequency spectrum can be obtained from $\theta n = k\pi$ where k represents the k-th eigenfrequency. The exact formulae valid for all eigenfrequencies of the Hencky- Bar-Chain model includes the one of the fundamental frequency:

$$\Omega_{k,n} = 4n^2\sin^2\left(\frac{k\pi}{2n}\right) \quad \Rightarrow \quad \Omega^2{}_{k,n} = 16n^4\sin^4\left(\frac{k\pi}{2n}\right) \tag{108}$$

The nonlocal continuous approximation obtained from Eringen's beam theory:

$$\Omega_{k,n}^2 = \frac{(k\pi)^4}{1+\frac{(k\pi)^2}{n^2}\left(\frac{l_c}{a}\right)^2} \tag{109}$$

The comparison between Eqs. (108) and (109) also shows that the best match for the fundamental frequency ($k = 1$) and the highest natural frequency ($k = n$) have a discrepancy:

$$k = 1 \quad \Rightarrow \quad e_0^2 = \left(\frac{l_c}{a}\right)^2 = \frac{1}{6}$$
$$\text{and} \quad k = n \quad \Rightarrow \quad e_0^2 = \left(\frac{l_c}{a}\right)^2 = \frac{\pi^2}{16} - \frac{1}{\pi^2} \tag{110}$$

The fundamental eigenfrequency ($k = 1$) can be calibrated from $e_0 = 1/\left(\sqrt{6}\right) \approx 0.408$, whereas $e_0 = \sqrt{\pi^2/16 - 1/\pi^2} \approx 0.718$ is valid for the highest frequencies (particularly $k = n$).

It is possible to consider the coupling between buckling and vibrations. In this case, the difference eigenvalue problem is governed by the following fourth-order linear difference equation:

$$w_{i+2} - 4w_{i+1} + 6w_i - 4w_{i-1} + w_{i-2} - \frac{\Omega^2}{n^4} w_i + \frac{\bar{\beta}}{n^2}\left(w_{i+1} - 2w_i + w_{i-1}\right) = 0 \tag{111}$$

Following the procedure detailed for the free vibration of the Hencky-Bar-Chain, the vibration mode can be assumed as $w_i = W\lambda^i$, and then by substituting the shape function into Eq. (111), one obtains the characteristic palindromic equation:

$$\left(\frac{1}{\lambda} + \lambda\right)^2 - \left(4 - \frac{\bar{\beta}}{n^2}\right)\left(\frac{1}{\lambda} + \lambda\right) + 4 - \frac{2\bar{\beta}}{n^2} - \frac{\Omega^2}{n^4} = 0 \tag{112}$$

Again, this quartic equation admits four solutions:

$$\lambda_{1,2} = \cos\theta \pm j\sin\theta$$
$$\text{and} \quad \lambda_{3,4} = 2 - \frac{\bar{\beta}}{2n^2} - \cos\theta \pm \sqrt{\left(2 - \frac{\bar{\beta}}{2n^2} - \cos\theta\right)^2 - 1}$$
$$\text{with} \quad \theta = \arccos\left(1 - \frac{\bar{\beta}}{4n^2} - \frac{1}{2n^2}\sqrt{\frac{\bar{\beta}^2}{4} + \Omega^2}\right) \tag{113}$$

These solutions include the free vibration case treated by Leckie and Lindberg [88] and Santoro and Elishakoff [89].

For simply supported boundary conditions, the eigenmodes could be assumed as the trigonometric shape function $w_i = B\sin(\theta i)$ and then we have

$$\sin(\theta n) = 0 \quad \Rightarrow \quad \theta = \frac{\pi}{n}$$

$$\Rightarrow \quad \frac{\bar{\beta}}{2} + \sqrt{\frac{\bar{\beta}^2}{4} + \Omega^2} = 4n^2 \sin^2 \frac{\pi}{2n} \tag{114}$$

The frequency-normal force equation can be equivalently reformulated in a Dunkerley type interaction formulae valid for the Hencky-Bar-Chain system:

$$\left[\frac{\Omega}{4n^2 \sin^2 \frac{\pi}{2n}}\right]^2 + \frac{\bar{\beta}}{4n^2 \sin^2 \frac{\pi}{2n}} = 1 \tag{115}$$

Dunkerley's line, or Melan's formula is known to relate linearly the square frequency and the load parameter (see Tarnai [90]). Even for the continuous beam problem, it is also known that the linearity between the square frequency and the load parameter valid for the simply supported boundary conditions does not exactly apply to other boundary conditions, even though it provides a closed approximation (see for instance Massonnet [91] and Galef [92]).

This exact relationship between load and frequency for the Hencky-Bar-Chain system can be generally approximated using the asymptotic expansion as follows

$$\left(\frac{\Omega}{\pi^2}\right)^2 = 1 - \frac{\pi^2}{6n^2} - \frac{\bar{\beta}}{\pi^2}\left(1 - \frac{\pi^2}{12n^2}\right) + o\left(\frac{1}{n^4}\right) \tag{116}$$

This last formulae valid for the Hencky-Bar-Chain system can now be matched to that obtained from Eringen's method based on the single length scale $l_c = e_0 a = e_0 L/n$ (see recently Wang et al. [79]):

$$\left(\frac{\Omega}{\pi^2}\right)^2 = \frac{1}{1 + e_0^2 \frac{\pi^2}{n^2}} - \frac{\bar{\beta}}{\pi^2} = 1 - e_0^2 \frac{\pi^2}{n^2} - \frac{\bar{\beta}}{\pi^2} + o\left(\frac{1}{n^4}\right) \tag{117}$$

The equivalent length scale coefficient of Eringen's model can be calculated from the discrete beam model, by comparing Eq. (116) with Eq. (117), thus leading to a load-dependent nonlocal length scale:

$$e_0^2 = \frac{1}{6} - \frac{\bar{\beta}}{12\pi^2} \tag{118}$$

In this chapter, the nonlocal length scale e_0 is calibrated from the exact relationship between normal force and frequency in the Hencky-Bar-Chain model, a method which differs from that in Wang et al. [79] who derived the same relationship by continualizing the difference equations of the lattice beam.

4 Discrete and Nonlocal Plates

4.1 *Hencky-Bar-Chain Net*

In this part, we propose a microstructured plate model (or Hencky-bar-net model) for the vibration of an axially loaded rectangular lattice plate with simply supported edges. The idea to investigate continuous plates with discrete elements is old, and is behind the development of modern numerical methods such as Finite Element Models (Absi and Prager [93]; Zienkiewicz and Taylor [94]). Wieghardt [95], Riedel [96] or Hrennikoff [97] used equivalent truss elements to investigate continuous structural problems such as plates or two-dimensional elasticity problems. The generalization of the Hencky-Bar-Chain model to plates with concentrated stiffnesses is probably more recent. Zaslavsky [72] and Wu [98] developed a beam grid net by connecting two Hencky-Bar-Chain systems. A complete Hencky-Bar-Net model (in the sense that the difference equations of the lattice plate model exactly coincide with the finite difference formulation of Kirchhoff-Love plate model) was initially developed at the end of the 1980s with the works of Wifi et al. [14], followed by El Naschie [15]. The governing equations for the Hencky-Bar-Net model are equivalent to the finite difference formulation of Kirchhoff-Love plate model. For sufficiently small spacing, Hencky-Bar-Net then asymptotically converges towards a Kirchhoff-Love plate model. The connection between Hencky-Bar-Net plate model and a continualized nonlocal Kirchhoff-Love plate model is more recent [99].

Some studies have been focused on modelling microstructured lattice membrane structures (Ari and Eringen [100]; Rosenau [64]; Andrianov and Awrejcewicz [48]; Lombardo and Askes [101]; Hérisson et al. [102]). In this last part of the chapter, we show that the Hencky-Bar-Net model (or the lattice plate model) can be reasonably well approximated by a nonlocal plate theory whose governing equation has a similar form as that of the Eringen's nonlocal plate theory. Exact solutions for buckling and vibration problems in the Hencky-Bar-Net model and the continualized nonlocal plate model are obtained. The small length scale coefficient e_0 of the Eringen's nonlocal model is calibrated for both the vibration and the buckling cases. It is found that e_0 is depend on the initial loading, effect of rotary inertia, mode shape and aspect ratio (width to length ratio) (see also Zhang et al. [103, 104]; Challamel et al. [99]). A continualized nonlocal plate model is then built from the difference equations of the Hencky-Bar-Net model. This alternative nonlocal plate theory possesses a length scale which is independent on the structural parameters of the problem.

Consider a rectangular lattice plate subjected to uniform compressive stress σ (see Fig. 5). The plate has four simply supported edges and of L and αL ($0 < \alpha \leq 1$) in length and width.

The bending deformation of the Hencky-Bar net plate model is made possible by rotational springs placed at individual nodes. Repetitive unit cells comprising four rigid beam segments in rectangular alignment connected by springs at center (Wifi et al. [14]; El Naschie [15]) are introduced for modelling the torsional deformation. The masses are lumped at the nodes of Hencky-Bar net plate model.

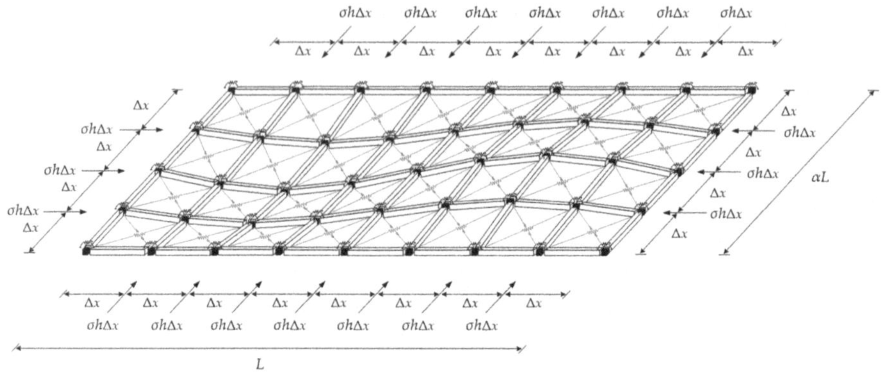

Fig. 5 Hencky-Bar-Net model under initial stress σ

The elastic stain energy potential due to the deformation of the rotational springs at joints is given by

$$U_b = \frac{D}{2}\sum_{i=1}^{n_x-1}\sum_{j=1}^{n_y-1}\left[(\Delta\theta_x^{i,j})^2\frac{\Delta y}{\Delta x} + (\Delta\theta_y^{i,j})^2\frac{\Delta x}{\Delta y} + 2\nu\Delta\theta_x^{i,j}\Delta\theta_y^{i,j}\right] \tag{119}$$

where n_x and n_y are the number of rigid segments in x and y directions, respectively. Δx and Δy are their lengths and they are given by $\Delta x = L/n_x$ and $\Delta y = \alpha L/n_y$. D is the equivalent bending stiffness of the elastic plate. ν is the Poisson's ratio. $\Delta\theta_x$ and $\Delta\theta_y$ are rotations at node (i, j) of the Hencky-Bar net plate model, which can be expressed by

$$\begin{aligned}\Delta\theta_x^{i,j} &= \frac{w_{i+1,j} - 2w_{i,j} + w_{i-1,j}}{\Delta x}\\ \Delta\theta_y^{i,j} &= \frac{w_{i,j+1} - 2w_{i,j} + w_{i,j-1}}{\Delta y}\end{aligned} \tag{120}$$

The total torsional energy in the Hencky-Bar-Net model is written as [15]:

$$U_t = \frac{D}{2}(1-\nu)\sum_{i=1}^{n_x}\sum_{j=1}^{n_y}\left[(\Delta\theta_{xy}^{i,j})^2\frac{\Delta x}{\Delta y} + (\Delta\theta_{yx}^{i,j})^2\frac{\Delta y}{\Delta x}\right] \tag{121}$$

where

$$\begin{aligned}\Delta\theta_{xy}^{i,j} &= \frac{w_{i,j+1} - w_{i+1,j+1}}{\Delta x} - \frac{w_{i,j} - w_{i+1j}}{\Delta x}\\ \Delta\theta_{yx}^{i,j} &= \frac{w_{i,j+1} - w_{i,j}}{\Delta y} - \frac{w_{i+1,j+1} - w_{i+1,j}}{\Delta y}\end{aligned} \tag{122}$$

The kinetic energy of the Hencky-Bar-Net model can be formulated, as

$$T = \frac{1}{2}(m_0 \Delta x \Delta y) \sum_{i=1}^{n_x-1} \sum_{j=1}^{n_y-1} (\frac{dw_{i,j}}{dt})^2 \tag{123}$$

where m_0 is the mass per unit cell area, i.e. $m_0 = M/(\Delta x \Delta y)$ and $M = \rho A h$ is the total mass of beam-grid. The work done by the axial loading in each direction is written

$$\begin{aligned} W_e &= \sigma h \Delta x \Delta y \sum_{i=1}^{n_x} \sum_{j=1}^{n_y-1} (1 - \cos\theta_x^{i,j}) + \sigma h \Delta x \Delta y \sum_{i=1}^{n_y} \sum_{j=1}^{n_x-1} (1 - \cos\theta_y^{i,j}) \\ &\approx \frac{1}{2}\sigma h \Delta x \Delta y \left[\sum_{i=1}^{n_x} \sum_{j=1}^{n_y-1} (\theta_x^{i,j})^2 + \sum_{i=1}^{n_y} \sum_{j=1}^{n_x-1} (\theta_y^{i,j})^2 \right] \end{aligned} \tag{124}$$

where

$$\begin{aligned} \theta_x^{i,j} &= \frac{w_{i+1,j} - w_{i,j}}{\Delta x} \\ \theta_y^{i,j} &= \frac{w_{i,j+1} - w_{i,j}}{\Delta y} \end{aligned} \tag{125}$$

By setting $\Delta y = \Delta x$, the total potential energy function is expressed as

$$\begin{aligned} W &= U - W_e \\ &= \frac{D}{2} \sum_{i=1}^{n_x-1} \sum_{j=1}^{n_y-1} (\Delta\theta_x^{i,j} + \Delta\theta_y^{i,j})^2 \\ &- D(1-\nu) \left[\sum_{i=1}^{n_x-1} \sum_{j=1}^{n_y-1} \Delta\theta_x^{i,j} \Delta\theta_y^{i,j} - \sum_{i=1}^{n_x} \sum_{j=1}^{n_y} (\Delta\theta_{xy}^{i,j})^2 \right] \\ &- \frac{1}{2}(\sigma h \Delta x^2) \sum_{i=1}^{n_x} \sum_{j=1}^{n_y-1} (\theta_x^{i,j})^2 - \frac{1}{2}(\sigma h \Delta x^2) \sum_{i=1}^{n_y} \sum_{j=1}^{n_x-1} (\theta_y^{i,j})^2 \end{aligned} \tag{126}$$

where $U = U_b + U_t$. By resorting to the Hamilton's principle

$$\int_{t_1}^{t_2} (\delta T - \delta W)\, dt = 0 \tag{127}$$

and considering harmonic motion, we have the governing equation for the two-dimensional microstructured beam-grid model given by

$$n_y^4 L^{i,j} - n_y^2 \lambda H_x^{i,j} - n_y^2 \lambda H_y^{i,j} - \hat{\Omega}^2 w_{i,j} = 0 \tag{128}$$

where

$$\begin{aligned}
L^{i,j} &= 20w_{i,j} + (w_{i,j-2} + w_{i,j+2} + w_{i-2,j} + w_{i+2,j}) \\
&\quad - 8(w_{i,j-1} + w_{i,j+1} + w_{i-1,j} + w_{i+1,j}) \\
&\quad + 2(w_{i+1,j-1} + w_{i+1,j+1} + w_{i-1,j-1} + w_{i-1,j+1}) \\
H_y^{i,j} &= w_{i,j-1} - 2w_{i,j} + w_{i,j+1} \\
H_x^{i,j} &= w_{i-1,j} - 2w_{i,j} + w_{i+1,j} \\
\lambda &= \frac{\sigma h(\alpha L)^2}{D} = \frac{N(\alpha L)^2}{D} \\
\hat{\Omega}^2 &= \frac{m_0 \omega^2 (\alpha L)^4}{D}
\end{aligned} \tag{129}$$

where $\hat{\Omega}$ is the vibration frequency for the Hencky-Bar-Net model.

For a Hencky-Bar-Net model with four simply supported edges, an exact solution can be thought for this difference boundary value problem of the Navier type, with a double Fourier sine series as

$$w_{i,j} = w_0 \sin \frac{m\pi i}{n_x} \sin \frac{n\pi j}{n_y} \tag{130}$$

In fact, the displacement field in Eq. (130) converges to the classical continuum solution for rectangular plates with simply supported edges when $n_x \to \infty$ and $n_y \to \infty$. The substitution of Eq. (130) into (128) yields

$$\sin \frac{m\pi i}{n_x} \sin \frac{n\pi j}{n_y} \left[-\frac{\hat{\Omega}^2}{n_y^4} + \frac{2\lambda(\gamma + \chi)}{n_y^2} + 4(\gamma + \chi)^2 \right] = 0 \tag{131}$$

where $\gamma = \cos \frac{m\pi}{n_x} - 1$ and $\chi = \cos \frac{n\pi}{n_y} - 1$. From Eq. (131), the vibration frequency can be expressed explicitly as

$$\hat{\Omega}^2 = 2n_y^2(\gamma + \chi)\left[2n_y^2(\gamma + \chi) + \lambda\right] \tag{132}$$

By assuming that $n_x = n_y/\alpha$ and then setting $n_y \to \infty$, Eq. (132) can be reformulated

$$\hat{\Omega}^2 = \pi^2(n^2 + m^2\alpha^2)\left[\pi^2(n^2 + m^2\alpha^2) - \lambda\right] \tag{133}$$

Equation (133) reduces to the classical solution of vibration frequency for transversely vibrating rectangular plates under uniform axial loadings (see for instance Rao [105]).

4.2 Eringen's Nonlocal Plate Model

In the following, we will try to capture the length scale effects of the micro-structured plate model with a phenomenological nonlocal model called Eringen's nonlocal model [3]. For Eringen's nonlocal plate, the nonlocal constitutive law is characterized by the following differential law between the moment and the curvature variables (see for instance Lu et al. [106]):

$$\begin{aligned} M_{xx} - (e_0 a)^2 \Delta M_{xx} &= D\left[\frac{\partial^2 w}{\partial x^2} + \upsilon \frac{\partial^2 w}{\partial y^2}\right] \\ M_{yy} - (e_0 a)^2 \Delta M_{yy} &= D\left[\frac{\partial^2 w}{\partial y^2} + \upsilon \frac{\partial^2 w}{\partial x^2}\right] \\ M_{xy} - (e_0 a)^2 \Delta M_{xy} &= D(1-\upsilon)\frac{\partial^2 w}{\partial x \partial y} \end{aligned} \tag{134}$$

where the length scale $e_0 a$ can be calibrated from lattice models. The equilibrium equations are in presence of uniform compressive stress:

$$\frac{\partial^2 M_{xx}}{\partial x^2} + 2\frac{\partial^2 M_{xy}}{\partial x \partial y} + \frac{\partial^2 M_{yy}}{\partial y^2} + N\left(\frac{\partial^2 w}{\partial x^2} + \frac{\partial^2 w}{\partial y^2}\right) - m_0 \omega^2 w = 0 \tag{135}$$

leading to the fourth-order partial differential equation, valid for the vibration problem of a nonlocal rectangular plate subjected uniform compressive stress

$$\begin{aligned} & D\left[\frac{\partial^4 w}{\partial x^4} + 2\frac{\partial^4 w}{\partial x^2 \partial y^2} + \frac{\partial^4 w}{\partial y^4}\right] + \sigma h\left(\frac{\partial^2 w}{\partial x^2} + \frac{\partial^2 w}{\partial y^2}\right) \\ & -m_0 \omega^2 w - (e_0 a)^2 \sigma h\left[\frac{\partial^4 w}{\partial x^4} + 2\frac{\partial^4 w}{\partial x^2 \partial y^2} + \frac{\partial^4 w}{\partial y^4}\right] \\ & +(e_0 a)^2 \omega^2 m_0\left(\frac{\partial^2 w}{\partial x^2} + \frac{\partial^2 w}{\partial y^2}\right) = 0 \end{aligned} \tag{136}$$

where the quantity w is the transverse displacement, ω is the vibration frequency, $D = Eh^3/[12(1-\nu^2)]$ denotes the flexural rigidity, h is the thickness of plate, ν is the Poisson ratio, E is Young's modulus. The quantity $m_0 = \rho h$ is the mass inertia and ρ denotes the mass density. Besides, e_0 is the small length scale coefficient, a is the internal characteristic length and $\sigma h = N$ the axial hydrostatic loading.

This equation valid for nonlocal Kirchhoff-Love plate theory can be reformulated as:

$$\left(D - (e_0 a)^2 N\right)\Delta\Delta w + \left(N + (e_0 a)^2 m_0 \omega^2\right)\Delta w - m_0 \omega^2 w = 0 \tag{137}$$

where $\Delta w = \left(\partial^2 w/\partial x^2 + \partial^2 w/\partial y^2\right)$ is the Laplacian operator. This last equation is given by Murmu and Pradhan [107] or Zhang et al. [104]. For the pure buckling problem, Eq. (137) has been obtained by Pradhan and Murmu [108]:

$$\left(D - (e_0 a)^2 N\right) \Delta\Delta w + N \Delta w = 0 \tag{138}$$

For the pure vibrations problem, Eq. (137) has been obtained by Lu et al. [106]:

$$D\Delta\Delta w + (e_0 a)^2 m_0 \omega^2 \Delta w - m_0 \omega^2 w = 0 \tag{139}$$

It is seen from Eq. (139) that the presented nonlocal plate theory reduces to a Kirchhoff plate theory with additional nonlocal rotary effects (as considered by Wang and Wang [109] for Kirchhoff-Love plates with rotary effects).

In case of simply supported boundary conditions, the general solution for the displacement w in Eq. (137) may be expressed as

$$w(x, y) = w_0 \sin \frac{m\pi x}{L} \sin \frac{n\pi y}{\alpha L} \tag{140}$$

where w_0 is an unknown constant, m and n are the number of half waves in x and y directions, respectively. The substitution of Eq. (140) into Eq. (137), generates the following vibration frequency expressed as,

$$\Omega^2 = \frac{\pi^4 (n^2 + m^2\alpha^2)^2 \left(\frac{\alpha L}{a}\right)^2}{\left[e_0^2 n^2 \pi^2 + e_0^2 m^2 \alpha^2 \pi^2 + \left(\frac{\alpha L}{a}\right)^2\right]} - \lambda \pi^2 (n^2 + m^2 \alpha^2) \tag{141}$$

where $\Omega^2 = \omega^2 m_0 (\alpha L)^4 / D$ is the square of dimensionless vibration frequency of nonlocal rectangular plate and $\lambda = \sigma h (\alpha L)^2 / D$ is the dimensionless buckling stress. When setting $e_0 = 0$, Eq. (141) can be simplified to

$$\Omega^2 = \pi^2 (n^2 + m^2 \alpha^2) \left[\pi^2 (n^2 + m^2 \alpha^2) - \lambda\right] \tag{142}$$

It can be checked that Eq. (142) reduces to the classical frequency solution for transversely vibrating rectangular plates under initial loading (Rao [105]).

The buckling load is calculated from $\Omega = 0$ in Eq. (141) and is equal to (see also Pradhan and Murmu [108]):

$$\lambda = \min_{(m,n)} \frac{(n\pi)^2 + (\alpha m \pi)^2}{1 + (e_0 a)^2 \left[\left(\frac{n\pi}{\alpha L}\right)^2 + \left(\frac{m\pi}{L}\right)^2\right]} \tag{143}$$

The frequency is calculated from $\lambda = 0$ in Eq. (141) and is equal to (see also Lu et al. [106]):

$$\Omega^2 = \frac{\left[(n\pi)^2 + (\alpha m\pi)^2\right]^2}{1 + (e_0 a)^2 \left[\left(\frac{n\pi}{\alpha L}\right)^2 + \left(\frac{m\pi}{L}\right)^2\right]} \tag{144}$$

The small length scale parameter existing in the Eringen's nonlocal model can be calculated based on the exact solution of the micro-structured plate model. For the pure buckling problem ($\Omega = 0$ in Eq. (144) as well as $\hat{\Omega} = 0$ in Eq. (132)), by setting their buckling stresses equal, the small length scale coefficient can be calibrated by assuming one half wave in y direction (i.e. $n = 1$) as

$$\lim_{n_y \to \infty} e_0^2 = \frac{1 + m^4\alpha^4}{12(1 + m^2\alpha^2)^2} \tag{145}$$

Note that the buckling stress for one half-wave buckling mode in y direction is $\lambda = \pi^2(1 + m^2\alpha^2)$. Equation (145) shows that the small length scale coefficient e_0 in nonlocal rectangular plates is relative to both the buckling mode m and the aspect ratio α. It is worth mentioning that $e_0 \to \sqrt{1/12} \approx 0.289$ when $\alpha \to 0$, which gives the same small length scale coefficient as that in the buckling of nonlocal beams.

A remarkable result is that the small length scale coefficient e_0 in the nonlocal rectangular plate varies with respect to both buckling modes and aspect ratios (see Fig. 6). It is not independent of the geometrical parameters of the structural problem. It means that Eringen's nonlocal plate model is not asymptotically derived from the lattice plate model, as opposed to the beam comparison problem. It is found that e_0 is in the range of 0.204 to 0.215. It is also confirmed from Fig. 6 that the calibrated value of $e_0 = \sqrt{1/24} \approx 0.204$ in nonlocal rectangular plate is consistent with the proposed Hencky-Bar-Net model subjected to uniform compression. For square plate ($\alpha = 1$) of 1st buckling mode ($m = 1$) rectangular plate $\alpha = 1/2$ of 2nd buckling mode ($m = 2$), we have the same length scale coefficient.

The dependence of the length scale factor to the geometry of the plate and the considered mode is also confirmed in the free vibration analysis. Regarding free vibration problem without axial loading, i.e. $\lambda = 0$, it is possible to show that

$$\lim_{n_y \to \infty} e_0^2 = \frac{1 + m^4\alpha^4}{6(1 + m^2\alpha^2)^2} \tag{146}$$

The small length scale coefficient e_0 in Eq. (146) reduces to $1/\sqrt{6}$ by taking $m = 1$ and $\alpha = 1$, which is also valid for free vibration of nonlocal Euler beam models. Figure 7 shows that varies with respect to both the mode shape m and the aspect ratio α. This discovery is different from the nonlocal beams where $e_0 = 1/\sqrt{6}$ is always a constant, irrespective of boundary conditions and mode shapes.

This apparent dependence of the length scale calibration to some other parameters of the problem can be explained by considereing an alternative nonlocal plate model built from the lattice difference equations.

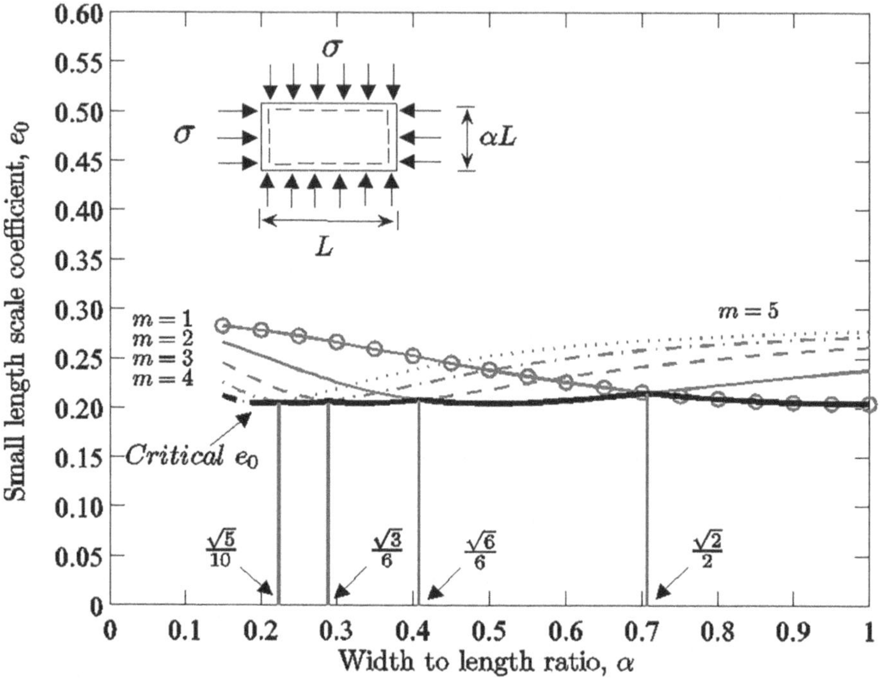

Fig. 6 Calibrated small length scale coefficient with respect to width to length ratio for buckling nonlocal rectangular plate of different modes m—Calibration with respect to the Hencky-Bar-Net model

4.3 Microstructure-Based Nonlocal Plate Model

Starting from the difference Eq. (128), it is possible to continualize the difference equations using pseudo-differential operators (as introduced for the beam lattice):

$$D\frac{16}{a^4}\left[\sinh^2\left(\frac{a}{2}\frac{\partial}{\partial x}\right)+\sinh^2\left(\frac{a}{2}\frac{\partial}{\partial y}\right)\right]^2 w$$
$$+N\frac{4}{a^2}\left[\sinh^2\left(\frac{a}{2}\frac{\partial}{\partial x}\right)+\sinh^2\left(\frac{a}{2}\frac{\partial}{\partial y}\right)\right] w - m_0\omega^2 w = 0 \qquad (147)$$

A Taylor-based asymptotic expansion of the pseudo-differential operators gives:

$$D\left[\Delta^2+\frac{a^2}{6}\Delta\left(\frac{\partial^4}{\partial x^4}+\frac{\partial^4}{\partial y^4}\right)\right] w$$
$$+N\Delta w + N\frac{a^2}{12}\left(\frac{\partial^4}{\partial x^4}+\frac{\partial^4}{\partial y^4}\right) w - m_0\omega^2 w = 0 \qquad (148)$$

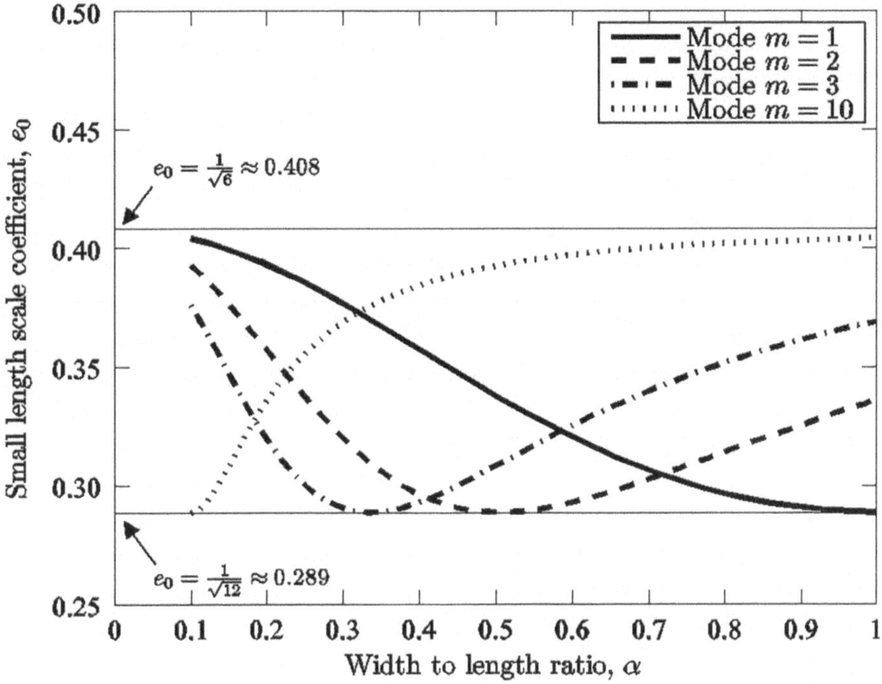

Fig. 7 Calibrated small length scale coefficient with respect to width to length ratio for free vibrating nonlocal rectangular plate of different modes m—Calibration with respect to the Hencky-Bar-Net modellabel

It is possible to multiply Eq. (148) by $1 - a^2/12\ \Delta$ and neglecting higher-order terms, thus leading to the alternative nonlocal equation:

$$D\Delta^2 w + \frac{a^2}{12} D\Delta \left[2\left(\frac{\partial^4}{\partial x^4} + \frac{\partial^4}{\partial y^4} \right) - \Delta^2 \right] w + N\Delta w$$
$$+N\frac{a^2}{12}\left[\frac{\partial^4}{\partial x^4} + \frac{\partial^4}{\partial y^4} - \Delta^2 \right] w - m_0\omega^2 \left(1 - \frac{a^2}{12}\Delta \right) w = 0 \qquad (149)$$

which can be equivalently re-expressed as (see also Challamel et al. [99]; Hache et al. [110])

$$D\Delta\Delta w + \left(N + \frac{a^2}{12} m_0\omega^2 \right) \Delta w - m_0\omega^2 w - \frac{a^2}{6} N \frac{\partial^4 w}{\partial x^2 \partial y^2}$$
$$+D\frac{a^2}{12}\left[\frac{\partial^6 w}{\partial x^6} - \frac{\partial^6 w}{\partial x^4 \partial y^2} - \frac{\partial^6 w}{\partial x^2 \partial y^4} + \frac{\partial^6 w}{\partial y^6} \right] = 0 \qquad (150)$$

This continualization scheme contains sixth-order spatial derivatives. It is possible to derive an alternative continualization scheme by multiplying Eq. (150) by $1 - a^2/12\ \Delta$, thus leading to:

$$D\Delta^2 w + N\left(1 - \frac{a^2}{12}\Delta\right)\Delta w - m_0\omega^2\left(1 - \frac{a^2}{6}\Delta\right)w$$
$$-\frac{a^2}{3}D\frac{\partial^4}{\partial x^2 \partial y^2}\Delta w - \frac{a^2}{6}N\frac{\partial^4 w}{\partial x^2 \partial y^2} = 0 \qquad (151)$$

This nonlocal model can be equivalently rewritten as

$$\left(D - (e_0 a)^2 N\right)\Delta\Delta w + \left(N + 2(e_0 a)^2 m_0\omega^2\right)\Delta w - m_0\omega^2 w$$
$$-4(e_0 a)^2 D\frac{\partial^4}{\partial x^2 \partial y^2}\Delta w - 2(e_0 a)^2 N\frac{\partial^4 w}{\partial x^2 \partial y^2} = 0 \qquad (152)$$

where $(e_0 a)^2 = a^2/12$. This nonlocal plate model is similar to an Eringen's nonlocal plate model, apart from the spatially coupled derivatives, highlighted by the last two terms in Eq. (152). Equation (152) has been also derived by Hache et al. [110]. It is seen here that the lattice-based nonlocal plate model slightly differs from a stress gradient Eringen's plate model.

The nonlocal beam model Eq. (73) is found from this nonlocal plate model when the spatially coupling terms of the continualized nonlocal plate model are vanishing:

$$\left[EI - (e_0 a)^2 P\right]\frac{\partial^4 w}{\partial x^4} + \left[P + 2(e_0 a)^2 m_0\omega^2\right]\frac{\partial^2 w}{\partial x^2} - m_0\omega^2 w = 0 \qquad (153)$$

These spatially coupling terms in Eq. (152) are caused by the difference between the Eringen's nonlocal plate model and the continualized nonlocal model based on Hencky-net plate model. This explains the geometrical dependence of the scaling factor of Eringen's nonlocal model, when we match the results of this model and the microstructured results. The corrected nonlocal model derived in Eq. (152) has a length scale independent on the mode and the geometry of the plate.

5 Conclusions

In this chapter, we have presented exact solutions of some lattice structural systems, including axial lattices, beam lattices and plate lattices. We have restricted our study to discrete linear elasticity. Lattice inelasticity is another branch of structural mechanics, which still needs to be explored. The exact solutions of the lattice structural problems are then compared to some approximated continuous models, built from the continualization of the linear difference equations for lattice structures. Nonlocal rod, beam and plate models have been derived based on a series of asymptotic expansions of the pseudo-differential operator of the discrete model. Nonlocal models continualized from one-dimensional lattices are typically consistent with the stress gradient model of Eringen [3], where the small length scale parameter was calibrated regarding the lattice spacing. For two-dimensional lattices such as lattice membranes or lattice plates (even if we mainly focus only on lattice plates in the last part of the chapter), some differences are highlighted between the two-dimensional stress gradient plate model and the continualized (or lattice-based) nonlocal plate model [99]. The same conclusions have been reached for lattice membrane (see for instance Hérisson et al. [102]).

This chapter focused on the building of nonlocal structural elements from physical discrete structural approaches. Nowadays, the applicability of nonlocal models (mostly strain-based integral nonlocal models) tends to be measured from purely mathematical point of view rather than their ability on reflecting the underlying discrete material behavior. We have shown, for some very elementary periodic microstructures, that a class of differential nonlocal models may be rigorously introduced from the governing equations of the lattice reference problem (for one-dimensional and two-dimensional lattice structural elements). The continuous nonlocal models are efficient in reproducing some scale effects inherent to the discrete approach.

The microstructure considered herein is based on concentrated microstructures (for instance, concentrated masses and concentrated stiffness for the Hencky-Bar-Chain model with localized rotational stiffness). It has been shown that such microstructures can be handled by the stress gradient nonlocal models with corresponding boundary conditions. However, other patterns of microstructures can also be considered as shown by Challamel et al. [23] for axial lattices. This is the case of distributed microstructure which may be associated to strain gradient elasticity models, with different macroscopic properties. It appears that the length scale dependence strongly depends on the type of microstructure that governs the mechanics of lattices. It is worth noting that the concluding remarks are limited to simple lattice structures (that is periodic spring networks, homogeneous Hencky-Bar Chain models or homogeneous Hencky-Bar Net models). Other microstucture may be alternatively used for each structural element (microstructured rods, microstructured beams or microstructured plates). More general lattices of two-dimension or three-dimension are under investigation for deriving new nonlocal elastic structural models.

References

1. Wang CM, Zhang H, Challamel N, Pan W (2020) Hencky-Bar-Chain/Net for structural analysis. World Scientific, Singapore
2. Eringen AC, Kim BS (1977) Relation between non-local elasticity and lattice dynamics. Cryst Lattice Defects 7:51–57
3. Eringen AC (1983) On differential equations of nonlocal elasticity and solutions of screw dislocation and surface waves. J Appl Phys 54:4703–4710
4. Eringen AC (2002) Nonlocal continuum field theories. Springer-Verlag, New York

5. Elishakoff I, Pentaras D, Dujat K, Versaci C, Muscolino G, Storch J, Bucas S, Challamel N, Natsuki T, Zhang YY, Wang CM, Ghyselinck G (2012) Carbon nanotubes and nanosensors: vibrations, buckling and ballistic impact. ISTE-Wiley, London
6. Gopalakrishnan S, Narendar S (2013) Wave propagation in nanostructures: nonlocal continuum mechanics formulations. Springer, New York
7. Karlic D, Murmu T, Adhikari S, McCarthy M (2016) Non-local structural mechanics. ISTE-Wiley, London
8. Ghavanloo E, Rafii-Tabar H, Fazelzadeh A (2019) Computational continuum mechanics of nanoscopic structures: nonlocal elasticity approaches. Springer, New York
9. Lagrange JL (1759) Recherches sur la nature et la propagation du son, Miscellanea Philosophico-Mathematica Societatis Privatae Taurinensis I, 2rd Pagination, i-112 (1759) (see also Œuvres, Tome 1, 39–148)
10. Lagrange JL (1853) Mécanique analytique (Paris, 1788) (3rd Edition, Mallet-Bachelier, Gendre et successeur de Bachelier, Imprimeur-Libraire du bureau des longitudes, de l'école Polytechnique, de lcole centrale des arts et manufactures, Paris, 1853)
11. Hencky H (1920) Über die angenäherte Lösung von Stabilitätsproblemen im Raummittels der elastischen Gelenkkette. Der Eisenbau 11:437–452
12. Wang CT (1951) Discussion on the paper of "Salvadori M.G., Numerical computation of buckling loads by finite difference". Trans ASCE 116:629–631
13. Wang CT (1953) Applied elasticity. McGraw-Hill, New York
14. Wi AS, Wu CW, Obeid KA (1989) A simple discrete element mechanical model for the stability analysis of elastic structures. In: Kabil YH, Said ME (eds). Current advances in mechanical design and production. Pergamon Press, 149–156 (1989)
15. El Naschie MS (1990) Stress, stability and chaos in structural engineering: an energy approach. McGraw-Hill, New York
16. Lerbet J, Challamel N, Nicot F, Darve F (2020) Stability of discrete non-conservative systems. ISTE Press, Elsevier, London
17. Zhang H, Wang CM, Challamel N, Gao RP (2018) Modelling vibrating nano-strings by lattice, finite difference and Eringen's nonlocal models. J Sound Vib 425:41–52
18. Born M, von Kármán T (1912) On fluctuations in spatial grids. Physikalishe Zeitschrift 13:297–309
19. Gantmacher F (1970) Lectures in analytical mechanics. Mir Publishers, Moscow
20. Thomson WT, Dahleh MD (1998) Theory of vibration with applications. Prentice-Hall, Englewood Cliffs, New Jersey
21. Blevins RD (2001) Formulas for natural frequency and mode shape. Krieger Publishing Company
22. Tong P, Pian THH, Bucciarelli LL (1971) Mode shapes and frequencies by finite element method using consistent and lumped masses. Comput Struct 1:623–638
23. Challamel N, Wang CM, Elishakoff I (2016) Nonlocal or gradient elasticity macroscopic models: a question of concentrated or distributed microstructure. Mech Res Commun 71:25–31
24. Challamel N, Picandet V, Collet B, Michelitsch T, Elishakoff I, Wang CM (2015) Revisiting finite difference and finite element methods applied to structural mechanics within enriched continua. Eur J Mech A/Sol 53:107–120
25. Luongo A, Zulli D (2011) Parametric, external and self-excitation of a tower under turbulent wind flow. J Sound Vib 330:3057–3069
26. Triantafyllidis N, Bardenhagen S (1993) On higher order gradient continuum theories in 1-D nonlinear elasticity. Derivation from and comparison to the corresponding discrete models. J Elast. 33(3):259–293
27. Hérisson B, Challamel N, Picandet V, Perrot A (2016) Nonlocal continuum analysis of a nonlinear uniaxial elastic lattice system under non-uniform axial load. Physica E 83:378–388
28. Gazis DC, Wallis RF (1965) Surface tension and surface modes in semi-infinite lattices. Surf Sci 3:19–32

29. Charlotte M, Truskinovsky L (2002) Linear elastic chain with a hyper-pre-stress. J Mech Phys Solids 50:217–251
30. Kruskal MD, Zabusky NJ (1964) Stroboscopic perturbation for treating a class of nonlinear wave equations. J Math Phys 5:231–244
31. Collins MA (1981) A quasicontinuum approximation for solitons in an atomic chain. Chem Phys Lett 77:342–347
32. Benjamin TB, Bona JL, Mahony JJ (1972) Model equations for long waves in nonlinear dispersive systems. Philos T R Soc A 272:47–78
33. Rosenau P (1986) Dynamics of nonlinear mass-spring chains near the continuum limit. Phys Lett A 118:222–227
34. Jaberolanssar H, Jr J (1981) Peddieson, On continuum representation of mechanical behaviour of discrete lattices. Mech Res Comm 8:251–257
35. Gul U, Aydogdu M, Gaygusuzoglu G (2017) Axial dynamics of a nanorod embedded in an elastic medium using doublet mechanics. Compos Struct 160:1268–1278
36. Aydogdu M (2009) Axial vibration of the nanorods with the nonlocal continuum rod model. Physica E 41:861–864
37. Challamel N, Rakotomanana L, Le Marrec L (2009) A dispersive wave equation using nonlocal elasticity. C R Méanique 337:591–595
38. Aydogdu M (2012) Axial vibration analysis of nanorods (carbon nanotubes) embedded in an elastic medium using nonlocal elasticity. Mech Res Comm 43:34–40
39. Aydogdu M (2014) Longitudinal wave propagation in multiwalled carbon nanotubes. Compos Struct 107:578–584
40. Challamel N, Aydogdu M, Elishakoff I (2018) Statics and dynamics of nanorods embedded in an elastic medium: Nonlocal elasticity and lattice formulations. Eur J Mech A/Sol 67:1–18
41. Myshkis AD (2005) Mixed functional differential equations. J Math Sci 129:4111–4226
42. Maugin GA (1999) Nonlinear waves in elastic crystals. Oxford University Press, Oxford
43. LeVeque RJ (2007) Finite difference methods for ordinary and partial differential equations: steady-state and time-dependent problems. Society for Industrial and Applied Mathematics, New York
44. Kivshar YS, Zhang F, Takeno S (1998) Nonlinear surface modes in monoatomic and diatomic lattices. Physica D 113:248–260
45. Goldberg S (1958) Introduction to difference equations with illustrative examples from economics, psychology and sociology. Dover Publications, New York
46. Salvadori MG (1951) Numerical computation of buckling loads by finite differences. Trans ASCE 116:590–624
47. Andrianov IV, Awrejcewicz J (2003) On the average continuous representation of an elastic discrete medium. J Sound Vib 264:1187–1194
48. Andrianov IV, Awrejcewicz J (2008) Continuous models for 2D discrete media valid for higher-frequency domain. Comput Struct 86:140–144
49. Andrianov IV, Awrejcewicz J, Weichert D (2010) Improved continuous models for discrete media. Math Probl Eng 986242:1–35
50. Baker GAJ, Graves-Morris P (1996) Padé approximants. Cambridge University Press, New York
51. Rosenau P (2003) Hamiltonian dynamics of dense chains and lattices: or how to correct the continuum. Phys Lett A 311:39–52
52. Love AEH (1927) A treatise on the mathematical theory of elasticity. Cambridge University Press, Cambridge
53. Mindlin RD (1964) Micro-structure in linear elasticity. Arch Rat Mech Anal 16:51–78
54. Polyzos D, Fotiadis DI (2012) Derivation of Mindlin's first and second strain gradient elastic theory via simple lattice and continuum models. Int J Solids Struct 49:470–480
55. Rayleigh JWSB (1877) The theory of sound. Macmillan Publications Co., Inc, London
56. Challamel N, Wang CM, Elishakoff I (2014) Discrete systems behave as nonlocal structural elements: bending, buckling and vibration analysis. Eur J Mech A/Sol 44:125–135

57. Challamel N, Wang CM, Zhang H, Kitipornchai S (2018) Exact and nonlocal solutions for vibration of axial lattices with direct and indirect neighbouring interactions. J Eng Mech 144:04018025
58. Challamel N, Zhang H, Wang CM, Kaplunov J (2019) Scale effect and higher-order boundary conditions for generalized lattices, with direct and indirect interactions. Mech Res Commun 97:1–7
59. Pipes LA (1966) Circulant matrices and the theory of symmetrical components. Matrix Tensor Quart 17:35–50
60. Chen FY (1970) Similarity transformation and the eigenvalue problem of certain far-coupled systems. Am J Phys 38:1036–1039
61. Chen FY (1971) On modeling and direct solution of certain free vibration systems. J Sound Vib 14:57–79
62. Eaton HC, Jr J (1973) Peddieson, On continuum description of one-dimensional lattice mechanics. J Ten Acad Sci 18:96–100
63. Brillouin L (1946) Wave propagation in periodic structures: electric filters and crystal lattices. Mc Graw-Hill, New York
64. Rosenau P (1987) Dynamics of dense lattices. Phys Rev B 36:5868–5876
65. Di Paola M, Failla G, Zingales M (2009) Physically-based approach to the mechanics of strong non-local linear elasticity theory. J Elast 97:103–130
66. Carcaterra A, Dell'Isola F, Esposito R, Pulvirenti M (2015) Macroscopic description of microscopically strongly inhomegenous systems : A mathematical basis for the synthesis of higher gradient metamaterials. Arch Rational Mech Anal 218:1239–1262
67. Tarasov VE (2015) Lattice model with nearest-neighbor and next-nearest-neighbor interactions for gradient elasticity. Discontinuity, Nonlinear Comp 4:11–23
68. Michelitsch T, Collet B, Nowakowski AF, Nicolleau FCGA (2015) Fractional Laplacian matrix on the finite periodic linear chain and its periodic Riesz fractional derivative continuum limit. J Phys A 48:295202
69. Ghavanloo E, Fazelzadeh SA, Rafii-Tabar H (2020) Formulation of an efficient continuum mechanics-based model to study wave propagation in one-dimensional diatomic lattices. Mech Res Commun 103:103467
70. Ghavanloo E, Fazelzadeh SA (2019) Wave propagation in one-dimensional infinite acoustic metamaterials with long-range interactions. Acta Mech 230(12):4453–4461
71. Silverman IK (1951) Discussion on the paper of "Salvadori M.G., Numerical computation of buckling loads by finite differences. Trans ASCE, 116, 590–636, 1951." Trans ASCE, 116:625–626
72. Zaslavsky A (1968) Models for elastic buckling behaviour. Israel J Tech 6:212–226
73. Challamel N, Reddy JN, Wang CM (2016) On Eringen's stress gradient model for bending of nonlocal beams. J Eng Mech 142:04016095
74. Wang CM, Zhang H, Challamel N, Duan WH (2017) On boundary conditions for buckling and vibration of nonlocal beams. Eur J Mech A/Soli 61:73–81
75. Wattis JAD (2000) Quasi-continuum approximations to lattice equations arising from the discrete non-linear telegraph equation. J Phys A 33:5925–5944
76. Challamel N (2013) Variational formulation of gradient or/and nonlocal higher-order shear elasticity beams. Compos Struct 105:351–368
77. Zhang YY, Wang CM, Challamel N (2010) Bending, buckling and vibration of hybrid nonlocal beams. J Eng Mech 136:562–574
78. Challamel N, Zhang Z, Wang CM (2015) Nonlocal equivalent continua for buckling and vibration analyses of microstructured beams. J Nanomech Micromech 5:A4014004
79. Wang CM, Zhang Z, Challamel N, Duan WH (2013) Calibration of Eringen's small length scale coefficient for initially stressed vibrating nonlocal Euler beams based on microstructured beam model. J Phys D 46:345501
80. Milton GW, Willis JR (2007) On modifications of Newton's second law and linear continuum elastodynamics. Proc Royal Soc A 463:855–880

81. Charlotte M, Truskinovsky L (2012) Lattice dynamics from a continuum viewpoint. J Mech Phys Sol 60:1508–1544
82. Domokos G (1993) Qualitative convergence in the discrete approximation of the Euler problem. Mech Struct Mach 21:529–543
83. Challamel N, Kocsis A, Wang CM (2015) Discrete and nonlocal elastica. Int J Nonlin Mech 77:128–140
84. Seide P (1975) Accuracy of some numerical methods for column buckling. J Eng Mech 101:549–560
85. Challamel N, Lerbet J, Wang CM, Zhang Z (2014) Analytical length scale calibration of nonlocal continuum from a microstructured buckling model. Z Angew Math Mech 94:402–413
86. Wang CM, Zhang YY, Ramesh SS, Kitipornchai S (2006) Buckling analysis of micro- and nano-rods/tubes based on nonlocal Timoshenko beam theory. J Phys D 39:3904–3909
87. Reddy JN, Pang SD (2008) Nonlocal continuum theories of beams for the analysis of carbon nanotubes. J Appl Phys 103:023511
88. Leckie FA, Lindberg GM (1963) The effect of lumped parameters on beam frequencies. Aeronaut Quart 14:224–240
89. Santoro R, Elishakoff I (2006) Accuracy of the finite difference method in stochastic setting. J Sound Vib 291:275–284
90. Tarnai T (1995) The southwell and the dunkerley theorems. In: Tarnai T (ed) Summation theorems in structural stability. Springer-Verlag, New York, pp 141–185
91. Massonnet C (1940) Les relations entre les modes normaux de vibration et la stabilité des systémes élastiques. Bulletin du CERES. Liège, 183–210. (see also C. Massonnet, Résistance des matériaux. Dunod, Paris, 1965)
92. Galef AE (1968) Bending frequencies of compressed beams. J Acoust Soc Am 44:643–643
93. Absi E, Prager W (1975) A comparison of equivalence and finite element methods. Comput Meth Appl Mech Eng 6:59–64
94. Zienkiewicz OC, Taylor RL (2000) The finite element method. Butterworth-Heinemann, New York
95. Wieghardt K (1906) Über einen Grenzübergang der Elastizitätslehre und seine Anwendung auf die Statik hochgradig statisch unbestimmter Fachwerke. in: Verhandtlungen des Vereinz z. Beförderung des Gewerbefleisses Abhandlungen 85:139–176
96. Riedel W (1928) Beträge zür lösung des ebenen Problem eines elastichen Körpers mittels er Airyschen Spannungsfunktion. Zeit. Für Angewandte Mathematik und Mechanik 8:159–160
97. Hrennikoff A (1941) Solutions of problems of elasticity by the framework method. J Appl Mech 8:A169–A175
98. Wu CW (1985) A Discrete element method for linear and nonlinear stress and bifurcation problems of elastic structures. Doctoral dissertation, New Mexico State University
99. Challamel N, Hache F, Elishakoff I, Wang CM (2016) Buckling and vibration of microstructured rectangular plates considering phenomenological and lattice-based nonlocal continuum models. Compos Struct 149:145–156
100. Ari N, Eringen AC (1983) Nonlocal stress field at Griffith crack. Cryst Latt Def Amorp 10:33–38
101. Lombado M, Askes H (2010) Elastic wave dispersion in microstructured membranes. Proc R Soc A 466:1789–1807
102. Hérisson B, Challamel N, Picandet V, Perrot A, Wang CM (2018) Static and dynamic behaviors of microstructured membranes within nonlocal mechanics. J Eng Mech 144:04017155
103. Zhang Z, Wang CM, Challamel N (2014) Eringen's length scale coefficient for buckling of nonlocal rectangular plates from microstructured beam-grid model. Int J Solids Struct 51:4307–4315
104. Zhang Z, Wang CM, Challamel N (2015) Eringen's length scale coefficient for vibration and buckling of nonlocal rectangular plates with simply supported edges. J Eng Mech 141:04014117
105. Rao SS (2007) Vibration of continuous systems. John Wiley & Sons Inc, New Jersey

106. Lu BP, Zhang PQ, Lee HP, Wang CM, Reddy JN (2007) Non-local elastic plate theories. Proc R Soc A 463:3225–3240
107. Murmu T, Pradhan SC (2009) Vibration analysis of nanoplates under uniaxial prestressed conditions via nonlocal elasticity. J Appl Phys 106:104301
108. Pradhan SC, Murmu T (2009) Small scale effect on the buckling of single-layered grapheme sheets under biaxial compression via nonlocal continuum mechanics. Comput Mat Sci 47:268–274
109. Wang CY, Wang CM (2013) Structural vibration: exact solutions for strings, membranes, beams and plates. CRC Press, Singapore
110. Hache F, Challamel N, Elishakoff I, Wang CM (2017) Comparison of nonlocal continualization schemes for lattice beams and plates. Arch Appl Mech 87:1105–1138

Eringen's Nonlocal Integral Elasticity and Applications for Structural Models

Constantinos Chr. Koutsoumaris and Konstantinos G. Eptaimeros

Abstract The requirement of studying the unsolved problems of continuum mechanics in the context of micro- and nano-dimensions imperatively arose during the second half of the 20th century. For instance, the concentration of stresses at the tip of a crack and the wave dispersion at atomic dimensions were some of the problems that concerned the scientific community for decades. Moreover, the mechanical response of micro- and nano-structures ranging from carbon nanotubes (CNTs), graphene sheets to microtubules is significant for designing structural elements, micro- and nano-devices (micro/nano-sensors, micro/nano-electromechanical systems (M/NEMS)), biomaterials and bioengineering structures, etc. Eringen's nonlocal theory, in particular, is widely used to explore problems of this scientific field. There are sheets of research papers investigating the mechanical response of CNTs, graphene sheets and microtubules by using the weak form (i.e., the differential form) of Eringen's theory, yet paradoxes and inconsistencies are raised in the majority of the practical engineering problems that studied. Recently, the great value of the strong form (i.e., the integral form) of Eringen's constitutive equation has been pointed up by many researchers, and this chapter will focus on this matter.

1 Introduction

There has been a significant development of advanced materials of micro/nano-dimensions in recent years. Scale dependent phenomena of materials on micro- and nano-structures have not been adequately explained by classical continuum theory.

C. Chr. Koutsoumaris (✉)
Department of Civil Infrastructure and Environmental Engineering,
Khalifa University of Science and Technology, Abu Dhabi, United Arab Emirates
e-mail: konstantinos.koutsoumaris@ku.ac.ae

K. G. Eptaimeros (✉)
Institute of Applied and Computational Mathematics, Foundation for Research and Technology,
Heraklion, Crete, Greece
e-mail: kosepta@gmail.com

© The Author(s), under exclusive license to Springer Nature Switzerland AG 2021
E. Ghavanloo et al. (eds.), *Size-Dependent Continuum Mechanics Approaches*,
Springer Tracts in Mechanical Engineering,
https://doi.org/10.1007/978-3-030-63050-8_2

However, the generalized continuum theories (GCT) suggest a way to handle the material size effect, since they bridge the gap between the classical continuum theory and lattice atomic theories as the long and the short range cohesive forces at an atomic scale are elucidated [1–6].

Micro-polar, couple stress, strain gradient and nonlocal elasticity theories are some of the most widely known and accepted GCT. The aforementioned theories have been successfully applied to numerous problems of engineering, such as the concentration of stresses at the tip of a crack, the wave dispersion at atomic dimensions and screw dislocations [7–17]. An overview of GCT has been presented by Maugin [18].

The nonlocal continuum field theories are put forward into several studies and they explore the physics of material bodies in which the behavior at a material point is dependent on the state of all the points of the body [11, 19–21]. The nonlocal field theory of elasticity, specifically advanced by Eringen [7, 10, 11, 20–25], suggests that the stress at a point is expressed by an integral constitutive equation by means of the convolution of an attenuation function and a macroscopic stress field as well [10, 11].

Both the gradient elasticity theory and the nonlocal elasticity theory were developed at the same time [26–28]. The former was considered as an alternative approach of handling size effect phenomena. Mindlin [29] subsequently assumed the cohesion within the aforementioned theories by introducing Taylor (expansion) series of the strain of the nonlocal integral constitutive equation leading to deducing the equations of the gradient elasticity theory.

The nonlocal integral constitutive equation is a first kind Fredholm integral equation that is hard to handle even in one-dimension. The use of either weakly singular or continuous kernels representing attenuation functions leads to ill-posed problems [30]. Eringen [10] noticed that the integral constitutive equation could be transformed into a differential equation in an infinite domain for a class of attenuation functions, considered a fundamental solution of Helmholtz type differential operator wherein a physical interpretation was attributed [10, 11, 31].

Meanwhile, the nonlocal differential constitutive equation was employed to investigate the beam models [32, 33]. Those works encouraged the use of the aforementioned models for the mechanical response of a CNT and a nanobeam [34–41], since the modeling of the CNT as a beam structural element had been suggested by experimental and theoretical researches in advance [42–46]. Studies [34–36, 47], looking into the dynamical response of the nonlocal differential beam model, concluded that the above model exhibits a softening response in comparison with that of the classic-local model for benchmark engineering applications. By this approach, however, the fundamental eigenfrequency of a cantilever beam was assigned higher values than that of the classic-local model. This was, of course, one of the paradoxes of the nonlocal differential form [35].

What is more, the researches into the static response of nonlocal differential beam models [37, 39, 41] suggested that the behavior of the nonlocal effect is more flexible than that of the classic-local beam model in the majority of boundary conditions (BCs) and loadings. Nevertheless, there are specific cases of BCs and loadings wherein paradoxes are raised [39, 41]. In particular, the nonlocal transverse deflection is identical to the classical one for the case of a cantilever beam subjected to a concentrated load at the free end, a clamped-clamped and a clamped-pinned beam subjected to a uniformly distributed load, respectively. In addition, the response of a nonlocal cantilever beam, with a uniformly distributed load, is more stiff in comparison with that of the classical one. As regards a cantilever beam with a concentrated load applied to an internal point, the nonlocal transverse deflection is identical to the classical one at the interval ranging from the clamped point of the beam to the load point, yet the nonlocal deflection is smaller than the classical one at the interval ranging from the load point to the free end of the beam [48].

All the paradoxes of a beam have drawn the researchers' attention [49, 50]. The last-mentioned studies have not address the main disadvantage of the nonlocal differential beam form, which is the inadequacy of producing energy quadratic functional [37]. This inadequacy is not consistent with Eringen's nonlocal integral theory, according to which it is proved that the energy quadratic functional can be formulated for a homogeneous and an isotropic solid [11, 51]. It goes without saying that the transformation of the integral form into the differential form is not an injective process in a finite domain, and as a consequence energy inconsistencies and paradoxes are raised [52–54].

Except for one-dimensional problems, the nonlocal differential constitutive equations have been used to formulate the equilibrium equations of shells and plates [40, 55–60]. These studies reached the conclusion that the nonlocal differential form demonstrates a flexible behavior compared to that of the classic-local form only for solvable cases.

The aforementioned inconsistencies of the nonlocal differential form are overcome on the condition that the integral stress form may be reconsidered. Eringen [61] and Altan [62] put forth the two phase nonlocal integral (TPNI) stress model regarded as the constitutive equation of a two phase elastic material. This model represents a convex combination of the local elasticity and the nonlocal integral elasticity. The TPNI stress model additionally transforms the constitutive equation into a second kind Fredholm integral equation being manageable in comparison with the first kind Fredholm one. Thereafter, Polizzotto [51] studied the variational principles and advanced the TPNI stress model, while Pisano and Fuschi [63] analytically solved the problem of a tensile bar by applying the last-mentioned model afterward.

After Khodabakhshi and Reddy's [64] research into the TPNI stress model and its applications, the researchers' interest centered on the nonlocal integral form of one-dimensional beam problems. Closed form and numerical solutions of static problems regarding the TPNI stress model were also proposed by Eptaimeros et al. [65], Wang et al. [66, 67], Koutsoumaris et al. [48], Zhu et al. [68], Zhu and Li [69]. Meanwhile, Fernández-Sáez and Zaera [70] pointed up the inadequacy of finding an accepted analytical solution to a beam problem from a physical point of view.

As regards dynamical problems, Eptaimeros et al. [71], Fernández-Sáez and Zaera [70], Zhu and Li [72, 73] as well as Mikhasev and Nobili [74] employed the TPNI stress model and investigated the free vibration problem of beams and rods. Not only did Eptaimeros et al. [71] investigate the free vibration Euler-Bernoulli (EB) beam problem by applying the FEM, but they also tackled the cantilever paradox for the first time. The FEM results were subsequently verified by the analytical results of Fernández-Sáez and Zaera [70].

A property of the TPNI stress model concerns the violation of the normalization condition in a finite body, especially on the boundary. Needless to say, a similar argument had been made by Eringen [61]. A modified type kernel suggested by Borino et al. [75] as well as Bažant and Jirásek [76] and satisfying all the kernel's properties of Eringen [10] is symmetric, well-defined and applied to the investigation of plasticity problems. A significant remark made by Bažant and Jirásek [76] is that the violation of the normalization condition gives rise to inconsistencies as a uniform straining of a body does not generate a uniform stress. Also, the normalization condition is satisfied if and only if the nonlocal attenuation function reverts to Dirac delta function (distribution) when the nonlocal parameter tends to zero at each point of a finite body. To elucidate, unlike the TPNI stress model, the modified kernel confirms that the nonlocal model reverts to the classic model as the nonlocality reduces [48].

Koutsoumaris et al. [48] pointed out the great significance of the attenuation function as a probability density function (PDF), and the satisfaction of the normalization condition in a finite domain too. Moreover, the physical interpretation was attributed to the modified kernel. As regards the nonlocal integral models (i.e., both modified kernel's and TPNI stress models), the flexibility of the response compared to that of the classic-local model for static EB beam problems was numerically presented. Furthermore, all the paradoxes deriving from the use of the nonlocal differential model were successfully tackled by the nonlocal integral models for the first time. Their numerical solutions and the closed form solutions of Wang et al. [66] additionally deduced similar results concerning the TPNI stress model.

The bending strain of beam problems was subsequently investigated by Koutsoumaris and Eptaimeros [54]. To be more specific, the modified kernel's and the TPNI stress models for Helmholtz and bi-Helmholtz types of kernel [77] were employed. The use of bi-Helmholtz type kernel led to strains' oscillations at the load points and on the boundary. This behavior is not encouranging for applying bi-Helmholtz type kernel in structural engineering problems. Another important conclusion was that the modified kernel successfully handles the boundary layer and is considered the optimal choice for a mixed local/nonlocal model.

In the meantime, the nonlocal integral stress models (i.e., the modified kernel's and the TPNI stress models) were used by Eptaimeros et al. [78, 79] for the investigation of the dynamical response of a nanobeam and a microtubule, embedded in an elastic medium, respectively. The responses of the nonlocal integral models appear to have a softening behavior in comparison with that of the classic model for all the investigated problems. It is crucial that no paradox is raised for the fundamental eigenfrequency of an embedded cantilever beam. A general conclusion is also that the nonlocality's effect attempts to prevail over the effect of the elastic medium and vice versa.

This chapter revolves around the contemporary advances in Eringen's nonlocal elasticity theory with emphasis on solving structural engineering problems. To be more specific, an overview of the literature is presented, and a discussion about the integral formulation (i.e., the strong form) and the differential formulation of the nonlocal elasticity theory then follows. What is more, we point out the correlation between the nonlocal elasticity theory and the strain gradient elasticity theory as introduced by Mindlin, presenting the above correlation in detail. Thereafter an exhaustive discussion regarding the suitability of the nonlocal modulus for the problems' solution in a finite body follows. Finally, applications for both static and dynamical engineering problems are investigated as well.

To recap, this chapter is structured as follows: Sect. 2 focuses on the governing equations of the nonlocal elasticity theory and the exhaustive discussion about the strong and the weak form of the aforementioned theory. We then go on correlating the strain gradient elasticity theory with the nonlocal elasticity theory. Section 3 contains a number of applications for static and dynamical benchmark engineering problems. Finally, the conclusions of this chapter are reiterated.

2 Eringen's Formulation

2.1 Governing Equations

The linear theory of the nonlocal elasticity is expressed by the following set of equations for a homogeneous and an isotropic solid [10, 11]:

$$t_{kl,k} + \rho(f_l - \ddot{u}_l) = 0 \tag{1}$$

$$t_{kl}(\mathbf{x}) = \int_V K(||\mathbf{x}' - \mathbf{x}||, \tau)\sigma_{kl}(\mathbf{x}')dv(\mathbf{x}') \tag{2}$$

$$\sigma_{kl}(\mathbf{x}') = \lambda e_{rr}(\mathbf{x}')\delta_{kl} + 2\mu e_{kl}(\mathbf{x}') \tag{3}$$

$$e_{kl}(\mathbf{x}') = \frac{1}{2}\left(\frac{\partial u_k(\mathbf{x}')}{\partial x'_l} + \frac{\partial u_l(\mathbf{x}')}{\partial x'_k}\right) \tag{4}$$

where λ and μ are the constants of Lamé and $||\cdot||$ is the Euclidean norm. The stress tensor, the mass density, the body force density and the displacement vectors at a reference point $\mathbf{x}$ of the body at time t are denoted by t_{kl}, ρ, f_l and u_l, respectively. Also, the classical stress tensor at $\mathbf{x}'$ at time t and the linear strain tensor at any point $\mathbf{x}'$ of the body at time t are denoted by $\sigma_{kl}(\mathbf{x}')$ and $e_{kl}(\mathbf{x}')$, respectively. The former is associated with the latter by means of Hooke's law. The volume integral of Eq. (2) is evaluated over the region V of the body as well.

Equation (2) expresses the contribution of the material points of the body to the stress field at a point $\mathbf{x}$ via the attenuation function K (nonlocal modulus) [10, 11]. The attenuation function is of dimension of $(\text{length})^{-3}$ as well. Ergo, it should be dependent on a characteristic length ratio $\tau := e_0 a/\ell$, where a is an internal characteristic length, i.e., a lattice parameter/bond length, ℓ is an external characteristic length, i.e., a wave length, and e_0 is also a dimensionless material constant.

The following properties are attributed to the nonlocal modulus K:

(i) K has to revert to the Dirac delta function (δ) as $\tau \to 0$, so that the classical elasticity can be approximated when the limit of the internal characteristic length vanishes.

$$\lim_{\tau \to 0} K(||\mathbf{x}' - \mathbf{x}||, \tau) = \delta(||\mathbf{x}' - \mathbf{x}||) \tag{5}$$

(ii) K is symmetric with respect to $\mathbf{x}$ as well.

(iii) The maximum of K is also attained at $\mathbf{x}' = \mathbf{x}$ and $K \to 0$ at large distances as well.

(iv) K satisfies the normalization condition:

$$\int_{\Omega} K(||\mathbf{x}||, \tau) d\Omega = 1 \tag{6}$$

where $\mathbf{x} \in \Omega$, Ω is an infinite domain, and K corresponds to a PDF [10, 48].

2.2 *A Discussion About Eringen's Nonlocal Stress Models*

2.2.1 Integral Equations, Fundamental Solutions and Green Functions

The nonlocal integral constitutive equation being generally hard to handle is a first kind Fredholm. The use of either weakly singular or continuous kernels representing attenuation functions leads to ill-posed problems [30].

Eringen [10] noticed that the nonlocal integral constitutive equation could be transformed into a differential one. Provided that K is a fundamental solution of a linear differential operator $\mathcal{L}$, it then holds in an infinite domain:

$$\mathcal{L}\left[K(||\mathbf{x}' - \mathbf{x}||, \tau)\right] = \delta(||\mathbf{x}' - \mathbf{x}||) \tag{7}$$

Applying the differential operator $\mathcal{L}$ to Eq. (2), it yields:

$$\mathcal{L} t_{kl} = \sigma_{kl} \tag{8}$$

Equation (8) is then called the nonlocal differential constitutive equation. Eringen attributed a physical interpretation to the nonlocal differential form [10, 11].

Although this fundamental solution was adequate for a number of engineering problems [7–10, 21], it is not suitable for a problem in a finite domain since it is essential to find the corresponding Green function [53].

2.2.2 Two Phase Nonlocal Integral Constitutive Model

The TPNI stress model suggested by Eringen [61] was based on Kröner's research [2]. Eringen underlined the advantages of the above model from a mathematical point of view, yet there was some doubt if this model was suitable from a physical point of view [61].

The TPNI stress model can be regarded as the constitutive equation of a two phase elastic material, where phase 1 complies with local elasticity and phase 2 with nonlocal elasticity. The aforementioned model transforms the constitutive equation into a second kind Fredholm integral equation that is manageable and a well-posed problem. Altan [62] and subsequently Polizzotto [51] advanced the TPNI stress model which is defined as follows:

$$t_{kl}(\mathbf{x}) := k_1\sigma_{kl}(\mathbf{x}) + k_2\int_V K(||\mathbf{x}' - \mathbf{x}||, \tau)\sigma_{kl}(\mathbf{x}')dv(\mathbf{x}') \tag{9}$$

where $k_1 + k_2 = 1$ and $k_1, k_2 > 0$.

The attenuation function corresponding to the TPNI stress model is defined as:

$$\tilde{K}(||\mathbf{x}' - \mathbf{x}||, \tau) := k_1\delta(||\mathbf{x}' - \mathbf{x}||) + k_2K(||\mathbf{x}' - \mathbf{x}||, \tau) \tag{10}$$

2.2.3 Two Phase Constitutive Model with the Modified Kernel

An essential feature of a nonlocal kernel is associated with the normalization condition. To elucidate, the normalization condition has to be such that attenuation function converges toward Dirac delta function as the nonlocal parameter tends to zero at each point of a finite body. This characteristic corroborates that the nonlocal model reverts toward the classic-local model at each point of a finite body. A PDF is defined by the attenuation function. The key point is therefore to find an appropriate transformation from an infinite domain into a finite one. Borino et al. [75] as well as Bažant and Jirásek [76] proposed an efficient transformation, which leads to a modified, symmetric attenuation function being well-defined and satisfying all the properties of a nonlocal kernel. The model of the modified kernel is defined as follows:

$$K_{\text{mod}}(||\mathbf{x}' - \mathbf{x}||, \tau) := \xi(\mathbf{x}, \tau)\delta(||\mathbf{x}' - \mathbf{x}||) + K(||\mathbf{x}' - \mathbf{x}||, \tau) \tag{11}$$

where $\xi(\mathbf{x}, \tau) := 1 - \int_V K(||\mathbf{x} - \zeta||, \tau)d\zeta$.

If the normalization condition is violated, the uniform straining of the finite body can not generate a uniform stress [76].

The nonlocal stresses at a reference point $\mathbf{x}$ are defined by a weighted average via a nonlocal modulus (PDF) and the macroscopic (Cauchy) stresses of all the points of the body. This definition introduces the nonlocal interactions between the atoms (or the molecules) without necessitating the determination of the cohesive forces which are responsible for. The nonlocal modulus defined in an infinite domain is symmetric with respect to **x** and it satisfies the normalization condition (Eq. (6)). The normalization condition corroborates that the sum of all the weights of Cauchy stresses is equal to 1 at each point **x** of the infinite body. The violation of the latter on the boundary of a finite domain is obvious, since no material points exist out of the body to contribute to the stresses [48].

To exemplify, let us consider a beam of length L with boundaries 0 and L in one-dimension. For $\varepsilon > 0$ and $\varepsilon \to 0$, the points found in the intervals $[0, \varepsilon)$ and $(L - \varepsilon, L]$ have no adjacent points found in the intervals $(-\infty, 0)$ and $(L, +\infty)$ to interact with each other. Ergo, the locality increases in these points and it exists together with the interactions between the adjacent points in the intervals $(0, \varepsilon)$ and $(L - \varepsilon, L)$.

To shed light on, the modified kernel gives rise to the locality when the interactions between the particles/points are absent. The latter consideration can be generalized in two- and three-dimensions. The modified kernel can be inserted into Eq. (2), and it can be then considered as a physical expansion of the nonlocal theory of Eq. (2) on a finite body [48].

The classic types of a kernel do not satisfy the normalization condition on the boundary of the body, because they are normalized in an infinite domain. The TPNI stress model appears both mathematical and physical inconsistencies when it is applied to a finite domain. This model additionally violates the normalization condition in a finite domain and especially on the body's boundary.

2.3 The Correlation Between Eringen's Nonlocal Model and Mindlin's Gradient Model

The correlation between the nonlocal elasticity theory of Eringen and the strain gradient elasticity theory of Mindlin has been an open problem. The next process following Mindlin's rationale [29] attempts to explain the aforementioned correlation.

Consider the nonlocal constitutive equation (Eq. (2)) with respect to the strains:

$$t_{ij}(x_m) = \int_V C_{ijkl} K(||x_m - x'_m||) e_{kl}(x'_m) d^3 x'_m \tag{12}$$

where $d^3 x'_m \equiv dV = dx'_i dx'_j dx'_k$ and C_{ijkl} is the stiffness tensor.

Defining that $C_{ijkl}(x_m - x'_m) := C_{ijkl} K(||x_m - x'_m||)$, Eq. (12) yields:

$$t_{ij}(x_m) = \int_V C_{ijkl}(x_m - x'_m) e_{kl}(x'_m) d^3 x'_m \tag{13}$$

Considering that $x'_m = x_m + \Delta x_m$, Taylor series of the strain $e_{kl}(x'_m)$ gives:

$$e_{kl}(x'_m) = e_{kl}(x_m) + \frac{\partial e_{kl}(x_m)}{\partial x_q} \Delta x_q + \frac{1}{2} \frac{\partial^2 e_{kl}(x_m)}{\partial x_q \partial x_n} \Delta x_q \Delta x_n + \ldots \tag{14}$$

Inserting Eq. (14) into Eq. (13), it yields:

$$t_{ij}(x_m) = \int_V C_{ijkl}(x_m - x'_m)\left[e_{kl}(x_m) + \frac{\partial e_{kl}(x_m)}{\partial x_q}\Delta x_q + \frac{1}{2}\frac{\partial^2 e_{kl}(x_m)}{\partial x_q \partial x_n}\Delta x_q \Delta x_n + \ldots\right] d^3x'_m \tag{15}$$

The stresses are ergo defined by selecting the first three terms of Taylor series of Eq. (15):

$$\begin{aligned}\bar{t}_{ij}(x_m) &= e_{kl}(x_m)\int_V C_{ijkl}(x_m - x'_m)d^3x'_m \\ &+ \frac{\partial e_{kl}(x_m)}{\partial x_q}\int_V C_{ijkl}(x_m - x'_m)\Delta x_q d^3x'_m \\ &+ \frac{1}{2}\frac{\partial^2 e_{kl}(x_m)}{\partial x_q \partial x_n}\int_V C_{ijkl}(x_m - x'_m)\Delta x_q \Delta x_n d^3x'_m\end{aligned} \tag{16}$$

We also consider the mathematical expressions:

$$\begin{aligned}\mathcal{D}_{ijkl} &:= \mathcal{D}_{ijkl}(x_m) = \int_V C_{ijkl}(x_m - x'_m)d^3x'_m = K_0, \\ \mathcal{B}_{ijklq} &:= \mathcal{B}_{ijklq}(x_m) = \int_V C_{ijkl}(x_m - x'_m)\Delta x_q d^3x'_m, \\ \mathcal{E}_{ijklqn} &:= \mathcal{E}_{ijklqn}(x_m) = \int_V C_{ijkl}(x_m - x'_m)\Delta x_q \Delta x_n d^3x'_m\end{aligned} \tag{17}$$

Thereby, Eq. (16) can be written as:

$$\bar{t}_{ij}(x_m) = \mathcal{D}_{ijkl}e_{kl}(x_m) + \mathcal{B}_{ijklq}e_{kl,q}(x_m) + \mathcal{E}_{ijklqn}e_{kl,qn}(x_m) \tag{18}$$

It will hold $\mathcal{B}_{ijklq} = 0$, provided that the material is centroid and symmetric. Equation (18) will therefore give:

$$\bar{t}_{ij}(x_m) = \mathcal{D}_{ijkl}e_{kl}(x_m) + \mathcal{E}_{ijklqn}e_{kl,qn}(x_m) \tag{19}$$

As regards the strain gradient elasticity theory, the equilibrium equation can be expressed as:

$$\sigma_{pq} = \tau_{pq} - \mu_{rpq,r} \tag{20}$$

where σ_{pq} are the total stresses, τ_{pq} are the Cauchy stresses according to Mindlin and $\mu_{rpq,r}$ are the double stresses.

Based on the simplified theory (i.e., the form II of the strain gradient elasticity theory of Mindlin), we have:

$$\mu_{rpq} = \ell^2 \tau_{pq,r} \tag{21}$$

where ℓ is a length scale which represents a material length associated with the surface elastic strain energy.

We thus obtain by combining Eq. (20) with Eq. (21):

$$\sigma_{pq} = \tau_{pq} - \ell^2 \tau_{pq,rr} \tag{22}$$

We next consider the simplified form:

$$\mathcal{E}_{ijklqn} = -\ell^2 \delta_{qn} \mathcal{D}_{ijkl} \tag{23}$$

where δ_{qn} is Kronecker delta.

Equation (19) can be then written by inserting Eq. (23):

$$\bar{t}_{ij}(x_m) = \mathcal{D}_{ijkl} e_{kl}(x_m) - \ell^2 \delta_{qn} \mathcal{D}_{ijkl} e_{kl,qn}(x_m) \tag{24}$$

or

$$\bar{t}_{ij}(x_m) = \mathcal{D}_{ijkl} e_{kl}(x_m) - \ell^2 \mathcal{D}_{ijkl} e_{kl,rr}(x_m) \tag{25}$$

Equation (25) will be identical to Eq. (22) provided that $\tau_{kl} = \mathcal{D}_{ijkl} e_{kl}(x_m)$.

3 Applications for One-Dimensional Problems

This section deals with engineering applications in one-dimension by using the structural beam element. Both static and dynamical beam problems are investigated, and the results deduced from the classic-local model, the nonlocal differential model and the nonlocal integral models are compared with each other.

The kernels of Eq. (10) and Eq. (11) are customized in one-dimension:

$$A_{\text{mod}}(|x-s|, \tau) := \xi(x, \tau)\delta(|x-s|) + A(|x-s|, \tau) \tag{26}$$

$$\tilde{A}(|x-s|, \tau) := k_1 \delta(|x-s|) + k_2 A(|x-s|, \tau) \tag{27}$$

where $k_1 + k_2 = 1$ and $k_1, k_2 > 0$.

The exponential kernel suggested by Eringen [10] is normalized in an infinite domain and it is employed to this work. This kernel's form is defined as follows:

$$A(|x-s|, e_0 a) = \frac{1}{2e_0 a} \exp\left(\frac{-|x-s|}{e_0 a}\right) \tag{28}$$

The kernel of Eq. (28) in a finite domain is depicted in Fig. 1, where the violation of the normalization condition is obvious.

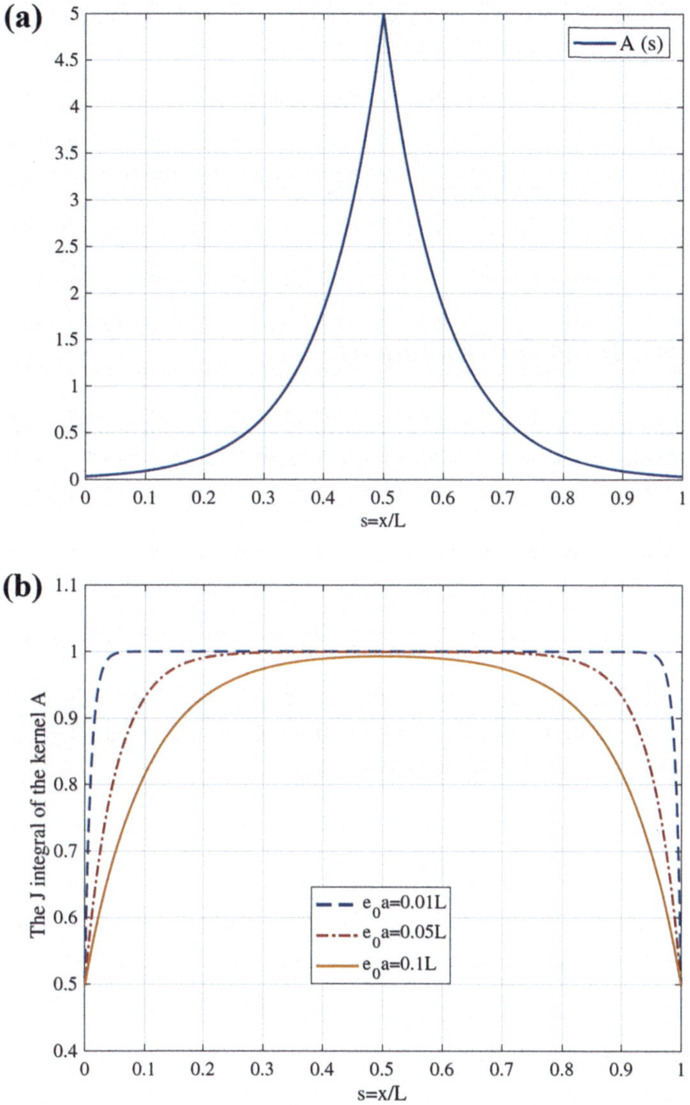

Fig. 1 **a** The nonlocal modulus A with respect to the normalized distance x/L with $e_0a = 0.1L$. **b** Values of the integral $J(x)$ of the nonlocal modulus A with respect to the normalized distance x/L, for different values of the nonlocal parameter, where $J(x) = \int_0^L A(|x-s|, e_0a)ds$

The one-dimensional differential operator corresponding to the exponential kernel is given by:

$$\mathcal{H} = 1 - (e_0 a)^2 \frac{d^2}{dx^2} \tag{29}$$

The nonlocal differential constitutive equation in one-dimension is then defined as:

$$\mathcal{H} t_{xx} = \sigma_{xx} \tag{30}$$

3.1 *Beam Equilibrium Equations*

Let us consider a beam of length L. The uniform cross section and the constant stiffness of the beam are denoted by $S\left(:= \int_S dS\right)$ and EI, respectively. Also, Young's modulus of the material and the second moment of an area about y-axis are respectively denoted by E and I, where $I := \int_S z^2 dS$. EB beam theory (EBBT) is based on the displacement field as well:

$$u_1 = u(x,t) - z\frac{\partial w(x,t)}{\partial x}, \; u_2 = 0, \; u_3 = w(x,t) \tag{31}$$

The axial and the transverse displacements of the point $(x, 0)$ on the middle plane of the beam as well as the axial and the longitudinal coordinate are denoted by u, w, x and z, respectively. The strain is, according to EBBT:

$$\varepsilon_{xx} = \frac{\partial u}{\partial x} - z\frac{\partial^2 w}{\partial x^2} \equiv \varepsilon^0_{xx} + z\kappa \tag{32}$$

where ε^0_{xx} is the axial strain and κ is the curvature. This work centers on the bending equation; thereby we assume that $\varepsilon_{xx} = -z\frac{\partial^2 w}{\partial x^2}$.

3.1.1 Integral Stress Models

The nonlocal integral motion equation of the Rayleigh prismatic beam is defined as follows [54]:

$$\frac{\partial^2}{\partial x^2}\left[EI\int_0^L \widehat{A}(|x-s|, e_0 a)\frac{\partial^2 w(s,t)}{\partial s^2}ds\right] + m_0\frac{\partial^2 w}{\partial t^2} - m_2\frac{\partial^4 w}{\partial x^2 \partial t^2} = q(x,t) \tag{33}$$

where $\widehat{A}$ is either A_{mod} or $\tilde{A}$. The distributed mass, the rotary inertia, the density, the moment of inertia, the cross section and the dynamical transverse distributed force of the beam are denoted by m_0, m_2, ρ, I, S and $q(x,t)$, respectively. The distributed mass and the rotary inertia are given by $m_0 = \rho S$ and $m_2 = \rho I$.

Furthermore, the BCs of Eq. (33), which are either the shear force or the transverse displacement and either the bending moment or the slope, follow:

$$Q^{NL} := \frac{\partial}{\partial x}\left[EI\int_0^L \widehat{A}(|x-s|, e_0a)\frac{\partial^2 w(s,t)}{\partial s^2}ds\right] - m_2\frac{\partial^3 w}{\partial x\partial t^2} = 0 \quad \text{or}$$
$$w = 0 \tag{34}$$

$$M^{NL} := EI\int_0^L \widehat{A}(|x-s|, e_0a)\frac{\partial^2 w(s,t)}{\partial s^2}ds = 0 \quad \text{or}$$
$$\frac{\partial w}{\partial x} = 0 \tag{35}$$

at $x = 0$ and $x = L$, where L is the beam's length.

3.1.2 Differential Stress Model

The nonlocal differential motion equation is defined as follows [37]:

$$-EI\frac{\partial^4 w}{\partial x^4} + (e_0a)^2\frac{\partial^2}{\partial x^2}\left(m_0\frac{\partial^2 w}{\partial t^2} - m_2\frac{\partial^4 w}{\partial x^2\partial t^2} - q\right)$$
$$-m_0\frac{\partial^2 w}{\partial t^2} + m_2\frac{\partial^4 w}{\partial x^2\partial t^2} + q = 0 \tag{36}$$

Equation (36) can be ergo written in a compressed form as:

$$EI\frac{\partial^4 w}{\partial x^4} + \mathcal{H}\left(m_0\frac{\partial^2 w}{\partial t^2} - m_2\frac{\partial^4 w}{\partial t^2\partial x^2} - q\right) = 0 \tag{37}$$

where the parenthesis is the variable of the operator $\mathcal{H}$ (Eq. (29)).

The boundary terms and their energy conjugated quantities are given by the following expressions:

$$Q := -EI\frac{\partial^3 w}{\partial x^3} + (e_0a)^2\frac{\partial}{\partial x}\left(m_0\frac{\partial^2 w}{\partial t^2} - m_2\frac{\partial^4 w}{\partial t^2\partial x^2} - q\right) + m_2\frac{\partial^3 w}{\partial t^2\partial x} = 0 \tag{38}$$

or $w = 0$ at the boundary either 0 or L,

$$M := -EI\frac{\partial^2 w}{\partial x^2} + (e_0a)^2\left(m_0\frac{\partial^2 w}{\partial t^2} - m_2\frac{\partial^4 w}{\partial t^2\partial x^2} - q\right) = 0 \tag{39}$$

or $\partial w/\partial x = 0$ at the boundary either 0 or L.

3.2 *Static Problems*

3.2.1 Direct Approach to the Nonlocal Integral Model

A direct handling of the nonlocal integral constitutive equation is to deduce the strains by vanishing the dynamical terms of Eq. (33) and then integrating by parts twice. The following equation concerning the strain is ergo obtained [80]:

$$\int_0^L \widehat{A}(|x-s|, e_0a)\frac{d^2w}{ds^2}ds = \tilde{q}(x) + C_1x + C_2 \tag{40}$$

where $\tilde{q}(x) := \frac{1}{EI}\int\left(\int_0^x q(y)dy\right)dx$ and $C_1, C_2 \in \Re$.

A cantilever and a simply supported beam subjected to two types of loading (i.e., a concentrated load applied to the beam's middle and a uniformly distributed load) are respectively studied. The essential BCs of a cantilever and a simply supported beam are respectively: $w(L) = dw(L)/dx = 0$ and $w(0) = w(L) = 0$.

The following values are selected for the concentrated load, the uniformly distributed load, the length, the bending stiffness, the radius as well as the thickness of the SWCNT/nanobeam: $P_0 = 1 \times 10^{-9}$N, $q_0 = 1 \times 10^{-9}$N/1×10^{-9}m, $L = 10 \times 10^{-9}$m, $EI = 4.41 \times 10^{-26}$Nm^2, $r = 0.34 \times 10^{-9}$m and $t = 0.34 \times 10^{-9}$m.

We note that $A \in L^2(0, L)$ (i.e., $\int_0^L\int_0^L |A(|x-s|, e_0a)|^2\,ds < \infty$), since the attenuation function is continuous and the integral operator is compact. This entails the solution's existence and uniqueness to the investigated problems. To solve numerically the integral Eq. (40), we use Gauss quadrature rule converging toward the analytical expression of the integral, since the points' number tends to infinity [81].

Owing to the existence of the boundary layer, the number of Gauss points selected without a significant computational cost is 3000. A linear system of equations $\mathbf{Ke} = \mathbf{g}$ is deduced. The coefficient matrix, the vector of unknowns and the vector of external loads are denoted by **K**, **e** and **g**, respectively.

The ratio of the beam's length to the beam's height is sufficiently large in all the above examples as well. Ergo, we assume that the thin beams' theory is appropriate, and the fact that the EBBT does not take into consideration the shear deformation does not produce a significance error.

The next results of figures exhibit the normalized strain $\varepsilon_{xx} := \frac{e_{xx}}{\max|e_{xx}^{cl}|}$, $e_{xx} := e_{xx}(x,z) = -z_{\max}\frac{d^2w}{dx^2}$, with respect to the length of the beam for each explored case. The classic strain is denoted by e_{xx}^{cl}. The following figures display the strain ε_{xx} of the classic-local model, the TPNI stress model and the model of the modified kernel for $e_0a = 0.01L$. To display the strain in all the figures, a representative set of 3000 Gauss points is indicatively selected.

Equation (40) can be alternatively written as:

$$M^{NL}(x) = EI\tilde{q}(x) + \tilde{C}_1x + \tilde{C}_2 \tag{41}$$

where $\tilde{C}_1, \tilde{C}_2 \in \Re$.

The strains are deduced by the following equations for the case of a cantilever beam subjected to a uniformly distributed load q_0 (BCs: $M^{NL}(0) = Q^{NL}(0) = 0$) [54]:

$$\frac{d^2w}{dx^2}\xi(x, e_0a) + \int_0^L A(|x-s|, e_0a)\frac{d^2w}{ds^2}ds = \frac{q_0x^2}{2EI} \tag{42}$$

$$k_1\frac{d^2w}{dx^2} + k_2\int_0^L A(|x-s|, e_0a)\frac{d^2w}{ds^2}ds = \frac{q_0x^2}{2EI} \tag{43}$$

What is more, the next equations give the strains for the case of a cantilever beam subjected to a concentrated load at an interior point x_0 (BCs: $M^{NL}(0) = Q^{NL}(0) = 0$) [54]:

$$\frac{d^2w}{dx^2}\xi(x, e_0a) + \int_0^L A(|x-s|, e_0a)\frac{d^2w}{ds^2}ds = \frac{P_0}{EI}(x-x_0)H(x-x_0) \tag{44}$$

$$k_1\frac{d^2w}{dx^2} + k_2\int_0^L A(|x-s|, e_0a)\frac{d^2w}{ds^2}ds = \frac{P_0}{EI}(x-x_0)H(x-x_0) \tag{45}$$

where $H(x-x_0)$ is the step/Heaviside function at point x_0. Furthermore, the strains are obtained by the following equations for the case of a simply supported beam subjected to a uniformly distributed load q_0 (BCs: $M^{NL}(0) = M^{NL}(L) = 0$) [54]:

$$\frac{d^2w}{dx^2}\xi(x, e_0a) + \int_0^L A(|x-s|, e_0a)\frac{d^2w}{ds^2}ds = \frac{q_0x}{EI}\left(\frac{x}{2}-1\right) \tag{46}$$

$$k_1\frac{d^2w}{dx^2} + k_2\int_0^L A(|x-s|, e_0a)\frac{d^2w}{ds^2}ds = \frac{q_0x}{EI}\left(\frac{x}{2}-1\right) \tag{47}$$

Moreover, the next equations yield the strains for the case of a simply supported beam subjected to a concentrated load at an interior point x_0 (BCs: $M^{NL}(0) = M^{NL}(L) = 0$) [54]:

$$\begin{aligned}&\frac{d^2w}{dx^2}\xi(x, e_0a) + \int_0^L A(|x-s|, e_0a)\frac{d^2w}{ds^2}ds\\&= \frac{P_0}{EI}(x-x_0)H(x-x_0) - \frac{P_0}{EI}\left(1-\frac{x_0}{L}\right)x\end{aligned} \tag{48}$$

$$\begin{aligned}&k_1\frac{d^2w}{dx^2} + k_2\int_0^L A(|x-s|, e_0a)\frac{d^2w}{ds^2}ds\\&= \frac{P_0}{EI}(x-x_0)H(x-x_0) - \frac{P_0}{EI}\left(1-\frac{x_0}{L}\right)x\end{aligned} \tag{49}$$

The discrete mathematical expressions of Eq. (42)–(49) take the form [54]:

$$f(t_i)\xi(t_i, e_0 a) + \sum_{j=1}^{G} c_j A(|t_i - t_j|, e_0 a) f(t_j) = g(t_i), \quad i = 1, \ldots, G \tag{50}$$

$$f(t_i)k_1 + k_2 \sum_{j=1}^{G} c_j A(|t_i - t_j|, e_0 a) f(t_j) = g(t_i). \quad i = 1, \ldots, G \tag{51}$$

A cantilever beam is subjected to a uniformly distributed load in Fig. 2. A difference between the nonlocal integral models and the classic model is indicated. Apart from an area near to the fixed point, the strains of the nonlocal integral models exceed that of the classic model (Fig. 2b). When k_1-parameter reduces, the integral models also tend to each other without including the area close to the fixed point (Fig. 2b). Moreover, the boundary layer concerning the TPNI stress model intensifies close to the fixed point of the beam (Fig. 2b). The boundary layer, in particular, decreases, yet its influence zone increases as k_1-parameter augments. The classic model is approached by the TPNI stress model when k_1-parameter increases as well (Fig. 2). Another conclusion is that the TPNI stress model converges toward the model of the modified kernel when k_1-parameter decreases, and the points of the body are not too close to the fixed point (Fig. 2). On the other hand, the boundary layer of the model of the modified kernel is indiscernible near to the fixed point. Additionally, the corresponding strain exceeds that of the classic model when the distance between an arbitrary point and the fixed point increases. What is more, the strain of the modified kernel model reduces and is exceeded by that of the classic model when the free end is approached.

A concentrated load is applied to the middle of a cantilever beam in Fig. 3. The stains of the nonlocal integral models are identical to that of the classic model except

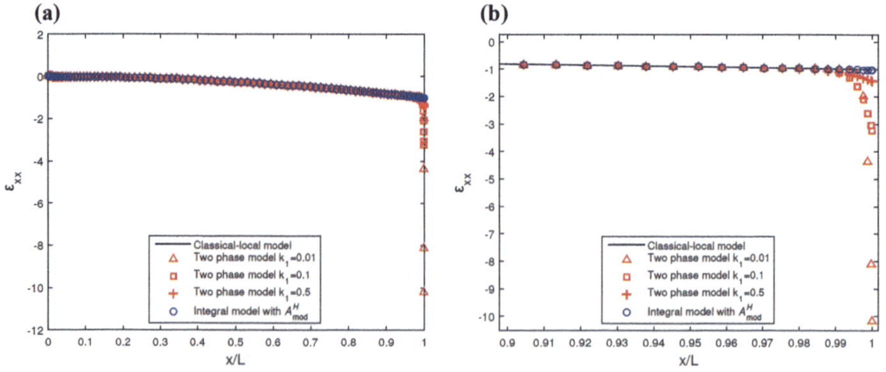

Fig. 2 **a** The normalized strain ε_{xx} of a cantilever beam, with a uniformly distributed load, for $e_0 a/L = 0.01$. **b** Details of an area close to the fixed point. ($A^H_{mod} = A_{mod}$)

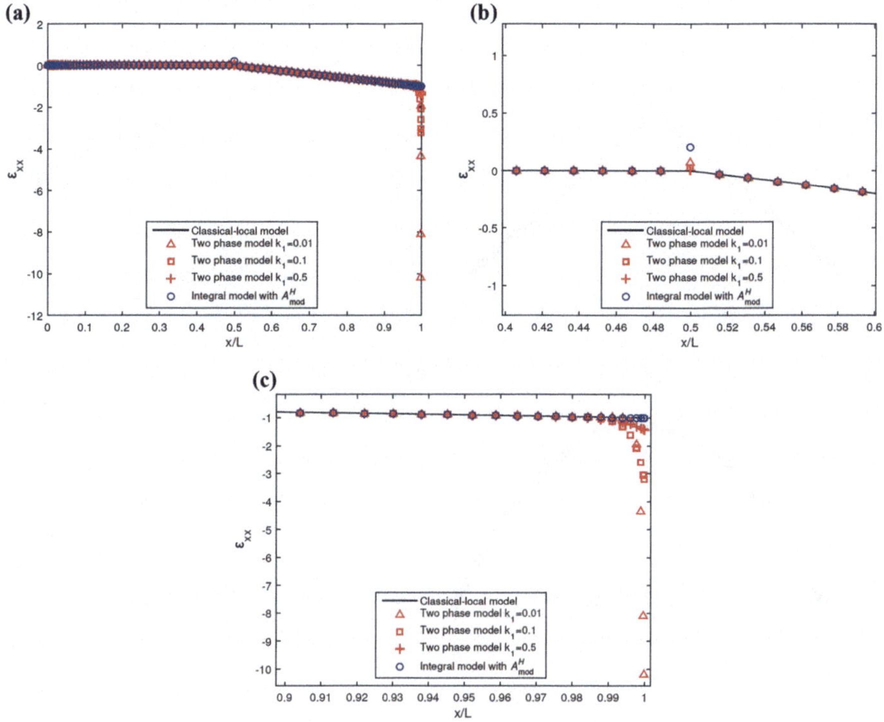

Fig. 3 **a** The normalized strain ε_{xx} of a cantilever beam, with a concentrated load applied to the beam's middle, for $e_0a/L = 0.01$. Details of an area of the load point **b** and an area close to the fixed point **c**

for the areas close to the load and the fixed points of the beam. When k_1-parameter reduces, the strain of the TPNI stress model decreases (Fig. 3a). As regards the TPNI stress model, a boundary layer, similar to the case of a cantilever beam with a uniformly distributed load, is developed near to the fixed point (Fig. 3b, c). To be more specific, the boundary layer decreases, but at the same time its influence zone increases as k_1-parameter augments. Taking into account the increase of k_1-parameter, the TPNI stress model tends to the classic model (Fig. 3). Nevertheless, the strain of the model of the modified kernel takes smaller values than those of the other models at the load points (Fig. 3b). Furthermore, the strain of the TPNI stress model tends to that of the model of the modified kernel at the load point when k_1-parameter decreases. A significant conclusion is that the model of the modified kernel does not appear to have a discernible boundary layer at the fixed point (Fig. 3c).

Another conclusion of importance is that the boundary layer depends only on the investigated governing equation, but not the exerted loading (Figs. 2b and 3c).

Two different types of loading are applied to a simply supported beam in Fig. 4a, c. Similarities are presented between the models which are explored. However, the boundary layer regarding the TPNI stress model is less sharp and the strain takes

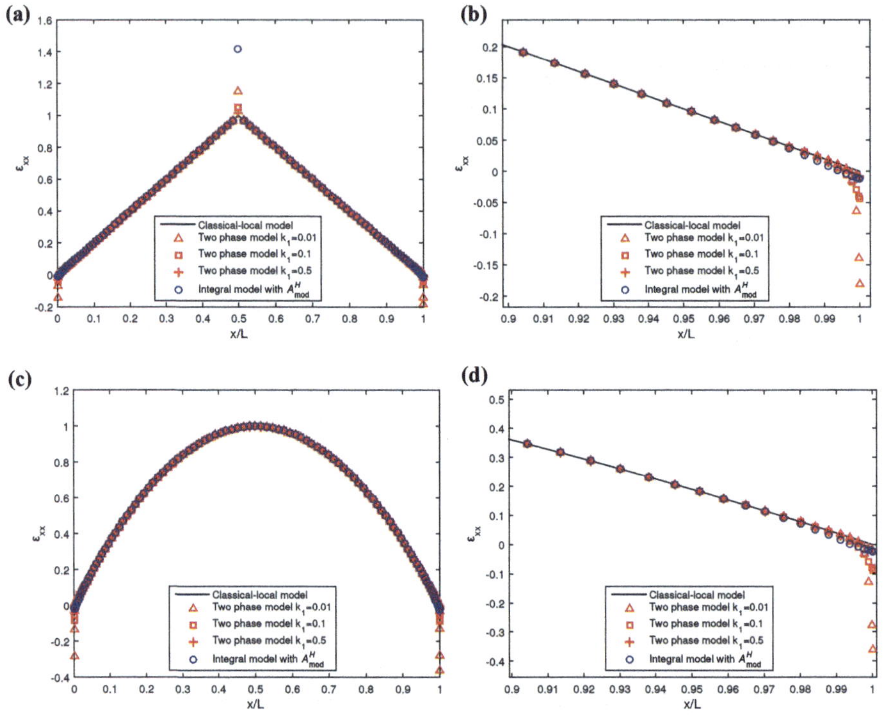

Fig. 4 The normalized strain ε_{xx} of a simply supported beam for $e_0a/L = 0.01$. **a** A concentrated load applied to the beam's middle. **b** Details of an area close to the fixed point. **c** A uniformly distributed load. **d** Details of an area close to the fixed point

smaller values than that of a cantilever beam in an area close to the fixed points (Fig. 4b, d). The influence zone of the boundary layer of the TPNI stress model enlarges as k_1-parameter increases, yet the boundary layer compared to that of the modified kernel and being close to the fixed points decreases (Fig. 4b, d). The strain of the model of the modified kernel, in particular, takes larger values than that of the TPNI stress model with $k_1 = 0.5$ in areas close to the fixed points. This indicates the difficulty of the model of the modified kernel to handle the boundary layer of a simply supported beam.

To check the stability of the linear system of equations, the condition number of the coefficient matrix **K** should be calculated. The condition number of a matrix **K** $n \times n$ is defined as $\kappa(\mathbf{K}) = ||\mathbf{K}|| \cdot ||\mathbf{K^{-1}}||$.

The definitions of the norm are given by $||\mathbf{K}||_\infty = \max_{1\le i\le n} \sum_{j=1}^{n} |K_{ij}|$, $||\mathbf{K}||_1 = \max_{1\le j\le n} \sum_{i=1}^{n} |K_{ij}|$ and $||\mathbf{K}||_2 = \left[\sum_{1\le i,j\le n}^{n} \left(K_{ij}\right)^2\right]^{1/2}$.

If **K** is a symmetric matrix, then $||\mathbf{K}||_2 = \rho(\mathbf{K})$, where $\rho(\mathbf{K}) = \max\{\lambda_i : \mathbf{K}\}$ and λ_i are the eigenvalues of the matrix **K**. In case that **K** is a symmetric and a normal matrix, it then holds that $\kappa(\mathbf{K}) = |\lambda_{\max}|/|\lambda_{\min}|$.

Table 1 Cantilever beam: Condition number $\left(\kappa(\mathbf{K}) = ||\mathbf{K}|| \cdot ||\mathbf{K}^{-1}||\right)$ of the matrix of coefficients, **K**

e_0a/L	Gauss points	A_{mod}	$k_1 = 0.01$	$k_1 = 0.1$	$k_1 = 0.2$	$k_1 = 0.5$
0.01	500	4875.431	100.9415	10.00192	4.984897	1.990991
	1000	2335.448	101.0362	10.00464	4.985585	1.991034
	2000	2240.871	101.1099	10.00701	4.986253	1.991106
	3000	1663.168	101.1350	10.00786	4.986504	1.991137
0.1 [54]	500	93.13096	97.12981	9.651288	4.82809	1.951197
	1000	92.71524	97.19989	9.653703	4.82881	1.951289
	2000	92.54316	97.25277	9.655588	4.829385	1.951368
	3000	92.51211	97.27045	9.65623	4.829583	1.951395

The linear system corresponding to the TPNI stress model is stable. The condition number of the coefficient matrix **K** tends to 1 when k_1-parameter increases. Also, the condition number is slightly affected by the nonlocal parameter (Table 1).

What is more, the condition number of the coefficient matrix **K** corresponding to the model of the modified kernel is acceptable for the case of $e_0a/L = 0.01$. When the nonlocal parameter takes such a small value, the model of the modified kernel approaches the first kind Fredholm integral equation. On the other hand, the condition number gives excellent results when $e_0a/L = 0.1$, and the linear system deriving is stable (Table 1).

3.2.2 Calculation of Static Deflection Through FEM

This section focuses on the transverse deflections of the nonlocal integral models for various types of loading and BCs. The static nonlocal integral EB beam bending equation derives from Eq. (33):

$$\frac{d^2}{dx^2}\left[EI\int_0^L \widehat{A}(|x-s|, e_0a)\frac{d^2w(s)}{ds^2}ds\right] = q(x) \tag{52}$$

The values of Sect. 3.2.1 have been taken into consideration for the numerical simulations. The selected values of e_0a also range from $0.01L$ to $0.04L$.

Each transverse deflection is normalized as $w^{norm} := w/w^{cl}_{max}$, where w is the delection of a nanobeam and ω^{cl}_{max} is the maximum deflection of the classical EBBT for all the problems explored below.

In the next figures, a comparison is drawn between the transverse deflections of the classic-local model, the nonlocal differential model, the TPNI stress model with different values of c_1-parameter and the model of the modified kernel of a nanobeam with respect to the nonlocal parameter.

The numerical transverse deflections regarding the nonlocal integral models are deduced by the following FEMs [48]:

$$\sum_{n=1}^{N}\left[\sum_{j=1}^{4} w_j^n \int_{x_{n-1}}^{x_n} \frac{d^2\phi_i^n}{dx^2} EI\xi(x, e_0a)\frac{d^2\phi_j^n}{dx^2}dx \right.$$
$$\left. + \int_{x_{n-1}}^{x_n} \frac{d^2\phi_i^n}{dx^2} EI \sum_{m=1}^{M}\sum_{j=1}^{4} w_j^m \int_{s_{m-1}}^{s_m} A(|x-s|, e_0a)\frac{d^2\phi_j^m}{ds^2} dsdx\right]$$
$$= \sum_{n=1}^{N}\int_{x_{n-1}}^{x_n} \phi_i^n(x)q(x)dx - \phi_i^N(L)\bar{Q}^{NL}(L) + \phi_i^1(0)\bar{Q}^{NL}(0)$$
$$+ \left(\phi_i^N\right)'(L)\bar{M}^{NL}(L) - \left(\phi_i^1\right)'(0)\bar{M}^{NL}(0) \tag{53}$$

$$\sum_{n=1}^{N}\left[k_1\sum_{j=1}^{4} w_j^n \int_{x_{n-1}}^{x_n} \frac{d^2\phi_i^n}{dx^2} EI\frac{d^2\phi_j^n}{dx^2}dx \right.$$
$$\left. + k_2\int_{x_{n-1}}^{x_n} \frac{d^2\phi_i^n}{dx^2} EI \sum_{m=1}^{M}\sum_{j=1}^{4} w_j^m \int_{s_{m-1}}^{s_m} A(|x-s|, e_0a)\frac{d^2\phi_j^m}{ds^2} dsdx\right]$$
$$= \sum_{n=1}^{N}\int_{x_{n-1}}^{x_n} \phi_i^n(x)q(x)dx - \phi_i^N(L)\tilde{Q}^{NL}(L) + \phi_i^1(0)\tilde{Q}^{NL}(0)$$
$$+ \left(\phi_i^N\right)'(L)\tilde{M}^{NL}(L) - \left(\phi_i^1\right)'(0)\tilde{M}^{NL}(0), \tag{54}$$

where the arbitrary finite elements of the beam are denoted by $V_n = (x_{n-1}, x_n)$ and $V_m' = (s_{m-1}, s_m)$ and the numbers of finite elements by N and M, respectively. Also, Hermite shape functions are denoted by $\phi_i^n(x)$ $(i = 1, \ldots, 4)$, the approximate transverse deflection by $w(x) = \sum_{j=1}^{4} w_j^n\phi_j^n(x)$ as well as the generalized displacements by w_j, respectively. Gauss quadrature rule is used for the integrals' calculation of Eq. (53) and Eq. (54) as well. $N = M = 100$ is also selected for all the below investigated problems.

(a) Cantilever beam with a concentrated load at the middle of the beam

The essential and the natural BCs of a cantilever beam subjected to a concentrated load at the middle of the beam are given by:

$$w(0) = \left.\frac{dw}{dx}\right|_{x=0} = 0 \tag{55}$$

$$\bar{M}(L) = \bar{Q}(L) = 0, \quad \tilde{M}(L) = \tilde{Q}(L) = 0 \tag{56}$$

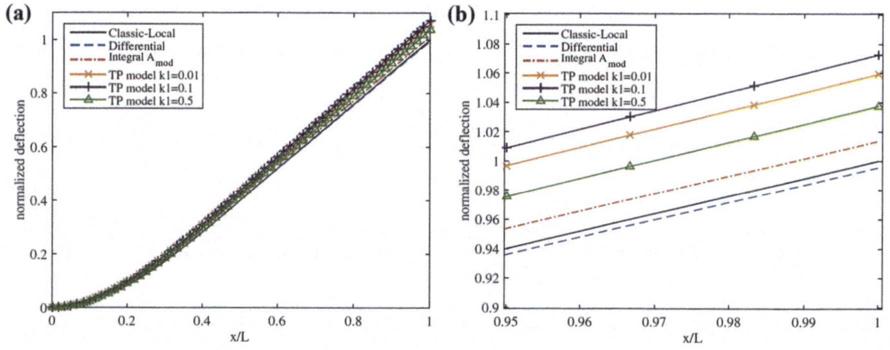

Fig. 5 **a** Normalized transverse deflections of a cantilever beam, with a concentrated load applied to the beam's middle, for $e_0a/L = 0.03$. **b** Detail of an area at the free end of the beam

The maximum classic-local transverse deflection takes the following value at the free end of the beam: $w_{max}^{cl} = 5P_0L^3/48EI$.

In Figs. 5 and 6, the responses of the nonlocal integral models present to be flexible in comparison with that of the classic-local model.

What is more, the transverse deflection of the nonlocal differential model is identical to that of the classic-local model at the interval $(0, L/2)$, yet it is smaller than that of the classic-local model at the interval $(L/2, L)$. In other words, the beam's stiffness increases in the interval ranging from the load point to the free end of the beam. It is critical to mention that the nonlocal differential and the nonlocal integral models present totally different results.

Furthermore, the deflection of the TPNI stress model exceeds that of the model of the modified kernel in Figs. 5 and 6. This behavior is expected by taking into consideration Fig. 3. The difference between their values increases as the nonlocal parameter augments (Fig. 6).

Based on Fig. 6a, the deflections of the model of the modified kernel linearly increase when $0.01 < e_0a/L < 0.03$, but their increase rate reduces when $0.03 < e_0a/L < 0.04$. The attenuation function widens as the nonlocal parameter increases. Its area close to the boundary and ergo the information's quantity decreases, yet the intervals in which the area of the attenuation function does not approach 1 increase (Fig. 1b). Hence, the created gap is occupied by the locality, and this is obvious in Fig. 6a.

(b) Cantilever beam with a concentrated load at the free end of the beam

The essential BCs of a cantilever beam with a concentrated load applied to the free end of the beam are given by Eq. (55) and the natural BCs by:

$$\bar{M}(L) = \tilde{M}(L) = 0, \quad \bar{Q}(L) = \tilde{Q}(L) = P_0 \tag{57}$$

The maximum classic-local transverse deflection takes the following value at the free end of the beam: $w_{max}^{cl} = P_0L^3/3EI$.

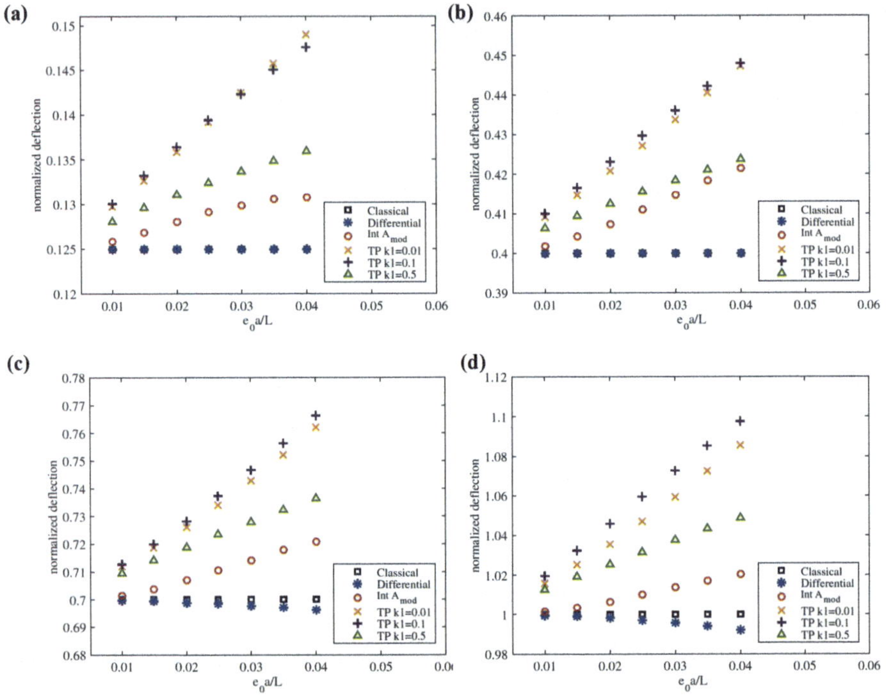

Fig. 6 Normalized transverse deflections of a cantilever beam, with a concentrated load applied to the beam's middle, for different values of e_0a/L at points **a** $0.25L$, **b** $0.5L$, **c** $0.75L$, **d** L

Based on Figs. 7 and 8, the responses of the nonlocal integral models appear to be flexible in comparison with that of the classic-local model. The transverse deflection of the TPNI stress model presents a monotonically increasing behavior as the nonlocal parameter augments.

What is more, the transverse deflection of the nonlocal differential model is identical to that of the classic-local model (Figs. 7 and 8). In other words, the nonlocal differential model is insensitive to the locality effect. It is worth mentioning that the nonlocal integral models appear to have a qualitatively different behavior from that of the nonlocal differential model.

Finally, the response of the model of the modified kernel of a cantilever beam with a concentrated load at the free end of the beam presents a similar behavior with that of a cantilever beam subjected to a concentrated load at the beam's middle close to the fixed point (Fig. 8a). This behavior is more intense than that presented in Fig. 6a.

(c) Cantilever beam with a uniformly distributed load

The essential and the natural BCs of a cantilever beam subjected to a uniformly distributed load are given by Eqs. (55) and (56).

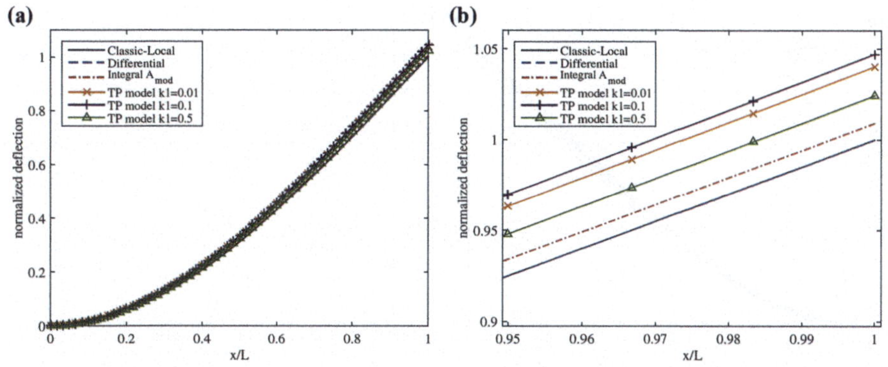

Fig. 7 **a** Normalized transverse deflections of a cantilever beam, with a concentrated load applied to the free end of the beam, for $e_0a/L = 0.03$. **b** Detail of an area at the free end of the beam

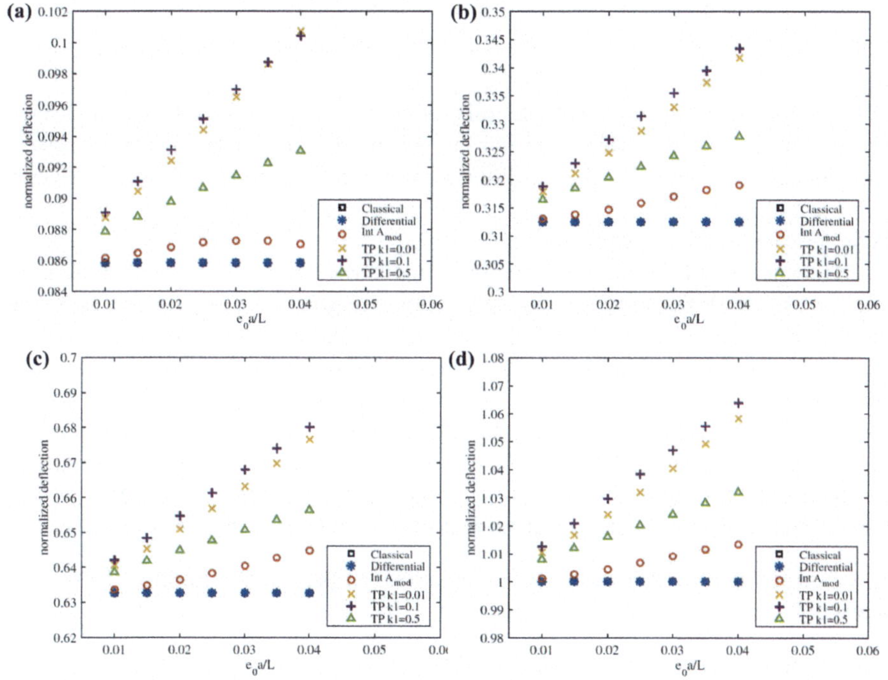

Fig. 8 Normalized transverse deflections of a cantilever beam, with a concentrated load applied to the free end of the beam, for different values of e_0a/L at points **a** $0.25L$, **b** $0.5L$, **c** $0.75L$, **d** L

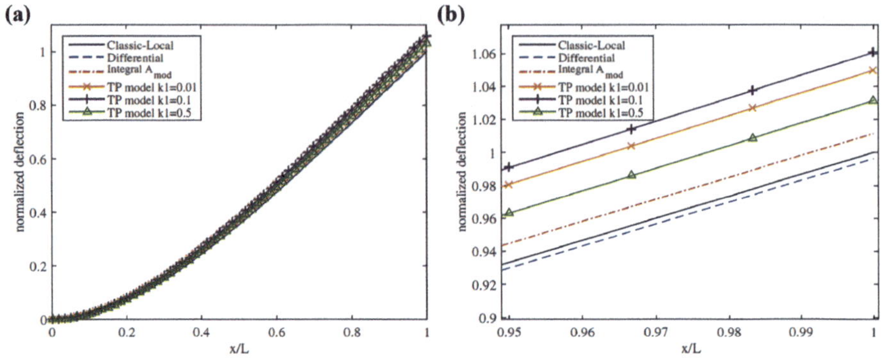

Fig. 9 **a** Normalized transverse deflections of a cantilever beam, with a uniformly distributed load, for $e_0a/L = 0.03$. **b** Detail of an area at the free end of the beam

The maximum classic-local transverse deflection takes the following value at the free end of the beam: $w_{max}^{cl} = q_0L^4/8EI$.

The responses of the nonlocal integral models present to be flexible in comparison with that of the classic-local model in Figs. 9 and 10. In Fig. 10a, the locality effect is evident for the model of the modified kernel. On the other hand, the transverse deflection of the TPNI stress model presents a monotonically increasing behavior as the nonlocal parameter augments.

The responses of the model of the modified kernel appear to have an extremum when $e_0a/L = 0.03$, since the locality's contribution increases as the nonlocal parameter increases (Fig. 10a). The deflection of the model of the modified kernel takes larger values than that of the classic-local model irrespective of the above behavior.

Moreover, the deflection of the classic-local model exceeds the deflection of the nonlocal differential model. In other words, the nonlocal differential model stimulates the stiffness of the beam. It is noteworthy that the nonlocal integral models present a qualitatively different behavior from that of the nonlocal differential model.

(d) Clamped-clamped beam with a concentrated load at the middle of the beam

The essential BCs of a clamped-clamped beam with a concentrated load applied to the middle of the beam are given by:

$$w(0) = \left.\frac{dw}{dx}\right|_{x=0} = 0, \quad w(L) = \left.\frac{dw}{dx}\right|_{x=L} = 0 \tag{58}$$

The maximum classic-local transverse deflection takes the following value at the middle of the beam: $w_{max}^{cl} = P_0L^3/192EI$.

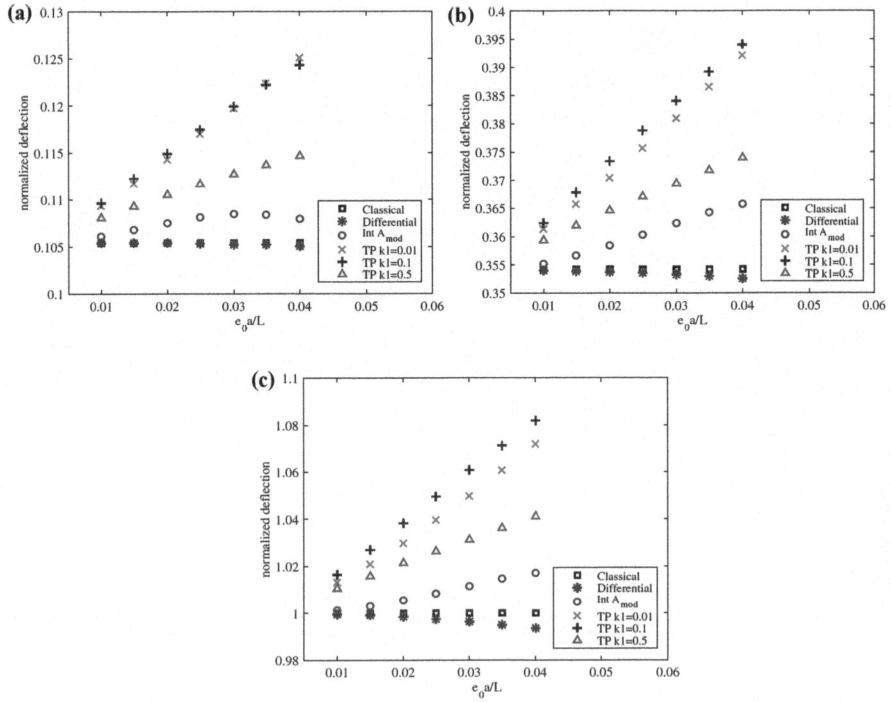

Fig. 10 Normalized transverse deflections of a cantilever beam, with a uniformly distributed load, for different values of e_0a/L at points **a** $0.25L$, **b** $0.5L$, **c** L

On the basis of Fig. 11, the responses of the nonlocal integral models appear to be flexible in comparison with those of the classic-local and the nonlocal differential models.

What is more, the transverse deflection of the model of the modified kernel, near to the fixed boundaries of the beam, takes smaller values than that of the TPNI stress model with both $k_1 = 0.01$ and $k_1 = 0.1$ (Fig. 11b). At the same time at the middle of the beam, the response of the model of the modified kernel appears to have a flexible behavior compared to the responses of the other explored models (Fig. 11c). Also, the model of the modified kernel is intensely affected by the locality added to the constitutive equations as the nonlocal parameter increases.

As regards the TPNI stress model, their deflections monotonically evolve with respect to k_1-parameter and when the nonlocal parameterer increases (Fig. 11b, c).

It is of importance that the nonlocal integral models do not appear to have a qualitatively different behavior from that of the nonlocal differential model.

(e) Clamped-clamped beam with a uniformly distributed load

The essential BCs of a clamped-clamped beam subjected to a uniformly distributed load are given by Eq. (58).

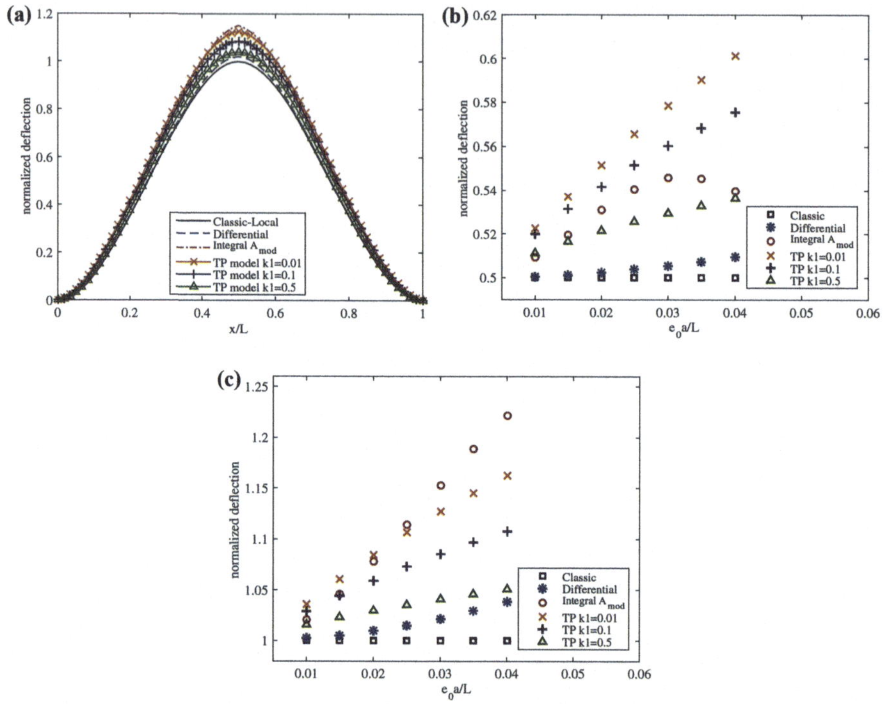

Fig. 11 **a** Normalized transverse deflections of a clamped-clamped beam, with a concentrated load applied to the beam's middle, for $e_0a/L = 0.03$. Normalized transverse deflections for different values of e_0a/L at points **b** $0.25L$, **c** $0.5L$

The maximum classic-local transverse deflection takes the following value at the middle of the beam: $w^{cl}_{max} = q_0L^4/384EI$.

In Fig. 12, the responses of the nonlocal integral models appear to have a flexible behavior in comparison with that of the classic-local model. Furthermore, the transverse deflection of the nonlocal differential model is identical to that of the classic-local model. The deflections concerning the TPNI stress model monotonically evolve with respect to k_1-parameter and when the nonlocal parameterer increases (Fig. 12c, d).

Moreover, the deflections of the TPNI stress model with both $k_1 = 0.01$ and $k_1 = 0.1$, close to the fixed boundaries of the beam, take larger values than that of the model of the modified kernel (Fig. 12c). At the same time at the middle of the beam, the response of the model of the modified kernel and the response of the TPNI stress model with $k_1 = 0.1$ compete with each other (Fig. 12d). By all the investigated models, the deflection of the TPNI stress model with $k_1 = 0.01$ takes the largest values (Fig. 12c, d).

What is more, the model of the modified kernel is intensely affected by the locality, which is added to the constitutive equations as the nonlocal parameter increases.

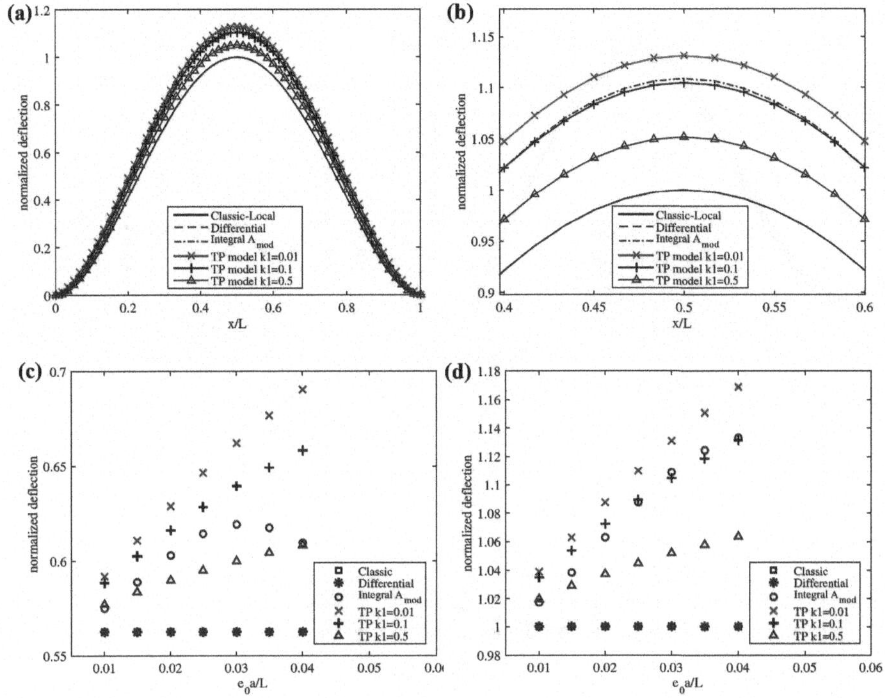

Fig. 12 **a** Normalized transverse deflections of a clamped-clamped beam, with a uniformly distributed load, for $e_0a/L = 0.03$. **b** Detail of an area at the middle of the beam. Normalized transverse deflections for different values of e_0a/L at points **c** $0.25L$, **d** $0.5L$

It is critical to mention that the nonlocal integral models present a qualitatively different behavior from that of the nonlocal differential model.

(f) Clamped-pinned beam with a concentrated load at the middle of the beam

The essential and the natural BCs of a clamped-pinned beam with a concentrated load applied to the middle of the beam are given by:

$$w(0) = \left.\frac{dw}{dx}\right|_{x=0} = 0, \quad w(L) = 0 \tag{59}$$

$$\bar{M}(L) = \tilde{M}(L) = 0 \tag{60}$$

The maximum classic-local transverse deflection takes the following value at the point of the beam $x = \left(1 - \sqrt{5}/5\right) L$: $w_{max}^{cl} = \sqrt{5}P_0L^3/240EI$.

Based on Fig. 13, the responses of the nonlocal integral models present to be flexible in comparison with those of the classic-local and the nonlocal differential models. What is more, the transverse deflection of the model of the modified kernel,

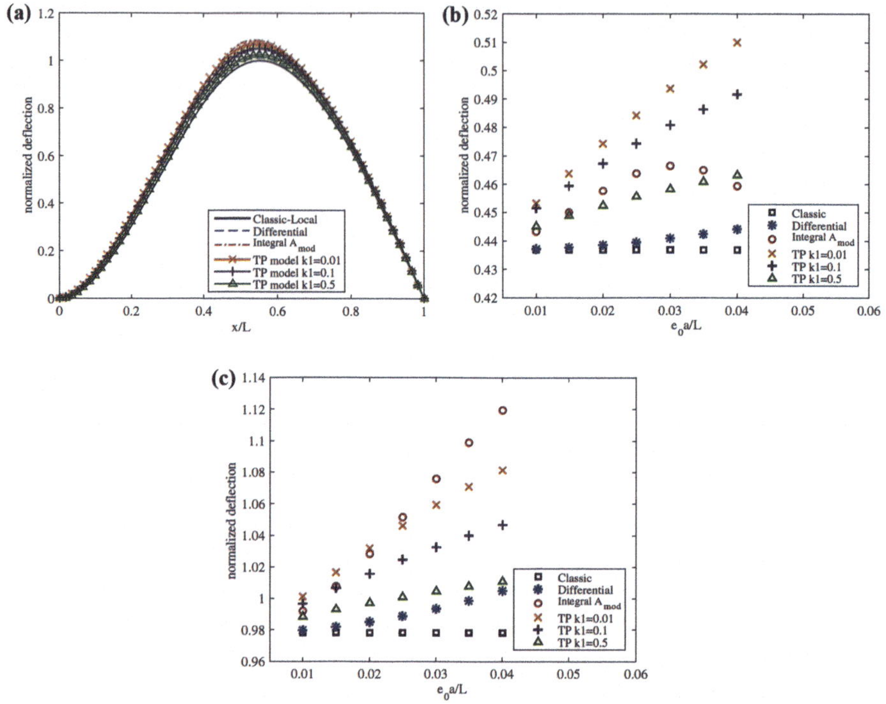

Fig. 13 **a** Normalized transverse deflections of a clamped-pinned beam, with a concentrated load applied to the beam's middle, for $e_0a/L = 0.03$. Normalized transverse deflections for different values of e_0a/L at points **b** $0.25L$, **c** $0.5L$

close to the fixed boundary of the beam, takes smaller values than that of the TPNI stress model with $k_1 = 0.01$ and $k_1 = 0.1$ too (Fig. 13b). At the same time at the middle of the beam, the response of the model of the modified kernel appears to have a flexible behavior compared to the responses of the other investigated models (Fig. 13c). Besides, the model of the modified kernel is intensely affected by the locality added to the constitutive equations as the nonlocal parameter increases. The deflections regarding the TPNI stress model monotonically evolve when the nonlocal parameter augments (Fig. 13b, c).

It is of significance that the behavior demonstrated by the nonlocal integral models is not qualitatively different from that of the nonlocal differential model.

(g) Clamped-pinned beam with a uniformly distributed load

The essential and the natural BCs of a clamped-pinned beam subjected to a uniformly distributed load are given by Eqs. (59) and (60).

The maximum classic-local transverse deflection takes the following value at the point of the beam $x = (15 - \sqrt{33})L/16$: $w_{max}^{cl} = (q_0L^4/EI)\left(\frac{39+55\sqrt{33}}{65536}\right)$.

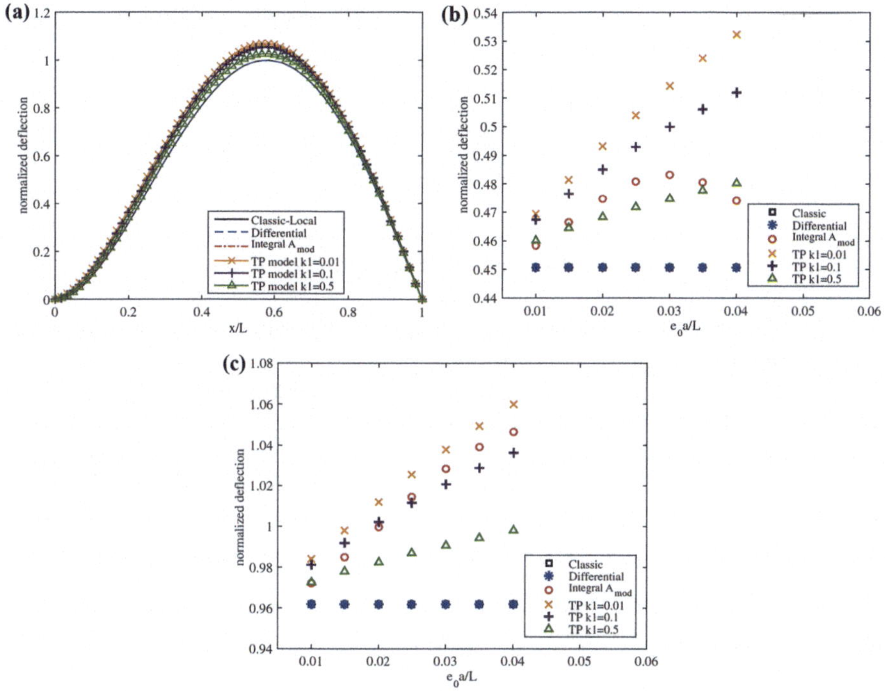

Fig. 14 **a** Normalized transverse deflections of a clamped-pinned beam, with a uniformly distributed load, for $e_0a/L = 0.03$. Normalized transverse deflections for different values of e_0a/L at points **b** $0.25L$, **c** $0.5L$

In Fig. 14, the responses of the nonlocal integral models appear to have a flexible behavior in comparison with that of the classic-local model. Furthermore, the transverse deflection of the nonlocal differential model is identical to that of the classic-local model. As regards the TPNI stress model, its deflections monotonically evolve when the nonlocal parameter increases (Fig. 14b, c).

Moreover, the deflections of the TPNI stress model with both $k_1 = 0.01$ and $k_1 = 0.1$, close to the fixed boundary of the beam, take larger values than that of the model of the modified kernel (Fig. 14b). At the same time at the middle of the beam, the response of the model of the modified kernel and the response of the TPNI stress model with $k_1 = 0.1$ compete with each other (Fig. 14c). By all the investigated models, the deflection of the TPNI stress model with $k_1 = 0.01$ takes the largest values (Fig. 14b, c). The model of the modified kernel is also affected by the locality, which is added to the constitutive equations as the nonlocal parameter increases.

It is critical to mention that the nonlocal integral models present a qualitatively different behavior from that of the nonlocal differential model.

(h) Simply supported beam with a concentrated load at the middle of the beam

The essential and the natural BCs of a simply supported beam with a concentrated load applied to the middle of the beam are given by:

$$w(0) = w(L) = 0 \quad (61)$$

$$\bar{M}(0) = \tilde{M}(0) = 0, \quad \bar{M}(L) = \tilde{M}(L) = 0. \quad (62)$$

The maximum classic-local transverse deflection takes the following value at the middle of the beam: $w^{cl}_{max} = P_0 L^3/48EI$.

In Fig. 15, the responses of the nonlocal integral models present to be flexible in comparison with that of the classic-local model.

Moreover, the transverse deflection of the model of the modified kernel takes larger values than those of the other models that studied. A extremum point is demonstrated in Fig. 15c as well. The extremum point is caused by the fact that the modified kernel adds locality to the gap created by the nonlocality on the boundary. The effect of the

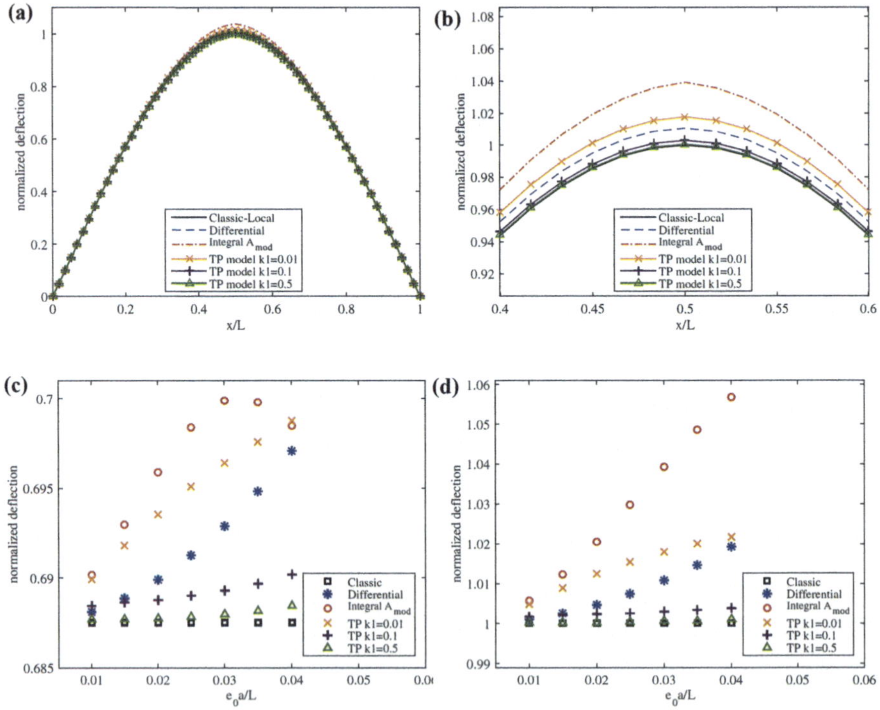

Fig. 15 **a** Normalized transverse deflections of a simply supported beam, with a concentrated load applied to the beam's middle, for $e_0a/L = 0.03$. **b** Detail of an area at the middle of the beam. Normalized transverse deflections for different values of e_0a/L at points **c** $0.25L$, **d** $0.5L$

modified kernel is therefore greater on points close to the boundary. Similar results are deduced for the case of a cantilever beam.

Besides, the deflections of the TPNI stress model monotonically evolve with respect to k_1-parameter and when the nonlocal parameterer increases in Fig.15c, d. The nonlocal deflection additionally tends to that of the classic-local model because of the augmentation of k_1-parameter.

The transverse deflection of the nonlocal differential model also appears to have a flexible response in comparison with that of the classic-local model. It is of significance that the nonlocal integral models present a qualitatively similar behavior to that of the nonlocal differential model.

(i) Simply supported beam with a uniformly distributed load

The essential and the natural BCs of a simply supported beam subjected to a uniformly distributed load are given by Eqs. (61) and (62).

The maximum classic-local transverse deflection takes the following value at the middle of the beam: $w^{cl}_{max} = 5q_0L^4/384EI$.

Based on Fig. 16, similar conclusions are drawn for the nonlocal models as in case of a concentrated load.

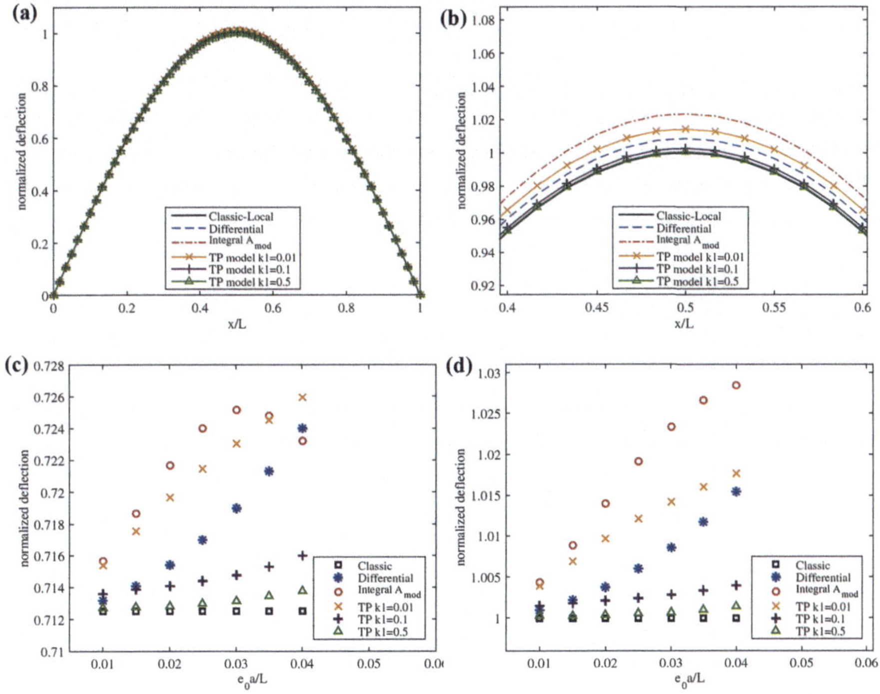

Fig. 16 **a** Normalized transverse deflections of a simply supported beam, with a uniformly distributed load, for $e_0a/L = 0.03$. **b** Detail of an area at the middle of the beam. Normalized transverse deflections for different values of e_0a/L at points **c** $0.25L$, **d** $0.5L$

The model of the modified kernel presents a flexible behavior in comparison with the other investigated models. When $e_0a/L > 0.03$, however, the aforementioned model is affected by the locality more than the corresponding model in case of a concentrated load.

In addition, the deflections of the TPNI stress model monotonically evolve with respect to k_1-parameter and when the nonlocal parameterer increases in Fig. 16c, d. The nonlocal deflection also tends to that of the classic-local model because of the augmentation of k_1-parameter.

3.3 Dynamical Problems

3.3.1 Free Vibration Problem of a Beam

This section revolves around the dynamical response of a cantilever, a clamped-clamped, a clamped-pinned and a simply supported nanobeam. The free vibration problem of an EB nanobeam (Eq. (33)), in particular, is solved by means of a numerical method.

The values of Sect. 3.2.1 have been taken into account for the numerical simulations. The following material constant is considered for the density: $\rho = 2300\,\text{kg/m}^3$. The selected values of e_0a also range from $0.01L$ to $0.04L$.

Each eigenfrequency is normalized as $\bar{\omega}_i := \omega_i/\omega_i^{clas}$, $i = \{1, 2, 3, 4\}$, where ω_i is the eigenfrequency of a nanobeam and ω_i^{clas} is the eigenfrequency of the classical EBBT of each mode for all the following investigated problems.

In the next figures, a comparison is drawn between the eigenfrequencies of the classic-local model, the nonlocal differential model, the TPNI stress model corresponding to different values of c_1-parameter and the model of the modified kernel of a nanobeam with respect to the nonlocal parameter.

The numerical eigenfrequencies concerning the nonlocal integral models are deduced by the following FEMs [71]:

$$\begin{aligned}
&\sum_{n=1}^{N}\Bigg[\sum_{j=1}^{4} c_j^n \int_{x_{n-1}}^{x_n} \frac{d^2\phi_i^n}{dx^2}(x) EI\xi(x, e_0a)\frac{d^2\phi_j^n}{dx^2}dx \\
&+\int_{x_{n-1}}^{x_n} \frac{d^2\phi_i^n}{dx^2} EI \sum_{m=1}^{M}\sum_{j=1}^{4} c_j^m \int_{s_{m-1}}^{s_m} A(|x-s|, e_0a)\frac{d^2\phi_j^m}{ds^2} ds dx \\
&-\omega^2 \sum_{j=1}^{4} c_j^n \int_{x_{n-1}}^{x_n} m_0\phi_i^n(x)\phi_j^n(x)dx\Bigg] \\
&= -\phi_i^N(L)\bar{Q}(L) + \phi_i^1(0)\bar{Q}(0) + \frac{d\phi_i^N}{dx}(L)\bar{M}(L) - \frac{d\phi_i^i}{dx}(0)\bar{M}(0) \qquad (63)
\end{aligned}$$

$$\sum_{n=1}^{N}\left[k_1\sum_{j=1}^{4}c_j^n\int_{x_{n-1}}^{x_n}\frac{d^2\phi_i^n}{dx^2}EI\frac{d^2\phi_j^n}{dx^2}dx\right.$$
$$+k_2\int_{x_{n-1}}^{x_n}\frac{d^2\phi_i^n}{dx^2}EI\sum_{m=1}^{M}\sum_{j=1}^{4}c_j^m\int_{s_{m-1}}^{s_m}A(|x-s|,e_0a)\frac{d^2\phi_j^m}{ds^2}dsdx$$
$$\left.-\,\omega^2\sum_{j=1}^{4}c_j^n\int_{x_{n-1}}^{x_n}m_0\phi_i^n(x)\phi_j^n(x)dx\right]$$
$$=-\phi_i^N(L)\tilde{Q}(L)+\phi_i^1(0)\tilde{Q}(0)+\frac{d\phi_i^N}{dx}(L)\tilde{M}(L)-\frac{d\phi_i^1}{dx}(0)\tilde{M}(0) \quad (64)$$

where the arbitrary finite elements of the beam are denoted by $V_n = (x_{n-1}, x_n)$ and $V_m' = (s_{m-1}, s_m)$ and the numbers of FEs by N and M, respectively. Hermite shape functions are also denoted by $\phi_i^n(x)$ $(i = 1, \ldots, 4)$, the approximate transverse deflection by $\psi(x) = \sum_{j=1}^{4} c_j^n\phi_j^n(x)$ as well as the generalized displacements by c_j, respectively. Gauss quadrature rule is used for the integrals' calculation of Eqs. (63) and (64). ω^2 (or $|\omega|$) is then calculated by $\det\left(\mathbf{K}^{NL} - \omega^2\mathbf{M}\right) = 0$. $N = M = 100$ is also selected for all the below investigated problems.

(a) Cantilever beam

The essential and the natural BCs of a cantilever nanobeam are given by Eqs. (55) and (56). In Fig. 17, the eigenfrequencies of the nonlocal models appear to have a softening response in comparison with those of the classic-local model, apart from the fundamental eigenfrequency of the nonlocal differential model. To be more specific, the fundamental eigenfrequency of the differential model increases as the nonlocal parameter augments (Fig. 17a). This remark has been obviously presented in the literature [35] and it is regarded as the paradox of the fundamental eigenfrequency of the nonlocal differential model of a cantilever beam.

What is more, the fundamental eigenfrequency of the model of the modified kernel takes larger values than those of the TPNI stress model for each k_1-parameter (Fig. 17a). However, that is not the case for the model of the modified kernel when the eigenfrequencies $\bar{\omega}_2$, $\bar{\omega}_3$, $\bar{\omega}_4$ are studied. In those cases, its eigenfrequencies take smaller values than the eigenfrequencies of all the other investigated models (Fig. 17b–d).

Furthermore, the fundamental eigenfrequency of the TPNI stress model appears a non-monotonic behavior with respect to k_1-parameter. Perhaps an explanation of this behavior centers on the fact that the TPNI stress model is unsuitable to successfully handle the boundary (Fig. 17a). On the other hand, the three remaining eigenfrequencies of the aforementioned model present a totally different behavior. To be more specific, the corresponding eigenfrequencies of the TPNI stress model evolve in a monotonic way and they tend to those of the classic-local model as k_1-parameter increases. Moreover, all the eigenfrequencies of the TPNI stress model exhibit a monotonic behavior with respect to the nonlocal parameter (Fig. 17b, c).

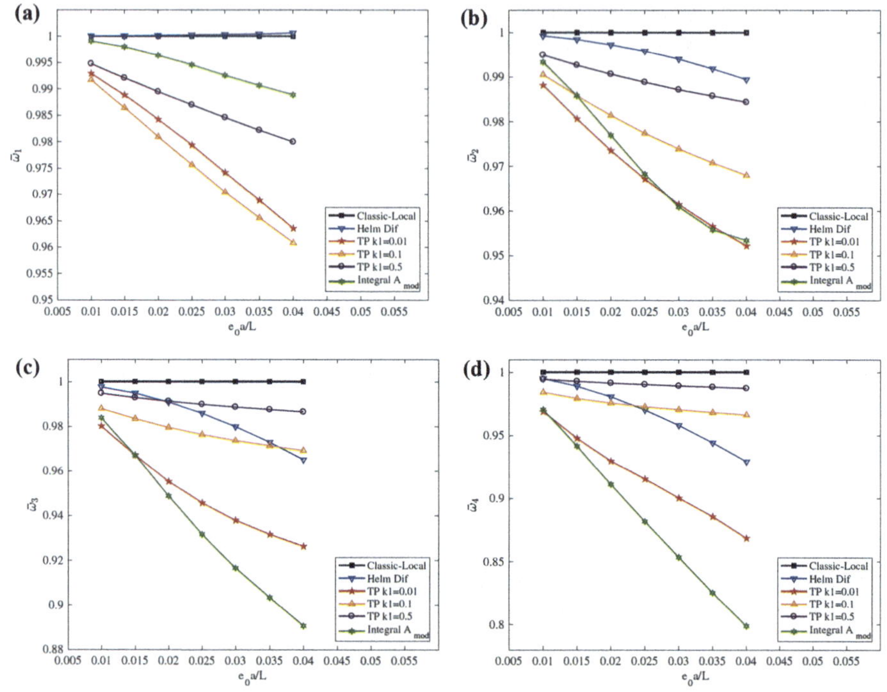

Fig. 17 The first four normalized eigenfrequencies of a cantilever nanobeam with respect to the nonlocal parameter. **a** $\bar{\omega_1}$, **b** $\bar{\omega_2}$, **c** $\bar{\omega_3}$, **d** $\bar{\omega_4}$

The third and the fourth eigenfrequencies of the nonlocal differential model demonstrate a softening response in comparison with those of the TPNI stress model for $k_1 = \{0.1, 0.5\}$ as well (Fig. 17c, d).

To recap, the responses of the nonlocal integral models present a softening behavior compared to those of the classic-local model. The last-mentioned results are in accordance with the results of a structural element in the context of a lattice atomic model [29].

(b) Clamped-clamped beam

The essential BCs of a clamped-clamped nanobeam are given by Eq. (58). The eigenfrequencies of the nonlocal models appear to have a softening response in comparison with those of the classic-local model in Fig. 18. Apart from the fundamental eigenfrequency, all the eigenfrequencies of the model of the modified kernel take smaller values than those of the other investigated models. As regards the fundamental eigenfrequency, the responses of the nonlocal integral models clearly differ from that of the nonlocal differential model (Fig. 18a).

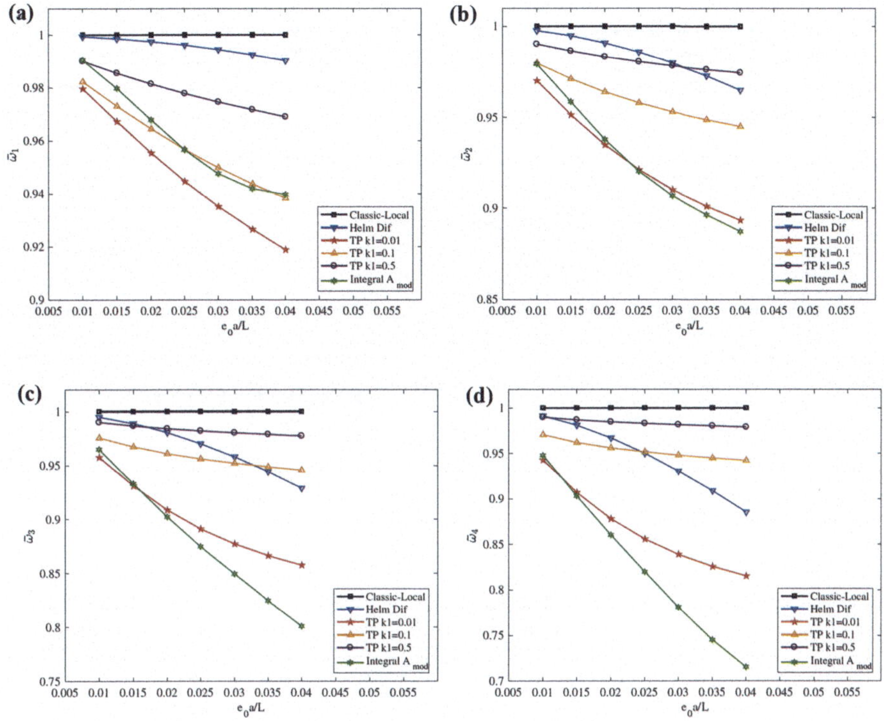

Fig. 18 The first four normalized eigenfrequencies of a clamped-clamped nanobeam with respect to the nonlocal parameter. **a** $\bar{\omega_1}$, **b** $\bar{\omega_2}$, **c** $\bar{\omega_3}$, **d** $\bar{\omega_4}$

What is more, the eigenfrequencies of the TPNI stress model monotonically evolve with respect to k_1-parameter. In other words, the aforementioned eigenfrequencies tend to those of the classic-local model as k_1-parameter increases. Besides, these eigenfrequencies decrease further when the nonlocal parameter augments.

Regarding the model of the modified kernel, its fundamental eigenfrequency diminishes up to a specific value of the nonlocal parameter (extremum), and when the nonlocal parameter exceeds this value, the eigenfrequency's behavior changes and starts to augment (Fig. 18a).

For the other three investigated eigenfrequencies, the model of the modified kernel and the TPNI stress model with $k_1 = 0.01$ present a competitive behavior with each other (Fig. 18b). The modified kernel's model also demonstrates a monotonically decreasing behavior with respect to the nonlocal parameter (Fig. 18b–d). The responses of the nonlocal differential model demonstrate a monotonically decreasing behavior as well.

(c) Clamped-pinned beam

The essential and the natural BCs of a clamped-pinned nanobeam are given by Eqs. (59) and (60). The responses of a clamped-pinned beam appear to have a similar

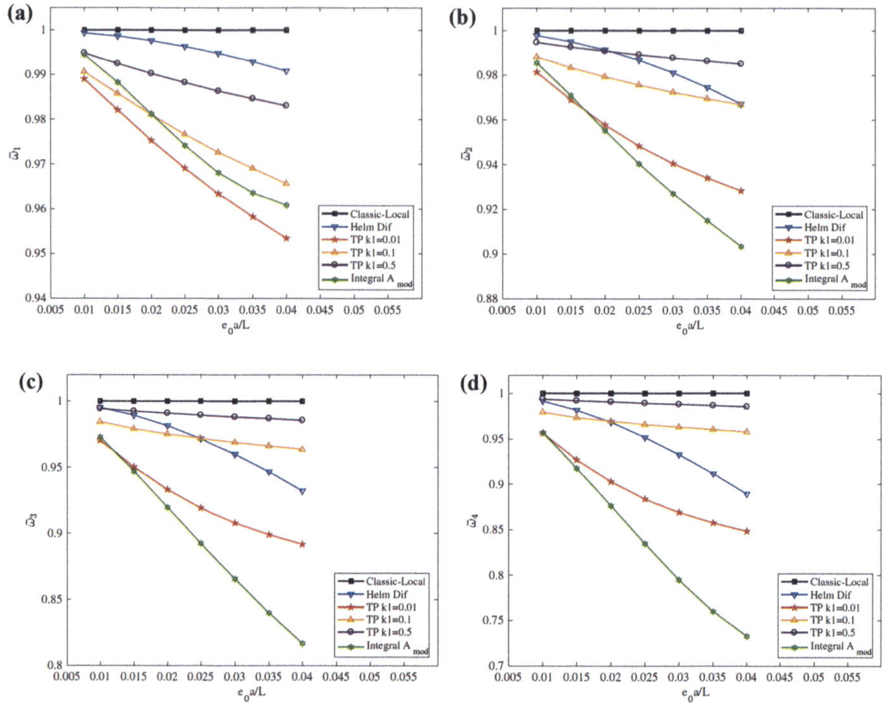

Fig. 19 The first four normalized eigenfrequencies of a clamped-pinned nanobeam with respect to the nonlocal parameter. **a** $\bar{\omega_1}$, **b** $\bar{\omega_2}$, **c** $\bar{\omega_3}$, **d** $\bar{\omega_4}$

behavior with those of a clamped-clamped beam in Fig. 19. Unlike the fundamental eigenfrequencies of a clamped-clamped beam, the fundamental eigenfrequencies of the nonlocal differential and the nonlocal integral models of a clamped-pinned beam do not present such a significant difference (Fig. 19a).

(d) Simply supported beam

The essential and the natural BCs of a simply supported nanobeam are given by Eqs. (61) and (62). The eigenfrequencies of the nonlocal models appear to have a softening response in comparison with those of the classic-local model in Fig. 20. Apart from the fundamental eigenfrequency when $e_0a/L > 0.03$ (Fig. 20a), all the eigenfrequencies of the model of the modified kernel take smaller values than those of the other investigated models.

What is more, the eigenfrequencies of the TPNI stress model monotonically evolve with respect to k_1-parameter. In other words, the above eigenfrequencies tend to those of the classic-local model as k_1-parameter increases. Besides, these eigenfrequencies decrease further when the nonlocal parameter augments.

As regards the model of the modified kernel, the decrease rate of its fundamental eigenfrequency reduces when the nonlocal parameter increases (Fig. 20a). It is critical to mention that the transverse deflections of the corresponding static problem present a similar behavior with respect to the nonlocal parameter. To elucidate, the increase of the contribution of the neighbouring particles to the nonlocal stresses at the interior points of the body is triggered by the augmentation of the nonlocal parameter. Nevertheless, the locality's contribution in stresses at an area close to the boundary is significant. The model of the modified kernel therefore suggests a behavior according to which the local attribute is of importance when the nonlocal parameter increases. Furthermore, the rest of the eigenfrequencies of the model of the modified kernel appear to have a linear, decreasing behavior with respect to the nonlocal parameter (Fig. 20b–d).

The responses of the nonlocal differential model demonstrate a monotonically decreasing behavior as well.

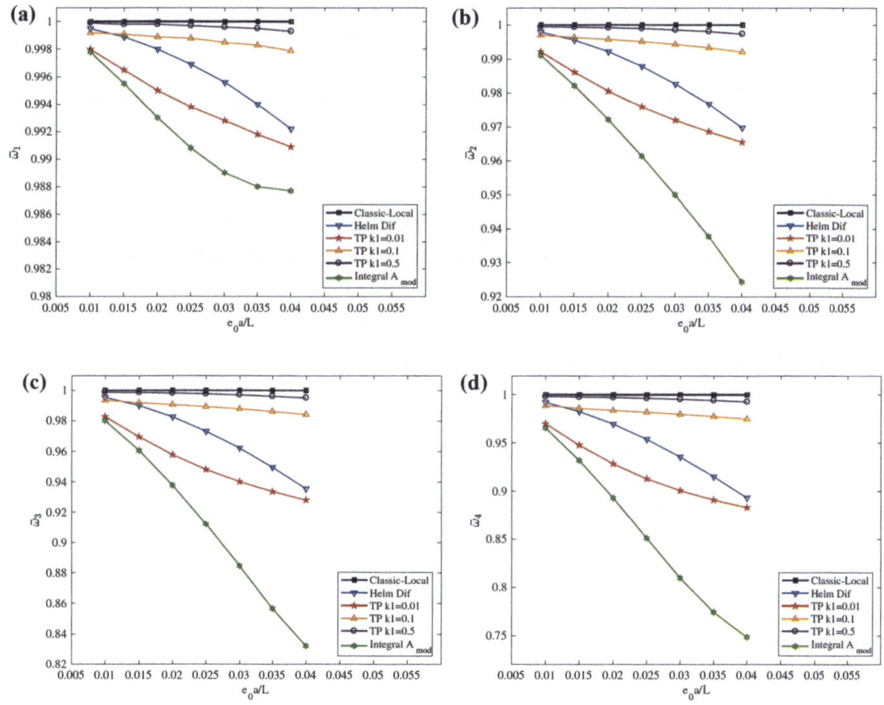

Fig. 20 The first four normalized eigenfrequencies of a simply supported nanobeam with respect to the nonlocal parameter. **a** $\bar{\omega}_1$, **b** $\bar{\omega}_2$, **c** $\bar{\omega}_3$, **d** $\bar{\omega}_4$

4 Conclusions

This work revolves around the contemporary advances in Eringen's nonlocal elasticity theory with emphasis on solving structural engineering problems.

An overview of the literature and a discussion about the integral formulation (i.e., the strong form) and the differential formulation of the nonlocal elasticity theory are presented. The correlation introduced by Mindlin [29] between the nonlocal elasticity theory and the strain gradient elasticity theory is pointed out. An attempt to explain the above correlation is performed by following Mindlin's rationale. An exhaustive discussion regarding the suitability of the nonlocal modulus for the problems' solution in a finite body is also presented. Applications for both static and dynamical engineering problems are investigated as well.

To be more specific, the bending problem and the free vibration problem of an EB nanobeam are explored. The strains are calculated by means of a direct approach to the nonlocal integral model. On the other hand, the static deflections and the eigenfrequencies are calculated by carrying out the FEM.

The following conclusions are drawn from the strain of a cantilever and a simply supported nanobeam, respectively. On the basis of the numerical results, the strain field exceeds the strain of the classic model at a fixed point, but that is not the case for points not too close to the above point in all the problems that studied. How flexible the beam's response is, it hinges on the boundary layer of the strain and the corresponding influence zone.

The boundary layer developed is successfully tackled by the model of the modified kernel for the case of a cantilever and a simply supported beam, respectively. It is critical to mention that the model of the modified kernel and the TPNI stress model appear to have a similar behavior except for points too close to the boundary. Owing to the boundary layer, the inadequacy of the TPNI stress model relating to the violation of the normalization condition in a finite domain is highlighted.

By all the investigated static problems, the responses of the nonlocal integral models appear to have a flexible behavior in comparison with that of the classic-local model.

The beam problems studied are successfully tackled by the nonlocal integral models (i.e., the model of the modified kernel and the TPNI stress model) and the FEM. We note that the model of the modified kernel almost counteracts the boundary layer appearing in the strain, unlike the TPNI stress model.

The results of the transverse deflections associated with the model of the modified kernel demonstrate a worth mentioning behavior as the nonlocal parameter increases. In particular in the investigated problems, the increase rate of the deflection is slower for points found close to the boundaries and when $e_0 a/L > 0.03$. There are also cases wherein the deflection of the model of the modified kernel approaches its greatest value when $e_0 a/L = 0.03$. This behavior can be explained by the constitutive equation. Owing to the fact that the body is finite and the nonlocal parameter increases, the information added by the attenuation function to the constitutive equation reduces.

Locality is then added to the created gap, and this triggers the nonlocal deflection to tend to the classic-local deflection.

However, that is not the case for the TPNI stress model, since its deflections monotonically evolve with respect to the nonlocal parameter in all the above explored problems. It is also critical to mention that the responses of the discrete lattice atomic models of a nanobeam appear to have a flexible behavior in comparison with the classical ones [82], unlike the responses of the strain gradient elasticity theory in one-dimensional problems [83–86].

As regards the free vibration problem, the eigenfrequencies of the nonlocal integral models appear to have a softening response in comparison with those of the classic-local model when the nonlocal parameter increases. Unlike the nonlocal differential model, no paradoxes are raised for the fundamental eigenfrequencies of the nonlocal integral models of a cantilever beam.

The responses of the nonlocal integral models differ from that of the nonlocal differential model as well. To be more specific, the eigenfrequencies of the TPNI stress model with $k_1 = 0.01$ and those of the model of the modified kernel generally take smaller values than the eigenfrequencies of the nonlocal differential model.

By all the investigated models, the eigenfrequencies of the model of the modified kernel generally take the smallest values with respect to the nonlocal parameter. Needless to say, the behavior of the fundamental eigenfrequency is noteworthy as well. The aforementioned behavior is particularly affected by the equilibrium between the locality and the nonlocality, imposed by the model of the modified kernel on each problem that studied. Why is the fundamental eigenfrequency more sensitive to this phenomenon than the other eigenfrequencies? It still remains an unanswered question.

A kind of convergence of the numerical solutions, accomplished by successively finer meshes, also suggests that the FEM is stable for all the problems that studied.

As regards the static and the dynamical problems, the use of the modified kernel in a finite domain can be regarded the optimal selection for a mixed-type local/nonlocal model applied to the nanodevice's design.

The overall conclusions drawn from this chapter are encouraging of triggering the study of complicated nanostructural problems such as nanoframes, nanoplates and nanoshells.

References

1. Kröner E (1967) Elasticity theory of materials with long range cohesive forces. Int J Solids Struct 3:731–742
2. Kröner E (1968) In Kröner (ed) Mechanics of Generalized Continua: Proceedings of the IUTAM-Symposium on The Generalized Cosserat Continuum and the Continuum Theory of Dislocations with Applications, Freudenstadt and Stuttgart (Germany) 1967. Springer Berlin Heidelberg, Berlin, Heidelberg, pp 330–340

3. Krumhansl JA (1968) Some considerations of the relation between solid state physics and generalized continuum mechanics. In: Kröner E (ed) Mechanics of generalized continua. Springer, Berlin Heidelberg, pp 298–311
4. Krumhansl JA (1965) In Lattice Dynamics, Wallis RF (eds) Generalized continuum field representations for lattice vibrations. Pergamon Press, Oxford
5. Kröner E, Datta BK (1966) Nichtlokale Elastostatik: Ableitung aus der Gittertheorie. Z Phys 196:203–211
6. Kunin IA (1966) Model of elastic medium with simple structure and space dispersion (in Russian). Prikl Mat Mekh 30:9
7. Eringen AC, Kim BS (1974) Stress concentration at the tip of crack. Mech Res Commun 1:233–237
8. Eringen AC, Kim BS (1974) On the problem of crack tip in nonlocal elasticity. In: Thoft-Christensen P (eds) Continuum Mechanics Aspects of Geodynamics and Rock Fracture Mechanics NATO Advanced Study Institutes Series (Series C–Mathematical and Physical Sciences) vol 12 Springer, Dordrecht
9. Eringen AC, Speziale CG, Kim BS (1977) Crack-tip problem in non-local elasticity. J Mech Phys Solids 25:339–355
10. Eringen AC (1983) On differential equations of nonlocal elasticity and solutions of screw dislocation and surface waves. J Appl Phys 54:4703–4710
11. Eringen AC (2002) Nonlocal continuum field theories. Springer-Verlag, New York
12. Georgiadis HG (2003) The mode III crack problem in microstructured solids governed by dipolar gradient elasticity: Static and dynamic analysis. J Appl Mech 70:517–530
13. Georgiadis HG, Vardoulakis I, Velgaki EG (2004) Dispersive Rayleigh-wave propagation in microstructured solids characterized by dipolar gradient elasticity. J. Elasticity 74:17–45
14. Lazar M, Maugin GA (2004) Defects in gradient micropolar elasticity-I: Screw dislocation. J Mech Phys Solids 52:2263–2284
15. Lazar M, Maugin GA (2005) Nonsingular stress and strain fields of dislocations and disclinations in first strain gradient elasticity. Int J Eng Sci 43:1157–1184
16. Lazar M, Agiasofitou E (2011) Screw dislocation in nonlocal anisotropic elasticity. Int J Eng Sci 49:1404–1414
17. Lazar M, Polyzos D (2015) On non-singular crack fields in Helmholtz type enriched elasticity theories. Int J Solids Struct 62:1–7
18. Maugin GA (2010) A historical perspective of generalized continuum mechanics. In: Advanced structured materials. Springer, Berlin, Heidelberg
19. Edelen DGB, Green AE, Laws N (1971) Nonlocal continuum mechanics. Arch Ration Mech Anal 43:36–44
20. Eringen AC, Edelen DGB (1972) On nonlocal elasticity. Int J Eng Sci 10:233–248
21. Eringen AC (1972) Linear theory of nonlocal elasticity and dispersion of plane waves. Int J Eng Sci 10:425–435
22. Eringen AC (1966) A unified theory of thermomechanical materials. Int J Eng Sci 4:179–202
23. Eringen AC (1974) Theory of nonlocal thermoelasticity. Int J Eng Sci 12:1063–1077
24. Eringen AC (1984) Theory of nonlocal piezoelectricity. J Math Phys 25:717–727
25. Eringen AC (2006) Nonlocal continuum mechanics based on distributions. Int J Eng Sci 44:141–147
26. Mindlin RD, Tiersten HF (1962) Effects of couple-stresses in linear elasticity. Arch Ration Mech Anal 11:415–448
27. Mindlin RD (1964) Micro-structure in linear elasticity. Arch Ration Mech Anal 16:51–78
28. Mindlin RD (1965) Second gradient of strain and surface tension in linear elasticity. Int J Solids Struct 1:417–438
29. Mindlin RD (1972) Elasticity, piezoelectricity and crystal lattice dynamics. J Elast 2:217–282
30. Kress R (2014) Linear integral equations. Springer-Verlag, New York
31. Shaat M, Abdelkefi A (2017) New insights on the applicability of Eringen's nonlocal theory. Int J Mech Sci 121:67–75

32. Peddieson J, Buchanan GR, McNitt RP (2003) Application of nonlocal continuum models to nanotechnology. Int J Eng Sci 41:305–312
33. Sudak LJ (2003) Column buckling of multiwalled carbon nanotubes using nonlocal continuum mechanics. J Appl Phys 94:7281–7287
34. Wang Q (2005) Wave propagation in carbon nanotubes via nonlocal continuum mechanics. J Appl Phys 98:124301
35. Lu P, Lee HP, Lu C, Zhang PQ (2006) Dynamic properties of flexural beams using a nonlocal elasticity model. J Appl Phys 99:073510
36. Wang Q, Varadan VK (2006) Vibration of carbon nanotubes studied using nonlocal continuum mechanics. Smart Mater Struct 15:659
37. Reddy JN (2007) Nonlocal theories for bending, buckling and vibration of beams. Int J Eng Sci 45:288–307
38. Wang CM, Zhang YY, He XQ (2007) Vibration of nonlocal Timoshenko beams. Nanotechnology 18:105401
39. Wang Q, Liew KM (2007) Application of nonlocal continuum mechanics to static analysis of micro- and nano-structures. Phys Lett A 363:236–242
40. Wang Q, Varadan VK (2007) Application of nonlocal elastic shell theory in wave propagation analysis of carbon nanotubes. Smart Mater Struct 16:178
41. Reddy JN, Pang SD (2008) Nonlocal continuum theories of beams for the analysis of carbon nanotubes. J Appl Phys 103:023511
42. Harik VM (2001) Ranges of applicability for the continuum beam model in the mechanics of carbon nanotubes and nanorods. Solid State Commun 120:331–335
43. Harik VM (2002) Mechanics of carbon nanotubes: Applicability of the continuum-beam models. Comput Mater Sci 24:328–342
44. Treacy MMJ, Ebbesen TW, Gibson JM (1996) Exceptionally high Young's modulus observed for individual carbon nanotubes. Nature 381:678–680
45. Wong EW, Sheehan PE, Lieber CM (1997) Nanobeam mechanics: Elasticity, strength, and toughness of nanorods and nanotubes. Science 277:1971–1975
46. Krishnan A, Dujardin E, Ebbesen TW, Yianilos PN, Treacy MMJ (1998) Young's modulus of single-walled nanotubes. Phys Rev B 58:14013–14019
47. Koutsoumaris CC, Vogiatzis GG, Thedorou DN, Tsamasphyros GJ (2015) Application of bi-Helmholtz nonlocal elasticity and molecular simulations to the dynamical response of carbon nanotubes. AIP Conf Proc 1702:190011
48. Koutsoumaris CC, Eptaimeros KG, Tsamasphyros GJ (2017) A different approach to Eringen's nonlocal integral stress model with applications for beams. Int J Solids Struct 112:222–238
49. Challamel N, Wang CM (2008) The small length scale effect for a nonlocal cantilever beam: A paradox solved. Nanotechnology 19:345703
50. Challamel N, Zhang Z, Wang CM, Reddy JN, Wang Q, Michelitsch T, Collet B (2014) On nonconservativeness of Eringen's nonlocal elasticity in beam mechanics: Correction from a discrete-based approach. Arch Appl Mech 84:1275–1292
51. Polizzotto C (2001) Nonlocal elasticity and related variational principles. Int J Solids Struct 38:7359–7380
52. Benvenuti E, Simone A (2013) One-dimensional nonlocal and gradient elasticity: Closed-form solution and size effect. Mech Res Commun 48:46–51
53. Mahmoud FF (2017) On the non-existence of a feasible solution in the context of the differential form of Eringen's constitutive model: A proposed iterative model based on a residual nonlocality formulation. Int J Appl Mech 9:17594
54. Koutsoumaris CC, Eptaimeros KG (2018) A research into bi-Helmholtz type of nonlocal elasticity and a direct approach to Eringen's nonlocal integral model in a finite body. Acta Mech 229:3629–3649
55. Hu YG, Liew KM, Wang Q, He XQ, Yakobson BI (2008) Nonlocal shell model for elastic wave propagation in single- and double-walled carbon nanotubes. J Mech Phys Solids 56:3475–3485
56. Ansari R, Arash B (2013) Nonlocal flügge shell model for vibrations of double-walled carbon nanotubes with different boundary conditions. J Appl Mech 80:021006

57. Liang Y, Han Q (2014) Prediction of the nonlocal scaling parameter for graphene sheet. Eur J Mech - A/Solids 45:153–160
58. Ghavanloo E, Fazelzadeh SA (2015) Nonlocal shell model for predicting axisymmetric vibration of spherical shell-like nanostructures. Mech Adv Mater Struct 22:597–603
59. Ke LL, Liu C, Wang YS (2015) Free vibration of nonlocal piezoelectric nanoplates under various boundary conditions. Physica E 66:93–106
60. Shaat M (2015) Iterative nonlocal elasticity for Kirchhoff plates. Int J Mech Sci 90:162–170
61. Eringen AC (1987) Theory of nonlocal elasticity and some applications. Res Mech 21:313–342
62. Altan SB (1989) Uniqueness of initial-boundary value problems in nonlocal elasticity. Int J Solids Struct 25:1271–1278
63. Pisano AA, Fuschi P (2003) Closed form solution for a nonlocal elastic bar in tension. Int J Solids Struct 40:13–23
64. Khodabakhshi P, Reddy JN (2015) A unified integro-differential nonlocal model. Int J Eng Sci 95:60–75
65. Eptaimeros KG, Koutsoumaris CC, Tsamasphyros GJ (2016) Nonlocal integral elasticity analysis of nanobeams by employing finite element method. AIP Conf Proc 1790:110009
66. Wang YB, Zhu XW, Dai HH (2016) Exact solutions for the static bending of Euler-Bernoulli beams using Eringen's two phase local/nonlocal model. AIP Adv 6:085114
67. Wang Y, Huang K, Zhu X, Lou Z (2018) Exact solutions for the bending of Timoshenko beams using Eringen's two-phase nonlocal model. Math Mech Solids 24:559–572
68. Zhu X, Wang Y, Dai HH (2017) Buckling analysis of Euler-Bernoulli beams using Eringen's two-phase nonlocal model. Int J Eng Sci 116:130–140
69. Zhu X, Li L (2019) A well-posed Euler-Bernoulli beam model incorporating nonlocality and surface energy effect. Appl Math Mech -Engl 40:1561–1588
70. Fernández-Sáez J, Zaera R (2017) Vibrations of Bernoulli-Euler beams using the two-phase nonlocal elasticity theory. Int J Eng Sci 119:232–248
71. Eptaimeros KG, Koutsoumaris CC, Tsamasphyros GJ (2016) Nonlocal integral approach to the dynamical response of nanobeams. Int J Mech Sci 115–116:68–80
72. Zhu X, Li L (2017) Longitudinal and torsional vibrations of size-dependent rods via nonlocal integral elasticity. Int J Mech Sci 133:639–650
73. Zhu X, Li L (2017) On longitudinal dynamics of nanorods. Int J Eng Sci 120:129–145
74. Mikhasev G, Nobili A (2020) On the solution of the purely nonlocal theory of beam elasticity as a limiting case of the two-phase theory. Int J Solids Struct 190:47–57
75. Borino G, Failla B, Parrinello F (2003) A symmetric nonlocal damage theory. Int J Solids Struct 40:3621–3645
76. Bažant ZP, Jirásek M (2002) Nonlocal integral formulations of plasticity and damage: Survey of progress. J Eng Mech-ASCE 128:1119–1149
77. Lazar M, Maugin GA, Aifantis EC (2006) On a theory of nonlocal elasticity of bi-Helmholtz type and some applications. Int J Solids Struct 43:1404–1421
78. Eptaimeros KG, Koutsoumaris CC, Dernikas IT, Zisis Th (2018) Dynamical response of an embedded nanobeam by using nonlocal integral stress models. Compos Part B Eng 150:255–268
79. Eptaimeros KG, Koutsoumaris CC, Karyofyllis IG (2020) Eigenfrequencies of microtubules embedded in the cytoplasm by means of the nonlocal integral elasticity. Acta Mech 231:1669–1684
80. Koutsoumaris CC, Eptaimeros KG, Zisis T, Tsamasphyros GJ (2016) A straightforward approach to Eringen's nonlocal elasticity stress model and applications for nanobeams. AIP Conf Proc 1790:150018
81. Atkinson KE (1997) The numerical solution of integral equations of the second kind. Cambridge monographs on applied and computational mathematics. Cambridge University Press, Cambridge
82. Challamel N, Reddy JN, Wang CM (2016) Eringen's stress gradient model for bending of nonlocal beams. J Eng Mech 142:04016095

83. Papargyri-Beskou S, Tsepoura KG, Polyzos D, Beskos DE (2003) Bending and stability analysis of gradient elastic beams. Int J Solids Struct 40:385–400
84. Eptaimeros KG (2018) Continuous interior penalty finite element method for a 6th-order bending gradient elastic (nano)beam equation. AIP Conf Proc 1978:470031
85. Koutsoumaris CC, Eptaimeros KG (2018) The gradient beam: A confrontation between the analytical closed type and numerical type solution. AIP Conf Proc 1978:470032
86. Eptaimeros KG, Koutsoumaris CC, Tsamasphyros GJ (2019) Interior penalty discontinuous Galerkin FEMs for a gradient beam and CNTs. Appl Numer Math 144:118–139

Nonlocal Mechanics in the Framework of the General Nonlocal Theory

Mohamed Shaat and Esmaeal Ghavanloo

Abstract A continuum is stable if and only if its body and inertia forces are balanced globally with surface tractions acting on it. The global balance requirement automatically postulates a local balance if a material particle interacts with its direct neighbors. However, when the interaction of the particle is nonlocal and extended to interactions between all particles of the continuum, the local balance is violated, and the material is still stable only if the restrictive global balance is satisfied. The latter continuum behavior can be described by a nonlocal theory. The nonlocal theory developed by Eringen postulates the global balance requirement by representing the stress field depends on the properties of the whole continuum. Eringen assumed a single nonlocal kernel function for all the material coefficients. Here, we show that utilizing a single attenuation function puts limits to the nonlocal model where it would break down when applied to materials operate at high frequency ranges. We also show that Eringen's nonlocal model is limited for slowly varying acoustic waves and low frequencies. To exceed this limit, the general nonlocal theory is presented. The general nonlocal theory outweighs Eringen's nonlocal theory for considering different nonlocal kernels for the different material coefficients. The mechanics of particles, the mechanics of linear elastic continua, and the materials dispersion are discussed in the framework of the general nonlocal theory. In addition, the reduction of the general nonlocal theory to the strain gradient theory is interpreted. In addition, we indicate that the strain gradient theory can capture the same phenomena as the general nonlocal theory as long as various strain gradients are considered.

M. Shaat (✉)
Mechanical Engineering Department, Abu Dhabi University,
P.O.BOX 1790 Al Ain, United Arab Emirates
e-mail: mohamed.i@adu.ac.ae; shaatscience@yahoo.com

E. Ghavanloo
School of Mechanical Engineering, Shiraz University, 71963-16548 Shiraz, Iran
e-mail: ghavanloo@shirazu.ac.ir

© The Author(s), under exclusive license to Springer Nature Switzerland AG 2021
E. Ghavanloo et al. (eds.), *Size-Dependent Continuum Mechanics Approaches*,
Springer Tracts in Mechanical Engineering,
https://doi.org/10.1007/978-3-030-63050-8_3

1 Introduction

Classical models of continuum mechanics have been well-established to describe the mechanics of various types of materials. In these models, it is assumed that the material's internal energy is only dependent on a strain field, which is generated upon local interactions of a material point with its direct neighbors. Thus, information on non-neighbor interactions of the material point with other points of the continuum is disregarded. The classical continuum mechanics is invariant with respect to time and length scales, and more notably size effects cannot be captured by these models [1]. With the recent advancement of advanced materials, these classical models have shown clear failure to capture the non-classical phenomena of these materials. To overcome this limitation, a series of size-dependent continuum approaches, such as micromorphic theory [2–5], microstretch theory [5], micropolar theory [5], first and second strain gradients theories [2, 6], couple-stress theories [7–9], and nonlocal theories [10, 11], have been proposed, which are usually pertinent to study phenomena at small scales. To model the mechanics of materials with independent microstructures, the micromorphic theories were developed. In the micromorphic theories, the material's internal energy depends on non-classical kinematical variables, which are independent of the conventional strain field. In the strain gradient theories, the internal energy depends on the strain and its gradients. The couple stress theories were developed to represent the material's internal energy as a function of the strain and the gradients of the rotation tensor.

The first attempt to model elastic continua with long-range cohesive forces was made by Kröner [12]. Then, Eringen and Edelen [10, 11] proposed a systematic representation of the nonlocal continuum field theories. In the original nonlocal model, the stress field was obtained from the integral of the strain field over the whole body. In 1983, Eringen [13] proposed a differential form of the nonlocal model, which is equivalent to the nonlocal integral model over unbounded continuous domains. The peridynamics theory was introduced by Silling [14] as a computational tool of nonlocal mechanics. The peridynamics nonlocal theory is integral-type nonlocal continuum theory which can account for both short-range and long-range interaction forces. A strain-difference based nonlocal elasticity model was proposed by Polizzotto et al. [15]. Among several nonlocal approaches, both integral and differential Eringen's nonlocal models have been extensively used to accommodate the influence of the small-scale on the mechanics of nanomaterials with various morphologies [16, 17].

By the end of the 2016, it seemed that the research on the nonlocal elasticity theories have reached a saturation point, while there exist several open questions and crucial understanding of the physics behind one of them. One important lingering question is the identification of the nonlocal parameters and length scales. Another unclarified issue in current literatures is how the nonlocal effect influences the equivalent stiffness of the nanostructures. Some researchers assert that the nonlocal models can capture the softening mechanisms in the material while some other references predict the hardening behavior. To overcome these obstacles in describing the physics of the nonlocal theories, the general nonlocal theory has been recently proposed by Shaat [18]. This theory has shed light on many aspects of nonlocal mechanics, especially when it has been used for several engineering applications [19, 20].

In this chapter, we present the general nonlocal theory and provide enough details in order to use it in future research. The theory expounded in this chapter follows the work by Shaat [18]. In the first section (Sect. 2), the concepts of nonlocal mechanics of particles and elastic continua are explained. In addition, the fundamental model early proposed by Eringen is interpreted showing its limitations. In Sect. 3, the general nonlocal theory is developed and compared to Eringen's nonlocal theory. In Sect. 4, we show that the general nonlocal theory can be reduced to the strain gradient and the couple stress theories. In Sect. 5, an approach is presented to identify the nonlocal parameters of the nonlocal theory and the material coefficients of the strain and couple stress theories. Along with these aspects, a critical assessment of the ability of the nonlocal theory to capture both hardening and softening behaviors of materials is provided.

2 Eringen's Nonlocal Theory

The nonlocal theory has been developed with the flavor of accounting for the effects of the long-range interatomic interactions on the material mechanics. Some nonlocal models were early developed [10–12]. The currently used model of the nonlocal theory is the one that has been developed by Eringen [10, 11]. In this section, the nonlocal mechanics of particles and the nonlocal mechanics of linear elastic continua are reformulated in the framework of the nonlocal theory. The dispersion relations are then derived on the basis of Eringen's nonlocal theory.

2.1 *Nonlocal Mechanics of Particles*

Let us consider a system of particles as shown in Fig. 1. According to the conventional mechanics of particles, a particle A only interacts with its neighbor particles within the local region. Using Newton's second law of motion, the equations of motion of the particle A can be written as:

$$\sum \mathbf{F}^{l} = m_A \ddot{\mathbf{u}}_A \tag{1}$$

$$\sum \mathbf{x}_A \times \mathbf{F}^{l} = \mathbf{x}_A \times m_A \ddot{\mathbf{u}}_A \tag{2}$$

where $\mathbf{F}^{l}$ denotes the local force vector, $\ddot{\mathbf{u}}_A$ is the particle's acceleration, and $\mathbf{x}_A$ is the particle's position vector. In addition, m_A denotes the mass of the particle.

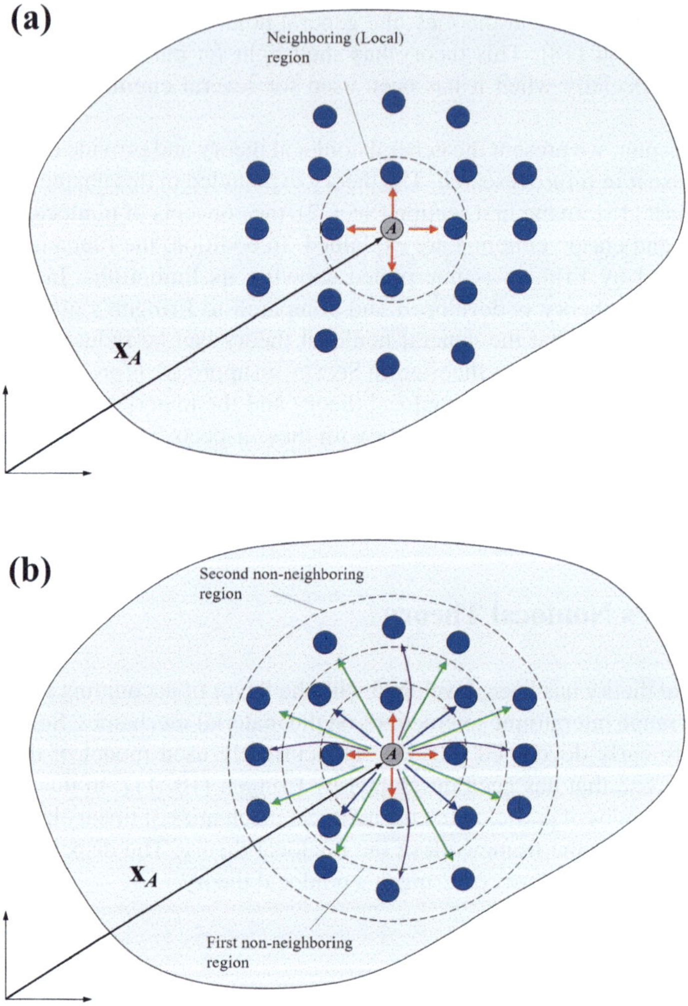

Fig. 1 Mechanics of particles in the context of **a** the classical mechanics and **b** the nonlocal mechanics

Some cases would require modeling the particle mechanics considering both neighbor and non-neighbor interactions. Accounting for the non-neighbor interactions, the dynamic equilibrium of particle A is conditional by its balance with all surrounding particles. A non-neighbor interaction depends on the distance between the particle and a non-neighbor particle. A weight function, $w\left(\left|\mathbf{x}_A - \mathbf{x}'\right|\right)$, can be introduced to define the intensity of the non-neighbor interaction between particle A and a particle at point $\mathbf{x}'$, where $\mathbf{x}_A$ is the position of particle A. Accordingly, the equations of motion for particle A can be rewritten as [18]:

$$w_0 \sum \mathbf{F}_0^{nl} + w_1 \sum \mathbf{F}_1^{nl} + w_2 \sum \mathbf{F}_2^{nl} + \cdots = m_A \ddot{\mathbf{u}}_A \tag{3}$$

$$w_0 \sum \mathbf{x}_A \times \mathbf{F}_0^{nl} + w_1 \sum \mathbf{x}_A \times \mathbf{F}_1^{nl} + w_2 \sum \mathbf{x}_A \times \mathbf{F}_2^{nl} + \cdots = \mathbf{x}_A \times m_A \ddot{\mathbf{u}}_A \tag{4}$$

where $\mathbf{F}_0^{nl}$ denotes the local force vector and, $\mathbf{F}_1^{nl}$ and $\mathbf{F}_2^{nl}$ are the nonlocal force vectors due to the interactions between particle A and the particles in the first and the second non-neighboring regions, respectively (Fig. 1b). In addition, w_i are the values of the weight function which satisfy the condition, i.e.,

$$\sum w_i = 1 \tag{5}$$

According to Eqs. (3) and (4), the force and moment equilibriums are globally satisfied in the context of the nonlocal mechanics of particles. This means that the local balance is violated when the interactions of the particles are nonlocal [10, 21].

2.2 *Nonlocal Continuum Mechanics*

Consider an elastic material with volume V and surface S. In the extreme limit, the material is a composition of an infinite number of infinitesimal particles. The physical properties of these particles and all the forces of interactions between them are described by smooth functions of space and time in the context of the theory of continuum mechanics [10]. Thus, the dynamic equilibrium of the material can be obtained based on the nonlocal theory integrating Eqs. (3) and (4) over the material volume, i.e.,

$$\int_V [\int_V w(|\mathbf{x}-\mathbf{x}'|)\mathbf{F}(\mathbf{x}')dV(\mathbf{x}')]dV(\mathbf{x}) + \int_S \mathbf{T}^{(n)}(\mathbf{x})dS(\mathbf{x}) = \int_V \rho\ddot{\mathbf{u}}(\mathbf{x})dV(\mathbf{x}) \tag{6}$$

$$\int_V [\int_V w(|\mathbf{x}-\mathbf{x}'|)(\mathbf{X}\times\mathbf{F}(\mathbf{x}'))dV(\mathbf{x}')]dV(\mathbf{x}) + \int_S (\mathbf{X}\times\mathbf{T}^{(n)}(\mathbf{x}))dS(\mathbf{x}) = \int_V \rho(\mathbf{X}\times\ddot{\mathbf{u}}(\mathbf{x}))dV(\mathbf{x}) \tag{7}$$

where ρ is the material's mass density, and $\mathbf{T}^{(n)}$ denotes the surface tractions, such that

$$T_i^{(n)} = t_{ji} n_j \tag{8}$$

where t_{ij} is a nonlocal stress tensor, which is formed accounting for the nonlocal interactions between the different material particles. n_j are components of the unit normal vector. Using the Green-Gauss theorem, we have

$$t_{ji,j} + f_i = \rho \ddot{u}_i \tag{9}$$

where f_i denotes the body forces and u_i are the Cartesian components of the infinitesimal displacement.

2.3 Eringen's Constitutive Equations

According to Eringen's nonlocal theory, equations of motion of a local particle are obtained from the global interaction of the particles with all other particles of the elastic body. These global interactions are considered in the constitutive equations, which gave the stress-based equations of motion identical to the one of the classical elasticity (Eq. (9)). To collect global interactions, the constitutive equations were expressed as follows (for homogeneous-isotropic linear elastic materials) [11, 16]:

$$t_{ij}(\mathbf{x}, t) = \int_V \alpha(|\mathbf{x} - \mathbf{x}'|) \left[\lambda \varepsilon_{qq}(\mathbf{x}', t)\delta_{ij} + 2\mu \varepsilon_{ij}(\mathbf{x}', t)\right] dV(\mathbf{x}') \tag{10}$$

where ε_{ij} is the infinitesimal strain tensor,

$$\varepsilon_{ij}(\mathbf{x}) = \frac{1}{2}[u_{i,j}(\mathbf{x}) + u_{j,i}(\mathbf{x})] \tag{11}$$

and λ and μ are the Lamé constants. δ_{ij} is the Kronecker delta and $\alpha(|\mathbf{x} - \mathbf{x}'|)$ is a nonlocal modulus. The basic features of the nonlocal kernel were discussed in Refs. [22, 23].

2.4 Nonlocal Field Equation

The nonlocal field equation can be obtained by substituting Eqs. (10) and (11) into Eq. (9), as follows:

$$\left(\int_V \alpha(|\mathbf{x} - \mathbf{x}'|) \left(\lambda u_{q,q}(\mathbf{x}', t)\delta_{ij} + \mu \left(u_{i,j}(\mathbf{x}', t) + u_{j,i}(\mathbf{x}', t)\right)\right) dV\left(\mathbf{x}'\right)\right)_{,j} + f_i(\mathbf{x}, t) = \rho \ddot{u}_i(\mathbf{x}, t) \tag{12}$$

Equation (12) is an integro-partial differential equation. Complications of solving the nonlocal field Eq. (12) have been early addressed in Ref. [13]. To avoid the difficulties of solving the integro-partial differential field equation, Eringen developed an equivalent model that gives the field equation in a differential form, as follows [13]:

$$(\lambda + \mu)u_{j,ij} + \mu u_{i,jj} + (f_i - \rho \ddot{u}_i) - \ell^2(f_{i,jj} - \rho \ddot{u}_{i,jj}) = 0 \tag{13}$$

where ℓ is a nonlocal parameter. The differential form of the nonlocal theory is equivalent to the nonlocal integral form only over unbounded elastic domains.

2.5 Dispersion Relations

Phonon dispersion is one of the fundamental properties of materials. Here, the propagation of a harmonic wave along [100] direction is modeled based on Eringen's nonlocal theory. By assuming a high-symmetric direction, the following one-dimensional harmonic functions can be considered:

$$u_r(x, t) = U_r \exp(i(kx - \omega t)) \tag{14}$$

where u_r is the vibration of a lattice site about its equilibrium position, k and ω are wave number and wave frequency, respectively. Substituting Eq. (14) into Eq. (13) and setting $f_i = 0$, we obtain the frequency-wavenumber relations as

$$\omega_L^2 = \frac{(\lambda + 2\mu)k^2}{\rho(1 + k^2\ell^2)} \tag{15}$$

$$\omega_T^2 = \frac{\mu k^2}{\rho(1 + k^2\ell^2)} \tag{16}$$

where ω_L and ω_T are the frequencies of longitudinal and transverse acoustic waves.

2.6 Limitations of Eringen's Nonlocal Theory

In this subsection, limitations of Eringen's nonlocal theory are addressed. It follows from Eq. (12) that Eringen's nonlocal theory assumes the same attenuation function for all the material coefficients. According to this theory, the nonlocal-dependence of the normal strain is identical to the nonlocal-dependence of the shear strain. However, this is not true for many materials. For example, the experimental dispersion curves of Si, Au, and Pt indicated that the decay of the longitudinal frequencies with the wavenumber is different from the decay of the transverse frequencies with the

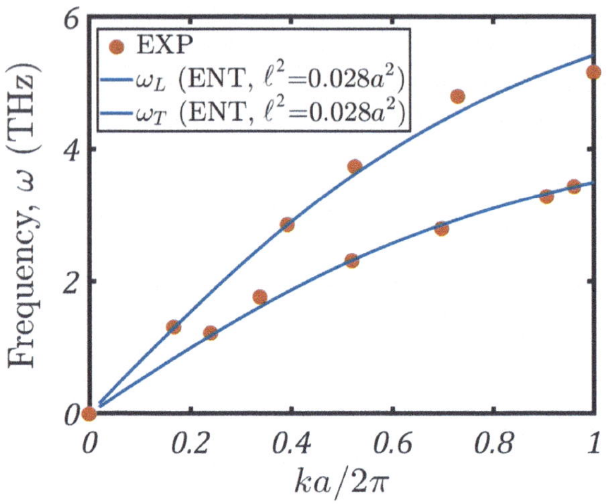

Fig. 2 Dispersion curves of Ag [100] based on Eringen's nonlocal theory (ENT)

wavenumber. For these materials, Eringen's nonlocal theory is inapplicable where it cannot properly fit their longitudinal and transverse dispersion curves [18, 19].

Figure 2 shows the applicability of Eringen's nonlocal theory to model the acoustic dispersion phonons of a material with nearly similar nonlocal characters in both the longitudinal and transverse directions. In this figure, the dispersion curves of Ag [100] predicted by Eringen's nonlocal model are plotted and compared with experimental results reported by Kamitakahara and Brcckhouse [24]. It is observed that the frequencies of both longitudinal and transverse acoustic waves are perfectly predicted by the nonlocal model for $\ell^2 = 0.028a^2$. Figure 3 shows the limitation of Eringen's nonlocal theory to model the acoustic dispersion phonons of a material with different nonlocal characters in the longitudinal and transverse directions. This figure shows the dispersion curves of Au [100] obtained from the nonlocal model and the experimental data given in Ref. [25]. These results indicate that Eringen's nonlocal theory has limitations to model Au. For Au [100], it can only predict one of the acoustic dispersion curves for specific value of the nonlocal parameter. For a detailed discussion about this topic, we refer the interested readers to Ref. [19]. The physical and material parameters used in the generation of these results based on Eringen's nonlocal theory are taken from Ref. [18] and listed in Table 1.

In addition to its limitations to model the dispersion curves of some materials, Eringen's nonlocal theory has limitations to reveal the nonlocal effects on the material's Poisson's ratio. It has been demonstrated that, due to nonlocal interatomic interactions, the Poisson's ratio ν of the material is generally non-constant and depend on the material dispersion [25], i.e.,

$$\nu = \frac{1 - 2R^2}{2\left(1 - R^2\right)} \tag{17}$$

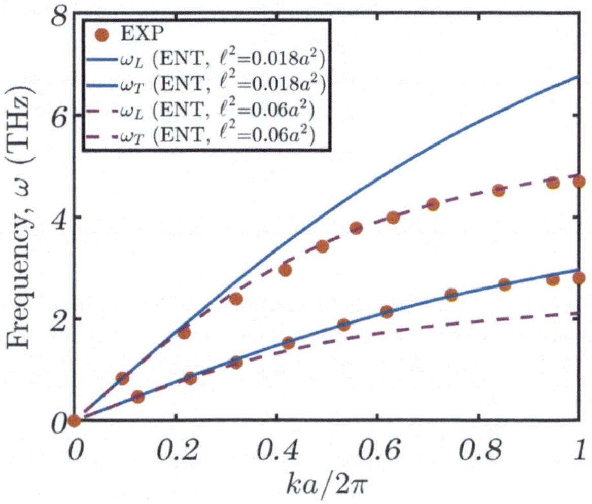

Fig. 3 Dispersion curves of Au [100] based on Eringen's nonlocal theory (ENT)

Table 1 Physical and material properties of Ag and Au

Material	Elastic moduli		Lattice constant	Density
	λ (GPa)	μ (GPa)	a (nm)	ρ (kg/m^3)
Silver (Ag)	18	45	0.4079	10490
Gold (Au)	154	48	0.4065	19300

where $R = \omega_T(k)/\omega_L(k)$. The substitution of the frequency-wavenumber relations (Eqs. (15) and (16)) into Eq. (17) gives Poisson's ratio with the form:

$$\nu = \frac{\lambda}{2(\lambda + \mu)} \tag{18}$$

which is the same Poisson's ratio as defined based on the classical theory of linear elasticity. This indicates that Poisson's ratio is independent of the nonlocal field in the context of Eringen's nonlocal theory. This, however, contradicts with the recent observations on the size-dependence of Poisson's ratio [26–29] and the microstructural topology-dependence of Poisson's ratio of metamaterials [26, 30–32].

3 General Nonlocal Theory

The general nonlocal theory [18, 19] was developed to remedy the drawbacks of Eringen's nonlocal theory. The general nonlocal theory considers independent nonlocal fields for the different material coefficients. This theory permits the simultaneous fitting of both the longitudinal and transverse acoustic dispersion curves of various materials [18, 19]. In addition, it effectively models the nonlocal character of Poisson's ratio of auxetic and non-auxetic materials [33, 34]. Here, the governing equations of the general nonlocal theory are presented. In addition, the superiority of the general nonlocal theory over Eringen's nonlocal theory is demonstrated.

3.1 Equilibrium and Constitutive Equations

In the context of the general nonlocal theory, an elastic solid consists of an infinite number of particles each of which is a mass point. The dynamic equilibrium of the solid is conditional by the global balance of its body forces, external surface tractions, and inertia forces.

The global balance (Eq. (9)) depends on a nonlocal stress tensor t_{ij}, which is expressed as a functional of the deformation gradients of all points of the elastic solid. According to the general nonlocal theory, the components of the nonlocal stress tensor are expressible as follows:

$$t_{ji}(\mathbf{x}) = t_{ij}(\mathbf{x}) = \int_V C_{ijkl}(\left|\mathbf{x}' - \mathbf{x}\right|, \ell)\varepsilon_{kl}(\mathbf{x}')dV\left(\mathbf{x}'\right) \tag{19}$$

In Eq. (19), the material coefficients $C_{ijkl}(\left|\mathbf{x}' - \mathbf{x}\right|, \ell)$ are functions of the distance $\left|\mathbf{x}' - \mathbf{x}\right|$ between a reference point $\mathbf{x}$ and any other point $\mathbf{x}'$ of the elastic solid. These material coefficients depend on the nonlocal character of the solid part, which is measured by a nonlocal parameter ℓ. The nonlocal parameter, ℓ, defines the intensity of the nonlocal field. According to the general nonlocal theory, for isotropic-linear elastic materials, $C_{ijkl}(\left|\mathbf{x}' - \mathbf{x}\right|, \ell)$ can be expressed as follows:

$$\begin{aligned} C_{ijkl}\left(\left|\mathbf{x}' - \mathbf{x}\right|, \ell\right) = {} & \bar{\lambda}\left(\left|\mathbf{x}' - \mathbf{x}\right|, \ell_\lambda\right)\delta_{ij}\delta_{kl} \\ & + \bar{\mu}\left(\left|\mathbf{x}' - \mathbf{x}\right|, \ell_\mu\right)\left(\delta_{ik}\delta_{jl} + \delta_{il}\delta_{jk}\right) \end{aligned} \tag{20}$$

where $\bar{\lambda}$ and $\bar{\mu}$ are material coefficients, which are functions of the distance $\left|\mathbf{x}' - \mathbf{x}\right|$. These material coefficients are defined in terms of the conventional Lame constants, λ and μ, as follows:

$$\bar{\lambda}\left(\left|\mathbf{x}' - \mathbf{x}\right|, \ell_\lambda\right) = \lambda\beta_\lambda\left(\left|\mathbf{x}' - \mathbf{x}\right|, \ell_\lambda\right) \tag{21a}$$

$$\bar{\mu}\left(\left|\mathbf{x}' - \mathbf{x}\right|, \ell_\mu\right) = \mu\beta_\mu\left(\left|\mathbf{x}' - \mathbf{x}\right|, \ell_\mu\right) \tag{21b}$$

where $\beta_\lambda\left(\left|\mathbf{x}' - \mathbf{x}\right|, \ell_\lambda\right)$ and $\beta_\mu\left(\left|\mathbf{x}' - \mathbf{x}\right|, \ell_\mu\right)$ are two independent nonlocal moduli, which are continuous functions that express the change of the nonlocal field with the distance $\left|\mathbf{x}' - \mathbf{x}\right|$. According to Eqs. (20)–(21), the nonlocal stress can be rewritten as:

$$t_{ji}(\mathbf{x}) = t_{ij}(\mathbf{x}) = \int_V \lambda \beta_\lambda\left(\left|\mathbf{x}' - \mathbf{x}\right|, \ell_\lambda\right) \varepsilon_{rr}\left(\mathbf{x}'\right) \delta_{ij} dV\left(\mathbf{x}'\right) + \int_V 2\mu \beta_\mu\left(\left|\mathbf{x}' - \mathbf{x}\right|, \ell_\mu\right) \varepsilon_{ij}\left(\mathbf{x}'\right) dV\left(\mathbf{x}'\right) \quad (22)$$

It is clear that the general nonlocal theory assumes two independent nonlocal fields for the different material coefficients. Two nonlocal parameters, ℓ_λ and ℓ_μ, are introduced as measures of the intensity of the nonlocal fields associated with the normal and shear strains. For the case, $\ell_\lambda = \ell_\mu$, the constitutive Eq. (22) reduces to the one of Eringen's nonlocal theory (Eq. (10)).

3.2 Nonlocal Moduli

Various continuous functions can be considered to express the variation of the nonlocal field with the distance $\left|\mathbf{x}' - \mathbf{x}\right|$. Eringen introduced the nonlocal mechanics using decaying (or attenuating) nonlocal moduli [13]. A nonlocal modulus as defined by Eringen is a continuous-attenuation function that satisfies the following conditions:

$$\lim_{\ell_\lambda, \ell_\mu \to 0} \left(\beta_\lambda\left(|\mathbf{x}|, \ell_\lambda\right), \beta_\mu\left(|\mathbf{x}|, \ell_\mu\right)\right) = \left(\delta(\mathbf{x}), \delta(\mathbf{x})\right) \quad (23a)$$

$$\lim_{|\mathbf{x}' - \mathbf{x}| \to \infty} \left(\beta_\lambda\left(\left|\mathbf{x}' - \mathbf{x}\right|, \ell_\lambda\right), \beta_\mu\left(\left|\mathbf{x}' - \mathbf{x}\right|, \ell_\mu\right)\right) = 0 \quad (23b)$$

where δ is the Dirac-delta function. These functions attenuate as $\left|\mathbf{x}' - \mathbf{x}\right|$ increases. In addition, these attenuation functions satisfy the condition that their integrals over the elastic domain give unity. In addition, when ℓ_λ, $\ell_\mu \to 0$, the attenuation functions become the Dirac delta functions indicating the transformation of the nonlocal model to the local-classical model.

Examples of the nonlocal attenuation functions that have been applied are [13, 35]:

$$\beta\left(|\mathbf{x}|, \ell\right) = \frac{1}{2\ell} \exp\left(-\frac{|\mathbf{x}|}{\ell}\right) \quad (24a)$$

$$\beta\left(|\mathbf{x}|, \ell\right) = \frac{5}{4\ell} \xi\left(|\mathbf{x}|, \ell\right) \quad (24b)$$

with

$$\xi\left(|\mathbf{x}|, \ell\right) = \begin{cases} 1 - 6\left(\frac{|\mathbf{x}|}{\ell}\right)^2 + 8\left(\frac{|\mathbf{x}|}{\ell}\right)^3 - 3\left(\frac{|\mathbf{x}|}{\ell}\right)^4 & |\mathbf{x}| \le \ell \\ 0 & |\mathbf{x}| > \ell \end{cases} \quad (25)$$

Attenuating nonlocal moduli reveal softening behaviors due to the nonlocal fields. These attenuation functions have been used to model the wave dispersions by various materials, the softening in structures' stiffnesses, and the decrease in the elastic moduli of the nanomaterials. For instance, Young's modulus of single crystalline silicon cantilever was observed monotonically decreasing from 169 to 53 GPa with a decrease in the cantilever's thickness from 300 to 12 nm [36].

Nonetheless, many cases would require non-attenuating nonlocal moduli, and instead *augmenting* nonlocal moduli would be required. One such a case is the increasing fringe frequency of a nondispersive interferometer and a synthetic quartz Wollaston prism spectrometer [37]. In addition, Young's modulus of ZnO films was increased by 22% upon decreasing the thickness from 8 to 1 nm [38]. Furthermore, Young's modulus of Al_2O_3 nano-films was intensively increased to ~776 GPa due to a decrease in the film thickness lower than 5 nm [39]. Therefore, nonlocal augmentation functions would be introduced to model the hardening behaviors of nanoscale structures, to explain the increase in the elastic moduli of some nanomaterials, and to predict the propagation of non-dispersive waves. Utilizing these augmentation functions, the nonlocal theory becomes more general and can capture both softening and hardening behaviors.

Here, we use a complex nonlocal parameter to obtain an augmenting nonlocal modulus. This augmenting nonlocal modulus satisfies the following conditions:

$$\lim_{\ell_\lambda, \ell_\mu \to 0} \left(\beta_\lambda \left(|\mathbf{x}|, \ell_\lambda\right), \beta_\mu \left(|\mathbf{x}|, \ell_\mu\right)\right) = \left(\delta\left(\mathbf{x}\right), \delta\left(\mathbf{x}\right)\right) \tag{26a}$$

$$\lim_{|\mathbf{x}'-\mathbf{x}| \to \infty} \left(\beta_\lambda \left(\left|\mathbf{x}'-\mathbf{x}\right|, \ell_\lambda\right), \beta_\mu \left(\left|\mathbf{x}'-\mathbf{x}\right|, \ell_\mu\right)\right) = \left(\mathcal{R}_\lambda\left(\mathbf{x}\right), \mathcal{R}_\mu\left(\mathbf{x}\right)\right) \tag{26b}$$

where $\mathcal{R}_\lambda\left(\mathbf{x}\right)$ and $\mathcal{R}_\mu\left(\mathbf{x}\right)$ are non-zero functions with values are higher than unity. These functions represent nonlocal sources that impose residual nonlocal interactions over the local interactions. In the next subsection, the variations of the wave frequencies of dispersive and aggregative waves with the wavenumber are determined.

3.3 Propagation of Dispersive and Aggregative Waves

When a wave propagates in a medium, it either extends or aggregates as it travels. The extension of the wave is common and known as wave dispersion, which is accompanied with a reduction in the wave frequency [40, 41]. The dispersion of waves in materials would be due to microstructural defects, heterogeneities, deformations, or friction. A non-neighbor interaction is another factor that would cause a wave extension, and hence a wave frequency reduction. On the other hand, waves would be aggregative in some special media and wave frequencies would rather increase as it travels.

The reduction or increase of the wave frequency can be effectively modeled by the general nonlocal theory. Here, the propagation of a harmonic wave along [100] direction is modeled based on the general nonlocal theory. Therefore, the one-dimensional harmonic functions defined in Eq. (14) are substituted into Eq. (11) gives the following non-zero strain components:

$$\begin{aligned}\varepsilon_{xx} &= ikU_x \exp\left(i\left(kx-\omega t\right)\right)\\ \varepsilon_{xy} &= \frac{1}{2}ikU_y \exp\left(i\left(kx-\omega t\right)\right)\\ \varepsilon_{xz} &= \frac{1}{2}ikU_z \exp\left(i\left(kx-\omega t\right)\right)\end{aligned}\tag{27}$$

Using the nonlocal modulus defined in Eq. (25), the non-zero components of the nonlocal stress can be written as follows:

$$\begin{aligned}t_{xx}(x) &= \left(\frac{5ik\lambda U_x}{4\ell_\lambda}\right)\int_x \left(\xi_\lambda\left(\left|x'-x\right|,\ell_\lambda\right)\exp\left(i\left(kx'-\omega t\right)\right)\right)dx'\\ &\quad+\left(\frac{5ik\mu U_x}{2\ell_\mu}\right)\int_x \left(\xi_\mu\left(\left|x'-x\right|,\ell_\mu\right)\exp\left(i\left(kx'-\omega t\right)\right)\right)dx'\\ t_{xy}(x) &= \left(\frac{5ik\mu U_y}{4\ell_\mu}\right)\int_x \left(\xi_\mu\left(\left|x'-x\right|,\ell_\mu\right)\exp\left(i\left(kx'-\omega t\right)\right)\right)dx'\\ t_{xz}(x) &= \left(\frac{5ik\mu U_z}{4\ell_\mu}\right)\int_x \left(\xi_\mu\left(\left|x'-x\right|,\ell_\mu\right)\exp\left(i\left(kx'-\omega t\right)\right)\right)dx'\end{aligned}\tag{28}$$

where

$$\begin{aligned}\xi_\lambda &= \begin{cases}1-6\left(\frac{|x'-x|}{\ell_\lambda}\right)^2+8\left(\frac{|x'-x|}{\ell_\lambda}\right)^3-3\left(\frac{|x'-x|}{\ell_\lambda}\right)^4 & \left|x'-x\right|\le\ell_\lambda\\ 0 & \left|x'-x\right|>\ell_\lambda\end{cases}\\ \xi_\mu &= \begin{cases}1-6\left(\frac{|x'-x|}{\ell_\mu}\right)^2+8\left(\frac{|x'-x|}{\ell_\mu}\right)^3-3\left(\frac{|x'-x|}{\ell_\mu}\right)^4 & \left|x'-x\right|\le\ell_\mu\\ 0 & \left|x'-x\right|>\ell_\mu\end{cases}\end{aligned}\tag{29}$$

Performing the integrations in Eq. (28), the components of the nonlocal stress are obtained in the form:

$$\begin{aligned}t_{xx} &= ik\left(\left(\frac{5\lambda}{4\ell_\lambda}\right)\mathcal{F}_\lambda U_x+\left(\frac{5\mu}{2\ell_\mu}\right)\mathcal{F}_\mu U_x\right)\exp\left(i\left(kx-\omega t\right)\right)\\ t_{xy} &= ik\left(\frac{5\mu}{4\ell_\mu}\right)\mathcal{F}_\mu U_y\exp\left(i\left(kx-\omega t\right)\right)\\ t_{xz} &= ik\left(\frac{5\mu}{4\ell_\mu}\right)\mathcal{F}_\mu U_z\exp\left(i\left(kx-\omega t\right)\right)\end{aligned}\tag{30}$$

where

$$\mathcal{F}_\lambda = \frac{48\,(2k\ell_\lambda + k\ell_\lambda \cos(k\ell_\lambda) - 3\sin(k\ell_\lambda))}{k^5\ell_\lambda^4}$$
$$\mathcal{F}_\mu = \frac{48\left(2k\ell_\mu + k\ell_\mu \cos\left(k\ell_\mu\right) - 3\sin\left(k\ell_\mu\right)\right)}{k^5\ell_\mu^4} \tag{31}$$

By substituting Eq. (30) into Eq. (9) and solving the obtained characteristic equations, the frequency-wavenumber relations are derived, i.e.,

$$\omega_L^2 = \frac{\left(\left(\frac{5\lambda}{4\ell_\lambda}\right)\mathcal{F}_\lambda + \left(\frac{5\mu}{2\ell_\mu}\right)\mathcal{F}_\mu\right)k^2}{\rho} \tag{32}$$

$$\omega_T^2 = \frac{\left(\frac{5\mu}{4\ell_\mu}\right)\mathcal{F}_\mu k^2}{\rho} \tag{33}$$

It should be noted that real positive nonlocal parameters, $\ell_\lambda > 0$ and $\ell_\mu > 0$, give attenuation nonlocal moduli. These attenuating nonlocal moduli give the wave frequency lower than that of the classical theory. On the other hand, complex nonlocal parameters give augmentation nonlocal moduli and the wave frequency higher than that of the classical theory. Using a complex nonlocal parameter gives the dispersion relations in the complex plane, and the real part of the frequency shows an increase with the wavenumber increase. Frequencies in the complex plan is one of the approaches and features of topological insulators and quantum matters [42]. Thus, the implementation of the complex nonlocal parameters gives the nonlocal theory a new trend to capture analogies with quantum matters.

Figure 4 shows the ability of the general nonlocal theory to reveal the softening of materials (Fig. 4a) and the hardening of materials (Fig. 4b). Using attenuating nonlocal moduli, the wave frequency of the nonlocal model is lower than that of the classical model, as shown in Fig. 4a. It is observed that, for a specific wavenumber, the wave frequency decreases with an increase in nonlocal parameters. This physically means that the general nonlocal model can predict the softening behavior when attenuating nonlocal moduli with real nonlocal parameters are used. In Fig. 4b, however, augmentation nonlocal moduli are used. It is observed that the wave frequency of the nonlocal model increases higher than the frequency of the classical model. This is mainly due to the nonlocal effects, which increases as the nonlocal parameter increases. Thus, the general nonlocal model can also predict the hardening behavior when complex nonlocal parameters are used.

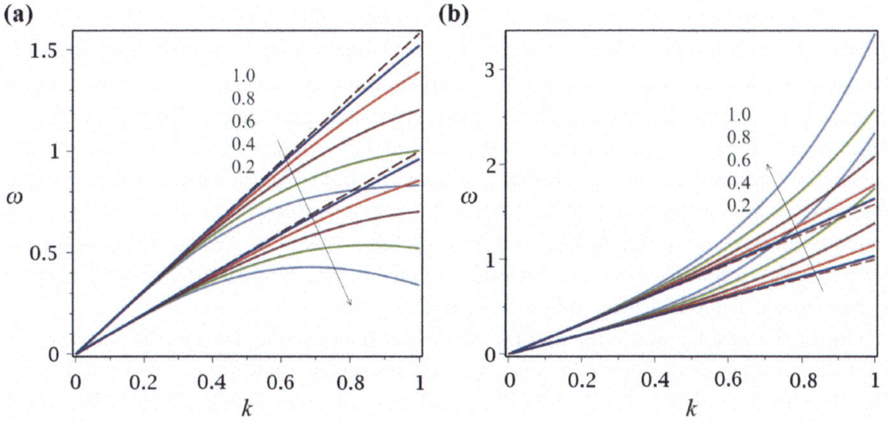

Fig. 4 Softening and hardening behaviors based the general nonlocal theory. **a** The variation of the longitudinal and transverse acoustic frequencies, ω (rad/s), as functions of the wave number, k, lower than the frequencies of the classical mechanic ($\lambda = 0.5$, $\mu = 1$, $\rho = 1$, $\ell_\lambda/2\pi = 0.2$, and $\ell_\mu/2\pi = 0.2 \rightarrow 1.0$). **b** The variation of the longitudinal and transverse acoustic frequencies, ω (rad/s), as functions of the wave number, k, higher than the frequencies of the classical mechanics ($\lambda = 0.5$, $\mu = 1$, $\rho = 1$, $\ell_\lambda/2\pi = i0.2$, and $\ell_\mu/2\pi = i0.2 \rightarrow i1.0$)

3.4 *Comparison to Eringen's Nonlocal Theory*

The wave frequency equations of the general nonlocal theory (Eqs. (32) and (33)) reduce to the ones of Eringen's nonlocal theory by setting $\ell_\lambda = \ell_\mu = \ell$:

$$\omega_L^2 = \frac{5\,(\lambda + 2\mu)\,\mathcal{F}k^2}{4\ell\rho} \tag{34}$$

$$\omega_T^2 = \frac{5\mu\mathcal{F}k^2}{4\ell\rho} \tag{35}$$

where

$$\mathcal{F} = \frac{48\,(2k\ell + k\ell\cos(k\ell) - 3\sin(k\ell))}{k^5\ell^4} \tag{36}$$

In addition, the wave frequency equations of the general nonlocal theory (Eqs. (32) and (33)) reduce to the ones of the classical theory of linear elasticity by setting $\ell_\lambda \rightarrow 0$ and $\ell_\mu \rightarrow 0$:

$$\omega_L^2 = \frac{(\lambda + 2\mu)\,k^2}{\rho} \tag{37}$$

$$\omega_T^2 = \frac{\mu k^2}{\rho} \tag{38}$$

We have detected the limitations of Eringen's nonlocal theory in modeling the dispersion of many materials, e.g., Au (Fig. 3). Eringen's nonlocal theory assumes the same nonlocal character for both the longitudinal and transverse acoustic waves. However, the general nonlocal theory outweighs Eringen's nonlocal theory by the ability to model the contrast between the nonlocal characters of the longitudinal and transverse acoustic waves. As shown in Eqs. (32) and (33), the transverse wave frequency depends on the nonlocal parameter, ℓ_μ, while the longitudinal wave frequency depends on two independent nonlocal parameters, ℓ_λ and ℓ_μ. With this new flavor, the general nonlocal theory can simultaneously fit the longitudinal and transverse dispersion curves of various materials [18, 19].

Unlike Eringen's nonlocal theory, the general nonlocal theory gives Poisson's ratio dependent on the nonlocal and the dispersion characteristics of the material. The substitution of Eqs. (32) and (33) into Eq. (17) gives Poisson's ratio based on the general nonlocal theory, as follows:

$$\nu = \frac{\lambda\left(\frac{\mathcal{F}_\lambda}{\ell_\lambda}\right)}{2\lambda\left(\frac{\mathcal{F}_\lambda}{\ell_\lambda}\right) + 2\mu\left(\frac{\mathcal{F}_\mu}{\ell_\mu}\right)} \tag{39}$$

In Eq. (18), Eringen's nonlocal theory gave the Poisson's ratio constant as assumed by the classical theory of elasticity. However, the general nonlocal theory gives the Poisson's ratio dependent on the material's nonlocal character, as shown in Eq. (39).

In comparison to the classical theory of elasticity, the general nonlocal theory models the distortion of the envelope of acoustic waves. The classical dispersion relations in Eqs. (37) and (38) indicate a constant phase velocity v_p, i.e., $\partial\omega/\partial k = v_p$ and $\partial^2\omega/\partial k^2 = 0$. Thus, according to the classical theory, acoustic dispersion curves are non-varying, and the envelope of an acoustic wave steadily varies in space as fast as its wavelength [18]. This is not true for all materials when operate at high frequencies. The envelope of an acoustic wave depends on the wavelength and is distorted as the wave travels. This can be revealed by the general nonlocal theory where the dispersion relations (Eqs. (32) and (33)) give the group velocity a function of the wave number, i.e., $\partial\omega/\partial k = v_g(k)$ and $\partial^2\omega/\partial k^2 \neq 0$.

4 Relation to Strain Gradient and Couple Stress Theories

Because the general nonlocal theory satisfies, both the force and the moment global balance conditions, the nonlocal stress tensor is a symmetric tensor with six independent components. The skew-symmetric part of the nonlocal stress tensor vanishes upon achieving the moment balance globally. However, if the moment balance is considered locally and is violated globally, the skew-symmetric part of the nonlocal stress tensor shall be non-vanishing. Given this observation, the balance equations of the general nonlocal theory can be written in three different forms, as follows:

Form I:

$$\begin{aligned} t_{ji,j} + f_i &= \rho \ddot{u}_i \\ e_{ijk} t_{jk} &= 0 \end{aligned} \tag{40}$$

Form II:

$$\begin{aligned} \sigma_{ji,j} + f_i &= \rho \ddot{u}_i \\ e_{ijk} t_{jk} &= 0 \end{aligned} \tag{41}$$

Form II:

$$\begin{aligned} t_{ji,j} + f_i &= \rho \ddot{u}_i \\ e_{ijk} \sigma_{jk} &= 0 \end{aligned} \tag{42}$$

where σ_{ij} and t_{ij} are the local and the nonlocal stress tensors, respectively. The stress tensors, σ_{ij} and t_{ij}, in Eqs. (40)–(42) are general tensors each of which has nine independent components. These tensors are defined by:

$$\sigma_{ij}(x) = \bar{C}_{ijkl} u_{k,l}(x) \tag{43}$$

$$t_{ij}(x) = \int_V C_{ijkl}\left(\left|x' - x\right|, \ell\right) u_{k,l}\left(x'\right) dV\left(x'\right) \tag{44}$$

where $\bar{C}_{ijkl}$ and C_{ijkl} are local and nonlocal stiffnesses, respectively. Form I (Eq. (40)) is a general form of the nonlocal theory, which achieves the force and the moment global balance requirements. This form is an equivalent form of the balance Eq. (9). According to the first equation of Eq. (40), the resultant force at a point is calculated as the sum of the local interaction of this point with its neighbors and the nonlocal interactions with all other points of the elastic domain. On the other hand, the second equation of Eq. (40) indicates that the resultant moment at a point is calculated as the sum of the moments of the local interaction of the point with its neighbors and the moments of the nonlocal interactions with all other points of the elastic domain.

Forms II and III are reduced versions of the general nonlocal theory (Form I). Form II achieves the force balance locally but violates it globally. In Form III, the moment balance is achieved locally, but it is violated globally. These different forms would lead to different versions of the nonlocal theory.

In the following, we show that Form I of the nonlocal theory can be reduced to the strain gradient theory. In addition, we show that Form II can be reduced to the couple stress theory.

4.1 Strain Gradient Theory

For weak nonlocal fields (i.e., weak nonlocality), the general nonlocal theory can be reduced to a strain gradient theory in which the deformation energy is a function of the strain and its gradients. The simplification of the general nonlocal theory to a strain gradient theory is achieved by considering Form I of the nonlocal model and expanding the strain field, $\varepsilon\left(\mathbf{x}'\right)$, about point $\mathbf{x}$ as follows [18]:

$$\begin{aligned}\varepsilon\left(\mathbf{x}'\right) &= \varepsilon\left(\mathbf{x}\right) + \left(\mathbf{x}' - \mathbf{x}\right).\nabla\varepsilon\left(\mathbf{x}\right) + \frac{1}{2}\left[\left(\mathbf{x}' - \mathbf{x}\right).\nabla\right]^2\varepsilon\left(\mathbf{x}\right)\\ &+ \frac{1}{6}\left[\left(\mathbf{x}' - \mathbf{x}\right).\nabla\right]^3\varepsilon\left(\mathbf{x}\right) + \frac{1}{24}\left[\left(\mathbf{x}' - \mathbf{x}\right).\nabla\right]^4\varepsilon\left(\mathbf{x}\right) + \ldots\end{aligned} \tag{45}$$

By the substitution of Eq. (45) into Eq. (22), the nonlocal stress is formed depending on the strain and its gradients, as follows:

$$\begin{aligned}t_{ji} = t_{ji} &= \left(\lambda\varepsilon_{rr} + a_3\varepsilon_{rr,ll} + a_5\varepsilon_{rr,llmm} + \ldots\right)\delta_{ij}\\ &+ \left(2\mu\varepsilon_{ij} + b_3\varepsilon_{ij,ll} + b_5\varepsilon_{ij,llmm} + \ldots\right)\end{aligned} \tag{46}$$

where

$$\begin{aligned}a_n &= \frac{\lambda}{\prod(n-1)}\int\limits_V\left[\left(\mathbf{x}' - \mathbf{x}\right)\right]^{n-1}\beta_\lambda\left(\left|\mathbf{x}' - \mathbf{x}\right|, \ell_\lambda\right)dV\left(\mathbf{x}'\right)\\ &\qquad\text{for } n = 3, 5, \ldots\\ b_n &= \frac{2\mu}{\prod(n-1)}\int\limits_V\left[\left(\mathbf{x}' - \mathbf{x}\right)\right]^{n-1}\beta_\mu\left(\left|\mathbf{x}' - \mathbf{x}\right|, \ell_\mu\right)dV\left(\mathbf{x}'\right)\\ &\qquad\text{for } n = 3, 5, \ldots\end{aligned} \tag{47}$$

Substituting Eq. (46) into Eq. (40) gives the equation of motion in the form:

$$\begin{aligned}&\left(\lambda\varepsilon_{rr,j} + a_3\varepsilon_{rr,llj} + a_5\varepsilon_{rr,llmmj} + \ldots\right)\delta_{ij}\\ &+ \left(2\mu\varepsilon_{ij,j} + b_3\varepsilon_{ij,llj} + b_5\varepsilon_{ij,llmmj} + \ldots\right) + f_i = \rho\ddot{u}_i\end{aligned} \tag{48}$$

The strain can be expressed in terms of the displacement vector, $\mathbf{u}$, as follows:

$$\varepsilon = \nabla\mathbf{u} - \frac{1}{2}\nabla\times\mathbf{u} \tag{49}$$

which gives the field equation with the form:

$$\left[\lambda + 2\mu + \sum_{n=3,\ 5,\ldots}^{\infty} (a_n + b_n)\,\nabla^{(n-1)}\right](\nabla\nabla.\mathbf{u})$$

$$-\frac{1}{2}\left[2\mu + \sum_{n=3,5,\ \ldots}^{\infty} b_n\nabla^{(n-1)}\right](\nabla \times \nabla \times \mathbf{u}) + \mathbf{F} = \rho\ddot{\mathbf{u}} \tag{50}$$

4.2 *Couple Stress Theory*

Form II of the nonlocal theory can be reduced to a couple stress theory. Form II indicates that the force resultant at a point is locally balanced while the moment resultant at a point is globally balanced. According to Eqs. (41) and (43), the skew-symmetric part of the nonlocal stress tensor t_{ij} is obtained for isotropic-linear elastic materials with the form:

$$e_{ijk}t_{jk}(\mathbf{x}) = \int_V 2\mu\beta_\mu\left(\left|\mathbf{x}' - \mathbf{x}\right|, \ell_\mu\right)\theta_{mn}\left(\mathbf{x}'\right)dV\left(\mathbf{x}'\right) \tag{51}$$

where θ_{mn} is the rotation tensor that expresses the rigid rotation of the elastic body. For a weak nonlocal moment field (i.e., weak nonlocality), Form II of the nonlocal theory can be reduced to the couple stress theory by expanding the rigid-rotation field $\theta\left(\mathbf{x}'\right)$ about point $\mathbf{x}$ as follows [18]:

$$\theta\left(\mathbf{x}'\right) = \theta(\mathbf{x}) + \left(\mathbf{x}' - \mathbf{x}\right).\nabla\theta(\mathbf{x}) + \frac{1}{2}\left[\left(\mathbf{x}' - \mathbf{x}\right).\nabla\right]^2\theta(\mathbf{x})$$

$$+ \frac{1}{6}\left[\left(\mathbf{x}' - \mathbf{x}\right).\nabla\right]^3\theta(\mathbf{x}) + \frac{1}{24}\left[\left(\mathbf{x}' - \mathbf{x}\right).\nabla\right]^4\theta(\mathbf{x}) + \ldots \tag{52}$$

The substitution of Eq. (52) into Eq. (51) yields:

$$e_{ijk}t_{jk}(\mathbf{x}) = 2\mu\theta_{jk}(\mathbf{x}) + b_3\theta_{jk,ll}(\mathbf{x}) + b_5\theta_{jk,llmm}(\mathbf{x}) + \ldots \tag{53}$$

where b_n are defined in Eq. (47). Equation (53) can be rewritten in terms of a rotation gradient tensor, R_{ij}, i.e.,

$$e_{ijk}t_{jk}(\mathbf{x}) = e_{jki}\left(2\mu\theta_i(\mathbf{x}) + b_3R_{li,l}(\mathbf{x}) + b_5R_{li,lmm}(\mathbf{x}) + \ldots\right) \tag{54}$$

where

$$R_{ji} = \theta_{i,j} \tag{55}$$

where $\theta_i = (e_{ijk}\theta_{kj})/2$ and $\theta_{ij} = e_{ijk}\theta_k$.

According to Eq. (41), the skew-symmetric part of the general nonlocal stress is zero, i.e., $e_{ijk}\ t_{jk}(x) = 0$. This gives the skew-symmetric part of the local stress tensor, $\sigma_{[ij]} = -e_{jik}(\mu\theta_k)$, with the form:

$$\sigma_{[ij]} = \frac{1}{2} e_{jik} \left(b_3 R_{lk,l} \left(\mathbf{x} \right) + b_5 R_{lk,lmm} \left(\mathbf{x} \right) + \ldots \right) \quad (56)$$

In light of the preceding equations, the balance equation of the couple stress theory is obtained in the following form by the substitution of Eq. (56) into the first equation of Eq. (41):

$$\sigma_{(ji),j} + \frac{1}{2} e_{jik} \left(b_3 R_{lk,lj} + b_5 R_{lk,lmmj} + \ldots \right) + f_i = \rho \ddot{u}_i \quad (57)$$

where $\sigma_{(ji)} = \lambda\varepsilon_{rr}\delta_{ij} + 2\mu\varepsilon_{ij}$ is the classical symmetric-stress tensor. The field equation can be obtained by substituting $R = (\nabla\nabla \times \mathbf{u})/2$ into Eq. (57):

$$(\lambda + 2\mu)(\nabla\nabla.\mathbf{u}) - \frac{1}{2} \left[2\mu + \frac{1}{2} \sum_{n=3,5,\ldots}^{\infty} b_n \nabla^{(n-1)} \right] (\nabla \times \nabla \times \mathbf{u}) + \mathbf{F} = \rho \ddot{\mathbf{u}} \quad (58)$$

4.3 Relation to Mindlin's Strain Gradient Theories

In the framework of Mindlin's strain gradient theory, the total strain energy is a function of the strain and its first and second gradients [2, 5, 43]. According to this theory, the field equation is expressed with the form [2]:

$$\begin{aligned} &(\lambda + 2\mu) \left[1 - l_1^2 \nabla^2 + l_2^4 \nabla^4 \right] (\nabla\nabla.\mathbf{u}) \\ &-\mu \left[1 - l_3^2 \nabla^2 + l_4^4 \nabla^4 \right] (\nabla \times \nabla \times \mathbf{u}) + \mathbf{F} = \rho \ddot{\mathbf{u}} \end{aligned} \quad (59)$$

where l_1, l_2, l_3, and l_4 are different length scales. By comparing Eqs. (59) to (50), the length scales, l_1, l_2, l_3, and l_4, can be related to the material coefficients, λ, a_n, μ, and b_n, as follows:

$$l_1^2 = -\frac{a_3 + b_3}{\lambda + 2\mu},\ l_2^4 = \frac{a_5 + b_5}{\lambda + 2\mu},\ l_3^2 = -\frac{b_3}{2\mu},\ l_4^4 = \frac{b_5}{2\mu} \quad (60)$$

The preceding discussion indicates that Mindlin's strain gradient is a nonlocal theory but for weak nonlocality. The strain gradient theory can model long-range interatomic/intermolecular interactions up to a certain number of neighbors depending on the gradients of the strain considered.

4.4 Relation to Couple Stress Theory

The couple stress theory [7, 8] was originally developed on the basis of the first-order gradient of the rotation tensor. This theory was then extended to the second rotation gradient theory [9] in which the internal energy depends on the strain and the first and the second gradients of the rotation. According to the couple stress theory, the field equation is expressed as follows [9]:

$$(\lambda + 2\mu)\,(\nabla\nabla.\mathbf{u}) - \mu\left[1 - l_5^2\nabla^2 + l_6^4\nabla^4\right](\nabla \times \nabla \times \mathbf{u}) + \mathbf{F} = \rho\ddot{\mathbf{u}} \tag{61}$$

where l_5 and l_6 are length scales. According to Eqs. (61) and (58), the length scales, l_5 and l_6, can be related to the material coefficients μ, and b_n, as follows:

$$l_5^2 = -\frac{b_3}{4\mu},\; l_6^4 = \frac{b_5}{4\mu} \tag{62}$$

It is clear that the couple stress theory, like the strain gradient theory, is a nonlocal theory but with weak nonlocality. The couple stress theory can model long-range interatomic/intermolecular interactions up to a certain number of neighbors depending on the considered gradients of the rotation tensor.

4.5 Wave Propagation

To derive the dispersion relations along [100] direction in the context of the strain gradient and couple stress theories, Eq. (14) is used. Equation (14) is substituted into the field Eq. (50), which gives the dispersion relations according to the strain gradient theory with the form:

$$\omega_L^2 = \left((\lambda + 2\mu)\,k^2 + \sum_{n=3,5,\ldots.}^{\infty} -(i)^{n+1}\,(a_n + b_n)\,k^{n+1}\right)/\rho \tag{63}$$

$$\omega_T^2 = \left(\mu k^2 + \frac{1}{2}\sum_{n=3,5,\ldots}^{\infty} -(i)^{n+1} b_n k^{n+1}\right)/\rho \tag{64}$$

Similarly, the dispersion relations according to the couple stress theory are derived in the following form by substituting Eq. (14) into Eq. (58):

$$\omega_L^2 = \left((\lambda + 2\mu) k^2 \right) / \rho \tag{65}$$

$$\omega_T^2 = \left(\mu k^2 + \frac{1}{4} \sum_{n=3,5,\ldots}^{\infty} - (i)^{n+1} b_n k^{n+1} \right) / \rho \tag{66}$$

5 Infeasibility of Nonlocal Strain Gradient Theory

An incorrect statement that circulates among investigators of the nonlocal mechanics is that the nonlocal theory only reveals the softening behaviors. However, the nonlocal theory can capture both softening and hardening behaviors. We demonstrated in Fig. 4 that the nonlocal theory can predict both the increase and the decrease in the wave frequency with the wavenumber. The softening behavior is captured using real nonlocal parameters while complex nonlocal parameters are used to capture the hardening behavior.

Another misunderstanding of the nonlocal mechanics is the claim that the strain gradient theory and couple stress theory only model the hardening behaviors of materials. It was claimed that the nonlocal theory and strain gradient/couple stress theory describe different material characteristics [44]. This motivated some researchers to combine the nonlocal theory with the strain gradient and/or the couple stress theory in a unified model, e.g., nonlocal strain gradient theory [44]. It was claimed that these unified models can give the flavor of combining the softening and hardening mechanisms. Here, these attempts are revisited and discussed according to their feasibility demonstrating the following facts:

- The general nonlocal theory can model both softening and hardening behaviors of materials.
- The strain gradient and couple stress theories can model both softening and hardening behaviors of materials.
- Neither the strain gradient theory nor the couple stress theory can be merged with the nonlocal theory in a unified model.
- The strain gradient theory is a nonlocal theory that can exactly emulate the nonlocal theory if more gradients of the strain are considered.
- The couple stress theory has a nonlocal character, which is promoted by considering more gradients of the rotation tensor.
- The newly claimed nonlocal strain gradient theory [44] is physically infeasible.

We demonstrate in Fig. 5 that each of the general nonlocal theory, the strain gradient theory, and the couple stress theory can model both the softening and hardening behaviors of materials. The wave frequency-wavenumber curves as obtained based on these three theories are depicted for different nonlocal parameters and length scales. In Fig. 5a, real positive nonlocal parameters, $\ell_\lambda > 0$ and $\ell_\mu > 0$, and complex length scales, $l_1^2 < 0$, $l_2^2 > 0$, $l_3^2 < 0$, $l_4^2 > 0$, $l_5^2 < 0$, and $l_6^2 > 0$, are defined to give the wave frequency decreasing lower than that of the classical theory with the

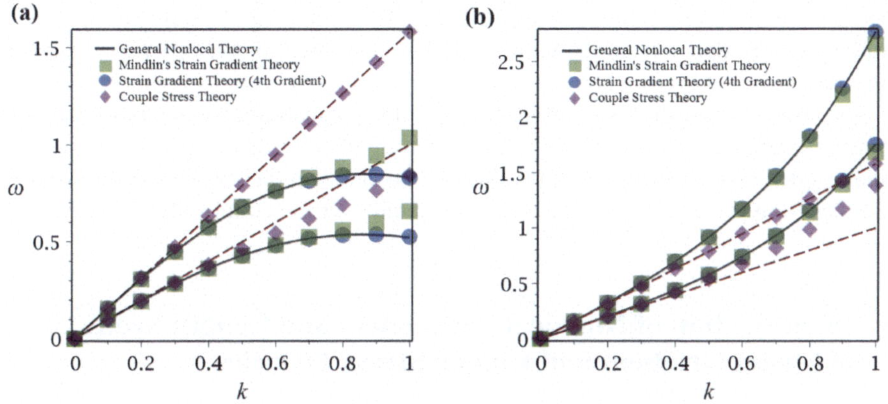

Fig. 5 Hardening and softening behaviors of nonlocal, strain gradient, and couple stress theories. **a** The variation of the longitudinal and transverse acoustic frequencies, ω (rad/s), as functions of the wavenumber, k, lower than the frequencies of the classical model. The results are represented based on the general nonlocal theory ($\ell_\lambda = 1.6\pi$ and $\ell_\mu = 1.6\pi$), Mindlin's strain gradient theory ($l_1 = i1.1$, $l_2 = 0.89$, $l_3 = i1.1$, and $l_4 = 0.91$), strain gradient theory up to 4th strain gradient ($a_3 = 0.6$, $a_5 = 0.32$, $a_7 = 0.1$, $a_9 = 0.02$, $b_3 = 2.4$, $b_5 = 1.67$, $b_7 = 0.39$, $b_9 = 0.08$), and the couple stress theory ($l_5 = i0.77$, and $l_6 = 0.65$). **b** The variation of the longitudinal and transverse acoustic frequencies, ω (rad/s), as functions of the wavenumber, k, higher than the frequencies of the classical model. The results are represented based on the general nonlocal theory ($\ell_\lambda = i1.6\pi$ and $\ell_\mu = i1.6\pi$), Mindlin's strain gradient theory ($l_1 = 1.1$, $l_2 = 0.89$, $l_3 = 1.1$, and $l_4 = 0.91$), strain gradient theory up to 4th strain gradient ($a_3 = -0.6$, $a_5 = 0.32$, $a_7 = -0.1$, $a_9 = 0.02$, $b_3 = -2.4$, $b_5 = 1.67$, $b_7 = -0.39$, $b_9 = 0.08$), and the couple stress theory ($l_5 = 0.77$, and $l_6 = 0.65$). ($\lambda = 0.5$, $\mu = 1$, and $\rho = 1$)

increase in the wavenumber. The decrease in the wave frequency indicates a material softening (or wave dispersion). Thus, with the nonlocal parameters and length scales as defined in Fig. 5a, the general nonlocal theory, Mindlin's strain gradient theory, and the couple stress theory can capture the softening behaviors of materials. In Fig. 5b, complex nonlocal parameters, $\ell_\lambda^2 < 0$ and $\ell_\mu^2 < 0$, and real positive length scales, $l_1^2 > 0$, $l_2^2 > 0$, $l_3^2 > 0$, $l_4^2 > 0$, $l_5^2 > 0$, and $l_6^2 > 0$, are used to demonstrate the ability of the general nonlocal theory, Mindlin's strain gradient theory, and the couple stress theory to capture materials' stiffness hardening. For the latter case, the wave frequency is obtained increasing higher than that of the classical theory with the increase in the wavenumber.

It follows from Fig. 5 that the strain gradient theory can perfectly emulate the nonlocal theory, and its nonlocal character is enhanced by considering more gradients of the strain. Mindlin's strain gradient theory, which accounts for up to second-order gradients of the strain, gives the same results of the general nonlocal theory for $k < 0.6$. However, discrepancies between the results of the two theories are seen for $k > 0.6$. When more gradients were considered (up to 4th-order gradient), the strain gradient theory gave the exact same results of the general nonlocal theory. Thus, in general, the strain gradient theory can reveal nonlocal effects similar to the nonlocal theory but within certain limits of applicability. The limit of applicability of the strain gradient theory depends on the number of the considered higher-order gradients of the strain.

The couple stress theory exhibits some nonlocal capabilities when modeling the transverse acoustic waves. Nonetheless, the couple stress theory gives the longitudinal acoustic waves identical to the ones of the classical theory. This demonstrates that the strain gradient theory would be preferred over the couple stress theory.

6 Identification of Nonlocal Parameters and Length Scales of Strain Gradient and Couple Stress Theories

Many material properties can be revealed by non-destructive tests. In a non-destructive test, the material is exposed into a wave and the velocity of the wave propagation through the material is measured. When a wave propagates in a material, it excites atomic/molecular vibrations. Phonons are the description of the modes of vibration of atoms/molecules at their equilibrium position. Thus, the prediction of these phonons is a nondestructive method of material characterization [18, 45–48]. Therefore, we propose identifying the nonlocal parameters along with the elastic properties of various materials by means of nondestructive testing of the material phonons, e.g., inelastic neutron scattering. The dispersion relations based on the nonlocal theory are determined and fitted to the experimentally measured phonon dispersions.

Acoustic phonons represent the translational modes of vibration. These phonons depend on neighbor and non-neighbor interactions. In metals, delocalized electrons connect atoms/molecules together and promote non-neighbor interactions. Because of the delocalized electrons, the stress at a point is generally nonlocal and functional of the deformation gradients of all points of the elastic solid. Therefore, the general nonlocal theory is used here to effectively model acoustic phonons of different materials. The longitudinal acoustic (LA) and transverse acoustic (TA) phonons according to the general nonlocal theory are determined in Eqs. (32) and (33). According to these equations, TA phonons only depend on the shear nonlocal effect while the LA phonons depend on both the dilatation and shear nonlocal effects. The acoustic dispersion relations of the general nonlocal theory are compared and fitted to the experimental acoustic phonons of different materials with the aim of determining their dilatation and shear nonlocal characteristics. The fitting is carried out for acoustic phonons along [100] direction of diamond, graphite, silicon (Si), silver (Ag), gold (Au), copper (Cu), and platinum (Pt) in Fig. 6. The experimental acoustic phonons of these materials are obtained from [25, 49–53]. The determined nonlocal parameters, ℓ_λ and ℓ_μ, are summarized in Table 2.

Upon identifying the nonlocal parameters, ℓ_λ and ℓ_μ, the length scales of the strain gradient and couple stress theories, l_i, can be determined based on Eqs. (60) and (62). The determined length scales of the strain gradient and couple stress theories

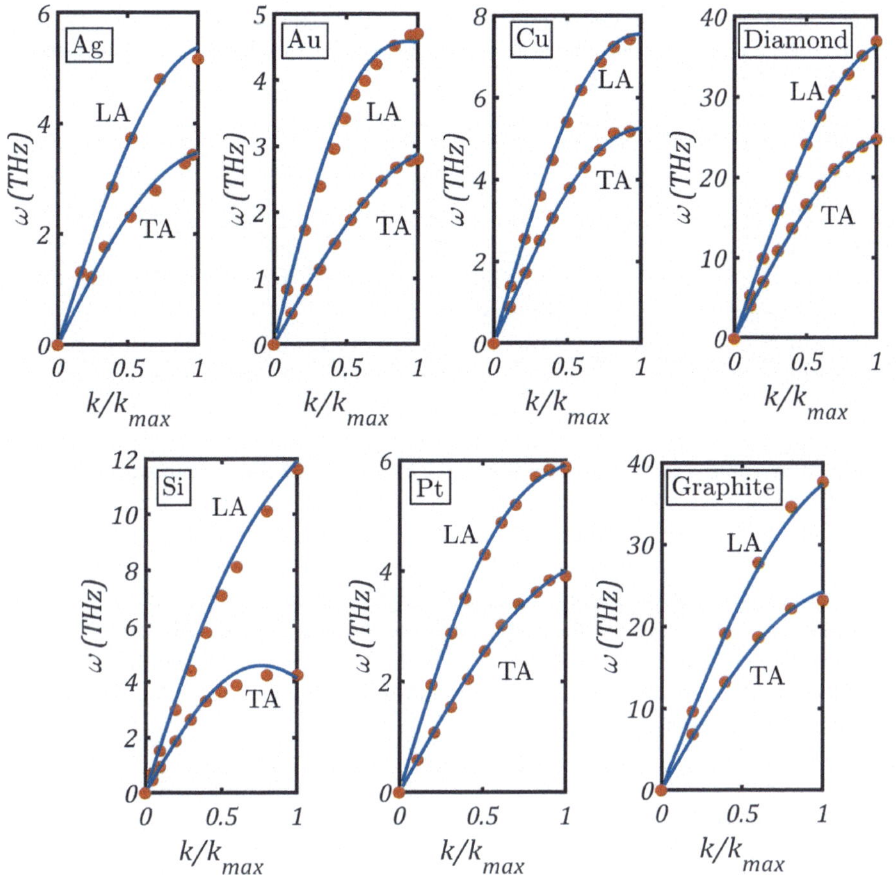

Fig. 6 Identification of nonlocal parameters, ℓ_λ and ℓ_μ. The general nonlocal theory (curves) fits the experimental acoustic phonons along [100] direction (closed circles) of Ag, Au, Cu, diamond, Si, Pt, and graphite

Table 2 Identification of nonlocal parameters, ℓ_λ and ℓ_μ for Ag, Au, Cu, diamond, Si, Pt, and graphite

Parameters	Diamond	Graphite	Si	Ag	Au	Cu	Pt
Elastic Moduli (λ and μ), Mass Density (ρ), and Lattice Parameter (a)							
λ (GPa)	87.463	30	74.672	18	154	25	140
μ (GPa)	537.4	160	63.624	45	48	78	95
ρ (kg/m^3)	3400	2266	2330	10490	19300	8960	21450
a (nm)	0.3567	0.246	0.543	0.4079	0.4065	0.3597	0.3912
Determined Nonlocal Parameters							
ℓ_λ (nm)	0.214	~0	~0	0.253	0.427	0.378	0.430
ℓ_μ (nm)	0.214	0.145	0.489	0.253	0.224	0.241	0.215

Table 3 Identification of the length scales ($l_1 \rightarrow l_4$) of the strain gradient theory using Eq. (60), and the length scales of the couple stress theory (l_5 and l_6) using Eq. (62) for Ag, Au, Cu, diamond, Si, Pt, and graphite

Parameters	Diamond	Graphite	Si	Ag	Au	Cu	Pt
Calculated Length Scales of the Strain Gradient Theory							
l_1 (m) $\times 10^{(-11)}$	$i4.67$	$i3.028$	$i8.4658$	$i5.518\,7$	$i7.9107$	$i5.7636$	$i7.0782$
l_2 (m) $\times 10^{(-11)}$	3.798	2.5188	7.7276	4.4883	6.7883	4.8803	6.2903
l_3 (m) $\times 10^{(-11)}$	$i4.67$	$i3.167$	$i10.664$	$i5.5187$	$i4.8788$	$i5.259$	$i4.6952$
l_4 (m) $\times 10^{(-11)}$	3.798	2.575	8.6732	4.4883	3.9679	4.2771	3.8185
Calculated Length Scales of the Couple Stress Theory							
l_5 (m) $\times 10^{(-11)}$	$i3.3022$	$i2.2396$	$i7.5408$	$i3.9023$	$i3.4498$	$i3.7187$	$i3.32$
l_6 (m) $\times 10^{(-11)}$	3.1938	2.166	7.2932	3.7742	3.3366	3.5966	3.211

are given in Table 3. It is clear that the strain gradient and the couple stress theories can capture the reduction of the wave frequency with the wave number and softening behaviors like the nonlocal theory. Some of the length scales are of complex values, which reflected a softening behavior. According to Mindlin, the strain energy density function is real-positive definite when the length scales are either real positive or complex [2, 7, 8]. The use of the length scales of the strain and couple stress theories was confined only for real positive values in the literature. These real length scales revealed hardening behaviors. However, complex length scales can be used to capture softening behaviors of materials.

7 Conclusions

In this chapter, the general nonlocal theory was interpreted. The limits of applicability of Eringen's nonlocal were discussed extending the assumptions of this theory to the general nonlocal theory. In addition, it was shown that, for weak nonlocality, the general nonlocal theory can be reduced to the strain gradient theory and the couple stress theory. Furthermore, we demonstrated that the general nonlocal theory can reflect both the softening and the hardening behaviors of materials. Finally, the nonlocal parameters of the general nonlocal theory and the material coefficients of the strain gradient and the couple stress theories were determined for diamond, graphite, silicon, silver, gold, copper, and platinum.

References

1. Askes H, Aifantis EC (2011) Gradient elasticity in statics and dynamics: an overview of formulations, length scale identification procedures, finite element implementations and new results. Int J Solids Struct 48:1962–1990
2. Mindlin RD (1964) Microstructure in linear elasticity. Arch Ration Mech Anal 16:51–78

3. Shaat M (2018) A reduced micromorphic model for multiscale materials and its applications in wave propagation. Compos Struct 201:446–454
4. Neff P, Ghiba ID, Madeo A, Placidi L, Rosi G (2014) A unifying perspective: the relaxed linear micromorphic continuum. Continuum Mech Therm 26:639–681
5. Eringen AC (1999) Microcontinuum field theories I: foundations and solids. Springer, New York
6. Mindlin R (1965) Second gradient of strain and surface-tension in linear elasticity. Int J Solids Struct 1:417–38
7. Mindlin R, Tiersten H (1962) Effects of couple stresses in linear elasticity. Arch Ration Mech Anal 11:415–48
8. Toupin RA (1962) Elastic materials with couple-stresses. Arch Ration Mech Anal 11:385–414
9. Shaat M, Abdelkefi A (2016) On a second-order rotation gradient theory for linear elastic continua. Int J Eng Sci 100:74–98
10. Eringen AC, Edelen DGB (1972) On nonlocal elasticity. Int J Eng Sci 10:233–248
11. Eringen AC (1972) Linear theory of nonlocal elasticity and dispersion of plane waves. Int J Eng Sci 10:425–435
12. Kröner E (1967) Elasticity theory of materials with long range cohesive forces. Int J Solids Struct 3:731–742
13. Eringen AC (1983) On differential equations of nonlocal elasticity and solutions of screw dislocation and surface waves. J Appl Phys 54:4703–4710
14. Silling SA (2000) Reformulation of elasticity theory for discontinuities and long-range forces. J Mech Phys Solids 48:175–209
15. Polizzotto C, Fuschi P, Pisano AA (2004) A strain-difference-based nonlocal elasticity model. Int J Solids Struct 41:2383–2401
16. Rafii-Tabar H, Ghavanloo E, Fazelzadeh SA (2016) Nonlocal continuum-based modeling of mechanical characteristics of nanoscopic structures. Phys Rep 638:1–97
17. Askari H, Younesian D, Esmailzadeh E, Cveticanin L (2017) Nonlocal effect in carbon nanotube resonators: a comprehensive review. Adv Mech Eng 9:1–24
18. Shaat M (2017) A general nonlocal theory and its approximations for slowly varying acoustic waves. Int J Mech Sci 130:52–63
19. Shaat M, Abdelkefi A (2017) New insights on the applicability of Eringen's nonlocal theory. Int J Mech Sci 121:67–75
20. Shaat M, Faroughi S (2019) Influence of long-range interatomic and interlayer interactions on dispersion of acoustic waves by multilayer graphene. Physica E 108:74–82
21. Eringen AC (2006) Nonlocal continuum mechanics based on distributions. Int J Eng Sci 44:141–147
22. Eringen AC (2002) Nonlocal continuum field theories. Springer, New York
23. Ghavanloo E, Rafii-Tabar H, Fazelzadeh SA (2019) Computational continuum mechanics of nanoscopic structures: nonlocal elasticity approaches. Springer, New York
24. Kamitakahara WA, Brcckhouse BN (1969) Crystal dynamics of silver. Phys Lett A 29:639–640
25. Lynn JW, Smith HG, Nicklow RM (1973) Lattice dynamics of gold. Phys Rev B 8:3493–3499
26. Shaat M (2019) Size-dependence of Young's modulus and Poisson's ratio: Effects of material dispersion. Mech Mater 133:111–119
27. Sun CT, Zhang H (2003) Size-dependent elastic moduli of platelike nanomaterials. J Appl Phys 93:1212–1218
28. Grima JN, Grech MC, Grima-Cornish JN, Gatt R, Attard D (2018) Giant auxetic behaviour in engineered graphene. Ann Phys 530:1–6
29. Ahadi A, Melin S (2016) Size dependence of the Poisson's ratio in single-crystal fcc copper nanobeams. Comput Mater Sci 111:322–327
30. Grima JN, Gatt R, Ellul B, Chetcuti E (2010) Auxetic behaviour in non-crystalline materials having star or triangular shaped perforations. J Non Cryst Solids 356:1980–1987
31. Jiang JW, Kim SY, Park HS (2016) Auxetic nanomaterials: Recent progress and future development. Appl Phys Rev 3:041101

32. Lakes RS (2017) Negative-Poisson's-ratio materials: Auxetic solids. Annu Rev Mater Res 47:63–81
33. Shaat M (2018) Correction of local elasticity for nonlocal residuals: application to Euler-Bernoulli beams. Meccanica 53:3015–3035
34. Faroughi S, Shaat M (2018) Poisson's ratio effects on the mechanics of auxetic nanobeams. Eur J Mech A/Solids 70:8–14
35. Zeng X, Chen Y, Lee JD (2006) Determining material constants in nonlocal micromorphic theory through phonon dispersion relations. Int J Eng Sci 44:1334–1345
36. Li X, Ono T, Wang Y, Esashi M (2003) Ultrathin single-crystalline-silicon cantilever resonators: Fabrication technology and significant specimen size effect on Young's modulus. Appl Phys Lett 83:3081–3083
37. Courtial J, Patterson BA, Hirst W, Harvey AR, Duncan AJ, Sibbett W, Padgett MJ (1997) Static Fourier-transform ultraviolet spectrometer for gas detection. Appl Opt 36:2813–2817
38. Cao GX, Chen X (2008) Size dependence and orientation dependence of elastic properties of ZnO nanofilms. Int J Solids Struct 45:1730–1753
39. Chen Y, Gao Q, Wang Y, An X, Liao X, Mai YW, Tan HH, Zou J, Ringer SP, Jagadish C (2015) Determination of Young's modulus of ultrathin nanomaterials. Nano Lett 15:5279–5283
40. Palik ED (1985) Handbook of optical constants of solids. Academic Press, New York
41. Gan F (1995) Optical properties of fluoride glasses: a review. J Non-Cryst Solids 184:9–20
42. Bacigalupo A, De Bellis ML, Gnecco G (2019) Complex frequency band structure of periodic thermo-diffusive materials by Floquet-Bloch theory. Acta Mech 230:3339–3363
43. Mindlin R, Eshel N (1968) On first strain-gradient theories in linear elasticity. Int J Solids Struct 4:109–24
44. Lim CW, Zhang G, Reddy J (2015) A higher-order nonlocal elasticity and strain gradient theory and its applications in wave propagation. J Mech Phys Solids 78:298–313
45. Schneider D, Shultrich B (1992) Determination of elastic modulus and thickness of surface layers by ultrasonic surface waves. Thin Solid Films 214:92–102
46. Eroglu A (2010) Wave propagation and radiation in gyrotropic and anisotropic media. Springer, New York
47. Sermeus J, Sinha R, Vanstreels K, Vereecken PM, Glorieux C (2014) Determination of elastic properties of a MnO_2 coating by surface acoustic wave velocity dispersion analysis. J Appl Phys 116:023503
48. Sas W, Gabryś K, Soból E, Szymański A (2016) Dynamic characterization of cohesive material based on wave velocity measurements. Appl Sci 6:49
49. Warren J, Wenzel R, Yarnell JL (1965) Inelastic scattering of neutrons. Vienna: Int Atomic Energy Agency 1:361–367
50. Dutton D, Brockhouse B, Miller AP (1972) Crystal dynamics of platinum by inelastic neutron scattering. Can J Phys 50:2915–2927
51. Cochran W (1973) The dynamics of atoms in crystals. Crane Russak, London
52. Jishi R, Venkataraman L, Dresselhaus M, Dresselhaus G (1993) Phonon modes in carbon nanotubules. Chem Phys Lett 209:77–82
53. Bian Q, Bose RC, Shukla SK (2008) Vibrational and thermodynamic properties of metals from a model embedded-atom potential. J Phys Chem Solids 69:168–181

Displacement Based Nonlocal Models for Size Effect Simulation in Nanomechanics

Gioacchino Alotta, Francesco P. Pinnola, and Marzia S. Vaccaro

Abstract The increasing number of nano and microscopic structural devices and their importance in several technological fields have pushed the research towards the formulation of mathematical models suitable for capturing mechanical small-scale effects. Usually, in these cases, an accurate modeling requires the simulation of the microstructure, the reproduction of intermolecular interactions and heterogeneity at the micro/nanoscale. Therefore, for this kind of mechanical modeling, classical local continuum theories fail in reproducing the real behavior at small-scale due to the inability to reproduce these scale-dependent effects. For this reason, an advanced theory is developed in the recent decades. Such formulation is known as nonlocal theory and it is able to take into account nonlocal effects. Among these nonlocal effects, there are long-range interactions, size-effects and heterogeneity of the material, strain localizations, and so on. Usually, in these advanced models, the nonlocal effects are reproduced by means of some additional terms in the governing equations. There are several kind of nonlocal models provided in literature. Among these various models, this chapter considers the displacement based nonlocal models which belong to the mechanically based nonlocality. Following this approach, the nonlocal effects are modeled as additional body forces acting on material volumes depending on their relative displacements. An overview of the main results of this theory and a summary of the other nonlocal models are reported in this chapter showing their differences and likeness.

G. Alotta (✉)
Department of Civil, Energy, Environment and Materials Engineering (DICEAM), Università Mediterranea di Reggio Calabria, Reggio Calabria, Italy
e-mail: gioacchino.alotta@unirc.it

F. P. Pinnola · M. S. Vaccaro
Department of Structures for Engineering and Architecture, University of Naples Federico II, Via Claudio 21, Ed. 6, 80125 Naples, Italy
e-mail: francescopaolo.pinnola@unina.it

M. S. Vaccaro
e-mail: marziasara.vaccaro@unina.it

© The Author(s), under exclusive license to Springer Nature Switzerland AG 2021
E. Ghavanloo et al. (eds.), *Size-Dependent Continuum Mechanics Approaches*, Springer Tracts in Mechanical Engineering,
https://doi.org/10.1007/978-3-030-63050-8_4

1 Introduction

In its first formulation, continuum mechanics theory was based on the idea that the matter mechanically interacts through local force interactions [1–3]. Following this approach, under certain assumptions, the solid continuum matter can be approximatively modeled as a set of discrete elements connected by elastic springs in which the parameters depend on the mechanical properties of the material which constitutes the matter. This approach allows to well describe the mechanical and thermodynamical behavior of several structures and materials providing well-posed formulations and accurate results. However, the local continuum mechanics fails in describing some structures at small-scale where size-effects, long range interactions and non-elastic forces influence the mechanical phenomena. Moreover, some mechanical effects, also at macro-scale, derive from micro- and nano-scale phenomena, such as stress-tip concentrations, dispersion of elastic waves, edge effects, shear bands. In this context, the local continuum theory may be inadequate in modeling these phenomena. For this reason, a new model able to take into account other mechanical effects was born in the last century. Pursuing this purpose the nonlocal model was formulated recently.

From the first formulations and the pioneering works [4–8], nonlocal theory is based on the idea that the stress at a given point of the continuum matter depends of on the entire local stress field of the domain. Nonlocal elasticity models provide the definition of an enriched continuum as compared to classical theories continuum definition. Such models are able to capture small-scale phenomena avoiding computational expensive procedures [9, 10].

Nonlocal approaches introduce a constitutive relation where long distance interactions exchanged within the body are described by internal parameters. Specifically, in many formulations of this kind, the long-range interaction is described by a convolution integral in the stress-strain relation. One of the first formulation has been provided by Eringen [11, 12]. In this model, the nonlocal stress is the output of a convolution integral between the entire elastic strain field and a particular attenuation function depending on an internal length-scale parameter. Since the input in the Eringen constitutive law is the strain field, this model is also called strain-driven approach. Substantially, this nonlocal theory provides a constitutive relation which is not pointwise but it is based on an integral average. Eringen's approach provides an accurate tool able to provide good results for screw dislocation and surface waves in unbounded domain. However, some issues appear when this model is applied to real structural problems in which, usually, the continuum domain must be considered bounded.

These issues are related to some meaningless boundary conditions which appear in a nonlocal Eringen finite domain. Specifically, from the integral formulation, it is possible to obtain the corresponding differential stress-strain relation by selecting a proper kernel. This relation is equal to the integral one only if proper boundary conditions are selected. By using the classical Eringen's approach, the integral formulation leads to an ill-posed differential problem in terms of meaningless elastic boundary conditions [13–16]. Several ways have been developed to overcome the ill-posedness of Eringen's nonlocal model for finite domain.

Specifically, in literature, there are some integral nonlocal formulations able to overcome this drawback, e.g. the two-phase local/nonlocal models [17–20], the strain-difference nonlocal elasticity [21–24], the stress-driven approach [15, 25]. Among these integral approaches, the latter still provides an integral formulation as in Eringen's approach but the nonlocal problem is set in a different way. Specifically, the output in the constitutive law is the nonlocal strain which is obtained as a convolution between an attenuation kernel and the entire stress field. In this way, the corresponding differential formulation avoids the boundary condition issues of Eringen's strain-driven one. By following this stress-driven approaches, several applications are performed [26–28] and other models are developed, such as local-nonlocal mixture stress-driven models [29] and two-phase models [30].

The aforementioned formulations provide an integral equation in the stress-strain relation. These approach are also known as strong nonlocality [31]. However, other nonlocal models exist and they are obtained following another approach known as weak nonlocality. In this context, we remember the gradient models proposed in Refs. [32–35]. Another class of nonlocal models which involves all the previous ones are the fractional-order models [36–38]. However, some of these formulations can not be considered a different approach but a kind of generalization. Specifically, as it has been shown in Refs. [39, 40], in all the integral nonlocal approaches if the convolution kernel is of power-law kind fractional-order operators may appear in the stress-strain relation. Moreover, such operators are characterized by a strong nonlocality due to their power-law kernels [41] and this peculiarity may be useful in several continuum mechanics applications [42, 43].

The aforementioned formulations are based on the assumption that the stress-strain relation of the nonlocal continuum is enriched by the introduction of additional contributions in terms of gradients or integrals of the strain and or stress fields. These other terms are able to take into account the presence of the effects of microstructures at small-scale. Some of the cited formulations presents some drawbacks. Specifically, in the strong nonlocality the selected kernel, that is the decaying function, must respects some geometrical constraints, whereas the weak nonlocality lack of an evident mechanical description. However, another class of different approach to take into account these effects exists, which avoid the restrictions of the previous approaches. This formulation is known as mechanically based nonlocal model and introduces nonlocal interactions among different locations of the body in terms of central long-range elastic [31, 36, 44] and/or viscoelastic body forces [45–47]. Such forces are proportional to the interactive volumes or masses of the solid. In other words, in comparison with the local continuum this model provides an enriched domain with nonlocal forces defined as dependent from the relative displacements between the volume elements and all the other volume elements in the domain [44, 48, 49]. For this reason this class of mechanically based model is known as displacement based nonlocal models. The model belongs to the class of integral or strong nonlocal models, and it can be seen as a modification of the peridynamic formulation provided by Silling [50].

This kind of model has some advantages. First of all, the nonlocal interactions are defined from a correspondent mechanical description, involving the presence of long-range springs connecting non-adjacent volume elements. For this reason, it is also known as mechanically based nonlocal model. Moreover, in presence of bounded domains, the nonlocal contributions to boundary conditions vanishes and this allows to enforce them as in a local continuum without any mathematical inconsistency. The displacement based nonlocal model has been successfully specialized to beam elements. In this context a finite element formulation with a closed formulation of non-local stiffness matrix elements has been obtained [44]. This approach is also suitable for modelling beams with viscoelastic long-range interactions both in quasi-static and dynamic conditions [46, 47].

Moreover, an analogous approach has been successfully developed in order to simulate non-local effects in the micromechanics of fluid flow, with the specific application of blood flowing along micro arterial vessels [51]. To this aim, the non-local forces are defined as dependent on the relative velocity instead of relative displacement. In this chapter, the mechanically based non-local model is introduced and its particularization to rods and beams are developed. Specifically, after an overview of the main nonlocal models in the Sect. 2, the Sect. 3 is devoted to the displacement based nonlocal model. In the latter nonlocal rod and beam models are derived and some applications in both static and dynamic conditions are provided.

2 An Overview on the Nonlocal Models

Considering the large number of contributions which have been published on nonlocal elasticity in the last decades, the present section aims to provide a synthesis of the main nonlocal formulations. Starting from Eringen's model in its strain-driven formulation, we recall the stress-driven approach, the gradient models and the mixture models.

2.1 Integral Models

Eringen's nonlocal integral formulation has been conceived to solve wave dispersion problems in unlimited domains [11, 12]. It is based on the idea that the stress $\boldsymbol{\sigma}$ at a point $\boldsymbol{x}$ of a nonlocal body $\mathcal{B}$ is the output of a convolution between the local response to the elastic strain field $\boldsymbol{\varepsilon}$ and a scalar kernel Φ_λ depending on a positive parameter λ. That is,

$$\boldsymbol{\sigma}(\boldsymbol{x}) = \int_{\mathcal{B}} \Phi_\lambda(\boldsymbol{x}, \bar{\boldsymbol{x}}) \boldsymbol{E}(\bar{\boldsymbol{x}}) \boldsymbol{\varepsilon}(\bar{\boldsymbol{x}}) d\bar{\boldsymbol{x}} \tag{1}$$

where $\boldsymbol{E}$ is the elastic stiffness tensor and $\boldsymbol{x}$, $\bar{\boldsymbol{x}}$ are position vectors. The integral law in Eq. (1) for the Bernoulli-Euler beam of length L leads to an integral nonlocal constitutive relation between the bending moment M and the elastic curvature χ,

$$M(x) = \int_0^L \Phi_\lambda(x, \bar{x}) k_f(\bar{x}) \chi(\bar{x}) d\bar{x} \tag{2}$$

where the beam axis coincides with the x-axis and $k_f(x)$ is the elastic bending stiffness. The attenuation kernel Φ_λ can be selected among exponential, Gaussian or power-law type functions and must satisfy the properties of symmetry, positivity and limit impulsivity. A frequent choice is the bi-exponential function defined as follows

$$\Phi_\lambda(x, \bar{x}) = \frac{1}{2\lambda L} \exp\left(-\frac{|x - \bar{x}|}{\lambda L}\right) \tag{3}$$

where λL is a characteristic length.

By choosing the kernel in Eq. (3), it can be proved [15] that the integral model expressed by Eq. (2) is equivalent to the second-order differential equation

$$\partial_x^2 M(x) - \frac{1}{(\lambda L)^2} M(x) = -\frac{EI}{(\lambda L)^2} \chi(x) \tag{4}$$

with the following constitutive boundary conditions (BCs)

$$\begin{aligned} \partial_x M(0) &= \frac{1}{\lambda L} M(0) \\ \partial_x M(L) &= -\frac{1}{\lambda L} M(L) \end{aligned} \tag{5}$$

being $k_f(x) = EI$, where E is the elastic modulus and I is the cross-sectional moment of inertia around the bending axis. Equations (5) are clearly in contrast with the static boundary conditions of most structural schemes. Therefore, the incompatibility between the equilibrium requirements and the constitutive nonlocal law reveals that applying the strain-driven model to bounded domains leads to ill-posed mechanical problems.

An efficient strategy to overcome the limits of Eringen's formulation has been developed in Refs. [15, 25]. It consists in a new nonlocal law, the stress-driven model, formally obtained by swapping the roles of stress and strain fields. Hence, the elastic strain tensor at a point $\boldsymbol{x}$ of the continuous body $\mathcal{B}$ is the output of the convolution integral between the local elastic strain field $\boldsymbol{\varepsilon}$ and the scalar kernel Φ_λ. That is,

$$\boldsymbol{\varepsilon}(\boldsymbol{x}) = \int_{\mathcal{B}} \Phi_\lambda(\boldsymbol{x}, \bar{\boldsymbol{x}}) \boldsymbol{C}(\bar{\boldsymbol{x}}) \boldsymbol{\sigma}(\bar{\boldsymbol{x}}) d\bar{\boldsymbol{x}} \tag{6}$$

where $\boldsymbol{C}$ is the elastic compliance $\boldsymbol{C} = \boldsymbol{E}^{-1}$.

Applied to a one-dimensional model, such as the Bernoulli-Euler beam introduced before, the stress-driven model gives

$$\chi(x) = \int_0^L \Phi_\lambda(x, \bar{x}) \frac{M(\bar{z})}{k_f(\bar{x})} d\bar{x} \tag{7}$$

The equivalent differential problem, obtained by assuming the attenuation function of Eq. (3) is expressed by a second-order differential equation

$$\partial_x^2 \chi(x) - \frac{1}{(\lambda L)^2} \chi(x) = -\frac{1}{EI(\lambda L)^2} M(x) \tag{8}$$

with the following constitutive BCs

$$\begin{aligned} \partial_x \chi(0) &= \frac{1}{\lambda L} \chi(0) \\ \partial_x \chi(L) &= -\frac{1}{\lambda L} \chi(L) \end{aligned} \tag{9}$$

where the signs in Eq. (9) are consequence of the fact that the first derivative of the special kernel is an odd function.

Unlike Eringen's formulation, the stress driven model provides exact solutions to nonlocal continuous problems and it is able to describe the actual behavior of micro- and nano-structures.

2.2 Gradient Models

Another nonlocal theory is the gradient model proposed in Ref. [35]. Such model overcomes the ill-posedness of the strain-driven model and it is able to simulate dispersion and wave propagation at atomic-scale [52, 53]. The approach combines the nonlocal strain-driven model by Eringen with the theory of strain gradient elasticity formulated by Mindlin [54, 55]. Nonlocal strain gradient model for a slender micro- or nano-beam is formulated by expressing the bending moment in terms of elastic flexural curvature χ and of its derivative $\partial_x \chi$. That is,

$$\begin{aligned} M(x) = &\int_0^L \alpha_0(x, \bar{x}, \lambda_0)\, k_f(\bar{x}) \chi(\bar{x}) d\bar{x} \\ &- l^2\, \partial_x \int_0^L \alpha_1(x, \bar{x}, \lambda_1)\, k_f(\bar{x}) \partial_{\bar{x}} \chi(\bar{x}) d\bar{x} \end{aligned} \tag{10}$$

The characteristic length l is introduced to make the two terms dimensionally compatible. The kernels α_0 and α_1 are attenuation functions depending on two nonlocal parameters λ_0, λ_1.

Following the treatment in Refs. [35] and [56], the attenuation kernels can be assumed to be coincident and equal to the bi-exponential function. Implicitly, it is also assumed the coincidence of nonlocal parameters $\lambda_0 = \lambda_1 = \lambda$. A second parameter, $\lambda_l = l/L$, may be associated with the characteristic length l of the strain gradient term.

As shown in Ref. [57], with the choice of the bi-exponential attenuation function, Eq. (10) is equivalent to the differential equation

$$(k_f \chi)(x) - l^2 \partial_x^2 (k \chi_f)(x) = M(x) - (\lambda L)^2 \partial_x^2 M(x) \tag{11}$$

equipped with the constitutive boundary conditions

$$\begin{aligned} \partial_x M(0) &= \frac{1}{\lambda L} M(0) + \frac{l^2}{(\lambda L)^2} \partial_x (k_f \chi)(0) \\ \partial_x M(L) &= -\frac{1}{\lambda L} M(L) + \frac{l^2}{(\lambda L)^2} \partial_x (k_f \chi)(L) \end{aligned} \tag{12}$$

Hence, the nonlocal strain gradient model is based on two parameters, λ and λ_l. As proved in Ref. [57], displacement solutions exhibit softening and stiffening responses for increasing nonlocal and gradient parameters, respectively. Thus, the nonlocal strain gradient law is able to model a wide class of small scale problems.

It has to be noted that for $\lambda \to 0$ and from Eq. (10), we recover the pure strain gradient law, by virtue of the kernel impulsivity property. When the characteristic length l approaches to zero, from the same equation, we get the fully nonlocal strain-driven model, which is ill-posed.

A stress gradient nonlocal model can be obtained by formally swapping the roles of stress and strain in Eq. (10). That is,

$$\chi(x) = \int_0^L \phi_\lambda(x, \bar{x}) \, \frac{M(\bar{x})}{k_f(\bar{x})} d\bar{x} - l^2 \, \partial_x \int_0^L \phi_\lambda(x, \bar{x}) \, \frac{\partial_{\bar{x}} M(\bar{x})}{k_f(\bar{x})} d\bar{x} \tag{13}$$

By selecting the bi-exponential attenuation function, the integral formulation Eq. (13) reverts to the differential equation

$$\chi(x) - (\lambda L)^2 \partial_x^2 \chi(x) = (k_f^{-1} M)(x) - l^2 \partial_x^2 (k_f^{-1} M)(x) \tag{14}$$

equipped with the constitutive boundary conditions

$$\begin{aligned} \partial_x \chi(0) &= \frac{1}{\lambda L} \chi(0) + \frac{l^2}{(\lambda L)^2} \partial_x (k_f^{-1} M)(0) \\ \partial_x \chi(L) &= -\frac{1}{\lambda L} \chi(L) + \frac{l^2}{(\lambda L)^2} \partial_x (k_f^{-1} M)(L) \end{aligned} \tag{15}$$

2.3 Mixture Models

Another strategy to formulate a well-posed nonlocal model is based on the two phases approach. It consists of a convex combination between local and nonlocal responses expressed as follows

$$M(x) = \alpha\,(k_f\,\chi)(x) + (1-\alpha)\int_0^L \Phi_\lambda(x,\bar{x})(k_f\,\chi)(\bar{x})d\bar{x} \tag{16}$$

where $\alpha \in [0, 1]$ and the nonlocal term is expressed by following the strain-driven approach. As shown in Ref. [30], for a fixed α, the response exhibits a softening behavior for increasing nonlocal parameter λ. When α approaches to zero, from Eq. (16), we recover the purely nonlocal strain-driven model, which is ill-posed. In Ref. [30], it has been shown that by choosing the kernel (3), it can be proved that the integral model expressed by Eq. (16) is equivalent to the second-order differential equation

$$\partial_x^2 M(x) - \frac{1}{(\lambda L)^2}M(x) = -\frac{k_f}{(\lambda L)^2}\chi(x) + \alpha k_f\,\partial_x^2\chi(x) \tag{17}$$

with the following constitutive boundary conditions (BCs)

$$\begin{aligned}
\partial_x M(0) &= \frac{1}{\lambda L}M(0) + \alpha k_f \partial_x \chi(0) - \frac{1}{\lambda L}k_f\,\chi(0) \\
\partial_x M(L) &= -\frac{1}{\lambda L}M(L) + \alpha k_f \partial_x \chi(L) + \frac{1}{\lambda L}k_f\,\chi(L)
\end{aligned} \tag{18}$$

A two phases mixture model can be also formulated by following the stress-driven approach, as shown below

$$\chi(x) = \alpha\left(\frac{M}{k_f}\right)(x) + (1-\alpha)\int_0^L \Phi_\lambda(x,\bar{x})\left(\frac{M}{k_f}\right)(\bar{x})d\bar{x} \tag{19}$$

In this case, for a fixed α, the response exhibits a stiffening behavior for increasing the nonlocal parameter λ and for any $\alpha \in [0, 1]$ the model is always well-posed.

The equivalent differential problem is expressed by the second-order differential equation as follow

$$\partial_x^2\chi(x) - \frac{1}{(\lambda L)^2}\chi(x) = -\frac{1}{k_f(\lambda L)^2}M(z) + \frac{\alpha}{k_f}\partial_x^2 M(x) \tag{20}$$

with the following constitutive BCs

$$\partial_x \chi(0) = \frac{1}{\lambda L}\chi(0) + \frac{\alpha}{k_f}\partial_x M(0) - \frac{\alpha}{k_f \lambda L} M(0)$$
$$\partial_x \chi(L) = -\frac{1}{\lambda L}\chi(L) + \frac{\alpha}{k_f}\partial_x M(L) + \frac{\alpha}{k_f \lambda L} M(L) \tag{21}$$

3 Displacement Based Nonlocal Model

A mechanically based approach to nonlocal elasticity is introduced in this section [31]. Specifically, we recall the displacement based model considered in Refs. [36, 45, 48, 49]. This formulation follows an alternative approach compared to the previous one. The local continuum is enriched by physically consistent long-range interactions depending on the relative displacements between non-adjacent volume elements. According to this model, the continuous domain is divided into volume elements exchanging contact and long distance interactions. Specifically, adjacent elements exchange classical contact forces while non-adjacent elements interact by means of long range forces. These nonlocal interactions depend on the relative displacements between two volume elements, on a distance-decaying function and on volumes product. Following this approach, the long-range force mutually exerted by two non adjacent infinitesimal volume elements is written as

$$df\,(\boldsymbol{x}, \boldsymbol{\xi}) = q\,(\boldsymbol{x}, \boldsymbol{\xi})\,dV(\boldsymbol{x})dV(\boldsymbol{\xi}) = \left[u(\boldsymbol{\xi}) - u(\boldsymbol{x})\right] g\,(\boldsymbol{x}, \boldsymbol{\xi})\,dV(\boldsymbol{x})dV(\boldsymbol{\xi}) \tag{22}$$

where $g\,(\boldsymbol{x}, \boldsymbol{\xi})$ is an attenuation function that rules the decaying of the long-range interactions. For the thermodynamics consistency, the model requires that $g\,(\boldsymbol{x}, \boldsymbol{\xi}) \geqslant 0$ and $g\,(\boldsymbol{x}, \boldsymbol{\xi}) = g\,(\boldsymbol{\xi}, \boldsymbol{x})$ [31, 48]. Considering the long-range interaction in Eq. (22) into the equilibrium equations, the nonlocal effects are captured. As it is shown in the following, the equilibrium problem is ruled by integro-differential equations in terms of the displacement field.

3.1 Nonlocal Rod

Now, a displacement based approach is applied with reference to a one-dimensional continuum for sake of simplicity. We consider the elastic nonlocal rod depicted in Fig. 1, and, as first step, we focus on unbounded domains by considering infinite length $L \to \infty$.

The bar is discretized into an infinite number of volume elements $V_j = A\Delta x$, $j = -\infty \ldots \infty$, where the $x-$axis coincides with the bar axis and A denotes the section area.

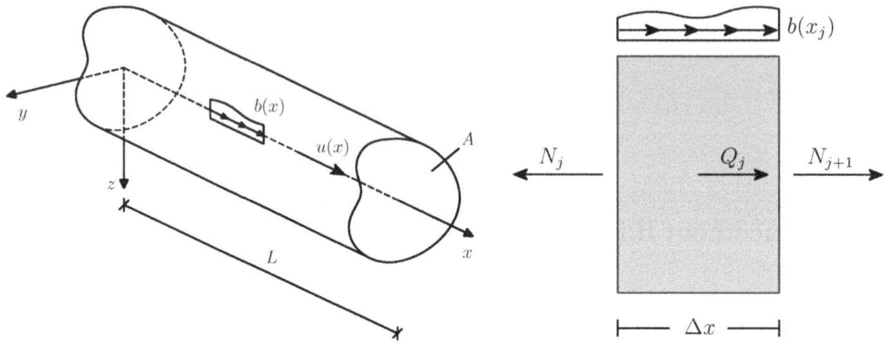

Fig. 1 Nonlocal elastic rod

Considering the Fig. 1, the equilibrium equation of the arbitrary volume element V_j is expressed as follows

$$\Delta N_j + Q_j + b(x_j)A\Delta x = 0 \tag{23}$$

where $b(x_j)$ is the axial volume force, ΔN_j is the difference between the contact forces and Q_j is the resultant of long-range interactions applied to the centroid of volume element V_j. That is,

$$Q_j = \sum_{h=-\infty, h\neq j}^{\infty} Q^{(h,j)} = \sum_{h=-\infty, h\neq j}^{\infty} q^{(h,j)}\, V_h\, V_j \tag{24}$$

where $q^{(h,j)}$ represents the specific long-range interaction between the volume elements V_h and V_j shown in Fig. 2. This force is expressed as

$$q^{(h,j)} = (u_h - u_j)\, g(x_h, x_j) \tag{25}$$

where u is the axial displacement of the rod and $g(x_h, x_j)$ is a positive and symmetric attenuation function. In virtue of Eq. (24), the equilibrium in Eq. (23) can be rewritten as

$$\Delta N_j + \sum_{h=-\infty, h\neq j}^{\infty} q^{(h,j)}\, (A\Delta x)^2 + b(x_j)\, \Delta x = 0 \tag{26}$$

Dividing Eq. (26) by Δx and taking the limit $\Delta x \to 0$ lead to the following integro-differential equation

$$A\partial_x \sigma_l(x) + A^2 \int_{-\infty}^{\infty} q(x, \xi)\, d\xi = -b(x) \tag{27}$$

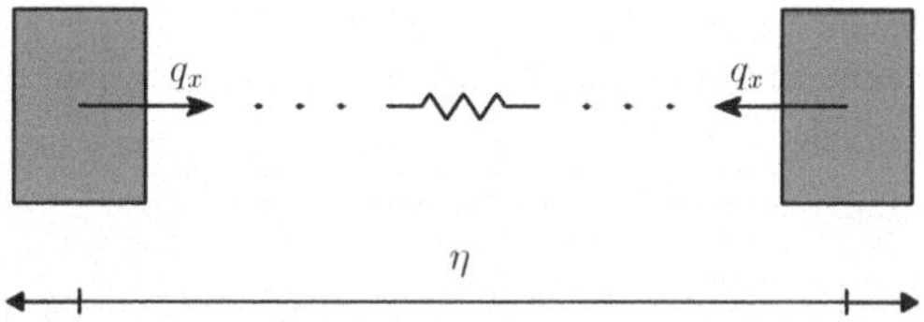

Fig. 2 Elastic long-range interaction

where σ_l denotes the local Cauchy stress while the second term at the left-hand side is related to the nonlocal interactions.

In Eq. (27), $q(x, \xi)$ is the long-rang force exerted by the volume at abscissa ξ on the element located at x. That is,

$$q(x, \xi) = [u(\xi) - u(x)]\, g(x, \xi) \tag{28}$$

which represents the nonlocal constitutive relation. By taking into account equation (28), the classical local stress-strain law and the strain-displacement relation $\varepsilon(x) = \frac{du}{dx}$, it is possible to rewrite the equilibrium equation (27) in terms of axial displacement u. That is,

$$EA\, \partial_x^2 u(x) + A^2 \int_{-\infty}^{\infty} [u(\xi) - u(x)]\, g(x, \xi)\, d\xi = -b(x) \tag{29}$$

If a rod of finite length L, divided into m volume elements $V_j = A\Delta x = AL/m$, is considered, the equilibrium equation (23) for $\Delta x \to 0$ takes the form

$$EA\, \partial_x^2 u(x) + A^2 \int_{0}^{L} [u(\xi) - u(x)]\, g(x, \xi)\, d\xi = -b(x) \tag{30}$$

3.1.1 Discrete Model of the Nonlocal Rod

An equivalent discrete mechanical model of the displacement based nonlocal rod is introduced in this section. The idea is that contact forces can be modeled by springs of stiffness $K^l = EA/\Delta x = EAm/L$, while the nonlocal forces can be represented by springs with distance-decaying stiffness. That is,

$$K_{jh}^{nl} = (A\Delta x)^2\, g(x_j, x_h) \tag{31}$$

Hence, the equilibrium equations of the point-spring model in Fig. 3 are expressed as follows

$$(\mathbf{K}^l + \mathbf{K}^{nl})\mathbf{u} = \mathbf{f} \tag{32}$$

where $\mathbf{K}^l$ is the following symmetric tridiagonal matrix.

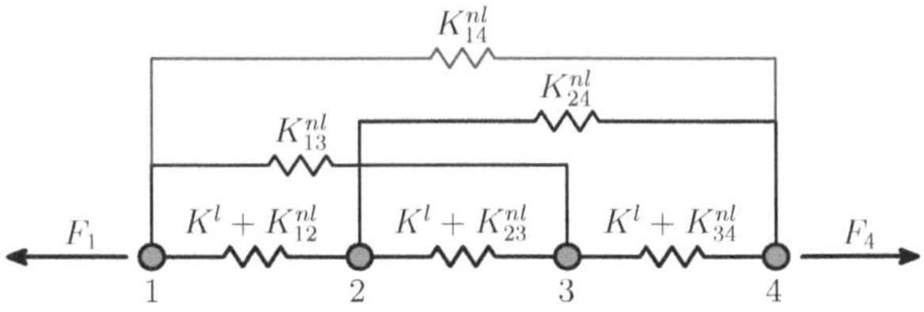

Fig. 3 Discrete elastic scheme of the local and nonlocal interactions

$$\mathbf{K}^l = \begin{bmatrix} K^l & -K^l & 0 & 0 & \dots & 0 \\ & 2K^l & -K^l & 0 & \dots & 0 \\ & & \dots & \dots & \dots & \dots \\ & & & & 2K^l & -K^l \\ & & & & & K^l \end{bmatrix} \tag{33}$$

while $\mathbf{K}^{nl}$ is a symmetric and full matrix. That is,

$$\mathbf{K}^{nl} = \begin{bmatrix} K_{11}^{nl} & -K_{12}^{nl} & -K_{13}^{nl} & \dots & \dots & -K_{1m}^{nl} \\ & K_{22}^{nl} & -K_{23}^{nl} & \dots & \dots & -K_{2m}^{nl} \\ & & \dots & \dots & \dots & \dots \\ & & & & K_{m-1m-1}^{nl} & -K_{m-1m}^{nl} \\ & & & & & K_{mm}^{nl} \end{bmatrix} \tag{34}$$

and $\mathbf{f}^T = \Delta x[F_1 \ \dots \ F_m]$ is the load vector in which $F_j = Ab(x_j)$ are the external nodal forces per unit length.

Kinematic and/or static boundary conditions can be imposed by assigning the first and last components of the displacement and load vectors.

By taking the limit $\Delta x \to 0$, Eq. (30) can be obtained from Eq. (32), that represents a discretized version of the previous model.

3.1.2 Differential Formulation of the Nonlocal Problem

In order to obtain an equivalent differential formulation of the integro-differential model of the nonlocal rod, Eq. (30) is firstly rewritten as

$$EA\,\partial_x^2 u(x) + A^2 \int_0^L u(\xi)\, g(x,\xi)\, d\xi - A^2 u(x)\gamma(x) = -b(x) \tag{35}$$

where

$$\gamma(x) = \int_0^L g(x,\xi)\, d\xi \tag{36}$$

Now, by selecting the bi-exponential attenuation function of the type of Eq. (3), the corresponding differential formulation of the Eq. (35) can be found in a similar way to the previous section [15]. Specifically, by assuming that

$$g(x, \xi) = \frac{E_{nl}}{2\lambda L^5} \exp\left(-\frac{|x-\xi|}{\lambda L}\right) \tag{37}$$

the following differential equation

$$-EA\partial_x^4 u(x) - \left[\frac{EA}{(\lambda L)^2} + A^2\gamma(x)\right]\partial_x^2 u(x) + 2A^2\partial_x\gamma(x)\partial_x u(x) + \\ + \left[\frac{1}{(\lambda L)^2}\left(\frac{EA^2}{L^4} - A^2\gamma(x)\right) + A^2\gamma(x)\right]u(x) = -\frac{b(x)}{(\lambda L)^2} \tag{38}$$

with the following BCs

$$\begin{aligned} &EA\partial_x^3 u(0) - A^2\left[\gamma(0)\partial_x u(0) + \partial_x\gamma(0)u(0)\right] = \\ &\frac{1}{\lambda L}\left[EA\partial_x^2 u(0) - A^2\gamma(0)u(0) + b(0)\right] \\ &EA\partial_x^3 u(L) - A^2\left[\gamma(L)\partial_x u(L) + \partial_x^4\gamma(L)u(L)\right] = \\ &-\frac{1}{\lambda L}\left[EA\partial_x^2 u(L) - A^2\gamma(L)u(L) + b(L)\right] \end{aligned} \tag{39}$$

is equivalent to the integro-differential equation in Eq. (35).

The integro-differential equation in Eq. (35) requires some specific numerical algorithms for its solution [58]. On the contrary, Eq. (38) with the BCs in Eq. (39) provides an alternative formulation of the displacement based nonlocal problem that can be readily solved. For example, consider a micro-rod constrained at $x = 0$ with length $L = 300\,\mu m$, rectangular cross section with width $b = 60\,\mu m$ and thickness $h = 25\,\mu m$, elastic modulus $E = 2.80$ GPa, $E_{nl} = 28$ TPa, and forced at $x = L$ by a point load force $F = 20$ N.

The effect of the nonlocal parameter λ on the mechanical response of the rod is investigated. Figure 4 shows the normalized axial displacement for different values of the nonlocal parameter. Each displacement function is normalized with respect to the related maximum value at $x = L$.

Numerical results show that the increasing of nonlocal parameter yields an increment in terms of rod stiffness. Specifically, Fig. 5 shows the maximum displacement $u_{nl}(L)$ normalized respect to the local one $u_l(L)$ for different values of nonlocal parameter. We can observe that the displacement decreases if the nonlocal parameter grows up. This behavior may be easily explained by focusing on the attenuation function (37). The nonlocal parameter λ plays a double role, it scales the whole attenuation function and at the same time modifies the velocity of decaying with the distance. An increase in the value of λ results in a larger scaling of the function, i.e. with lower peak, and in a slower decaying with the distance. From the numerical application of Figs. 4 and 5, it may be asserted that the influence of λ in the

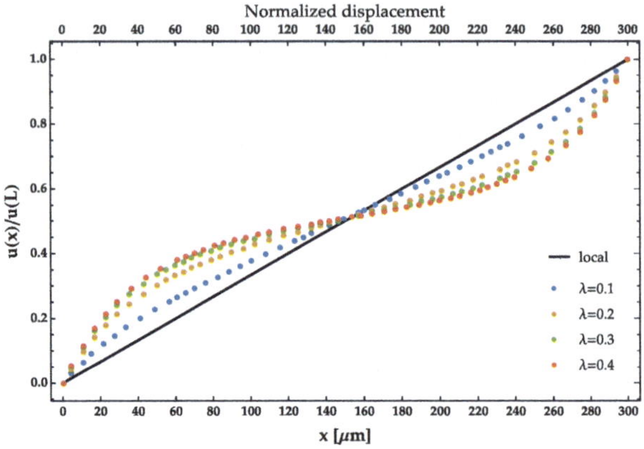

Fig. 4 Normalized displacement of nonlocal rod for different values of λ

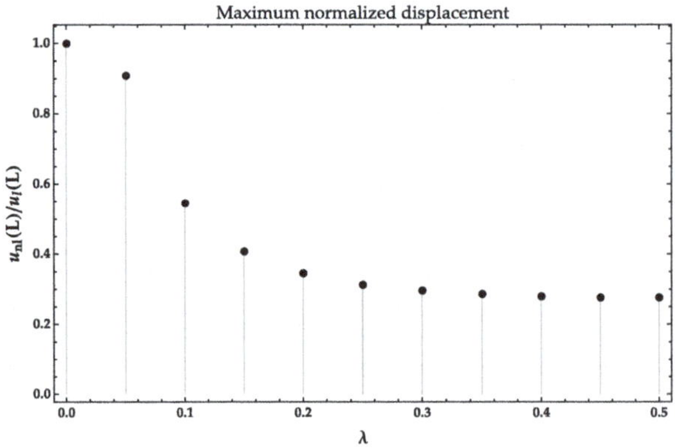

Fig. 5 Maximum normalized displacement for different values of λ

velocity of the attenuation function decaying is predominant and, as expected, the rod stiffness increases with λ. Further, a strain localization effect at the rod ends occurs. This effect is more significant for larger values of λ, and is due to the fact that volume elements in the neighborhood of the rod ends are involved in a smaller amount of nonlocal interactions compared with volume elements in the central part of the rod. The behavior of the nonlocal rod may be further manipulated by varying also the nonlocal modulus E_{nl} and/or introducing coefficients weighting the local and nonlocal contributions (similarly to Eq. (16)), resulting in a very flexible model.

3.2 Dynamical Problem of Nonlocal Beam Model

As shown in the previous section, long-range interactions can be modeled with a displacement based approach. With this approach, the nonlocal rod is readily obtained, but this mechanically based model can be applied also to more complex structural systems. In this section, we consider the dynamics of the nonlocal Timoshenko

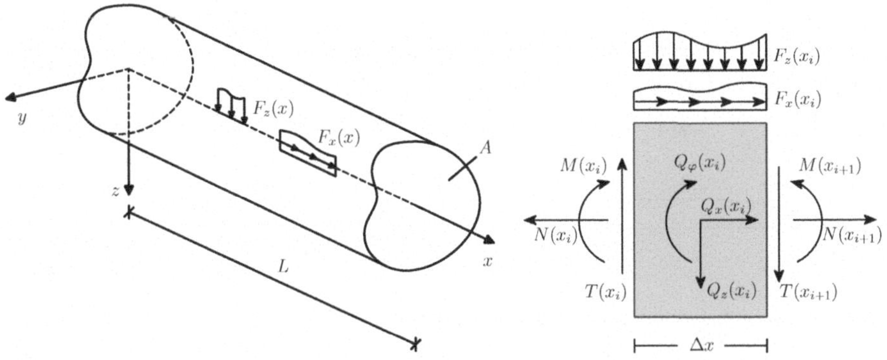

Fig. 6 Nonlocal elastic Timoshenko beam

(TM) beam model, derived below. The problem formulation is obtained following the previous approach, that is by introducing long-range interactions. However, to avoid that long-distance forces arise from relative displacements induced by a rigid motion, pure deformation modes are introduced.

We consider an isotropic and linearly elastic TM beam of length L, with centroidal axis x as depicted in Fig. 6.

In this case, two arbitrary volume elements $\Delta V(x_i)$ and $\Delta V(x_k)$ interact each other according to the pure deformation modes [59]. For the TM beam, the modes are expressed in terms of pure axial, pure bending and pure shear deformations. That is,

$$\begin{aligned}\eta(x_i, x_k) &= u(x_k) - u(x_i)\\ \theta(x_i, x_k) &= \varphi(x_k) - \varphi(x_i)\\ \psi(x_i, x_k) &= \left[\frac{v(x_k) - v(x_i)}{x_k - x_i} - \varphi(x_k)\right] + \left[\frac{v(x_k) - v(x_i)}{x_k - x_i} - \varphi(x_i)\right]\end{aligned} \tag{40}$$

where u denotes the axial displacement, φ is the rotation and v is the transverse displacement. The time dependence of the previous functions has been omitted for brevity. The axial mode is related to the rod and has been depicted in Fig. 2, while the other two are reported in Fig. 7.

According to the displacement based nonlocal model, the long-range forces and moments mutually exerted by unitary volume elements $\Delta V(x_i) = 1$ and $\Delta V(x_k) = 1$ are expressed by

$$q_x^{(i,k)} = g_x(x_i, x_k)\, \eta(x_i, x_k) \tag{41}$$

which represents the volume axial forces due to the pure axial deformation,

$$q_{\varphi,\varphi}^{(i,k)} = g_\varphi(x_i, x_k)\, \theta(x_i, x_k) \tag{42}$$

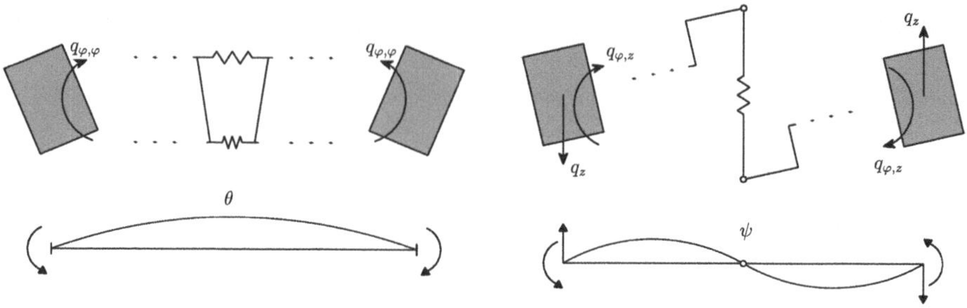

Fig. 7 Pure deformation modes for Timoshenko beam

representing the volume moments due to the pure bending mode, and finally

$$
\begin{aligned}
q_{\varphi,z}^{(i,k)} &= g_z(x_i, x_k)\, \psi(x_i, x_k) \\
q_z^{(i,k)} &= \frac{2}{x_k - x_i} g_z(x_i, x_k)\, \psi(x_i, x_k)
\end{aligned}
\tag{43}
$$

which are the long-range moments and transverse forces related to pure shear mode. The functions $g_x(x_i, x_k) \neq g_\varphi(x_i, x_k) \neq g_z(x_i, x_k)$ are independent attenuation functions governing the distance decay of the nonlocal interactions.

Let us consider a discretization of the TM beam into n volumes $V(x_i) = A\Delta x$, where $x_i = i\Delta x$ for $i = 0, ...n-1$. We introduce external forces per unit length $F_x(x, t)$, $F_z(x, t)$ and denote with $Q_x(x, t)$, $Q_z(x, t)$, $Q_\varphi(x, t)$ the resultants of the long-range forces and moments. The dynamic equilibrium equations of the arbitrary volume element $V(x_i)$ are expressed as follows

$$N(x_i + \Delta x, t) - N(x_i, t) + Q_x(x_i, t) + F_x(x_i, t)\, \Delta x - m(x_i)\partial_t^2 u(x_i, t)\Delta x = 0 \tag{44a}$$

$$T(x_i + \Delta x, t) - T(x_i, t) + Q_z(x_i, t) + F_z(x_i, t)\, \Delta x - m(x_i)\partial_t^2 v(x_i, t)\Delta x = 0 \tag{44b}$$

$$M(x_i + \Delta x, t) - M(x_i, t) - T(x_i)\Delta x - Q_\varphi(x_i, t) + I_p(x_i)\partial_t^2 \varphi(x_i, t)\Delta x = 0 \tag{44c}$$

where $m(x)$ is the mass per unit length and $I_p(x)$ is the rotational inertia per unit length. By performing simple manipulation of Eqs. (44) and taking the limit $\Delta x \to 0$ we get

$$EA\, \partial_x^2 u(x, t) + F_x(x, t) + A^2 \int_0^L g_x(x, \xi)\eta(x, \xi, t)d\xi - m(x)\partial_t^2 u(x, t) = 0 \tag{45a}$$

$$K_S GA[\partial_x^2 v(x,t) - \partial_x \varphi(x,t)] + F_z(x,t) + A^2 \int_0^L \frac{2}{\xi - x)} g_z(x,\xi)\psi(x,\xi,t)d\xi - m(x)\partial_t^2 v(x,t) = 0 \tag{45b}$$

$$K_S GA[\partial_x v(x,t) - \varphi(x,t)] + EI\partial_x^2 \varphi(x,t) + A^2 \int_0^L g_z(x,\xi)\psi(x,\xi,t)d\xi + A^2 \int_0^L g_\varphi(x,\xi)\theta(x,\xi,t)d\xi - I_p(x)\partial_t^2 \varphi(x,t) = 0 \tag{45c}$$

where the mechanical boundary conditions hold the classical form of local theory [36]. Obviously, in order to evaluate the solution of the problem in Eq. (45), a numerical procedure must be adopted. For this reason, next section shows a way to apply the finite element method to this kind of nonlocal problem.

3.2.1 Finite Element Formulations

Let us consider a beam subdivided in a mesh of n finite elements. The nodes of the i-th element are denoted as x_i and x_{i+1} and $\mathbf{d}_i$ is the vector of nodal displacements/rotation of the element. The displacement field along the finite element is expressed by the following vector

$$\mathbf{u}_i(x,t) = \mathbf{N}_i(x)\boldsymbol{d}_i(t) \qquad i = 1, ..., n \tag{46}$$

where $\mathbf{u}_i(x,t) = [u(x,t), v(x,t), \varphi(x,t)]^T$, $\boldsymbol{d}_i(t)$ is the vector collecting the nodal displacements of the i-th element and $\mathbf{N}_i(x)$ is the matrix of shape functions, taken as the standard first-order (axial problem) and third-order (bending problem) polynomial expressions. $\mathbf{N}_i^T(x)$ is given by

$$\mathbf{N}_i^T(x) = \begin{bmatrix} \frac{x_{i+1}-x}{l} & 0 & 0 \\ 0 & \frac{(l-y_i)(l^2(1+12\Omega)+(l-2y_i)y_i)}{l^3(1+12\Omega)} & \frac{6y_i(-l+y_i)}{l^3(1+12\Omega)} \\ 0 & \frac{(l-y_i)(l+6l\Omega-y_i)y_i}{l^2(1+12\Omega)} & \frac{(l+12l\Omega-3y_i)(l-y_i)}{l^2(1+12\Omega)} \\ \frac{x-x_i}{l} & 0 & 0 \\ 0 & \frac{y_i(12l^2\Omega+3ly_i-2y_i^2)}{l^3(1+12\Omega)} & \frac{6(l-y_i)y_i}{l^3(1+12\Omega)} \\ 0 & -\frac{(l-y_i)(6l\Omega+y_i)y_i}{l^2(1+12\Omega)} & \frac{2l(-1+6\Omega)+3y_i)y_i}{l^2(1+12\Omega)} \end{bmatrix} \tag{47}$$

where $y_i = x - x_i$, l is the length of the finite elements and $\Omega = EI/GAl^2$. Furthermore, $\mathbf{d}_i$ can be expressed as

$$\boldsymbol{d}_i(t) = \mathbf{C}_i \boldsymbol{d}(t) \tag{48}$$

where $\mathbf{C}_i$ is the connectivity matrix and $\boldsymbol{d}(t)$ denotes the nodal displacements vector of the whole beam.

With the standard Galerkin approach, we get the equation of motion in the form

$$\mathbf{M}\partial_t^2 \boldsymbol{d}(t) + [\mathbf{K}^l + \mathbf{K}^{nl}]\boldsymbol{d}(t) = \mathbf{F}(t) \tag{49}$$

where $\mathbf{M}$ is $3(n+1) \times 3(n+1)$ mass matrix and $\mathbf{F}(t)$ is the nodal load vector. $\mathbf{K}^l + \mathbf{K}^{nl}$ is $3(n+1) \times 3(n+1)$ global stiffness matrix, i.e.:

$$\mathbf{K}^l + \mathbf{K}^{nl} = \sum_{i=1}^{n} \mathbf{K}_i^l + \sum_{i=1}^{n} \mathbf{K}_i^{nl} \tag{50}$$

In particular, $\mathbf{K}_i^l$ is the classical local stiffness matrix, given by

$$\mathbf{K}_i^l = \int_{x_i}^{x_{i+1}} [\mathbf{B}_i(x)\mathbf{C}_i]^T \, \mathbf{D} \, \mathbf{B}_i(x)\mathbf{C}_i \, dx \tag{51}$$

where $\mathbf{D} = \mathrm{Diag}[EA \;\; EI \;\; GK_sA]$ and $\mathbf{B}_i^T(x)$ is the following matrix

$$\mathbf{B}_i^T(x) = \begin{bmatrix} -\frac{1}{l} & 0 & 0 \\ 0 & \frac{6(-2l^2\Omega - ly_i + y_i^2)}{l^3(1+12\Omega)} & -\frac{6(l-2y_i)}{l^3(1+12\Omega)} \\ 0 & \frac{l^2(1+6\Omega) - 4(l+3l\Omega)y_i + 3y_i^2}{l^2(1+12\Omega)} & \frac{-4(l+3l\Omega)+6y_i}{l^2(1+12\Omega)} \\ \frac{1}{l} & 0 & 0 \\ 0 & \frac{6(2l^2\Omega + (l-y_i)y_i)}{l^3(1+12\Omega)} & \frac{6(l-2y_i)}{l^3(1+12\Omega)} \\ 0 & \frac{-6l^2\Omega + 2l(-1+6\Omega)y_i + 3y_i^2}{l^2(1+12\Omega)} & \frac{2(l(-1+6\Omega)+3y_i)}{l^2(1+12\Omega)} \end{bmatrix} \tag{52}$$

In addition, $\mathbf{K}_i^{nl}$ is the nonlocal stiffness matrix, depending on three contributions related to the corresponding pure modes, i.e.: $\mathbf{K}_i^{nl,\eta}$, $\mathbf{K}_i^{nl,\theta}$, $\mathbf{K}_i^{nl,\psi}$. That is,

$$\mathbf{K}_i^{nl} = \mathbf{K}_i^{nl,\eta} + \mathbf{K}_i^{nl,\theta} + \mathbf{K}_i^{nl,\psi} = \sum_{j=1}^{n} \mathbf{K}_{ij}^{nl,\eta} + \sum_{j=1}^{n} \mathbf{K}_{ij}^{nl,\theta} + \sum_{j=1}^{n} \mathbf{K}_{ij}^{nl,\psi} \tag{53}$$

where the nonlocal matrices $\mathbf{K}_{ij}^{nl,\eta}$, $\mathbf{K}_{ij}^{nl,\theta}$, $\mathbf{K}_{ij}^{nl,\psi}$ are given as

$$\mathbf{K}_{ij}^{nl,\eta} = \frac{A^2}{2}\int_{x_i}^{x_{i+1}}\int_{x_j}^{x_{j+1}} [\mathbf{N}_j^{(u)}(\xi)\mathbf{C}_j - \mathbf{N}_i^{(u)}(x)\mathbf{C}_i]^T g_x(x,\xi) \times[\mathbf{N}_j^{(u)}(\xi)\mathbf{C}_j - \mathbf{N}_i^{(u)}(x)\mathbf{C}_i] dx d\xi \tag{54a}$$

$$\mathbf{K}_{ij}^{nl,\theta} = \frac{A^2}{2}\int_{x_i}^{x_{i+1}}\int_{x_j}^{x_{j+1}} [\mathbf{N}_j^{(\varphi)}(\xi)\mathbf{C}_j - \mathbf{N}_i^{(\varphi)}(x)\mathbf{C}_i]^T g_\varphi(x,\xi) \times[\mathbf{N}_j^{(\varphi)}(\xi)\mathbf{C}_j - \mathbf{N}_i^{(\varphi)}(x)\mathbf{C}_i] dx d\xi \tag{54b}$$

$$\mathbf{K}_{ij}^{nl,\psi} = \frac{A^2}{2}\int_{x_i}^{x_{i+1}}\int_{x_j}^{x_{j+1}} \left[2\frac{\mathbf{N}_j^{(v)}(\xi)\mathbf{C}_j - \mathbf{N}_i^{(v)}(x)\mathbf{C}_i}{\xi - x} - \mathbf{N}_j^{(\varphi)}(\xi)\mathbf{C}_j - \mathbf{N}_i^{(\varphi)}(x)\mathbf{C}_i\right]^T g_z(x,\xi)\left[2\frac{\mathbf{N}_j^{(v)}(\xi)\mathbf{C}_j - \mathbf{N}_i^{(v)}(x)\mathbf{C}_i}{\xi - x} - \mathbf{N}_j^{(\varphi)}(\xi)\mathbf{C}_j - \mathbf{N}_i^{(\varphi)}(x)\mathbf{C}_i\right] dx d\xi \tag{54c}$$

In Eqs. (54), $\mathbf{N}_i^{(u)}, \mathbf{N}_i^{(v)}, \mathbf{N}_j^{(\varphi)}$ are the row vectors of the matrix $\mathbf{N}_i$ of shape functions. Hence, the nonlocal matrices $\mathbf{K}_{ij}^{nl,\eta}$, $\mathbf{K}_{ij}^{nl,\theta}$, $\mathbf{K}_{ij}^{nl,\psi}$ express the contributions due to the long-range interactions between the differential volumes $dV(x)$ inside the i-th element, $x_i \leqslant x \leqslant x_{i+1}$, and the differential volumes $dV(\xi)$ inside the j-th element, $x_j \leqslant \xi \leqslant x_{j+1}$.

Finally, it is important to remark that unlike the local stiffness matrix, the nonlocal one is fully-populated and each element of $\mathbf{K}^{nl}$ can be obtained in closed-form by choosing an appropriate attenuation function such as exponential or power-law functions [44].

3.2.2 Dynamical Analysis of Nonlocal Beams

In this section, the dynamics of the nonlocal beam in bending is investigated. The quasi-static behavior of the nonlocal beam in bending mirrors the axial behavior described in Sect. 3.2.1 and is not analyzed for brevity. Numerical simulation are performed in order to the obtain the natural frequencies and the modal shapes of the displacement based beam model described above. The finite element formulation described in Sect. 3.2.1 has been adopted to model the beam, discretized with 40 elements. The consistent mass matrix is adopted, derived assuming the third order kinematic model of Eq. (47) and density equal to $\rho = 1000\,\text{kg/m}^3$. Natural frequency and modal shape are evaluated by means of a standard eigen-analysis.

The geometry and parameters of the beam are equal to those of the rod considered in the numerical application of Figs. 4 and 5. Two boundary conditions are considered: clamped-free beam and simply-supported beam. Natural frequencies and modal shapes are evaluated for different values of the nonlocal parameter, that is $\lambda = 0,\ 0.05,\ 0.10,\ 0.15,\ 0.20$.

In Fig. 8, the first four modal frequencies for the clamped-free beam are plotted as functions of the nonlocal parameter λ. It is shown that all the natural frequencies considered increase with λ. Further, the behavior of all frequencies with λ is non-linear and characterized by a decreasing rate with increasing λ. These results are in agreement with the mechanics of the model described in this work, since the introduction of long-range springs generates an increase in the stiffness of the beam. As expected, this stiffening effect is more evident for higher values of λ.

Fig. 8 First four natural frequencies of the clamped-free beam versus λ

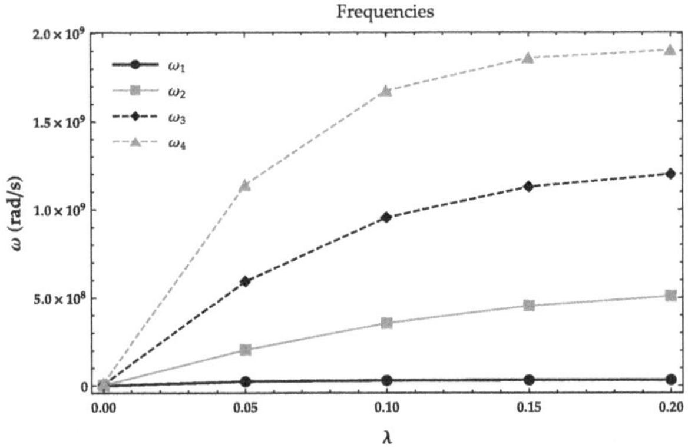

In Fig. 9, the first two modal shapes for the clamped-free beam are plotted for various values of λ. From this figure, it may be asserted that the general features of modal shapes are not drastically modified by the nonlocal interactions. However, close inspection of Fig. 9 reveals that as λ increases a deformation concentration in the neighborhood of the clamped end of the beam is observed. This is due to the fact that the volume elements in the central part of the beam are stiffened by a larger amount of nonlocal interactions than those in the neighborhood of the beam ends. Similar comments hold for the second modal shape in Fig. 9. Nevertheless, the global effect of increasing λ is a stiffening effect, as evidenced by analyzing the natural frequency trend in Fig. 8.

Figure 10 shows the first four natural frequencies of a simply supported beam as a function of the nonlocal parameter λ. As expected also in this case, the nonlocal parameter λ governs the stiffening effect of the nonlocal interactions on the beam. In Fig. 11, the first two modal shapes for the simply supported beam are depicted for various values of λ. Similar to the case of clamped-free beam, the modal shapes are not drastically modified, but a deformation concentration is observed close to the beam ends.

From the numerical results of Figs. 8, 9, 10 and 11, it may be concluded that the displacement based approach to nonlocal elasticity analyzed in this chapter has a predictable behavior also in dynamical conditions. Moreover, it has to be noted that in this numerical applications only the influence of the nonlocal parameter λ on the response has been investigated. Indeed, varying also the nonlocal modulus E_{nl} and/or the local elastic modulus allows for a wide range of nonlocal behaviors, making the displacement based nonlocal models a flexible tool in various nanomechanics applications.

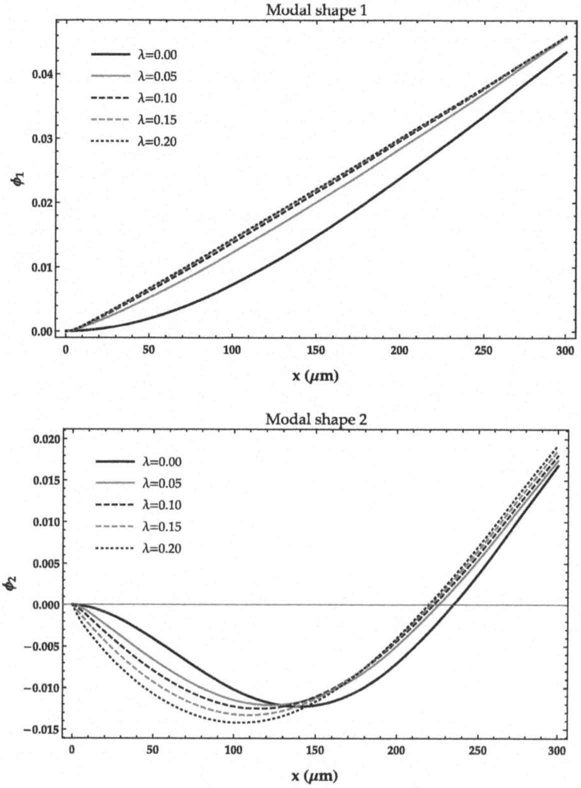

Fig. 9 Clamped-free beam: modal shapes for various values of λ

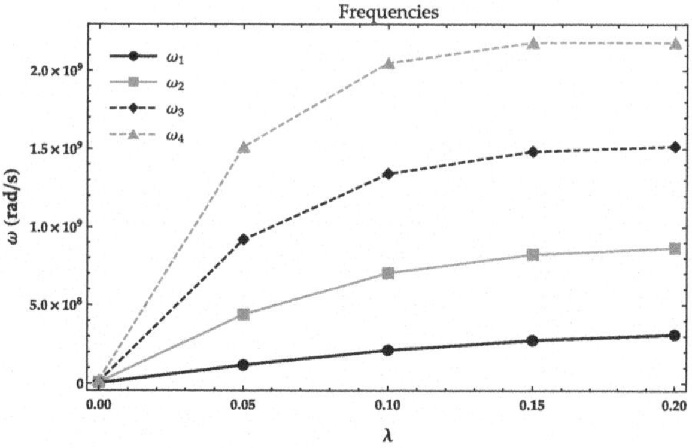

Fig. 10 First four natural frequencies of the simply supported beam versus λ

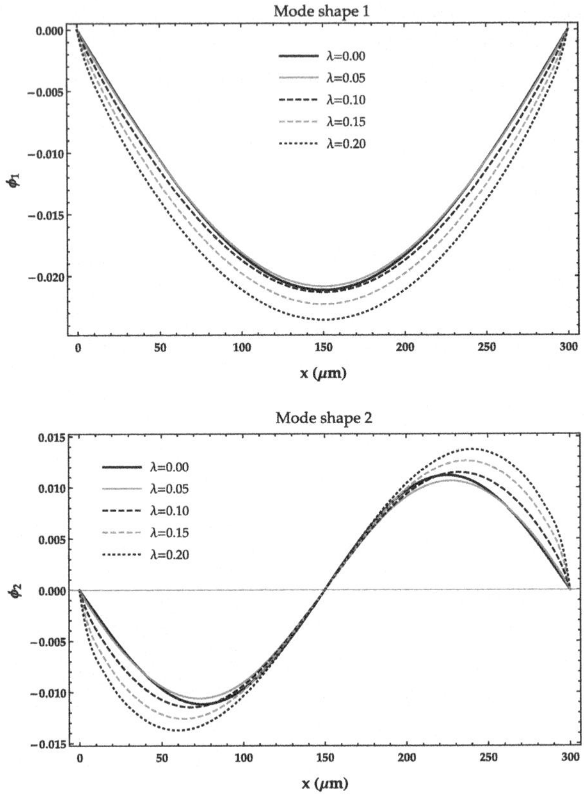

Fig. 11 Simply supported beam: modal shapes for various values of λ

4 Concluding Remarks

In this chapter, the so-called displacement based nonlocal model has been discussed. Differently from the other integral approaches to nonlocality, this formulation is based on a clear mechanical representation of nonlocal forces. More specifically, the nonlocal forces mutually exerted between two non-adjacent volume elements are thought as linearly depending on the product between the volumes, the relative displacement and an attenuation function accounting for the distance between the volume elements. The discussion about the model has been specialized to axial and bending behavior of a Timoshenko beam model. In this context it is shown that the model is suitable for dynamical problem formulation of nonlocal beams, as well as for a finite element formulation with closed form of the elements of the nonlocal stiffness matrices. Finally, it has been shown that the integro-differential equation ruling the elasto-static problem for a displacement based rod may be reverted to an equivalent differential equation with additional constitutive boundary conditions.

References

1. Navier CLMH (1827) Mémoire sur les lois de l'équilibre et du mouvement des corps solides élastiques. Mem Ac Sc In Fr 7:375–393
2. Poisson CLMH (1828) Mémoire sur l'équilibre et du mouvement des corps élastiques. Mem Ac Sc In Fr 8:357–570
3. Cauchy AL (1829) Sur l'équilibre et le mouvement intérieur des corps considérés comme des masses continues. Exer Math 4:293–319
4. Rogula D (1965) Influence of spatial acoustic dispersion on dynamical properties of dislocations. Bull Acad Pol Sci Ser Sci Tech 13:337–343
5. Kröner E, Datta BK (1966) Nichtlokal Elastostatik: Ableitung aus der Gittertheorie. Z Phys 196:203–21
6. Kunin IA (1966) Model of elastic medium with simple structure and space dispersion. Prikl Mat Mekh 30:542–550
7. Kröner E (1967) In Mechanics of Generalized Continua: Proceedings of the IUTAM-Symposium on The Generalized Cosserat Continuum and the Continuum. Theory of Dislocations with Applications, Freudenstadt and Stuttgart (Germany), E Kroner, Editor. Springer Berlin Heidelberg, Berlin, Heidelberg 1968, 330–340
8. Rogula D (1892) Introduction to nonlocal theory of material media. In: Rogula D (ed) Nonlocal theory of material media. CISM courses and lectures, vol 268. Springer, Wien, pp 125–222
9. Lakes RS (1991) Experimental micro mechanics methods for conventional and negative Poissons ratio cellular solids as Cosserat continua. J Eng Mater Tech 113:148–155
10. Arash B, Wang Q (2012) A review on the application of nonlocal elastic models in modeling of carbon nanotubes and graphenes. Comput Mater Sci 51:303–313
11. Eringen AC (1972) Linear theory of nonlocal elasticity and dispersion of plane waves. Int J Eng Sci 10:425–435
12. Eringen AC (1983) On differential equations of nonlocal elasticity and solutions of screw dislocation and surface waves. J Appl Phys 54:4703–4710
13. Challamel N, Wang CM (2008) The small length scale effect for a non-local cantilever beam: a paradox solved. Nanotechnology 19:345703
14. Fernández-Sáez J, Zaera R, Loya JA, Reddy JN (2016) Bending of Euler-Bernoulli beams using Eringen's integral formulation: A paradox resolved. Int J Eng Sci 99:107–116
15. Romano G, Barretta R, Diaco M, Marotti de Sciarra F (2017) Constitutive boundary conditions and paradoxes in nonlocal elastic nano-beams. Int J Mech Sci 121:151–156
16. Mahmoud FF (2017) On the non-existence of a feasible solution in the context of the differential form of Eringen's constitutive model: A proposed iterative model based on a residual nonlocality formulation. Int J Appl Mech 9:17594
17. Benvenuti E, Simone A (2013) One-dimensional nonlocal and gradient elasticity: closed-form solution and size effects. Mech Res Commun 48:46–51
18. Polizzotto C (2001) Nonlocal elasticity and related variational principles. Int J Solids Struct 38:7359–80
19. Pisano AA, Fuschi P (2003) Closed form solution for a nonlocal elastic bar in tension. Int J Solids Struct 40:13–23
20. Marotti de Sciarra F (2008) On non-local and non-homogeneous elastic continua. Int J Solids Struct 46:651–676
21. Borino G, Failla B, Parrinello F (2003) A symmetric nonlocal damage theory. Int J Solids Struct 40:3621–45
22. Polizzotto C, Fuschi P, Pisano AA (2004) A strain-difference-based nonlocal elasticity model. Int J Solids Struct 41:2383–2401
23. Polizzotto C, Fuschi P, Pisano AA (2006) A non homogeneous nonlocal elasticity model. Eur J Mech A Solids 25:308–333
24. Fuschi P, Pisano AA, Polizzotto C (2019) Size effects of small-scale beams in bending addressed with a strain-difference based nonlocal elasticity theory. Int J Mech Sci 151:661–671

25. Romano G, Barretta R (2017) Nonlocal elasticity in nanobeams: the stress-driven integral model. Int J Eng Sci 115:14–27
26. Apuzzo A, Barretta R, Luciano R, Marotti de Sciarra F, Penna R (2017) Free vibrations of Bernoulli-Euler nano-beams by the stress-driven nonlocal integral model. Compos Part B 123:105–111
27. Barretta R, Canadija M, Feo L, Luciano R, Marotti de Sciarra F, R Penna R (2018) Exact solutions of inflected functionally graded nano-beams in integral elasticity. Compos Part B 142:273–286
28. Barretta R, Fazelzadeh SA, Feo L, Ghavanloo E, Luciano R (2018) Nonlocal inflected nano-beams: A stress-driven approach of bi-Helmholtz type. Compos Struct 200:239–245
29. Barretta R, Caporale A, Faghidian SA, Luciano R, Marotti de Sciarra F, Medaglia CM (2019) A stress-driven local-nonlocal mixture model for Timoshenko nano-beams. Compos. Part B 164:590–598
30. Barretta R, Fabbrocino F, Luciano R, Marotti de Sciarra F (2018) Closed-form solutions in stress-driven two-phase integral elasticity for bending of functionally graded nano-beams. Physica E 97:13–30
31. Di Paola M, Failla G, Pirrotta A, Sofi A, Zingales M (2013) The mechanically based non-local elasticity: an overview of main results and future challenges. Philos T R Soc, A 371:20120433
32. Aifantis EC (1994) Gradient effects at macro, micro and nano scales. J Mech Behav Mater 5:355–375
33. Aifantis EC (2003) Update on a class of gradient theories. Mech Mater 35:259–280
34. Tarasov VE, Aifantis EC (2015) Non-standard extensions of gradient elasticity: Fractional nonlocality, memory and fractality. Commun Nonlinear Sci Numer Simulat 22:197–227
35. Lim CW, Zhang G, Reddy JN (2015) A higher-order nonlocal elasticity and strain gradient theory and its applications in wave propagation. J Mech Phys Solids 78:298–313
36. Di Paola M, Zingales M (2008) Long-range cohesive interactions of nonlocal continuum mechanics faced by fractional calculus. Int J Solids Struct 45:5642–5659
37. Carpinteri A, Cornetti P, Sapora A (2011) A fractional calculus approach to nonlocal elasticity. Eur Phys J 193:193–204
38. Drapaca CS, Sivaloganathan S (2012) A fractional model of continuum mechanics. J Elast 107:105–123
39. Challamel N, Zorica D, Atanackovič TM, Spasič DT (2013) On the fractional generalization of Eringen's nonlocal elasticity for wave propagation. C R Mécanique 341:298–303
40. Carpinteri A, Cornetti P, Sapora A (2014) Nonlocal elasticity: an approach based on fractional calculus. Meccanica 49:2551–2569
41. Tarasov VE (2018) No nonlocality. No fractional derivative. Commun Nonlinear Sci Numer Simul 62:157–163
42. Cottone G, Di Paola M, Zingales M (2009) Elastic waves propagation in 1D fractional non-local continuum. Physica E 42:95–103
43. Sapora A, Cornetti P, Carpinteri A (2013) Wave propagation in nonlocal elastic continua modelled by a fractional calculus approach. Commun Nonlinear Sci Numer Simul 18:63–74
44. Alotta G, Failla G, Zingales M (2014) Finite element method for a nonlocal Timoshenko beam model. Finite Elem Anal Des 89:77–92
45. Di Paola M, Failla G, Zingales M (2013) Non-local stiffness and damping models for shear-deformable beams. Eur J Mech A-Solids 40:69–83
46. Alotta G, Failla G, Pinnola FP (2017) stochastic analysis of a nonlocal fractional viscoelastic bar forced by gaussian white noise. ASCE-ASME J Risk Uncertain Eng Syst Part B 3:030904-030904-7 (2017)
47. Alotta G, Di Paola M, Failla G, Pinnola FP (2018) On the dynamics of non-local fractional viscoelastic beams under stochastic agencies. Compos Part B 137:102–110
48. Di Paola M, Pirrotta A, Zingales M (2010) Mechanically-based approach to non-local elasticity: variational principles. Int J Solids Struct 47:539–548
49. Failla G, Sofi A, Zingales M (2015) A new displacement-based framework for non-local Timoshenko beams. Meccanica 50:2103–2122

50. Silling SA (2000) Reformulation of elasticity theory for discontinuities and long-range forces. J Mech Phys Solids 48:175–209
51. Alotta G, Di Paola M, Pinnola FP, Zingales M (2020) A fractional nonlocal approach to non-linear blood flow in small-lumen arterial vessels. Meccanica 55:891–906
52. Fafalis DA, Filopoulos SP, Tsamasphyros GJ (2012) On the capability of generalized continuum theories to capture dispersion characteristics at the atomic scale. Eur J Mech A/Solids 36:25–37
53. Tsamasphyros GJ, Koutsoumaris CC (2016) Mixed nonlocal-gradient elastic materials with applications in wave propagation of beams. AIP Conf Proc 1790:150031
54. Mindlin RD (1963) Micro-structure in linear elasticity. Arch Ration Mech Anal 16:51–78
55. Mindlin RD (1965) Second gradient of strain and surface-tension in linear elasticity Int. J Solids Struct 1:414–438
56. Li L, Li X, Hu Y (2016) Free vibration analysis of nonlocal strain gradient beams made of functionally graded material. Int J Eng Sci 102:77–92
57. Barretta R, Marotti de Sciarra F (2018) Constitutive boundary conditions for nonlocal strain gradient elastic nano-beams. Int J Eng Sci 130:187–198
58. Press WH, Teukolsky SA, Vetterling et al WT (1997) Numerical recipes in fortran 77: the art of scientific computing. Cambridge University Press
59. Fuchs MB (1991) Unimodal beam elements. Int J Solids Struct 27:533–45

One-Dimensional Well-Posed Nonlocal Elasticity Models for Finite Domains

Mohammad Ali Maneshi, Esmaeal Ghavanloo, and S. Ahmad Fazelzadeh

Abstract Nonlocal modeling of physical phenomena is a very long history. Researchers in scientific and engineering communities increasingly recognize that the nonlocality is essential in realistic mathematical models of physical phenomena. Eringen's nonlocal elasticity is one of the most well-known nonlocal continuum theories. Eringen proposed two distinct forms of the nonlocal elasticity theory, namely the nonlocal integral form and the nonlocal differential form. In recent years, several inconsistencies and paradoxical results in the existing solutions of the nonlocal elasticity approaches have been revealed and they are not completely understood. To solve some paradoxical results, in this chapter, a well-posed nonlocal differential model for finite domains is developed and its applicability to predict the static behaviour of nanorods and nanobeams is investigated. We indicate that the proposed integral and differential nonlocal models are equivalent to each other over bounded continuous domains and the corresponding elastic problems are well-posed. In addition, some static problems are solved and we show that the paradoxical results disappear by using the present model.

M. A. Maneshi · E. Ghavanloo (✉) · S. A. Fazelzadeh
School of Mechanical Engineering, Shiraz University, 71963-16548 Shiraz, Iran
e-mail: ghavanloo@shirazu.ac.ir

M. A. Maneshi
e-mail: m.a.maneshi@shirazu.ac.ir

S. A. Fazelzadeh
e-mail: fazelzad@shirazu.ac.ir

S. A. Fazelzadeh
Visiting Professor in Department of Aerospace Science and Technology Politecnico di Milano, Via La Masa 34, 20156 Milan, Italy

© The Author(s), under exclusive license to Springer Nature Switzerland AG 2021
E. Ghavanloo et al. (eds.), *Size-Dependent Continuum Mechanics Approaches*,
Springer Tracts in Mechanical Engineering,
https://doi.org/10.1007/978-3-030-63050-8_5

1 Introduction

Eringen's nonlocal elasticity theory is one of the most computationally efficient approaches for modeling nanoscopic structures [1]. Theory of nonlocal elasticity has been proposed with two different approaches [2, 3]: nonlocal integral approach and nonlocal differential approach. Compared to the integral approach, the differential one is mostly utilized to model the mechanical characteristic of nanoscopic structures due to its simplicity [1].

In recent years, several inconsistencies and paradoxical results in the existing solutions of the nonlocal elasticity approaches have been revealed. In year 2003, the first attempt to solve the bending problem of beams on the basis of Eringen's nonlocal differential model was carried out by Peddieson et al. [4]. Their results showed that the nonlocal transvers deflections of clamped-free beam under a concentrated force are identical to the local ones. It is the first paradoxical results reported in the literature. The similar results were reported and discussed by Challamel and Wang [5] and Li et al. [6]. In addition, it was shown in the literature [3, 7, 8] that the local and the nonlocal solutions are identical for the following problems: clamped-clamped beams subjected to uniform or linear distributed loads, fixed-fixed rods subjected to uniform and/or linear distributed axial loads, fixed-free rods subjected to constant distributed axial loads.

The second paradox in differential-based nonlocal models appears when the free vibration of clamped-free beams is investigated. It was shown in several studies [9–11] that the fundamental frequency of a cantilever beam increases due to the nonlocal field while the other higher-mode frequencies decrease due to the nonlocal field. Another paradoxes is observed when the nonlocal integral form and the nonlocal differential form are compared to each other. Fernández-Sáez et al. [12] showed discrepancies between the nonlocal differential and the nonlocal integral beam models for various boundary conditions. They also claimed that these nonlocal models are not equivalent.

Various attempts have been made during the last decade to overcome the mentioned paradoxes. As the first strategy, application of the nonlocal integral type is a feasible option. However, in this case, there is only trivial solution and so the problem is not well-posed [13, 14]. To eliminate the ill-posedness of the fully nonlocal integral model, hybrid local-nonlocal model or two-phase nonlocal model has been used in the literature [5, 15–17]. Romano et al. [18] demonstrated that the hybrid local-nonlocal model may also be converted to ill-posed model when it is near to the pure nonlocal component. In another attempt, Challamel and co-workers [19, 20] developed lattice-based nonlocal model to overcome the paradoxes. In addition, Khodabakhshi and Reddy [21] suggested a general finite element formulation in conjunction with a two-phase integro-differential nonlocal model to solve the nonlocal paradoxes. Their model did not fully overcome the cantilever paradoxes and they also introduced the simply supported beam paradox. Shaat [22] demonstrated that the paradoxes associated with nonlocal differential models are eliminated using the iterative nonlocal residual approach.

Challamel et al. [23] attributed the nonlocal paradoxes to the fact that it is impossible to construct the underlying quadratic energy functional for such nonlocal beam models, as noticed first by Reddy [24]. Benvenuti and Simone [7] proved that the one-dimensional nonlocal differential model is not equivalent to the nonlocal integral model over the finite domain. Furthermore, Romano et al. [18] have introduced the concept of constitutive boundary conditions. They demonstrated that the transformation of the nonlocal integral form to its differential counterpart is ill-posed unless the extra constitutive boundary conditions (the constitutive boundary conditions were first represented by Benvenuti and Simone [7]) are fulfilled. In addition, they indicated that the inconsistency between the equilibrium conditions and the constitutive boundary conditions leads to the existing paradoxes between the integral and differential forms of the nonlocal theory. Furthermore, Koutsoumaris et al. [25] and Challamel [26] pointed out that many nonlocal strain measures developed in the literature violate the nonlocal invariance of the uniform strain field (normalization condition) and the paradoxical results are due to the violation of normalization condition of nonlocal kernel.

By reviewing the mentioned studies, it can be concluded that one of the most sources for the nonlocal paradoxical results is unsuitable selection of the nonlocal kernel. Motivated by this fact, in this chapter, we developed a one-dimensional well-posed nonlocal elasticity models by using a normalized kernel in a finite domain and proper transformation of the integral problem to the differential one. In addition, the applicability of the proposed model for static behavior of rods and beams is investigated.

2 One-Dimensional Well-Posed Nonlocal Elasticity Theory for Finite Domains

2.1 Nonlocal Integral Constitutive Equation

According to Eringen's nonlocal elasticity theory, one-dimensional constitutive equation for homogeneous-isotropic linear materials is expressed as follows [27]

$$\sigma (x) = \int_a^b \alpha(|x - \chi| , e_0 l_i) E\varepsilon (\chi) \, d\chi \tag{1}$$

where σ denotes the nonlocal stress, ε is strain, E is Young's modulus, α is known as nonlocal kernel and a and b are end-points of domain. In addition, e_0 and l_i are the nonlocal parameter and internal characteristic length, respectively. The following conditions must be satisfied by the nonlocal kernel [3]:

(a) The kernel function must reach a maximum at $x = \chi$ and kernel's value decreases monotonically to zero at large distances as well.

(b) The nonlocal kernel assume to be symmetric function with respect to x and χ.

(c) The kernel must converted to the Dirac delta function, δ, when the nonlocal parameter approaches zero.

(d) In order to preserve the uniform local strain field would produce a uniform nonlocal stress field, the kernel function must satisfy the normalization condition:

$$\int_a^b \alpha(|x-\chi|, e_0 l_i) d\chi = 1 \tag{2}$$

Polizzotto [28] and Borino et al. [29] proposed a nonlocal kernel function which satisfies all the mentioned conditions in a finite domain. The proposed kernel is

$$\begin{aligned} \alpha_{\text{mod}}(|x-\chi|, e_0 l_i) = [1 - \int_a^b \alpha(|x-\kappa|, e_0 l_i) d\kappa]\delta(|x-\chi|) \\ + \alpha(|x-\chi|, e_0 l_i) \end{aligned} \tag{3}$$

where $\alpha(|x-\chi|, e_0 l_i)$ is a kernel which satisfies all the properties of the nonlocal kernel in an infinite domain. Equation (3) has received scant attention and has only been employed by limited number of research groups. In this chapter, the bi-exponential kernel function, which is one of the most famous kernels in the infinite domains, is employed. This kernel is defined as follows.

$$\alpha(|x-\chi|, e_0 l_i) = \frac{1}{2e_0 l_i} \exp(-\frac{|x-\chi|}{e_0 l_i}) \tag{4}$$

Thus, the substitution of Eqs. (3) and (4) into Eq. (1), we have

$$\sigma(x) = g(x) E\varepsilon(x) + \frac{1}{2e_0 l_i} \int_a^b \exp(-\frac{|x-\chi|}{e_0 l_i}) E\varepsilon(\chi) d\chi \tag{5}$$

where

$$g(x) = \frac{1}{2}\left(\exp(\frac{a-x}{e_0 l_i}) + \exp(\frac{x-b}{e_0 l_i})\right) \tag{6}$$

2.2 Differential Constitutive Equation and Its Boundary Conditions

In this subsection, we transform the nonlocal integral equation, Eq. (5), into a nonlocal differential equation and corresponding boundary conditions. For this purpose, Eq. (5) is rewritten as

$$\sigma(x) = g(x)E\varepsilon(x) + \frac{1}{2e_0 l_i}\int_a^x \exp(\frac{\chi - x}{e_0 l_i})E\varepsilon(\chi)d\chi + \frac{1}{2e_0 l_i}\int_x^b \exp(\frac{x - \chi}{e_0 l_i})E\varepsilon(\chi)d\chi \tag{7}$$

Differentiating Eq. (7) with respect to x yields

$$\frac{d\sigma(x)}{dx} = E\left(\frac{dg(x)}{dx}\varepsilon(x) + g(x)\frac{d\varepsilon(x)}{dx}\right) - \frac{1}{2e_0^2 l_i^2}\int_a^x \exp(\frac{\chi - x}{e_0 l_i})E\varepsilon(\chi)d\chi + \frac{1}{2e_0^2 l_i^2}\int_x^b \exp(\frac{x - \chi}{e_0 l_i})E\varepsilon(\chi)d\chi \tag{8}$$

In order to simplify Eq. (8), we can again apply differentiation with respect to x. The resultant equation can be written in the form:

$$\frac{d^2\sigma(x)}{dx^2} = E\left(\frac{d^2g(x)}{dx^2}\varepsilon(x) + 2\frac{dg(x)}{dx}\frac{d\varepsilon(x)}{dx} + g(x)\frac{d^2\varepsilon(x)}{dx^2}\right) - \frac{1}{e_0^2 l_i^2}E\varepsilon(x) + \frac{1}{2e_0^3 l_i^3}\left(\int_a^x \exp(\frac{\chi - x}{e_0 l_i})E\varepsilon(\chi)d\chi + \int_x^b \exp(\frac{x - \chi}{e_0 l_i})E\varepsilon(\chi)d\chi\right) \tag{9}$$

Using Eqs. (5) and (6), we obtain

$$\frac{d^2\sigma(x)}{dx^2} - \frac{1}{e_0^2 l_i^2}\sigma(x) = Eg(x)\frac{d^2\varepsilon(x)}{dx^2} + 2E\frac{dg(x)}{dx}\frac{d\varepsilon(x)}{dx} + E\left(\frac{d^2g(x)}{dx^2} - \frac{g(x)}{e_0^2 l_i^2} - \frac{1}{e_0^2 l_i^2}\right)\varepsilon(x) \tag{10}$$

In addition, differentiating Eq. (6) with respect to x twice yields

$$\frac{d^2g(x)}{dx^2} = \frac{1}{e_0^2 l_i^2}g(x) \tag{11}$$

Substituting Eq. (11) into Eq. (10) and multiplying the resultant equation by $e_0^2 l_i^2$, we obtain the nonlocal differential constitutive equation

$$\sigma(x) - e_0^2 l_i^2\frac{d^2\sigma(x)}{dx^2} = E\varepsilon(x) - 2e_0^2 l_i^2 E\frac{dg(x)}{dx}\frac{d\varepsilon(x)}{dx} - e_0^2 l_i^2 Eg(x)\frac{d^2\varepsilon(x)}{dx^2} \tag{12}$$

which must be supplemented by the following boundary conditions

$$\frac{d\sigma(a)}{dx} = E\left(\frac{dg(a)}{dx}\varepsilon(a) + g(a)\frac{d\varepsilon(a)}{dx}\right) + \frac{1}{2e_0^2 l_i^2}\int_a^b \exp(\frac{a-\chi}{e_0 l_i})E\varepsilon(\chi)d\chi \tag{13}$$

$$\frac{d\sigma(b)}{dx} = E\left(\frac{dg(b)}{dx}\varepsilon(b) + g(b)\frac{d\varepsilon(b)}{dx}\right) - \frac{1}{2e_0^2 l_i^2}\int_a^b \exp(\frac{\chi-b}{e_0 l_i})E\varepsilon(\chi)d\chi \tag{14}$$

Now, let us eliminate the integral terms from Eqs. (13) and (14). Therefore, the boundary conditions are rewritten as

$$\sigma(a) - e_0 l_i \frac{d\sigma(a)}{dx} = E\left(\varepsilon(a) - e_0 l_i g\,(a)\,\frac{d\varepsilon(a)}{dx}\right) \tag{15}$$

$$\sigma(b) + e_0 l_i \frac{d\sigma(b)}{dx} = E\left(\varepsilon(b) + e_0 l_i g\,(b)\,\frac{d\varepsilon(b)}{dx}\right) \tag{16}$$

It should be noted that the same equations can be derived in Ref. [26]. Equations (15) and (16) are higher-order boundary conditions which known as constitutive boundary conditions [18, 30]. Note that the nonlocal differential constitutive equation (Eq. (12)) with boundary conditions (Eqs. (15) and (16)) is equivalent to the nonlocal integral constitutive equation (Eq. (5)) in the finite domain. Furthermore, since the constitutive boundary conditions, do not contrast with equilibrium requirements, the present nonlocal model leads to well-posed elastostatic problems [30]. Generally, a mathematical problem is well-posed if it satisfies two conditions: (1) the problem has unique solution for a specific boundary condition and (2) the solution depends continuously on its parameters including the boundary conditions.

3 Equilibrium Equation and Boundary Conditions

3.1 Governing Equation of a Nanorod

Consider a uniform rod of length L and cross-section A which is subjected to axial distributed force $p(x)$, and boundary forces F_0 and F_L(Fig. 1). The governing equation for longitudinal deformation of the rod is given by

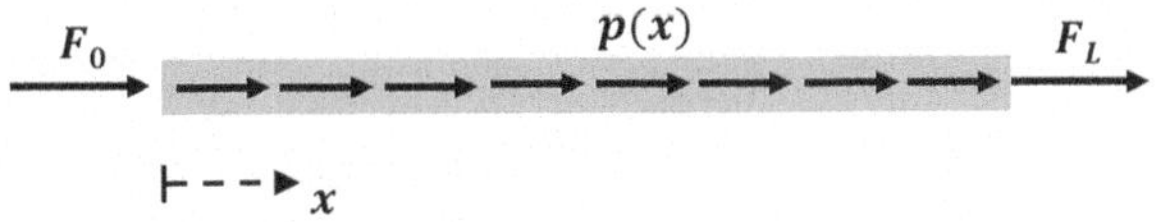

Fig. 1 Schematic illustration of a rod

$$\frac{\partial N(x)}{\partial x} + p(x) = 0 \tag{17}$$

where N is stress resultant which can be defined as

$$N = \int_A \sigma dA \tag{18}$$

For small longitudinal deformation of the rod, the strain component is

$$\varepsilon(x) = \frac{\partial U(x)}{\partial x} \tag{19}$$

where U denotes the longitudinal displacement. Using Eqs. (18) and (19) and upon integrating Eqs. (12), (15) and (16) over the rod cross section, we have

$$\begin{aligned} N(x) - e_0^2 l_i^2 \frac{d^2 N(x)}{dx^2} &= EA\frac{dU(x)}{dx} \\ &- 2e_0^2 l_i^2 EA\frac{dg(x)}{dx}\frac{d^2U(x)}{dx^2} - e_0^2 l_i^2 EAg(x)\frac{d^3U(x)}{dx^3} \end{aligned} \tag{20}$$

$$N(0) - e_0 l_i \frac{dN(0)}{dx} = EA\left(\frac{dU(0)}{dx} - e_0 l_i g(0)\frac{d^2U(0)}{dx^2}\right) \tag{21}$$

$$N(L) + e_0 l_i \frac{dN(L)}{dx} = EA\left(\frac{dU(L)}{dx} + e_0 l_i g(L)\frac{d^2U(L)}{dx^2}\right) \tag{22}$$

Here we set $a = 0$ and $b = L$. In addition, differentiating both sides of Eq. (17) with respect to x and substituting the resultant equation into Eq. (20) yields

$$\begin{aligned} N(x) = EA&\left(\frac{dU(x)}{dx} - 2e_0^2 l_i^2 \frac{dg(x)}{dx}\frac{d^2U(x)}{dx^2} - e_0^2 l_i^2 g(x)\frac{d^3U(x)}{dx^3}\right) \\ &- e_0^2 l_i^2 \frac{dp(x)}{dx} \end{aligned} \tag{23}$$

Now, substituting Eq. (23) into Eq. (17), we obtain

$$e_0^2 l_i^2 g(x) \frac{d^4 U(x)}{dx^4} + 3 e_0^2 l_i^2 \frac{dg(x)}{dx} \frac{d^3 U(x)}{dx^3} + (2g(x) - 1) \frac{d^2 U(x)}{dx^2} + \frac{e_0^2 l_i^2}{EA} \frac{d^2 p(x)}{dx^2} - \frac{p(x)}{EA} = 0 \tag{24}$$

Furthermore, using Eqs. (21–22), the higher-order boundary conditions are obtained in terms of the displacement, i.e.,

$$EA\left((e_0 l_i g(0) - 2 e_0^2 l_i^2 \frac{dg(0)}{dx}) \frac{d^2 U(0)}{dx^2} - e_0^2 l_i^2 g(0) \frac{d^3 U(0)}{dx^3} \right) - e_0^2 l_i^2 \frac{dp(0)}{dx} + e_0 l_i p(0) = 0 \tag{25}$$

$$-EA\left((e_0 l_i g\,(L) + 2 e_0^2 l_i^2 \frac{dg(L)}{dx}) \frac{d^2 U(L)}{dx^2} + e_0^2 l_i^2 g(L) \frac{d^3 U(L)}{dx^3} \right) - e_0^2 l_i^2 \frac{dp(L)}{dx} - e_0 l_i p(L) = 0 \tag{26}$$

Furthermore, we have two classical boundary conditions. For the case of the fixed-free rod, the classical boundary conditions are

$$U(0) = 0 \tag{27}$$

$$N(L) = EA(\frac{dU(L)}{dx} - 2 e_0^2 l_i^2 \frac{dg(L)}{dx} \frac{d^2 U(L)}{dx^2} - e_0^2 l_i^2 g(L) \frac{d^3 U(L)}{dx^3}) - e_0^2 l_i^2 \frac{dp(L)}{dx} = F_L \tag{28}$$

Introducing the dimensionless parameters

$$\lambda = \frac{e_0 l_i}{L},\ \xi = \frac{x}{L},\ u = \frac{U}{L},\ \bar{p} = \frac{pL}{EA},$$
$$f_1 = \frac{F_L}{EA},\ \bar{g}\,(\xi) = \frac{1}{2}\left(\exp(\frac{-\xi}{\lambda}) + \exp(\frac{\xi - 1}{\lambda}) \right) \tag{29}$$

the dimensionless governing equation is

$$\lambda^2 \bar{g} \frac{\partial^4 u}{\partial \xi^4} + 3\lambda^2 \frac{d\bar{g}}{d\xi} \frac{\partial^3 u}{\partial \xi^3} + (2\bar{g} - 1) \frac{\partial^2 u}{\partial^2 \xi} + \lambda^2 \frac{\partial^2 \bar{p}}{\partial \xi^2} - \bar{p} = 0 \tag{30}$$

and the dimensionless boundary conditions are

$$u(0) = 0 \tag{31}$$

$$(\lambda\bar{g}(0) - 2\lambda^2\frac{d\bar{g}(0)}{d\xi})\frac{d^2u(0)}{d\xi^2} - \lambda^2\bar{g}(0)\frac{d^3u(0)}{d\xi^3} - \lambda^2\frac{d\bar{p}(0)}{d\xi} + \lambda\bar{p}(0) = 0 \quad (32)$$

$$(\lambda\bar{g}(1) + 2\lambda^2\frac{d\bar{g}(1)}{d\xi})\frac{d^2u(1)}{d\xi^2} + \lambda^2\bar{g}(1)\frac{d^3u(1)}{d\xi^3} + \lambda^2\frac{d\bar{p}(1)}{d\xi} + \lambda\bar{p}(1) = 0 \quad (33)$$

$$\frac{du(1)}{d\xi} - 2\lambda^2\frac{d\bar{g}(1)}{d\xi}\frac{d^2u(1)}{d\xi^2} - \lambda^2\bar{g}(1)\frac{d^3u(1)}{d\xi^3} - \lambda^2\frac{d\bar{p}(1)}{d\xi} = f_1 \quad (34)$$

It should be noted that Eqs. (30)–(34) are reduced to the nonlocal differential model if the function $\bar{g}(\xi) = 0$. In addition, for the case $\lambda = 0$ and $\bar{g}(\xi) = 0$, Eqs. (30)–(34) correspond to the local differential model. Since it is difficult to obtain closed-form solutions for the static problems, in this chapter, the differential equation along with the boundary conditions are solved numerically using the Bvp4c Matlab solver.

3.2 Governing Equation of a Nanobeam

Here we consider a straight homogeneous isotropic nanobeam of length L, cross-section A, area moment of inertia I and Young's modulus E. It is assumed that the beam is subjected to a transverse distributed load $q(x)$, axial force F, boundary forces V_0 and V_L, and boundary moments M_0 and M_L (Fig. 2). On the basis of the Euler-Bernoulli beam theory, the equilibrium equation of the beam is

$$\frac{d^2M(x)}{dx^2} + F\frac{d^2W(x)}{dx^2} + q(x) = 0 \quad (35)$$

where M is moment resultant which can be defined as

$$M = \int_A \sigma z dA \quad (36)$$

where z is the distance of any point of the cross section of the beam to the neutral axis. Since the deformations of the beam take place in the xz plane, the displacement field can be written as

$$U_x(x, z) = -z\frac{dW}{dx}$$
$$U_z(x, z) = W(x) \quad (37)$$

where U_x and U_z denote respectively the axial and transverse displacements of a typical point of the beam and W is the transverse displacement of the reference line

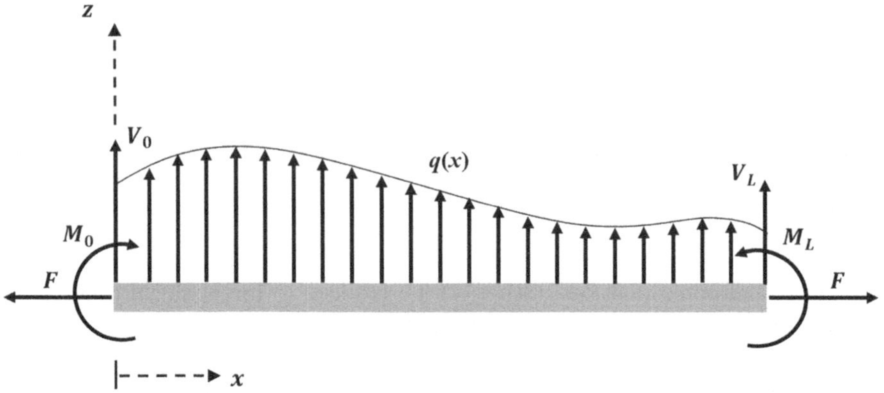

Fig. 2 Schematic illustration of a beam under various load conditions

of the beam (i.e., $z = 0$). Assuming small deformations, the nonzero component of strain is

$$\varepsilon = \frac{\partial U_x}{\partial x} = -z\frac{\partial^2 W}{\partial x^2} \tag{38}$$

Using Eqs. (36) and (38), multiplying Eqs. (12), (15) and (16) by z and then integrating the resultant equations over the cross section of the beam, we obtain

$$M(x) - {e_0}^2{l_i}^2\frac{d^2M(x)}{dx^2} = EI{e_0}^2{l_i}^2 g(x)\frac{d^4W(x)}{dx^4} + 2EI{e_0}^2{l_i}^2\frac{dg(x)}{dx}\frac{d^3W(x)}{dx^3} - EI\frac{d^2W(x)}{dx^2} \tag{39}$$

$$M(0) - e_0 l_i\frac{dM(0)}{dx} = EIe_0 l_i g(0)\frac{d^3W(0)}{\partial x^3} - EI\frac{d^2W(0)}{dx^2} \tag{40}$$

$$M(L) + e_0 l_i\frac{dM(L)}{dx} = -EIe_0 l_i g(L)\frac{d^3W(L)}{dx^3} - EI\frac{d^2W(L)}{dx^2} \tag{41}$$

Furthermore, using Eqs. (35) and (39), the moment resultant is obtained as

$$M(x) = EI\left({e_0}^2{l_i}^2 g(x)\frac{d^4W(x)}{dx^4} + 2{e_0}^2{l_i}^2\frac{dg(x)}{dx}\frac{d^3W(x)}{dx^3} - \frac{d^2W(x)}{dx^2}\right) - {e_0}^2{l_i}^2\left(F\frac{d^2W(x)}{dx^2} + q(x)\right) \tag{42}$$

Now, by two times differentiating Eq. (42) with respect to x and substituting the resultant equation into Eq. (35) the governing equation of the beam is obtained as

$$e_0{}^2l_i{}^2EIg(x)\frac{d^6W(x)}{dx^6}+4e_0{}^2l_i{}^2EI\frac{dg(x)}{dx}\frac{d^5W(x)}{dx^5}$$
$$+EI(5g(x)-1)\frac{d^4W(x)}{dx^4}+2EI\frac{dg(x)}{dx}\frac{d^3W(x)}{dx^3}+F\frac{d^2W(x)}{dx^2}$$
$$-e_0{}^2l_i{}^2(F\frac{d^4W(x)}{dx^4}+\frac{d^2q(x)}{dx^2})+q(x)=0 \tag{43}$$

Furthermore, using Eqs. (40–42), the higher-order boundary conditions are derived as

$$e_0{}^2l_i{}^2EIg(0)\frac{d^4W(0)}{dx^4}+2e_0{}^2l_i{}^2EI\frac{dg(0)}{dx}\frac{d^3W(0)}{dx^3}-EI\frac{d^2W(0)}{dx^2}$$
$$-e_0{}^2l_i{}^2g(0)EI\frac{d^5W(0)}{dx^5}-3e_0{}^2l_i{}^2EI\frac{dg(0)}{dx}\frac{d^4W(0)}{dx^4}$$
$$-2EIg(0)\frac{d^3W(0)}{dx^3}+EI\frac{d^3W(0)}{dx^3}-e_0{}^2l_i{}^2\left(F\frac{d^2W(0)}{dx^2}+q(0)\right)$$
$$+e_0{}^3l_i{}^3\left(F\frac{d^3W(0)}{dx^3}+\frac{dq(0)}{dx}\right)$$
$$=EIe_0l_ig(0)\frac{d^3W(0)}{\partial x^3}-EI\frac{d^2W(0)}{dx^2} \tag{44}$$

$$e_0{}^2l_i{}^2EIg(L)\frac{d^4W(L)}{dx^4}+2e_0{}^2l_i{}^2EI\frac{dg(L)}{dx}\frac{d^3W(L)}{dx^3}-EI\frac{d^2W(L)}{dx^2}$$
$$+e_0{}^2l_i{}^2EIg(L)\frac{d^5W(L)}{dx^5}+3e_0{}^2l_i{}^2EI\frac{dg(L)}{dx}\frac{d^4W(L)}{dx^4}$$
$$+2EIg(L)\frac{d^3W(L)}{dx^3}-EI\frac{d^3W(L)}{dx^3}-e_0{}^2l_i{}^2\left(F\frac{d^2W(L)}{dx^2}+q(L)\right)$$
$$-e_0{}^3l_i{}^3\left(F\frac{d^3W(L)}{dx^3}+\frac{dq(L)}{dx}\right)$$
$$=-EIe_0l_ig(L)\frac{d^3W(L)}{dx^3}-EI\frac{d^2W(L)}{dx^2} \tag{45}$$

In addition, there are four classical boundary conditions. For the case of the clamped-free beam, the classical boundary conditions are

$$W(0)=0 \tag{46}$$

$$\frac{dW(0)}{dx}=0 \tag{47}$$

$$V_L - F\frac{dW(L)}{dx} + e_0{}^2 l_i{}^2 (F\frac{d^3W(L)}{dx^3} + \frac{dq(L)}{dx}) - e_0{}^2 l_i{}^2 EIg(L)\frac{d^5W(L)}{dx^5}$$
$$-3e_0{}^2 l_i{}^2 EI\frac{dg(L)}{dx}\frac{d^4W(L)}{dx^4} - EI(2g(L) - 1)\frac{d^3W(L)}{dx^3} = 0 \tag{48}$$

$$M_L - e_0{}^2 l_i{}^2 (F\frac{d^2W(L)}{dx^2} + q(L)) + e_0{}^2 l_i{}^2 EIg(L)\frac{d^4W(L)}{dx^4}$$
$$+2e_0{}^2 l_i{}^2 EI\frac{dg(L)}{dx}\frac{d^3W(L)}{dx^3} - EI\frac{d^2W(L)}{dx^2} = 0 \tag{49}$$

The foregoing four conditions together with Eqs. (44) and (45) complete the boundary conditions needed to solve the sixth-order differential equation (43). To simplify the analysis, the following dimensionless parameters are defined.

$$\lambda = \frac{e_0 l_i}{L},\ \xi = \frac{x}{L},\ w = \frac{W}{L}$$
$$\bar{g}(\xi) = \frac{1}{2}\left(\exp(\frac{-\xi}{\lambda}) + \exp(\frac{\xi - 1}{\lambda})\right),\ \bar{q} = \frac{qL}{EI},\ f = \frac{FL^2}{EI}$$
$$v_1 = \frac{V_L L^2}{EI},\ m_1 = \frac{M_L L}{EI} \tag{50}$$

Substitution of Eq. (50) into Eq. (43) gives

$$\lambda^2\bar{g}\frac{d^6w}{d\xi^6} + 4\lambda^2\frac{d\bar{g}}{d\xi}\frac{d^5w}{d\xi^5} + \left(5\bar{g} - \lambda^2 f - 1\right)\frac{d^4w}{d\xi^4} + 2\frac{d\bar{g}}{d\xi}\frac{d^3w(\xi)}{d\xi^3}$$
$$+f\frac{d^2w}{d\xi^2} - \lambda^2\frac{d^2\bar{q}}{d\xi^2} + \bar{q}(\xi) = 0 \tag{51}$$

Furthermore, the dimensionless boundary conditions are

$$w(0) = 0 \tag{52}$$

$$\frac{dw(0)}{d\xi} = 0 \tag{53}$$

$$\lambda^2\bar{g}(1)\frac{d^5w(1)}{d\xi^5} + 3\lambda^2\frac{d\bar{g}(1)}{d\xi}\frac{d^4w(1)}{d\xi^4} + (2\bar{g}(1) - 1 - \lambda^2 f)\frac{d^3w(1)}{d\xi^3}$$
$$+f\frac{dw(1)}{d\xi} - \lambda^2\frac{d\bar{q}(1)}{d\xi} - v_1 = 0 \tag{54}$$

$$\lambda^2\bar{g}(1)\frac{d^4w(1)}{d\xi^4} + 2\lambda^2\frac{d\bar{g}(1)}{d\xi}\frac{d^3w(1)}{d\xi^3}$$
$$-(\lambda^2 f + 1)\frac{d^2w(1)}{d\xi^2} - \lambda^2\bar{q}(1) + m_1 = 0 \tag{55}$$

$$\lambda^3\bar{g}(0)\frac{d^5w(0)}{d\xi^5}+\left(3\lambda^3\frac{d\bar{g}(0)}{d\xi}-\lambda^2\bar{g}(0)\right)\frac{d^4w(0)}{d\xi^4}$$
$$-\left(2\lambda^2\frac{d\bar{g}(0)}{d\xi}-3\lambda\bar{g}(0)+f\lambda^3+\lambda\right)\frac{d^3w(0)}{d\xi^3}$$
$$+f\lambda^2\frac{d^2w(0)}{d\xi^2}+\lambda^2\bar{q}(0)-\lambda^3\frac{d\bar{q}(0)}{d\xi}=0 \tag{56}$$

$$\lambda^3\bar{g}(1)\frac{d^5w(1)}{d\xi^5}+\left(3\lambda^3\frac{d\bar{g}(1)}{d\xi}+\lambda^2\bar{g}(1)\right)\frac{d^4w(1)}{d\xi^4}$$
$$+\left(2\lambda^2\frac{d\bar{g}(1)}{d\xi}+3\lambda\bar{g}(1)-f\lambda^3-\lambda\right)\frac{d^3w(1)}{d\xi^3}$$
$$-f\lambda^2\frac{d^2w(1)}{d\xi^2}-\lambda^2\bar{q}(1)-\lambda^3\frac{d\bar{q}(1)}{d\xi}=0 \tag{57}$$

It is obvious that it is difficult to obtain closed-form solutions for above boundary value differential equations. As a result, the problem must be solved numerically. By setting the axial force parameter $f = 0$, Eqs. (51)–(57) reduce to the governing equation and boundary conditions for the bending of the well-posed nonlocal Euler beam under distributed transverse load $\bar{q}$. In addition, the equilibrium equation of buckling can be obtained by neglecting the transverse load term $\bar{q}$ in Eqs. (51)–(57).

4 Numerical Results

This section presents numerical results of static deformation of the nanorods and the nanobeams under different loading conditions. The numerical results are obtained by using the present model and are compared with the local and nonlocal differential models.

4.1 Static Deformation of Nanorods

In order to elucidate the effectiveness of the well-posed nonlocal elasticity model for predicting the static behavior of the nanorods, three problems which have technical interest in nanotechnology have been studied. In the first case, a nanorod is only subjected to end-point load $f_1 = 1$. The variations of strain along the axis of the nanorod is shown in Fig. 3. It is seen from this figure that the strain obtained from three models is uniform along the rod. This finding is not paradoxical and is correct from both physical and mathematical point of views because uniform straining generates a uniform stress at the domain and its boundaries.

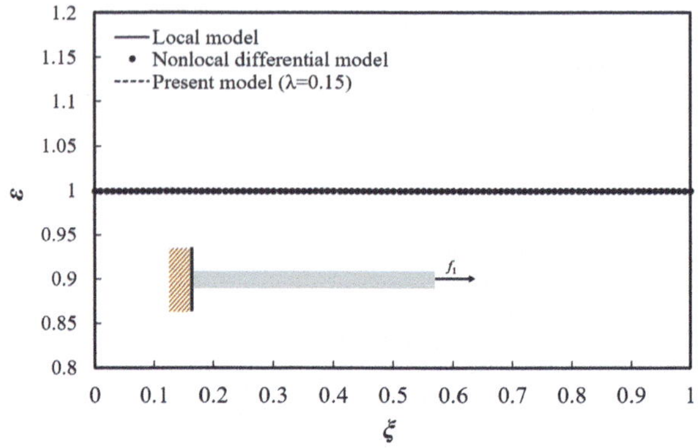

Fig. 3 The variation of strain along the rod axis for load condition $f_1 = 1$

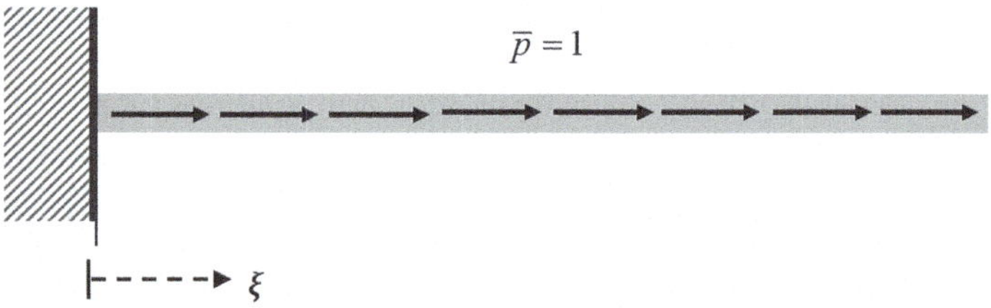

Fig. 4 Nanorod under uniform axial load $\bar{p} = 1$

As the second case, we consider a nanorod under a uniform axial load $\bar{p} = 1$ (Fig. 4). Figure 5 shows the variations of the strain and the deformation along the axis of the nanorod for several values of nonlocal parameter. It can be seen from the figure that the results of the nonlocal differential and local models are identical to each other and the nonlocal effect is not present in the nonlocal differential solutions. These paradoxical results were previously reported by Barretta et al. [8]. In addition, it is observed that the displacements predicted by the present model are greater than the local one except at the boundary points, and therefore, the model displays a softening behavior in term of the nonlocal parameter λ. It is of interest that the maximum displacement $u(1)$ of the rod is independent of the nonlocal parameter and the maximum axial displacement is identical to the one of the classical local. This finding is consistent with those reported in the literature [31]. One of the main advantage of the present model is that it avoids giving rise to paradoxical results for all values of the nonlocal parameter.

As the third case, the results obtained from the local, the nonlocal differential and the present model, for the nanorod subjected to linear distributed load (Fig. 6) are compared. The results of this comparison is shown in Fig. 7 that shows the variations of the axial deformation along the axis of the nanorod for different values of nonlocal parameter.

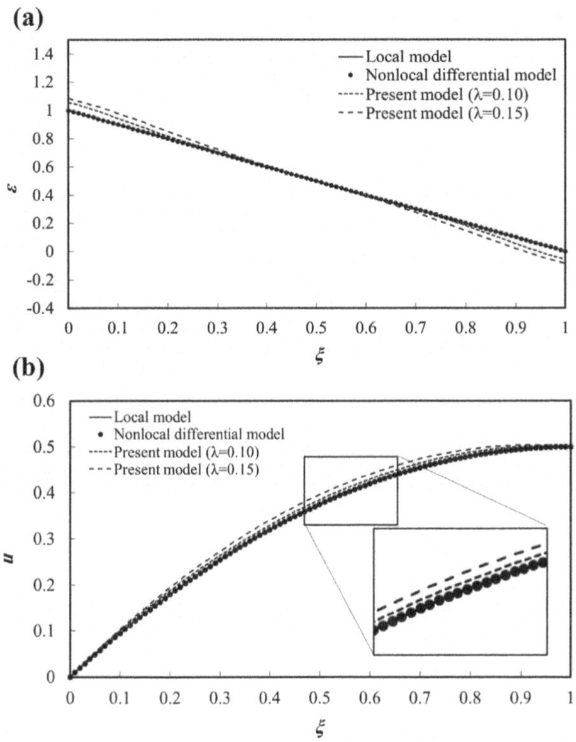

Fig. 5 Fixed- free nanorod subjected to uniform load; **a** Strain and **b** Displacement

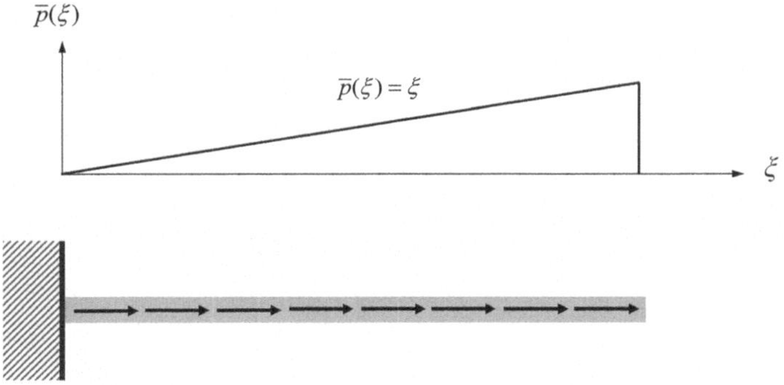

Fig. 6 Nanorod subjected to linear axial load

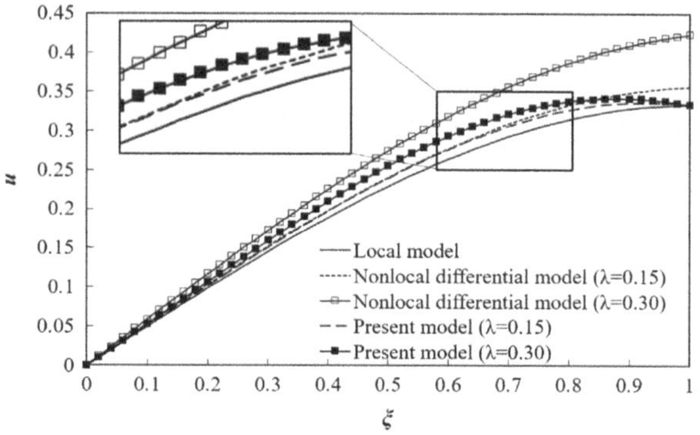

Fig. 7 Axial deformation of fixed-free rod loaded by a linear distributed load

Figure 7 shows that the axial displacements of nanorod obtained from the present and the nonlocal differential models are greater than the local ones. This physically implies that the well-posed model presents a softening behavior in term of the non-local parameter λ. It should be noted that the maximum displacement $u(1)$ of the nanorod does not depend on the nonlocal parameter.

4.2 Bending of Nanobeams

In this subsection, several bending problems, including the cantilever beam under end-point load, the cantilever beam loaded by uniform transverse load and the cantilever beam under linear distributed force, are considered. The cantilever beam under a point load is a paradigmatic static case. The effect of the nonlocal parameter on the transverse deflection of the nanobeam is depicted in Fig. 8.

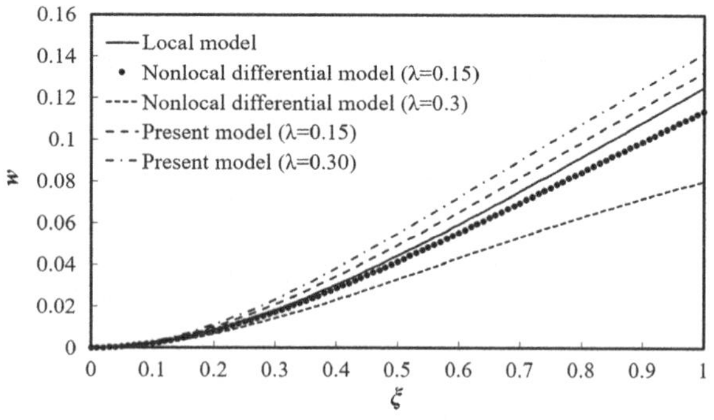

Fig. 8 Displacements of cantilever beam subjected to concentrated transverse force at the free end

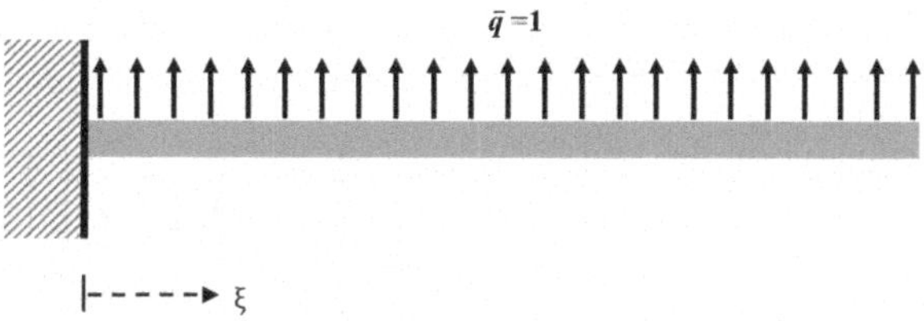

Fig. 9 Cantilever beam loaded by a uniform transverse load $\bar{q} = 1$

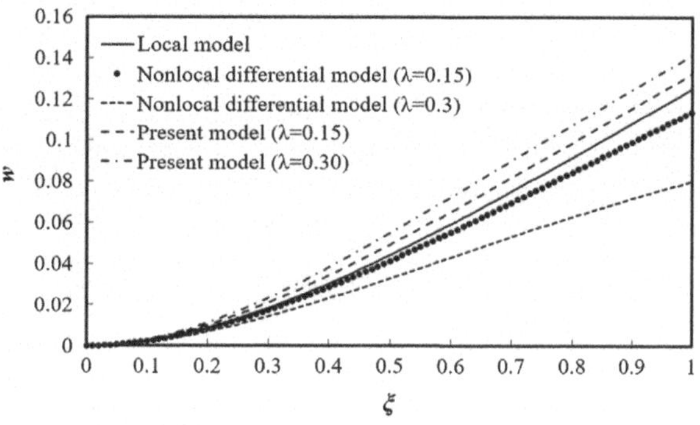

Fig. 10 Transverse deflection of a cantilever nanobeam subjected to a uniform distributed load

As expected, a paradox is observed where the bending solutions of the nonlocal differential model are identical to the local solutions. However, the deflections obtained from the present model are larger than those of the local one. Furthermore, it is observed that the deflections increase with an increase in the nonlocal coefficient λ. This means physically that the present model exhibits a softening behavior. As seen from Fig. 8, the paradox is resolved by the well-posed nonlocal model presented in this chapter.

Next, a nonlocal cantilever beam loaded by uniform transverse load is considered, as shown by Fig. 9. Variations of the deflection of the beam are obtained from the local, the nonlocal differential and the present models and plotted in Fig. 10. As can be seen in Fig. 10, the present nonlocal model shows that the stiffness of the beam decreases whereas in the differential nonlocal one the stiffness increases.

Finally, a comparison is made between the results obtained from the local, the nonlocal differential and the present models, for the beam subjected to linear distributed load (Fig. 11) and two different values of the nonlocal parameter. The deflection and slope of the beam shown in Fig. 12. The results indicate that the deflections obtained from the present are greater than the local one, whereas the deflections obtained from the nonlocal differential model are smaller than the local. One interesting result is that the slope of the end of the beam, according to the well-posed nonlocal model, does not depend on the nonlocal parameter and is identical to the slope of the local model.

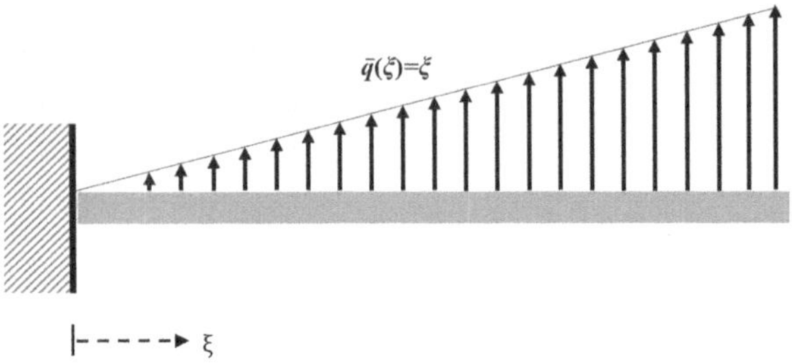

Fig. 11 Transverse deflection of a cantilever nanobeam subjected to a uniform distributed load

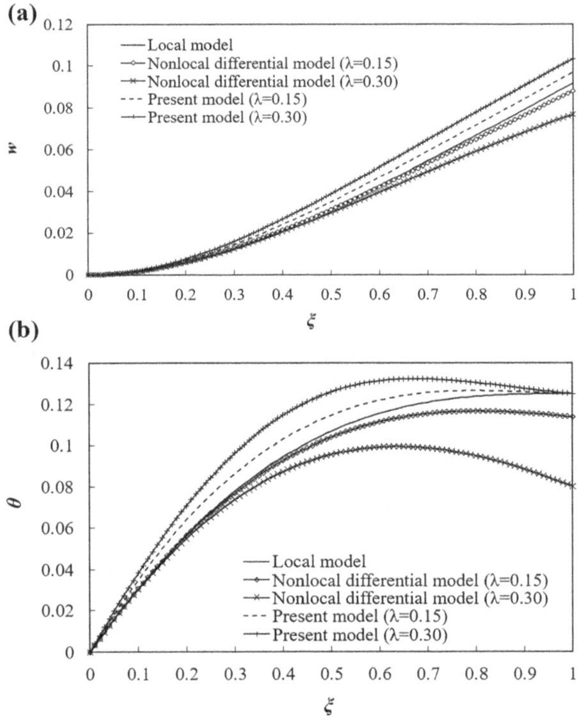

Fig. 12 Response of cantilever beam under linear distributed load; **a** Displacement and **b** Slope

4.3 Buckling of Nanobeams

In this subsection, the dimensionless buckling loads of the cantilever beam will be given for different models and nonlocal parameters. Table 1 lists the critical buckling loads for different values of nonlocal parameters from zero to 0.25. As can be inferred from this table, the nonlocal parameter decreases the buckling loads in comparison with those of the local results. Furthermore, the buckling loads of the nonlocal differential model are smaller than ones predicted by the present model. It should be noted that the negative sign shows the dimensionless load f is compressive.

Table 1 The buckling loads of the cantilever beam

λ	Local model	Nonlocal differential model	Present model
0.00	−2.4674	−2.4674	−2.4674
0.05	−2.4674	−2.4523	−2.4540
0.10	−2.4674	−2.4080	−2.4213
0.15	−2.4674	−2.3376	−2.3803
0.20	−2.4674	−2.2458	−2.3401
0.25	−2.4674	−2.1377	−2.3065

5 Conclusions

In this chapter, we developed a modified strain-based integral model for finite domains and utilize it to predict the static deformation of nanorods and the static behavior of nanobeams (bending and buckling). Numerical results for several static problems were solved and compared with the local and nonlocal differential models. The main conclusions of this chapter are as follows:

- The developed models obey the locality recovery condition, which implies that the classical Hooke's law is recovered in the presence of a uniform strain field, no matter the value of the length scale parameter.
- Using the well-posed nonlocal elastic models, the nonlocal paradoxical results are resolved.
- Elastic displacement solutions of the studied nanorods and nanobeams, formulated according to the well-posed models, indicate the softening behavior when the nonlocal parameter is increased.

References

1. Rafii-Tabar H, Ghavanloo E, Fazelzadeh SA (2016) Nonlocal continuum-based modeling of mechanical characteristics of nanoscopic structures. Phys Rep 638:1–97
2. Polizzotto C (2001) Nonlocal elasticity and related variational principles. Int J Solids Struct 38:7359–7380
3. Ghavanloo E, Rafii-Tabar H, Fazelzadeh SA (2019) Computational continuum mechanics of nanoscopic structures: nonlocal elasticity approaches. Springer
4. Peddieson J, Buchanan GR, McNitt RP (2003) Application of nonlocal continuum models to nanotechnology. Int J Eng Sci 41:305–312
5. Challamel N, Wang CM (2008) The small length scale effect for a non-local cantilever beam: a paradox solved. Nanotechnology 19:345703
6. Li C, Yao L, Chen W, Li S (2015) Comments on nonlocal effects in nano-cantilever beams. Int J Eng Sci 87:47–57
7. Benvenuti E, Simone A (2013) One-dimensional nonlocal and gradient elasticity: closed-form solution and size effect. Mech Res Commun 48:46–51

8. Barretta R, Marotti de Sciarra F, Diaco M (2014) Small-scale effects in nanorods. Acta Mech 225:1945–1953
9. Lu P, Lee HP, Lu C (2006) Dynamic properties of flexural beams using a nonlocal elasticity model. J Appl Phys 99:073510
10. Eltaher M, Alshorbagy AE, Mahmoud F (2013) Vibration analysis of Euler-Bernoulli nanobeams by using finite element method. Appl Math Model 37:4787–4797
11. Tashakorian M, Ghavanloo E, Fazelzadeh SA, Hodges DH (2018) Nonlocal fully intrinsic equations for free vibration of Euler-Bernoulli beams with constitutive boundary conditions. Acta Mech 229:3279–3292
12. Fernández-Sáez J, Zaera R, Loya JA, Reddy JN (2016) Bending of Euler-Bernoulli beams using Eringen's integral formulation: a paradox resolved. Int J Eng Sci 99:107–116
13. Romano G, Barretta R (2017) Stress-driven versus strain-driven nonlocal integral model for elastic nano-beams. Compos Part B 114:184–188
14. Fernández-Sáez J, Zaera R (2017) Vibrations of Bernoulli-Euler beams using the two-phase nonlocal elasticity theory. Int J Eng Sci 119:232–248
15. Wang YB, Zhu XW, Dai HH (2016) Exact solutions for the static bending of Euler-Bernoulli beams using Eringen's two-phase local/nonlocal model. AIP Adv 6:085114
16. Zhu X, Li L (2017) Longitudinal and torsional vibrations of size-dependent rods via nonlocal integral elasticity. Int J Mech Sci 133:639–650
17. Meng L, Zou D, Lai H, Guo Z, He X, Xie Z, Gao C (2018) Semi-analytic solution of Eringen's two-phase local/nonlocal model for Euler-Bernoulli beam with axial force. Appl Math Mech 39:1805–1824
18. Romano G, Barretta R, Diaco M, Marotti de Sciarra F (2017) Constitutive boundary conditions and paradoxes in nonlocal elastic nanobeams. Int J Mech Sci 121:151–156
19. Challamel N, Wang C, Elishakoff I (2014) Discrete systems behave as nonlocal structural elements: Bending, buckling and vibration analysis. Eur J Mech A/Solids 44:125–135
20. Hache F, Challamel N, Elishakoff I, Wang CM (2017) Comparison of nonlocal continualization schemes for lattice beams and plates. Arch Appl Mech 87:1105–1138
21. Khodabakhshi P, Reddy J (2015) A unified integro-differential nonlocal model. Int J Eng Sci 95:60–75
22. Shaat M (2018) Correction of local elasticity for nonlocal residuals: application to Euler-Bernoulli beams. Meccanica 53:3015–3035
23. Challamel N, Zhang Z, Wang CM, Reddy JN, Wang Q, Michelitsch T, Collet B (2014) On nonconservativeness of Eringen's nonlocal elasticity in beam mechanics: correction from a discrete-based approach. Arch Appl Mech 84:1275–1292
24. Reddy JN (2007) Nonlocal theories for bending, buckling and vibration of beams. Int J Eng Sci 45:288–307
25. Koutsoumaris C, Eptaimeros K, Tsamasphyros G (2017) A different approach to Eringen's nonlocal integral stress model with applications for beams. Int J Solids Struct 112:222–238
26. Challamel N (2018) Static and dynamic behaviour of nonlocal elastic bar using integral strain-based and peridynamic models. C. R. Mécanique 346:320–335
27. Eringen AC (2002) Nonlocal continuum field theories. Springer
28. Polizzotto C (2002) Remarks on some aspects of nonlocal theories in solid mechanics. Proceedings, 6th National Congress SIMAI, Chia Laguna, Italy
29. Borino G, Failla B, Parrinello F (2002) A symmetric formulation for nonlocal damage models. In: Proceedings, 5th World Congress on Computational Mechanics, Vienna, Austria
30. Barretta R, Marotti de Sciarra F (2018) Constitutive boundary conditions for nonlocal strain gradient elastic nano-beams. Int J Eng Sci 130:187–198
31. Barretta R, Čanadija M, Marotti de Sciarra F (2019) Modified nonlocal strain gradient elasticity for nano-rods and application to carbon nanotubes. Appl Sci 9:514

Iterative Nonlocal Residual Elasticity

Mohamed Shaat

Abstract Motivated by the existing complications of finding solutions of Eringen's nonlocal model, an alternative model is developed here. The new formulation of the nonlocal elasticity is centered upon expressing the dynamic equilibrium requirements based on a nonlocal residual stress field. This new nonlocal elasticity is explained from the lattice mechanics and continuum mechanics points of view. Boundary value problems obtained based on the new nonlocal elasticity are solved following a proposed iterative procedure. This iterative procedure is centered upon correcting the solution of the classical field problem for the nonlocal residual field of the elastic domain. Convergence analyses are presented to show the convergence of the iterative procedure to the solution of the nonlocal field problem. The iterative procedure is an integrated part of the proposed nonlocal elasticity. Therefore, the newly developed nonlocal elasticity is given the name "iterative nonlocal residual elasticity".

1 Introduction

In classical mechanics, a body consists of an infinite number of particles each of which is a mass point. Each particle undergoes interactions only with the nearest particles (neighbor particles). Whereas the interactions between non-neighbor particles are weaker than interactions between neighbor particles, these non-neighbor interactions are exist and may contribute to the continuum mechanics in certain occasions. Therefore, the nonlocal theory was developed to propose a generalized theory, which models the particle exhibits interactions with its neighbors and non-neighbors.

According to Eringen [1, 2], the dynamic equilibrium of a solid elastic body is conditional by the global balance of its body forces, external surface tractions, and inertia forces. The nonlocal theory, as early proposed by Eringen, postulates the condition of the global balance, thus [2]:

M. Shaat (✉)
Mechanical Engineering Department, Abu Dhabi University,
P.O.BOX 1790, Al Ain, United Arab Emirates
e-mail: mohamed.i@adu.ac.ae; shaatscience@yahoo.com

© The Author(s), under exclusive license to Springer Nature Switzerland AG 2021
E. Ghavanloo et al. (eds.), *Size-Dependent Continuum Mechanics Approaches*,
Springer Tracts in Mechanical Engineering,
https://doi.org/10.1007/978-3-030-63050-8_6

$$t_{ji,j} + F_i + \mathcal{F}_i = \rho \frac{\partial^2 u_i}{\partial t^2} \tag{1}$$

where ρ is the mass density, u_i is the displacement field of a material particle, and t is the time. F_i is a local-type body force, which is conjugate to neighbor interactions. $\mathcal{F}_i$ is a nonlocal body force. The nonlocal body force ($\mathcal{F}_i$) is a residual field, which is generated due to non-neighbor interactions. This nonlocal residual body force yields to the condition [1, 2]:

$$\int_V \mathcal{F}_i dV = 0 \tag{2}$$

where V is the volume of the elastic body. For a homogeneous-isotropic-linear elastic material, $\mathcal{F}_i = 0$ [1,2].

The balance laws of nonlocal continua (Eq. (1)) depend on a total stress field (t_{ij}), which sums neighbor interactions and non-neighbor interactions at a specific point $\boldsymbol{x}$ belongs to the elastic body. This total stress field (t_{ij}) was expressed as a functional of the deformation gradients of all points of the elastic body [1, 2]. Thus, for a homogeneous-isotropic-linear elastic material, the constitutive equations were determined based on some invariance and thermodynamics requirements with the form [1, 2]:

$$t_{ij}(\boldsymbol{x}) = \int_V \lambda'\left(\left|\boldsymbol{x}' - \boldsymbol{x}\right|\right) \varepsilon_{rr}\left(\boldsymbol{x}'\right) \delta_{ij} + 2\mu'\left(\left|\boldsymbol{x}' - \boldsymbol{x}\right|\right) \varepsilon_{ij}\left(\boldsymbol{x}'\right) dV\left(\boldsymbol{x}'\right) \tag{3}$$

where ε_{ij} is the infinitesimal strain. λ' and μ' are Lame moduli, which are functionals of $\left|\boldsymbol{x}' - \boldsymbol{x}\right|$. The integration over the elastic domain, V, was involved to collect all non-neighbor interactions.

With Eringen's manipulation of the nonlocal theory, the balance laws are identical to those of the classical continuum mechanics [1, 2]. However, the constitutive law is different where the total stress field (t_{ij}) was introduced to model neighbor and non-neighbor interactions via integral operators (Eq. (3)). This total stress field is commonly known as "nonlocal stress field".

Complications of finding solutions of nonlocal elasticity problems have been early discussed [3]. According to Eq. (3), the field equation of the nonlocal problem is an integro-partial differential equation. Analytical solutions for integro-partial differential equations are difficult to be determined, especially for mixed boundary value problems [3]. This motivated Eringen [3] to formulate the nonlocal elasticity in terms of singular differential equations. By assuming that (i) the Lame moduli exhibit the same attenuation with the distance $\left|\boldsymbol{x}' - \boldsymbol{x}\right|$ according to (ii) a Green's function of a differential operator with constant coefficients, the nonlocal field problem was formulated as the following partial differential equation [3]:

$$\sigma_{ji,j} + \mathcal{L}\left(F_i - \rho\frac{\partial^2 u_i}{\partial t^2}\right) = 0 \tag{4}$$

where $\mathcal{L}$ is the differential operator as defined by Eringen [3], i.e. $\mathcal{L}t_{ij} = \sigma_{ij}$. σ_{ij} is the local stress of the classical elasticity.

Mathematically speaking, solutions can be easily obtained using the differential form of the nonlocal elasticity (Eq. (4)). Nonetheless, the ability of this form to secret exact solutions, which should be consistent with the solutions of the integral nonlocal model is controversial. Discrepancies were observed between solutions of the differential nonlocal field problems and integral nonlocal field problems [4–7]. Details on the paradoxes of the differential nonlocal elasticity can be found in Ref. [4–15]. Therefore, the current focus of research on nonlocal mechanics depends on the integral nonlocal model [16–18].

Moreover, the "ill-posedness" of Eringen's nonlocal elasticity was discussed in Ref. [15]. It was revealed that the transformation of the integral constitutive law (Eq. (3)) into a differential nonlocal constitutive law secretes inconsistencies between the natural boundary conditions of the nonlocal equilibrium and the boundary conditions of the constitutive model generated upon the transformation [15].

An example of the paradox and inconsistency of the differential nonlocal elasticity can be exhibited by considering a nonlocal problem of an elastic body under a body force equals its inertia force, in particular $F_i - \rho\partial^2 u_i/\partial t^2 = 0$. An example of the latter case is a nonlocal static problem with a zero body force. For this case, Eq. (4) reduces to:

$$\sigma_{ji,j} = 0 \tag{5}$$

which gives the stress field identical to that of the classical elasticity with no dependency on the nonlocal character of the body. This is a clear violation of the concept of the nonlocal mechanics.

The illustrated example and the paradoxes and inconsistencies expressed in previous studies imply that the transformation of the integral nonlocal constitutive law (Eq. (3)) into a differential nonlocal constitutive law as proposed by Eringen [3] may be not convenient to express nonlocal boundary value problems, especially in bounded domains [15]. Thus, the transformation of the integral nonlocal constitutive law into a differential nonlocal constitutive law is an injective process in infinite domains, yet it is not in finite domains [16]. However, with the deep inspection of the situation, one can realize the fact that the transformation into the differential constitutive law proposed in Ref. [3] is mathematically consistent. So, the question is: what are the reasons behind the paradoxes of Eringen's nonlocal elasticity?

Motivated by the aforementioned complications and debates regarding Eringen's nonlocal elasticity presented in Eqs. (1)–(4), an alternative form of the nonlocal theory is developed in this chapter. This new form of the nonlocal elasticity depends on splitting the total nonlocal stress field (t_{ij}) into a local stress (σ_{ij}) and a nonlocal residual stress (τ_{ij}), i.e. $t_{ij} = \sigma_{ij} + \tau_{ij}$. The local stress ($\sigma_{ij}$) is a stress field that models the material stiffness due to neighbor (local) interactions. The nonlocal residual stress

(τ_{ij}) models the material stiffness due to non-neighbor (nonlocal) interactions. In the context of the new nonlocal elasticity, the balance laws are similar to those of the classical elasticity of an elastic continuum with a residual stress field. The proposed nonlocal theory is centered upon modifying the classical elasticity for an inherited nonlocal residual field. At the boundary of the elastic body, the surface tractions depend on both the local stress and the nonlocal residual stress.

The new manipulation of the nonlocal theory proposed here permits implementing an iterative procedure to extract solutions of nonlocal field problems. This iterative procedure is centered upon correcting the solution of the classical field problem for the nonlocal residual field of the elastic domain. In the context of the proposed approach, the classical field problem of an elastic domain with a pre-formed nonlocal residual stress is solved. Convergence analyses are presented to show the convergence of the iterative procedure to the solution of the nonlocal field problem formed based on the proposed nonlocal theory. It should be mentioned that the iterative procedure is an integrated part of the theory. Therefore, the newly developed form of the nonlocal theory will be give the name "iterative nonlocal residual elasticity".

2 Nonlocal Residual Elasticity

In the classical mechanics, a balance law is valid not only for the entire elastic body but also for each point belongs to the elastic domain [2]. However, in the nonlocal mechanics, the balance is achieved globally for the entire body. Thus, a nonlocal balance law is violated at a local point while satisfied globally because of the nonlocal interactions [2]. The nonlocal mechanics distinguishes between a local-type field and a nonlocal-type field. A nonlocal field (e.g., nonlocal stress field) is the sum of a local field, which accounts for the direct-neighbor interactions, and a nonlocal residual field. The nonlocal residual field is the sum of all the nonlocal interactions at a point, as shown in Fig. 1.

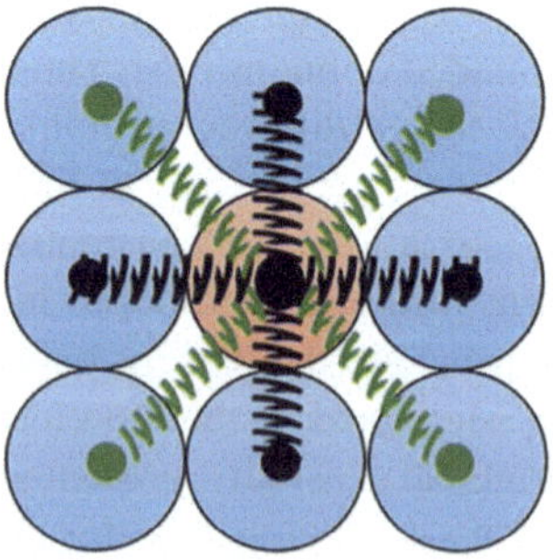

Fig. 1 Nonlocal lattice mechanics. Dark interactions are direct-neighbor interactions while the green ones are non-neighbor interactions only up to the next-nearest interaction

2.1 Nonlocal Mechanics of Particles

To give more insights into the nonlocal elasticity developed here, the system of particles presented in Fig. 1 is considered. As shown in the figure, a set of interaction forces is generated between a particle (n) and its neighbor particles. In the classical mechanics, these direct interactions are exist. Each one of these interactions forms a local force between two neighbors. From lattice mechanics, the local force field $\boldsymbol{F}_{nm}$, between two neighbor particles (n and m) can be defined as follows:

$$\boldsymbol{F}_{nm} = \boldsymbol{K} \cdot (\boldsymbol{u}_n - \boldsymbol{u}_m) \tag{6}$$

where $\boldsymbol{K}$ is an equivalent stiffness of an interaction between two neighbors. $\boldsymbol{u}$ is the displacement field expressing the translational motion of the particle. $(\cdot)$ is the single-dot operator of the dot production of the tensor $\boldsymbol{K}$ and the vector $(\boldsymbol{u}_n - \boldsymbol{u}_m)$.

In addition to the direct neighbor interactions, interactions are generated between a particle (n) and all other non-neighbor particles. A non-neighbor interaction, $\mathcal{F}$, depends on the distance between the two particles. Thus, the non-neighbor interaction between two non-neighbors (n and j) is a functional of the distance between them $|\boldsymbol{x}_n - \boldsymbol{x}_j|$, as follows:

$$\mathcal{F}_{nj} = \boldsymbol{k}_r \left(|\boldsymbol{x}_n - \boldsymbol{x}_j|\right) . \left(\boldsymbol{u}_n - \boldsymbol{u}_j\right) \tag{7}$$

where $\boldsymbol{k}_r \left(|\boldsymbol{x}_n - \boldsymbol{x}_j|\right)$ is a stiffness that is equivalent to the interaction between two non-neighbor particles.

Under the influence of the neighbor and non-neighbor interactions, the condition of the dynamic equilibrium of a particle (n) secretes the following balance equations:

$$\begin{gathered} \sum_{m=1}^{N_m} \boldsymbol{F}_{nm} + \sum_{j=1}^{N_j} \mathcal{F}_{nj} = \frac{\partial^2}{\partial t^2} (\bar{m}\boldsymbol{u}) \\ \sum_{m=1}^{N_m} (\boldsymbol{X} \times \boldsymbol{F}_{nm}) + \sum_{j=1}^{N_j} \left(\boldsymbol{X} \times \mathcal{F}_{nj}\right) = \frac{\partial^2}{\partial t^2} (\boldsymbol{X} \times \bar{m}\boldsymbol{u}) \end{gathered} \tag{8}$$

where $\boldsymbol{X} \times \boldsymbol{F}_{nm}$ and $\boldsymbol{X} \times \mathcal{F}_{nj}$ are moment fields. N_m is the number of neighbors of particle (n) while N_j is the number of its non-neighbor particles. $\bar{m}$ is the particle's mass.

We introduce in Eqs. (6)–(8) a new form of the nonlocal theory in which the nonlocal model is formulated depending on the nonlocal residual field ($\sum_{j=1}^{N_j} \mathcal{F}_{nj}$), which is the sum of all the non-neighbor (nonlocal) interactions of the particle under consideration. As a crucial requirement of nonlocal mechanics, the balance Eqs. (8) are obtained from the global statement of equilibrium.

In Eringen's form of the nonlocal theory, the balance equations were formed depending on a total field that merges the nonlocal residual field with the local field.

However, in the context of the proposed nonlocal elasticity, the nonlocal residual field is formed separately, and it is not merged with the local field. In addition, two types of constitutive equations are needed. In addition to the local-type constitutive equations (which are used to form the local field), constitutive equations are used to form the nonlocal residual field.

2.2 Nonlocal Continuum Mechanics: Balance Laws

The global balance laws of a nonlocal continuum with a volume V and bounded by a surface S can be derived according to Eq. (8), as follows:

$$\int_V \boldsymbol{F} dV + \int_V \boldsymbol{\mathcal{F}} dV + \int_S \boldsymbol{T} dS = \int_V \rho \ddot{\boldsymbol{u}} dV \tag{9}$$

$$\int_V (\boldsymbol{X} \times \boldsymbol{F}) dV + \int_V (\boldsymbol{X} \times \boldsymbol{\mathcal{F}}) dV + \int_S (\boldsymbol{X} \times T) dS = \int_V (\boldsymbol{X} \times \rho \ddot{\boldsymbol{u}}) dV \tag{10}$$

where ρ is the mass density. $\boldsymbol{F}$ denotes the local body force. $\boldsymbol{\mathcal{F}}$ is a nonlocal residual body force. $\boldsymbol{T}$ denotes a surface traction. A local body force $\boldsymbol{F}$ generates due to external forces acting on the system of particles form the elastic domain. However, the nonlocal body force $\boldsymbol{F}$ generates due to non-neighbor interactions that act on the system of particles form the elastic domain. The nonlocal body force is internal and satisfy the condition (Eq. (2)).

Like the classical mechanics, the external surface traction ($\boldsymbol{T}$) is balanced with the body forces and the inertia of the elastic domain (Eqs. (9) and (10)). However, in the proposed nonlocal elasticity, the surface traction $\boldsymbol{T}$ is formed depending on two tractions:

$$\boldsymbol{T} = \boldsymbol{n} \cdot \boldsymbol{\sigma} + \boldsymbol{n} \cdot \boldsymbol{\tau} \tag{11}$$

where $\boldsymbol{n}$ is the unit normal vector. $\boldsymbol{\sigma}$ is the local stress, and $\boldsymbol{\tau}$ is a nonlocal residual stress. The local stress ($\boldsymbol{\sigma}$) is conjugate to the direct neighbor interactions while the nonlocal residual stress ($\boldsymbol{\tau}$) is conjugate to the non-neighbor interactions. It should be observed that the stress tensors, $\boldsymbol{\sigma}$ and $\boldsymbol{\tau}$, are general tensors, which have symmetric and skew-symmetric parts.

According to Eq. (11) and the divergence theorem, Eqs. (9) and (10) can be rewritten as follows:

$$\boldsymbol{F} + \boldsymbol{\mathcal{F}} + \nabla \cdot \boldsymbol{\sigma} + \nabla \cdot \boldsymbol{\tau} = \rho \ddot{\boldsymbol{u}} \tag{12}$$

$$\Xi : (\sigma + \tau) = 0 \tag{13}$$

where Ξ is the permutation tensor, and (:) is the operator of the double-dot product. Equation (13) indicates that the skew-symmetric parts of the local stress tensor, $\boldsymbol{\sigma}$, and the nonlocal residual stress tensor, $\boldsymbol{\tau}$, vanish. Therefore, the balance equations can be written in the form:

$$\boldsymbol{F} + \boldsymbol{\mathcal{F}} + \nabla \cdot \boldsymbol{\sigma}^{sym} + \nabla \cdot \boldsymbol{\tau}^{sym} = \rho \ddot{\boldsymbol{u}} \tag{14}$$

where the balance equations depend on the symmetric parts of the stress tensors, $\boldsymbol{\sigma}^{sym}$ and $\boldsymbol{\tau}^{sym}$. In the rest of the chapter, the superscript"sym" is omitted for convenient writing.

2.3 Nonlocal Continuum Mechanics: Constitutive Model

The balance laws of Eringen's model of the nonlocal linear elasticity depend on a total stress field that merges the local stress with the nonlocal residual stress. This total-nonlocal stress field was formed as a functional depending on the motions histories of all points of the body. However, the balance laws of the proposed nonlocal elasticity (Eq. (14)) depend on a nonlocal residual stress field, which is not merged with the local stress field. In this section, constitutive equations are derived for the nonlocal residual stress field.

Eringen expressed a postulation [1] that states, "*For the linear theory of nonlocal elastic materials, whose natural state is free of nonlocal effects, the nonlocal body force vanishes.*", i.e. $\mathcal{F} = 0$. In light of this postulation, the balance laws of a homogeneous-isotropic-linear elastic solid material can be written in the form:

$$\sigma_{ji,j} + \tau_{ji,j} + F_i = \rho \ddot{u}_i \tag{15}$$

where $\sigma_{ij} = \sigma_{ji}$ and $\tau_{ij} = \tau_{ji}$, which are the components of the local stress tensor and the nonlocal residual stress tensor, respectively. For a linear elastic material, the infinitesimal strain is the fundamental measure of deformation:

$$\varepsilon_{ij} = \frac{1}{2}\left(u_{i,j} + u_{j,i}\right) \tag{16}$$

The local stress, σ_{ij}, has the same form as the stress field of the classical mechanics:

$$\sigma_{ji} = \lambda_0 \varepsilon_{rr} \delta_{ij} + 2\mu_0 \varepsilon_{ij} \tag{17}$$

where λ_0 and μ_0 are the Lamé constants as defined in the classical mechanics. Following Eringen's definition of nonlocal fields, the following constitutive equation can be proposed for the nonlocal residual stress (τ_{ij}):

$$\tau_{ij}(\boldsymbol{x}) = \int\limits_V \lambda_r\left(\left|\boldsymbol{x}' - \boldsymbol{x}\right|\right)\varepsilon_{rr}\left(\boldsymbol{x}'\right)\delta_{ij} + 2\mu_r\left(\left|\boldsymbol{x}' - \boldsymbol{x}\right|\right)\varepsilon_{ij}\left(\boldsymbol{x}'\right)dV\left(\boldsymbol{x}'\right) \quad (18)$$

where λ_r and μ_r are Lamé moduli, which model the stiffness of a non-neighbor interaction. The magnitudes of these Lamé moduli depend of the distance between two non-neighbor particles within the elastic domain ($|\boldsymbol{x}' - \boldsymbol{x}|$).

2.4 Boundary Value Problem and Solution Procedure: Iterative Nonlocal Residual Elasticity

The boundary value problem of the homogeneous-isotropic-linear elastic material based on the proposed nonlocal elasticity constitutes the following system of equations:

1. Equation of Motion:

$$\sigma_{ji,j}(\boldsymbol{x}) + \tau_{ji,j}(\boldsymbol{x}) + F_i(\boldsymbol{x}) = \rho\ddot{u}_i(\boldsymbol{x}) \qquad \forall\boldsymbol{x} \in V$$

2. Boundary Conditions:

$$\sigma_{ji}(\boldsymbol{x})n_j + \tau_{ji}(\boldsymbol{x})n_j = T_i(\boldsymbol{x}) \quad \text{or} \quad u_i(\boldsymbol{x}) = U_i(\boldsymbol{x}) \quad \forall\boldsymbol{x} \in S$$

3. Constitutive Equations:

$$\sigma_{ji}(\boldsymbol{x}) = \lambda_0\varepsilon_{rr}(\boldsymbol{x})\delta_{ij} + 2\mu_0\varepsilon_{ij}(\boldsymbol{x}) \qquad \forall\boldsymbol{x} \in V$$

4. Constitutive Equations of the Residual Nonlocal Field:

$$\tau_{ij}(\boldsymbol{x}) = \int\limits_V \lambda_r\left(\left|\boldsymbol{x}' - \boldsymbol{x}\right|\right)\varepsilon_{rr}\left(\boldsymbol{x}'\right)\delta_{ij}dV\left(\boldsymbol{x}'\right)$$
$$+ \int\limits_V 2\mu_r\left(\left|\boldsymbol{x}' - \boldsymbol{x}\right|\right)\varepsilon_{ij}\left(\boldsymbol{x}'\right)dV\left(\boldsymbol{x}'\right) \qquad \forall\boldsymbol{x} \in V$$

5. Kinematical Variables:

$$\varepsilon_{ij}(\boldsymbol{x}) = \frac{1}{2}\left(u_{i,j}(\boldsymbol{x}) + u_{j,i}(\boldsymbol{x})\right) \qquad \forall\boldsymbol{x} \in V \quad (19)$$

where $U_i(\boldsymbol{x})$ is a prescribed displacement field at the boundary.

It should be mentioned that the boundary value problem (Eq. (19)) is similar to the one represented based on Eringen's nonlocal elasticity [1, 2, 19]. However, the new form of the nonlocal boundary value problem presented in Eq. (19) depends on splitting the total nonlocal stress field into a local stress field and a nonlocal residual stress field.

In this section, an iterative procedure is proposed to solve the nonlocal boundary value problem (Eq. (19)). With the implementation of the iterative procedure proposed here, the nonlocal field equation is converted into a local-type field equation with an imposed nonlocal residual field. In each iteration, the nonlocal residual stress is known. This nonlocal residual stress is formed based on the stress field obtained in a previous iteration. Thus, the implementation of this iterative procedure gives the merit of forming the nonlocal residual stress iteratively based on a pre-determined local-type field. In the context of this iterative procedure, a local-type boundary value problem is corrected for the nonlocal field and then solved.

The iterative procedure implemented in this chapter is pioneered from an iterative-finite element-based model of nonlocal elasticity proposed in Ref. [19] and implemented in Refs. [20, 21] for beams and plates. In this chapter, a new iterative procedure is implemented to extract solutions of the proposed boundary value problem (Eq. (19)).

In an iteration (k), the nonlocal boundary value problem (19) is converted into a local boundary value problem of an elastic body exposed to a residual stress field, as follows:

1. Equation of Motion:
$$\sigma_{ji,j}^{(k)}(\boldsymbol{x}) + \tau_{ji,j}^{(k)}(\boldsymbol{x}) + F_i(\boldsymbol{x}) = \rho \ddot{u}_i^{(k)}(\boldsymbol{x}) \qquad \forall \boldsymbol{x} \in V$$
2. Boundary Conditions:
$$\sigma_{ji}^{(k)}(\boldsymbol{x})\, n_j = T_i(\boldsymbol{x}) - \tau_{ji}^{(k)}(\boldsymbol{x})\, n_j \quad \text{or} \quad u_i^{(k)}(\boldsymbol{x}) = U_i(\boldsymbol{x}) \quad \forall \boldsymbol{x} \in S$$
3. Constitutive Equations:
$$\sigma_{ji}^{(k)}(\boldsymbol{x}) = \lambda_0 \varepsilon_{rr}^{(k)}(\boldsymbol{x})\, \delta_{ij} + 2\mu_0 \varepsilon_{ij}^{(k)}(\boldsymbol{x}) \qquad \forall \boldsymbol{x} \in V$$
4. Constitutive Equations of the Residual Nonlocal Field:
$$\tau_{ij}^{(k)}(\boldsymbol{x}) = \int_V \lambda_r\left(\left|\boldsymbol{x}' - \boldsymbol{x}\right|\right) \varepsilon_{rr}^{(k-1)}\left(\boldsymbol{x}'\right) \delta_{ij} dV\left(\boldsymbol{x}'\right)$$
$$+ \int_V 2\mu_r\left(\left|\boldsymbol{x}' - \boldsymbol{x}\right|\right) \varepsilon_{ij}^{(k-1)}\left(\boldsymbol{x}'\right) dV\left(\boldsymbol{x}'\right) \qquad \forall \boldsymbol{x} \in V$$
5. Kinematical Variables:
$$\varepsilon_{ij}^{(k)}(\boldsymbol{x}) = \frac{1}{2}\left(u_{i,j}^{(k)}(\boldsymbol{x}) + u_{j,i}^{(k)}(\boldsymbol{x})\right) \qquad \forall \boldsymbol{x} \in V \qquad (20)$$

Equations (20) form a local boundary value problem of an elastic domain exposed to a pre-defined residual stress field, τ_{ij}. A procedure to derive a solution of the nonlocal boundary value problem (Eq. (20)) is shown in Fig. 2.

The proposed PROCEDURE presents a simple but effective approach to determine the solution of a nonlocal field problem formed based on the proposed iterative nonlocal residual elasticity. In an iteration k, the determined fields of a previous iteration $k-1$ ($\sigma_{ji}^{(k-1)}(\boldsymbol{x})$ and $\varepsilon_{ij}^{(k-1)}(\boldsymbol{x})$) are used to form the nonlocal residual stress, $\tau_{ji}^{(k)}(\boldsymbol{x})$, according to Eq. $(20)_4$. Then, the determined nonlocal residual stress is substituted into the local boundary value problem $(20)_1$ and $(20)_2$. This local boundary value problem is then solved for the fields $\sigma_{ji}^{(k)}(\boldsymbol{x})$ and $\varepsilon_{ij}^{(k)}(\boldsymbol{x})$. The procedure is then repeated where the solution converges to the nonlocal solution with the iterations.

PROCEDURE: in iteration k,

(I) Nonlocal residual stress, $\tau_{ji}^{(k)}(\mathbf{x})$, formation:

→ *Recall* $\varepsilon_{ij}^{(k-1)}(\mathbf{x})$, *the determined strain field of the previous iteration* $k-1$.

→ *Form* $\tau_{ji}^{(k)}(\mathbf{x})$ *using* Eq. (6.20)₄.

→ *Save* $\tau_{ji}^{(k)}(\mathbf{x})$.

(II) Solving the local boundary value problem:

→ *Recall* $\tau_{ji}^{(k)}(\mathbf{x})$.

→ *Form the local boundary value problem* (Eqs. $(6.20)_1$ and $(6.20)_2$).

→ *Solve the local boundary value problem for* $\sigma_{ji}^{(k)}(\mathbf{x})$ *and* $\varepsilon_{ij}^{(k)}(\mathbf{x})$.

→ *Save* $\sigma_{ji}^{(k)}(\mathbf{x})$ *and* $\varepsilon_{ij}^{(k)}(\mathbf{x})$.

End Procedure

Fig. 2 Procedure to determine a solution of the nonlocal boundary value problem Eq. (20)

It follows from the presented PROCEDURE that a solution is guaranteed in each iteration where the local boundary value problem is the one that is solved (a local boundary value problem is well-posed and admits a unique solution). The convergence of the iterative procedure to the nonlocal solution is conditional by the correct formation of the nonlocal residual stress. Indeed, with the new form of the nonlocal elasticity proposed here, the paradoxes and the inconsistencies associated with Eringen's nonlocal elasticity are effectively resolved. Examples of different boundary value problems are considered in the next section to demonstrate this fact.

3 Application to Euler-Bernoulli Beams

In this section, a set of nonlocal elasticity problems of the Euler-Bernoulli beams is solved using the proposed iterative nonlocal residual elasticity.

Because the nonlocal residual stress only models non-neighbor interactions, its Lame moduli (λ_r and μ_r) can be considered as follows:

$$\begin{aligned} \lambda_r(|\boldsymbol{x}|) &= \lambda_0\left(\beta_\lambda(|\boldsymbol{x}|) - \delta(|\boldsymbol{x}|)\right) \\ \mu_r(|\boldsymbol{x}|) &= \mu_0\left(\beta_\mu(|\boldsymbol{x}|) - \delta(|\boldsymbol{x}|)\right) \end{aligned} \tag{21}$$

with

$$\beta_\lambda(|\boldsymbol{x}|) = \frac{1}{2\ell_\lambda}\exp\left(-\frac{|\boldsymbol{x}|}{\ell_\lambda}\right)$$
$$\beta_\mu(|\boldsymbol{x}|) = \frac{1}{2\ell_\mu}\exp\left(-\frac{|\boldsymbol{x}|}{\ell_\mu}\right) \tag{22}$$

where $\delta(|\boldsymbol{x}|)$ is the Dirac-delta function. The substitution of Eqs. (21) and (22) into Eq. (18) yields:

$$\begin{aligned}\tau_{ij}(\boldsymbol{x}) = &\int_V \lambda_0\beta_\lambda\left(\left|\boldsymbol{x}'-\boldsymbol{x}\right|\right)\varepsilon_{rr}\left(\boldsymbol{x}'\right)\delta_{ij}dV\left(\boldsymbol{x}'\right)\\ &+\int_V 2\mu_0\beta_\mu\left(\left|\boldsymbol{x}'-\boldsymbol{x}\right|\right)\varepsilon_{ij}\left(\boldsymbol{x}'\right)dV\left(\boldsymbol{x}'\right)\\ &-\lambda_0\varepsilon_{rr}(\boldsymbol{x})\delta_{ij}-2\mu_0\varepsilon_{ij}(\boldsymbol{x})\end{aligned} \tag{23}$$

which is an explicit form of the nonlocal residual stress (τ_{ij}). This constitutive law depends on $\beta_\lambda\left(\left|\boldsymbol{x}'-\boldsymbol{x}\right|\right)$ and $\beta_\mu\left(\left|\boldsymbol{x}'-\boldsymbol{x}\right|\right)$, which are two independent kernels that define nonlocal effects. These kernels depend on two independent nonlocal parameters ℓ_λ and ℓ_μ. As previously demonstrated by the author, a general form of the nonlocal theory with expanded applicability is the one that considers different nonlocal kernels for the different material coefficients [22].

According to Eqs. (20) and (23), the boundary value problem of the elastostatic equilibrium of the bending of a beam subjected to a transverse load distribution $q(x)$ is derived as follows:

1. Equilibrium Equation:
$$\frac{d^2M^{(k)}(x)}{dx^2}+\frac{d^2\mathcal{M}^{(k)}(x)}{dx^2}=q(x)\quad \forall x\in[0,L]$$
2. Boundary Conditions:
$$\frac{dM^{(k)}}{dx}=\bar{V}-\frac{d\mathcal{M}^{(k)}}{dx}\quad \text{or}\quad w=\bar{w}\quad \text{at } x=0 \text{ or } L$$
$$M^{(k)}=\bar{M}-\mathcal{M}^{(k)}\quad \text{or}\quad w_{,x}=\bar{w}_{,x}\quad \text{at } x=0 \text{ or } L$$
3. Constitutive Equations:
$$M^{(k)}(x)=-(\lambda_0+2\mu_0)\frac{bh^3}{12}\frac{d^2w^{(k)}(x)}{dx^2}\quad \forall x\in[0,L]$$
4. Constitutive Equations of the Residual Nonlocal Field:
$$\mathcal{M}^{(k)}(x)=\int_0^L \xi_\lambda\beta_\lambda\left(\left|x'-x\right|\right)M^{(k-1)}\left(x'\right)dx'$$
$$+\int_0^L \xi_\mu\beta_\mu\left(\left|x'-x\right|\right)M^{(k-1)}\left(x'\right)dx'-M^{(k-1)}(x)\quad \forall x\in[0,L]$$

5. Kinematical Variables:

$$\varepsilon_{xx}^{(k)}(x) = -z\frac{d^2 w^{(k)}(x)}{dx^2} \quad \forall x \in [0, L]$$

where

$$\xi_\lambda = \frac{\lambda_0}{\lambda_0 + 2\mu_0}, \quad \xi_\mu = \frac{2\mu_0}{\lambda_0 + 2\mu_0}$$

$$\beta_\lambda\left(\left|x' - x\right|\right) = \frac{1}{2\ell_\lambda}\exp\left(-\frac{\left|x' - x\right|}{\ell_\lambda}\right)$$

$$\beta_\mu\left(\left|x' - x\right|\right) = \frac{1}{2\ell_\mu}\exp\left(-\frac{\left|x' - x\right|}{\ell_\mu}\right) \tag{24}$$

where $M(x)$ is the local-type bending moment, and $\mathcal{M}(x)$ is the nonlocal residual bending moment. $w(x)$ is the beam deflection. L, b and h are the beam length, width and height, respectively. $\bar{V}$, $\bar{M}$, $\bar{w}$, and $\bar{w}_{,x}$ are, respectively, prescribed shear force, bending moment, deflection, and slope at the boundary.

The proposed PROCEDURE is used to obtain the solution of the nonlocal boundary value problem (Eq. (24)). The procedure starts such that the nonlocal residual stress $\mathcal{M}(x)$ is zero. In this case, a solution of the boundary value problem (Eqs. $(24)_1$ and $(24)_2$) gives the local solution of the bending field $M(x)$. Then, the obtained bending field ($M(x)$) is substituted into the constitutive law $(24)_4$ to form the nonlocal residual stress ($\mathcal{M}(x)$). Afterwards, $\mathcal{M}(x)$ is substituted into the boundary value problem (Eqs. $(24)_1$ and $(24)_2$), which is then solved for the bending field $M(x)$. This process, when repeated, secretes the bending field $M(x)$ updated in each iteration. After a few iterations, both, the bending field $M(x)$ and the nonlocal residual bending field $\mathcal{M}(x)$ converge to their targeted solutions.

It should be mentioned that the bending field $M(x)$ is a local-type field, which accounts for the direct-neighbor interactions. This local-type field is adapted for the inclusion of the nonlocal residual fields of the elastic domain. The bending field $M(x)$ when added to the nonlocal residual bending $\mathcal{M}(x)$ forms a total bending field, which forms the constitutive law of Eringen's nonlocal elasticity, i.e.,

$$M_{total}(x) = M(x) + \mathcal{M}(x) \tag{25}$$

It should also be mentioned that the total bending field $M_{total}(x)$ formed by the sum of the bending fields $M(x)$ and $\mathcal{M}(x)$ is exactly the same as the one obtained from the equilibrium equation of Eringen's nonlocal elasticity. This total bending satisfies the requirement of the global balance.

Figures 3, 4 and 5 show the nonlocal residual bending moment $\mathcal{M}(x)$, the curvature, and the deflection of cantilever, simply supported, and clamped-clamped beams. The boundary conditions for the different beam configurations are defined in each iteration (k), as follows:

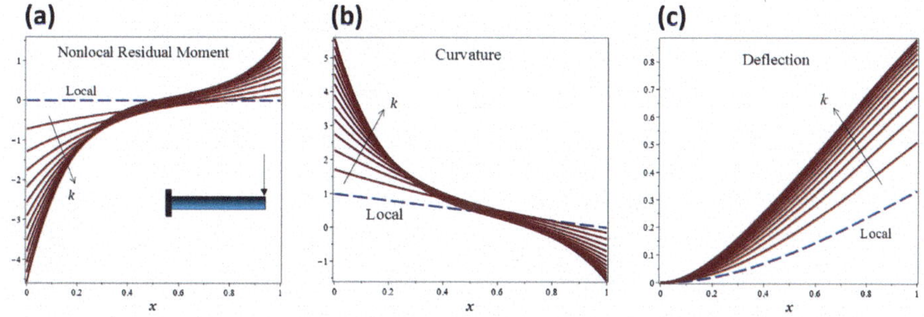

Fig. 3 Cantilever beam under a point load ($F = 1$): **a** The nonlocal residual moment ($\mathcal{M}(x)$), **b** the curvature ($d^2w/dx^2 = 12M/(\lambda_0 + 2\mu_0)\,bh^3$), and **c** the deflection (w) at each iteration (k). The broken curves represent the local solutions. $\xi_\lambda = 1/3$, $\xi_\mu = 2/3$, $\ell_\lambda = \ell_\mu = 0.5$, $L = 1$

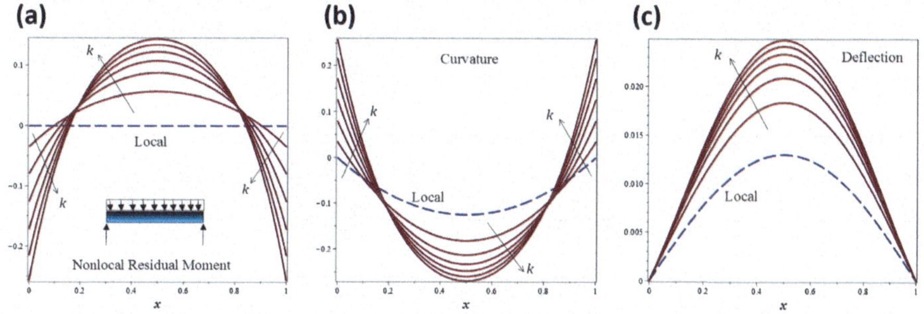

Fig. 4 Simply supported beam under a uniform load ($q = 1$): **a** The nonlocal residual moment ($\mathcal{M}(x)$), **b** the curvature ($d^2w/dx^2 = 12M/(\lambda_0 + 2\mu_0)\,bh^3$), and **c** the deflection (w) at each iteration (k). The broken curves represent the local solutions. $\xi_\lambda = 1/3$, $\xi_\mu = 2/3$, $\ell_\lambda = \ell_\mu = 0.4$, $L = 1$

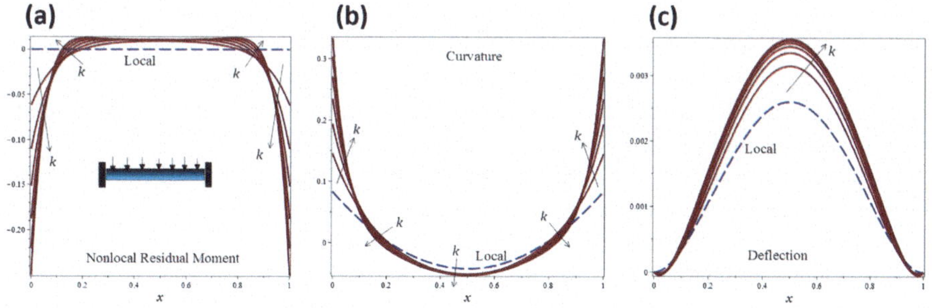

Fig. 5 Clamped-clamped beam under a uniform load (q): **a** The nonlocal residual moment ($\mathcal{M}(x)$), **b** the curvature ($d^2w/dx^2 = 12M/(\lambda_0 + 2\mu_0)\,bh^3$), and **c** the deflection (w) at each iteration (k). The broken curves represent the local solutions. $\xi_\lambda = 1/3$, $\xi_\mu = 2/3$, $\ell_\lambda = \ell_\mu = 0.1$, $L = 1$

Cantilever beam with a point load (F) at its free-end:

$$\begin{aligned}
&\frac{dM^{(k)}(L)}{dx} = -F - \frac{d\mathcal{M}^{(k)}(L)}{dx} \\
&M^{(k)}(L) = -\mathcal{M}^{(k)}(L) \\
&w(0) = 0 \\
&\frac{dw(0)}{dx} = 0
\end{aligned} \tag{26}$$

Simply supported beam under a uniform load (q):

$$\begin{aligned}
&M^{(k)}(0) = -\mathcal{M}^{(k)}(0) \\
&M^{(k)}(L) = -\mathcal{M}^{(k)}(L) \\
&w(0) = 0 \\
&w(L) = 0
\end{aligned} \tag{27}$$

Clamped-clamped beam under a uniform load (q):

$$\begin{aligned}
&M^{(k)}(0) = \frac{qL^2}{12} - \mathcal{M}^{(k)}(0) \\
&M^{(k)}(L) = \frac{qL^2}{12} - \mathcal{M}^{(k)}(L) \\
&w(0) = 0 \\
&w(L) = 0
\end{aligned} \tag{28}$$

As one of the main drawbacks of Eringen's nonlocal elasticity, it secretes stress fields identical to those of the classical elasticity. This is because it only forms the total stress field, which is always identical to the stress field of the classical elasticity due to the equilibrium requirements. Thus, Eringen's nonlocal elasticity totally hides the role of the nonlocal residual field and gives no information about it. This leads to the paradoxes and inconsistencies previously discussed.

The proposed iterative nonlocal residual elasticity outweighs Eringen's nonlocal elasticity where it distinguishes between nonlocal residual fields, local-type fields, and total nonlocal fields. The proposed nonlocal elasticity can effectively form nonlocal residual fields (this is for the first time). In addition, it secretes a stress field dependent on the nonlocal character of the elastic domain. This stress field is the local-type stress field ($M(x)$).

Figures 3, 4 and 5 show the effectiveness of the proposed iterative nonlocal residual elasticity to model nonlocal problems. In the first iteration, a local solution is obtained (the local solution is represented by broken-blue curves in Figs. 3, 4 and 5). After a few iterations, the solution converges to the targeted solution. The convergence is evident in Figs. 3, 4 and 5. For a detailed convergence analysis of the iterative nonlocal residual elasticity, readers can refer to Ref. [21].

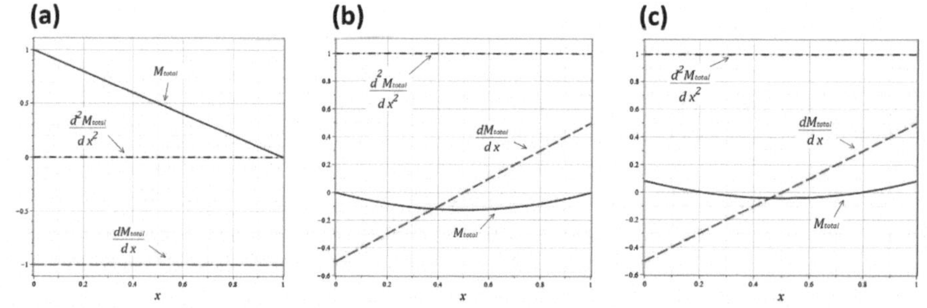

Fig. 6 Fulfillment of the global balance requirements: The total nonlocal bending $M_{total}(x)$ (Eq. (25)), the shear force $dM_{total}(x)/dx$, and the loading $d^2M_{total}(x)/dx^2$ fields of **a** cantilever under a unit point load, **b** simply supported beam under a unit-uniform load, and **c** clamped-clamped beam under a unit-uniform load. ($L = 1$)

The nonlocal residual moment ($\mathcal{M}(x)$) is zero in the first iteration to give the local solution. With the iterations, the nonlocal moment grows and converges to its complete value. This nonlocal residual moment adapt the boundary value problem to give updated solutions of the different fields of the beam. For the first time, nonlocal residual fields can be obtained by the proposed iterative nonlocal residual elasticity.

The obtained results in Figs. 3 , 4 and 5 are consistent. All fields represented in these figures for the different beam configurations reflect the inherited softness of the nonlocal elasticity. The nonlocal curvatures and deflections are obtained larger than those of the local elasticity.

In addition, the obtained solutions fulfill the requirement of the global balance. The proposed iterative approach is operated based on fulfilling the equilibrium requirements globally in each iteration. This can be demonstrated by depicting the total nonlocal bending $M_{total}(x)$ (Eq. (25)), shear force $dM_{total}(x)/dx$, and loading $d^2M_{total}(x)/dx^2$ fields in Fig. 6 for the different beam configurations. For a cantilever beam with a length L and under a point load of F, the global elastostatic equilibrium requires the total moment M_{total} is zero at the free-end and increases linearly to FL at the fixed end, and a constant shear force distribution over the beam. It is clear that the depicted results in Fig. 6a perfectly match these requirements. In addition to the cantilever beam, the equilibrium requirements of the other beams are effectively fulfilled, as shown in Fig. 6b, c. Furthermore, the external loading distribution is obtained when differentiating the total moment twice ($d^2M_{total}(x)/dx^2$), as shown in Fig. 6. These observations demonstrate the effectiveness of the proposed iterative nonlocal residual elasticity.

4 Conclusions

In this chapter, a new nonlocal elasticity, which completely forms the nonlocal residual fields, was developed. This formulation was motivated by the existing complications of finding solutions of the conventional model of the nonlocal elasticity early proposed by Eringen. The new nonlocal elasticity was given the name "iterative nonlocal residual elasticity". This new nonlocal elasticity is centered upon expressing the dynamic equilibrium requirements based on a nonlocal residual stress field. The nonlocal residual stress was formulated to model the material stiffness due to non-neighbor interactions. Solutions of the boundary value problems formed based on the new nonlocal elasticity are determined using an iterative procedure. When the iterative procedure is implemented, the nonlocal boundary value problem is converted into a local boundary value problem of an elastic domain exposed to a residual field. This residual field is a nonlocal-type field, which grows with the iterations and is used to correct the local field problem for the nonlocal fields of the elastic domain. This iterative procedure is an integrated part of the new nonlocal elasticity.

The iterative nonlocal residual elasticity was applied to Euler-Bernoulli beams to verify the convergence to the targeted nonlocal solution and the fulfillment of the global balance requirements. It was shown that the iterative nonlocal residual elasticity decomposes the total nonlocal stress field into a local-type stress field and a nonlocal residual stress field. Thus, for the first time, information about nonlocal residual stress fields can be obtained using the proposed nonlocal elasticity. Findings presented in this study came to demonstrate that the complications of Eringen's nonlocal elasticity are effectively resolved when using the iterative nonlocal residual elasticity.

References

1. Eringen AC (1972) Linear theory of nonlocal elasticity and dispersion of plane waves. Int J Eng Sci 10:425–435
2. Eringen AC, Edelen DGB (1972) On nonlocal elasticity. Int J Eng Sci 10:233–48
3. Eringen AC (1983) On differential equations of nonlocal elasticity and solutions of screw dislocation and surface waves. J Appl Phys 54:4703–4710
4. Benvenuti E, Simone A (2013) One-dimensional nonlocal and gradient elasticity: Closed-form solution and size effect. Mech Res Commun 48:46–51
5. Challamel N, Wang CM (2008) The small length scale effect for a non-local cantilever beam: a paradox solved. Nanotechnology 19:345703
6. Peddieson J, Buchanan GR, McNitt RP (2003) Application of nonlocal continuum models to nanotechnology. Int J Eng Sci 41:305–312
7. Fernández-Sáez J, Zaera R, Loya JA, Reddy JN (2016) Bending of Euler-Bernoulli beams using Eringen's integral formulation: a paradox resolved. Int J Eng Sci 99:107–116
8. Tuna M, Kirca M (2016) Exact solution of Eringen's nonlocal integral model for bending of Euler-Bernoulli and Timoshenko beams. Int J Eng Sci 105:80–92
9. Tuna M, Kirca M (2016) Exact solution of Eringen's nonlocal integral model for vibration and buckling of Euler-Bernoulli beam. Int J Eng Sci 107:54–67

10. Challamel N, Zhang Z, Wang CM, Reddy JN, Wang Q, Michelitsch T, Collet B (2014) On nonconservativeness of Eringen's nonlocal elasticity in beam mechanics: correction from a discrete-based approach. Arch Appl Mech 84:1275–1292
11. Wang CM, Kitipornchai S, Lim C, Eisenberger M (2008) Beam bending solutions based on nonlocal Timoshenko beam theory. J Eng Mech 134:475–481
12. Wang Q, Liew KM (2007) Application of nonlocal continuum mechanics to static analysis of micro-and nano-structures. Phys Lett A 363:236–242
13. Li C, Yao L, Chen W, Li S (2015) Comments on nonlocal effects in nano-cantilever beams. Int J Eng Sci 87:47–57
14. Rafii-Tabar H, Ghavanloo E, Fazelzadeh SA (2016) Nonlocal continuum-based modeling of mechanical characteristics of nanoscopic structures. Phys Rep 638:1–97
15. Romano G, Barretta R, Diaco M, de Sciarra FM (2017) Constitutive boundary conditions and paradoxes in nonlocal elastic nanobeams. Int J Mech Sci 121:151–156
16. Koutsoumaris CC, Eptaimeros KG (2018) A research into bi-Helmholtz type of nonlocal elasticity and a direct approach to Eringen's nonlocal integral model in a finite body. Acta Mech 229:3629–3649
17. Mahmoud FF (2017) On the non-existence of a feasible solution in the context of the differential form of Eringen's constitutive model: a proposed iterative model based on a residual nonlocality formulation. Int J Appl Mech 9:17594
18. Pour EM, Hosseini-Hashemi SH, Faghidian SA (2018) Nonlinear vibration analysis of FG nano-beams resting on elastic foundation in thermal environment using stress-driven nonlocal integral model. Appl Math Model 57:302–315
19. Polizzotto C (2017) Nonlocal elasticity and related variational principles. Int J Solids Struct 38:7359–7380
20. Shaat M (2018) Correction of local elasticity for nonlocal residuals: application to Euler-Bernoulli beams. Meccanica 53:3015–3035
21. Shaat M (2015) Iterative nonlocal elasticity for Kirchhoff plates. Int J Mech Sci 90:162–170
22. Shaat M (2017) A general nonlocal theory and its approximations for slowly varying acoustic waves. Int J Mech Sci 130:52–63

Nonlocal Gradient Mechanics of Elastic Beams Under Torsion

Francesco P. Pinnola, S. Ali Faghidian, Marzia S. Vaccaro, Raffaele Barretta, and Francesco Marotti de Sciarra

Abstract Nonlocal gradient mechanics of elastic beams subject to torsion is established by means of a variationally consistent methodology, equipped with suitable functional spaces of test fields. The proposed elasticity theory is the generalization of size-dependent models recently contributed in literature to assess size-effects in nano-structures, such as modified nonlocal strain gradient and strain- and stress-driven local/nonlocal elasticity formulations. General new ideas are elucidated by examining the torsional behavior of elastic nano-beams. Equivalence between nonlocal integral convolutions and differential problems subject to variationally consistent boundary conditions is demonstrated for special averaging kernels. The variational procedure leads to well-posed engineering problems in nano-mechanics. Elasto-static responses and free vibrations of nano-beams under torsion are analyzed applying an effective analytical solution technique. Nonlocal strain- and stress-driven gradient models of elasticity can efficiently predict both stiffening and softening structural responses, and thus, notably characterize small-scale phenomena in structures exploited in modern Nano-Electro-Mechanical-Systems (NEMS).

F. P. Pinnola · M. S. Vaccaro · R. Barretta (✉) · F. Marotti de Sciarra
Department of Structures for Engineering and Architecture, University of Naples Federico II, Via Claudio 21, 80125 Naples, Italy
e-mail: rabarret@unina.it

F. P. Pinnola
e-mail: francescopaolo.pinnola@unina.it

M. S. Vaccaro
e-mail: marziasara.vaccaro@unina.it

F. Marotti de Sciarra
e-mail: marotti@unina.it

S. A. Faghidian
Department of Mechanical Engineering, Science and Research Branch, Islamic Azad University, Tehran, Iran
e-mail: faghidian@srbiau.ac.ir

© The Author(s), under exclusive license to Springer Nature Switzerland AG 2021
E. Ghavanloo et al. (eds.), *Size-Dependent Continuum Mechanics Approaches*, Springer Tracts in Mechanical Engineering,
https://doi.org/10.1007/978-3-030-63050-8_7

1 Introduction

Nano-engineered materials are nowadays widely exploited as fundamental constituents of modern Nano-Electro-Mechanical-Systems (NEMS) due to their outstanding physical features [1–5]. Nano-structures are well-recognized to demonstrate size-dependent mechanical responses at small-scales which cannot be technically modeled by the local elasticity theory. The thematic concerning with the analysis of size-effects in advanced materials and structures has stimulated a great deal of interest in the current literature, such as nonlocal elasticity [6–12], local/nonlocal mixture [13–15], strain gradient elasticity [16–19], nonlocal strain gradient theory [20–22] and variational nonlocal gradient elasticity [23, 24]. Recent review on size-dependent elasticity models can be found in Ref. [25].

Nonlocal constitutive law associated with the strain-driven nonlocal elasticity, originally exploited by Eringen [26], is extensively applied for the investigation of size-effects in nano-structures [27, 28]. Inapplicability of strain-driven nonlocal integral elasticity to nano-beams of technical interest, involving bounded structural domains has been discussed and acknowledged in the recent literature [29]. On the contrary, the stress-driven nonlocal elasticity, conceived in [30], leads to mathematically well-posed nonlocal problems. Pure and two-phase stress-driven nonlocal models have been effectively utilized to capture size-effects in nano-structures subject to both static and dynamic phenomena, see e.g. [31, 32]. Nonlocal strain gradient model [33, 34] is also widely utilized to tackle small-scale effects in nano-continua while employing unnecessary higher-order boundary conditions. There is, however, a dispute in the literature on the suitable choice of non-standard boundary conditions required to close the nonlocal strain gradient problem [20, 21, 35, 36]. The constitutive boundary conditions, naturally stemmed from the nonlocal integral constitutive law, have been recently addressed in the framework of modified nonlocal strain gradient elasticity [35]. The consistent variational scheme, with suitably selected functional spaces describing test fields, is lately conceived for nonlocal gradient inflected beams [36]. The well-posed strain- and stress-driven approaches of nonlocal gradient elasticity are able to efficiently demonstrate both softening and stiffening nonlocal responses in the flexure of elastic nano-beams [23, 36].

In view of the importance of examining scale phenomena in torsional elements of modern NEMS, the motivation of the present chapter is to generalize the variational nonlocal gradient approach to the torsion of elastic nano-beams. The outline of the present chapter is as follows. Preliminary notions of kinematics and equilibrium of elastic beams under torsion are briefly recalled in Sect. 2. Nonlocal strain- and stress-driven gradient formulations of elasticity for torsion of nano-beams are developed in Sect. 3. Section 4 is devoted to the elasto-static and -dynamic analysis of the torsional behavior of nano-beams where numerical results are also provided and commented upon. Concluding remarks are drawn in Sect. 5.

2 Local Elastic Beams

A straight beam of length L subject to torsion, with circular cross-section Ξ, is considered. The abscissa x is taken along the beam axis, orthogonal to the cross-sectional plane including the axes y and z. Motivated by the Saint-Venant's problem solution [37, 38], the displacement field $\mathbf{u}$ of the beam, up to an inessential additional rigid body motion, writes as

$$\mathbf{u}\,(\mathbf{r}, x) = \theta\,(x)\,\mathbf{Rr} \tag{1}$$

with $\theta : [0, L] \mapsto \Re$ torsional cross-sectional rotation function. The tensor $\mathbf{R}$ is the rotation by $\pi/2$ counter clock-wise in Ξ and position vector of a cross-sectional point with respect to the centroid is represented by $\mathbf{r} = (y, z)$. The shear strain vector $\boldsymbol{\gamma} = \left(\gamma_{yx}, \gamma_{zx}\right)$, kinematically compatible with the displacement field $\mathbf{u}$, is detected as

$$\boldsymbol{\gamma}\,(\mathbf{r}, x) = \chi\,(x)\,\mathbf{Rr} = \partial_x\theta\,(x)\,\mathbf{Rr} \tag{2}$$

with the geometric torsional curvature $\chi : [0, L] \mapsto \Re$ being the first derivative of the torsional rotation along the axial abscissa x. Furthermore, introducing the mass polar moment of inertia J_ρ and torsional stiffness J_G is useful in the formulation of the torsional problem

$$\begin{aligned} J_\rho &= \iint\limits_{\Xi} \rho\,(\mathbf{r}\cdot\mathbf{r})\,dA \\ J_G &= \iint\limits_{\Xi} G\,(\mathbf{r}\cdot\mathbf{r})\,dA \end{aligned} \tag{3}$$

with ρ and G material density and shear elastic modulus, respectively. The dot into the integrals stands for inner product between vectors.

The loading system on the beam is assumed to consist of distributed torque per unit length $m : [0, L] \mapsto \Re$ and concentrated couples T_0 and T_L at the end cross-sections. The principle of virtual work can be applied to prescribe the dynamic equilibrium condition as

$$\int_0^L \left(m - J_\rho\partial_{tt}\theta\right)\delta\theta dx + \left[T_0\delta\theta\,(0) + T_L\delta\theta\,(L)\right] = \int_0^L T\delta\chi dx \tag{4}$$

for any virtual torsional rotation field $\delta\theta : [0, L] \mapsto \Re$ fulfilling homogeneous kinematic boundary conditions. The twisting resultant moment T is defined by

$$T = \iint\limits_{\Xi} \boldsymbol{\tau}\cdot(\mathbf{Rr})\,dA \tag{5}$$

with $\boldsymbol{\tau} = \left(\tau_{yx}, \tau_{zx}\right)$ shear stress vector field.

As a result of employing a standard localization procedure based on the integration by parts, differential and standard boundary conditions associated with the variational formulation Eq. (4) are expressed by

$$\partial_x T + m = J_\rho \partial_{tt} \theta \tag{6}$$

$$(T + T_0)\,\delta\theta|_{x=0} = (T - T_L)\,\delta\theta|_{x=L} = 0 \tag{7}$$

3 Nonlocal Gradient Elastic Beams

The nonlocal gradient model introduced in Ref. [36] for elastic inflected nano-beams is extended in this section to formulate nonlocal gradient nano-beams under torsion. A consistent variational formulation for both Nonlocal strain-driven Gradient (NstrainG) and Nonlocal stress-driven Gradient (NstressG) models of elasticity is introduced. The conceived NstrainG and NstressG elasticity theories are shown to result in well-posed nonlocal problems in bounded structural domains of nano-engineering interest.

The definition of integral convolution between a smoothing kernel φ_c and a scalar field f is preliminarily recalled for conciseness sake

$$(\varphi_c * f)(x) := \int_0^L \varphi_c(x - \bar{x}) f(\bar{x}) d\bar{x} \tag{8}$$

with x and $\bar{x}$ being the points of the structural interval $[0, L]$. The length-scale parameter demonstrating nonlocal effects is denoted by $c \in]0, \infty[$. The smoothing kernel φ_c is selected to meet positivity, symmetry, normalization and impulsivity properties [29].

3.1 Nonlocal Strain-Driven Gradient (NstrainG) Elasticity

The abstract formulation of nonlocal gradient beams under torsion consistent with NstrainG is governed by the following elastic energy, Π_{NstrainG},

$$\Pi_{\text{NstrainG}}(\chi) := \frac{1}{2} J_G \int_0^L \left[\alpha\chi^2 + (1-\alpha)(\varphi_c * \chi)\chi + \ell^2 (\varphi_c * \partial_x \chi)\partial_x \chi\right] dx \tag{9}$$

with $\alpha \in [0, 1]$ mixture parameter and $\ell \in [0, \infty[$ gradient length-scale parameter. The relevant nonlocal gradient twisting moment T is provided by the variational condition

$$\langle T, \delta\chi \rangle := \int_0^L T(x)\delta\chi(x)\,dx = \langle d\Pi_{\text{NstrainG}}(\chi), \delta\chi \rangle \tag{10}$$

for any virtual torsional curvature field $\delta\chi \in C_0^1([0, L]; \Re)$ having compact support in the structural domain. To determine the NstrainG constitutive law defining the twisting moment T in terms of torsional curvature χ, the directional derivative of the elastic energy Π_{NstrainG} is evaluated while integrating by parts

$$\begin{aligned}
\langle d\Pi_{\text{NstrainG}}(\chi), \delta\chi \rangle &= J_G \int_0^L [\alpha\chi\delta\chi + (1-\alpha)(\varphi_c * \chi)\,\delta\chi]\,dx \\
&+ J_G \int_0^L \left[\ell^2 (\varphi_c * \partial_x\chi)(\partial_x\delta\chi)\right] dx \\
&= J_G \int_0^L \left[\alpha\chi + (1-\alpha)(\varphi_c * \chi) - \ell^2 \partial_x (\varphi_c * \partial_x\chi)\right]\delta\chi dx \\
&+ J_G \ell^2 \left[(\varphi_c * \partial_x\chi)\,\delta\chi|_{x=L} - (\varphi_c * \partial_x\chi)\,\delta\chi|_{x=0}\right]
\end{aligned} \tag{11}$$

The choice of assuming virtual torsional curvature fields to have compact supports leads to vanishing boundary values $\delta\chi|_{x=0}$ and $\delta\chi|_{x=L}$. As a result of implementing a standard localization procedure, the sought nonlocal gradient constitutive law is detected via prescription of the variational condition Eq. (10) as

$$T(x) = J_G \left[\alpha\chi(x) + (1-\alpha)(\varphi_c * \chi)(x) - \ell^2 \partial_x (\varphi_c * \partial_x\chi)(x)\right] \tag{12}$$

Helmholtz bi-exponential kernel, fulfilling positivity, symmetry, normalization and impulsivity conditions, is a well-accepted choice for the special smoothing kernel φ_c

$$\varphi_c(x) := \frac{1}{2c} \exp\left(-\frac{|x|}{c}\right) \tag{13}$$

Following the proposition 3.1 of [36], the NstrainG integro-differential law can be demonstrated to be equivalent to an appropriate differential constitutive problem equipped with suitable constitutive boundary conditions.

Proposition 7.1. Constitutive equivalency for NstrainG

The nonlocal gradient constitutive law Eq. (12), on a bounded structural interval $[0, L]$, with bi-exponential kernel Eq. (13) for nano-beams subject to torsion is equivalent to the differential constitutive equation

$$\frac{1}{c^2} T(x) - \partial_{xx} T(x) = \frac{1}{c^2} J_G \chi(x) - J_G \left(\alpha + \frac{\ell^2}{c^2}\right) \partial_{xx}\chi(x) \tag{14}$$

subject to two constitutive boundary conditions (CBCs)

$$\partial_x T(0) - \frac{1}{c} T(0) = -J_G \frac{\alpha}{c} \chi(0) + J_G \left(\alpha + \frac{\ell^2}{c^2}\right) \partial_x \chi(0)$$
$$\partial_x T(L) + \frac{1}{c} T(L) = J_G \frac{\alpha}{c} \chi(L) + J_G \left(\alpha + \frac{\ell^2}{c^2}\right) \partial_x \chi(L) \tag{15}$$

The general formulation of nonlocal gradient elasticity comprises some well-established special models adopted in mechanics of nano-structures. The nonlocal strain gradient constitutive law equipped with suitable CBCs [24] can be recovered via vanishing the mixture parameter $\alpha \to 0$ as

$$\frac{1}{c^2} T(x) - \partial_{xx} T(x) = \frac{1}{c^2} J_G \chi(x) - J_G \frac{\ell^2}{c^2} \partial_{xx} \chi(x)$$
$$\partial_x T(0) - \frac{1}{c} T(0) = J_G \frac{\ell^2}{c^2} \partial_x \chi(0)$$
$$\partial_x T(L) + \frac{1}{c} T(L) = J_G \frac{\ell^2}{c^2} \partial_x \chi(L) \tag{16}$$

The two-phase local/nonlocal strain-driven model and associated CBCs can be also obtained as the gradient characteristic length approaches zero $\ell \to 0$ [32]

$$\frac{1}{c^2} T(x) - \partial_{xx} T(x) = \frac{1}{c^2} J_G \chi(x) - J_G \alpha \partial_{xx} \chi(x)$$
$$\partial_x T(0) - \frac{1}{c} T(0) = -J_G \frac{\alpha}{c} \chi(0) + J_G \alpha \partial_x \chi(0)$$
$$\partial_x T(L) + \frac{1}{c} T(L) = J_G \frac{\alpha}{c} \chi(L) + J_G \alpha \partial_x \chi(L) \tag{17}$$

3.2 Nonlocal Stress-Driven Gradient (NstressG) Elasticity

While the roles of stress and strain fields can be readily swapped in the framework of local elasticity, two distinct nonlocal gradient formulations should be prescribed based on the physical interpretation of source and output elastic fields. NstressG can be introduced by converting the source and output fields of the integral convolution with respect to NstrainG model. Accordingly, the elastic potential of nano-beams under torsion Π_{NstressG} associated with NstressG is defined by

$$\Pi_{\text{NstressG}}(T) := \frac{1}{2} \frac{1}{J_G} \int_0^L \left[\alpha T^2 + (1-\alpha)(\varphi_c * T) T\right] dx$$
$$+ \frac{1}{2} \frac{1}{J_G} \int_0^L \left[\ell^2 (\varphi_c * \partial_x T) \partial_x T\right] dx \tag{18}$$

The torsional curvature χ of NstressG is established by the variational constitutive condition

$$\langle \chi, \delta T \rangle := \int_0^L \chi(x) \delta T(x) \, dx = \langle d\Pi_{\text{NstressG}}(T), \delta T \rangle \tag{19}$$

for any virtual twisting moment field $\delta T \in C_0^1([0, L]; \Re)$ having compact support in the structural domain. The directional derivative of the elastic potential along a virtual twisting moment can be determined by introducing the expression of Π_{NstressG} and integrating by parts

$$\begin{aligned}
\langle d\Pi_{\text{NstressG}}(T), \delta T \rangle &= \frac{1}{J_G} \int_0^L [\alpha T \delta T + (1-\alpha)(\varphi_c * T)\delta T] \, dx \\
&+ \frac{1}{J_G} \int_0^L \left[\ell^2 (\varphi_c * \partial_x T)(\partial_x \delta T)\right] dx \\
&= \frac{1}{J_G} \int_0^L \left[\alpha T + (1-\alpha)(\varphi_c * T) - \ell^2 \partial_x (\varphi_c * \partial_x T)\right] \delta T dx \\
&+ \frac{1}{J_G} \ell^2 \left[(\varphi_c * \partial_x T) \delta T|_{x=L} - (\varphi_c * \partial_x T) \delta T|_{x=0} \right]
\end{aligned} \tag{20}$$

The boundary terms in Eq. (20) are disappeared due to assuming virtual test fields to have compact supports in the structural domain. A standard localization procedure then provides the NstressG torsional curvature χ in terms of twisting moment T while applying the variational condition Eq. (19)

$$\chi(x) = \frac{1}{J_G} \left[\alpha T(x) + (1-\alpha)(\varphi_c * T)(x) - \ell^2 \partial_x (\varphi_c * \partial_x T)(x)\right] \tag{21}$$

The equivalent differential constitutive problem with the corresponding new CBCs in the framework of NstressG is similarly determined by assuming the smoothing kernel to be the Helmholtz bi-exponential function.

Proposition 7.2. Constitutive equivalency for NstressG

The nonlocal gradient constitutive relation Eq. (21), endowed with the Helmholtz bi-exponential kernel Eq. (13), is equivalent to the differential constitutive equation

$$\frac{1}{c^2} \chi(x) - \partial_{xx} \chi(x) = \frac{1}{c^2} \frac{1}{J_G} T(x) - \frac{1}{J_G} \left(\alpha + \frac{\ell^2}{c^2}\right) \partial_{xx} T(x) \tag{22}$$

equipped with the constitutive boundary conditions

$$\partial_x \chi(0) - \frac{1}{c}\chi(0) = -\frac{1}{J_G}\frac{\alpha}{c}T(0) + \frac{1}{J_G}\left(\alpha + \frac{\ell^2}{c^2}\right)\partial_x T(0)$$
$$\partial_x \chi(L) + \frac{1}{c}\chi(L) = \frac{1}{J_G}\frac{\alpha}{c}T(L) + \frac{1}{J_G}\left(\alpha + \frac{\ell^2}{c^2}\right)\partial_x T(L) \tag{23}$$

The stress-driven type formulation of NstressG includes well-known particular elasticity models of nano-mechanics. Setting the gradient characteristic length zero $\ell \to 0$, the two-phase local/nonlocal stress-driven model and associated CBCs can be recovered [32]

$$\frac{1}{c^2}\chi(x) - \partial_{xx}\chi(x) = \frac{1}{c^2}\frac{1}{J_G}T(x) - \frac{1}{J_G}\alpha\partial_{xx}T(x)$$
$$\partial_x \chi(0) - \frac{1}{c}\chi(0) = -\frac{1}{J_G}\frac{\alpha}{c}T(0) + \frac{1}{J_G}\alpha\partial_x T(0)$$
$$\partial_x \chi(L) + \frac{1}{c}\chi(L) = \frac{1}{J_G}\frac{\alpha}{c}T(L) + \frac{1}{J_G}\alpha\partial_x T(L) \tag{24}$$

As expected, the stress-driven purely nonlocal formulation of elastic torsion can be obtained via vanishing the mixture parameter $\alpha \to 0$ and the gradient characteristic length $\ell \to 0$ as [32]

$$\frac{1}{c^2}\chi(x) - \partial_{xx}\chi(x) = \frac{1}{c^2}\frac{1}{J_G}T(x)$$
$$\partial_x \chi(0) - \frac{1}{c}\chi(0) = 0$$
$$\partial_x \chi(L) + \frac{1}{c}\chi(L) = 0 \tag{25}$$

4 Nonlocal Gradient Nano-Structures Under Torsion

The established variationally consistent nonlocal strain- and stress-driven gradient models of elasticity are exploited in this section to examine size-dependent torsional responses of structural schemes of nano-mechanical interest: cantilever and fully-clamped nano-beams. In elasto-static torsional analysis, a nano-beam of length L is assumed to be subjected to uniformly distributed couples $\bar{m}$ per unit length. The non-dimensional parameters: axial abscissa $\bar{x}$, nonlocal characteristic parameter λ, gradient characteristic parameter μ, torsional rotation $\bar{\theta}$ and fundamental torsional frequency $\bar{\omega}$ are introduced as

$$\bar{x} = \frac{x}{L}, \quad \lambda = \frac{c}{L}, \quad \mu = \frac{\ell}{L}, \quad \bar{\theta}(\bar{x}) = \theta(x)\frac{J_G}{\bar{m}L^2}, \quad \bar{\omega}^2 = \left(\frac{L^2 J_\rho}{\pi^2 J_G}\right)\omega^2 \tag{26}$$

4.1 Elastostatic Torsion

Inertia terms are absent in the elasto-static analysis, and thus, the differential condition of equilibrium Eq. (6) can be integrated to detect the twisting moment T in terms of an integration constant Λ_1 as

$$T(x) = -\int_0^x m(\zeta)\, d\zeta + \Lambda_1 \tag{27}$$

The torsional curvature χ is subsequently determined via solving the constitutive differential equation of NstrainG Eq. (14) in terms of integration constants Λ_2 and Λ_3

$$\begin{aligned}\chi(x) = {} & \Lambda_2 \exp\left(-\frac{x}{\sqrt{\alpha c^2+\ell^2}}\right) + \Lambda_3 \exp\left(\frac{x}{\sqrt{\alpha c^2+\ell^2}}\right) \\ & + \frac{1}{2J_G\sqrt{\alpha c^2+\ell^2}} \exp\left(-\frac{x}{\sqrt{\alpha c^2+\ell^2}}\right) \\ & \cdot \int_0^x \exp\left(\frac{\xi}{\sqrt{\alpha c^2+\ell^2}}\right)\left(T(\xi) - c^2 \partial_{\xi\xi} T(\xi)\right) d\xi \\ & - \frac{1}{2J_G\sqrt{\alpha c^2+\ell^2}} \exp\left(\frac{x}{\sqrt{\alpha c^2+\ell^2}}\right) \\ & \cdot \int_0^x \exp\left(-\frac{\eta}{\sqrt{\alpha c^2+\ell^2}}\right)\left(T(\eta) - c^2 \partial_{\eta\eta} T(\eta)\right) d\eta \end{aligned} \tag{28}$$

Similarly, the constitutive differential equation of NstressG Eq. (22) can be solved to determine the torsional curvature field in terms of integration constants Λ_2 and Λ_3

$$\begin{aligned}\chi(x) = {} & \Lambda_2 \exp\left(-\frac{x}{c}\right) + \Lambda_3 \exp\left(\frac{x}{c}\right) \\ & + \frac{1}{2cJ_G} \exp\left(-\frac{x}{c}\right) \int_0^x \exp\left(\frac{\xi}{c}\right)\left(T(\xi) - \left(\alpha c^2+\ell^2\right) \partial_{\xi\xi} T(\xi)\right) d\xi \\ & - \frac{1}{2cJ_G} \exp\left(\frac{x}{c}\right) \int_0^x \exp\left(-\frac{\eta}{c}\right)\left(T(\eta) - \left(\alpha c^2+\ell^2\right) \partial_{\eta\eta} T(\eta)\right) d\eta \end{aligned} \tag{29}$$

Lastly, the torsional rotation field θ can be evaluated integrating the differential condition of kinematic compatibility $\chi = \partial_x \theta$ in terms of the integration constant Λ_4 as

$$\theta(x) = \int_0^x \chi(\zeta) d\zeta + \Lambda_4 \tag{30}$$

The integration constants Λ_k ($k = 1,\ldots,4$) can be detected by prescribing two Constitutive Boundary Conditions of NstrainG Eq. (15) or NstressG Eq. (23), which are independent of the considered boundary kinematic constraints, in addition to two standard kinematic and static boundary conditions (BCs) specialized here for the examined case-studies.

In case of the cantilever nano-beam subject to uniformly distributed couples, the torsional solution field has to fulfill the (standard) classical boundary conditions

$$\theta(0) = 0, \quad T(L) = 0 \tag{31}$$

The (standard) essential kinematic boundary conditions in case of a fully-clamped nano-beam subject to uniformly distributed couples are also given by

$$\theta(0) = 0, \quad \theta(L) = 0 \tag{32}$$

The proposed analytical approach provides exact analytical solutions in consequence of integrating differential equations of lower order. In the sequel, the acronyms LOC, NstrainG and NstressG, respectively, denote the local beam model, nonlocal strain-driven gradient model and nonlocal stress-driven gradient model.

To visibly demonstrate the effects of the gradient characteristic parameter on the torsional responses of cantilever nano-beams, numerical values of torsional rotations are evaluated in the mid-span. The normalized torsional rotations at the mid-span of the cantilever nano-beam associated with NstrainG and NstressG under uniformly distributed couples are exhibited in Figs. 1 and 2.

Likewise, the normalized maximum torsional rotations of fully-clamped nano-beams consistent with NstrainG and NstressG under uniformly distributed couples are illustrated in Figs. 3 and 4. Detected torsional rotation fields are also normalized exploiting the corresponding torsional rotation of the local beam model $\bar{\theta}^{LOC}$. In Figs. 1, 2, 3 and 4, while the nonlocal characteristic parameter λ is ranging in the interval $]0, 1[$, the gradient characteristic parameter μ is ranging in the set of

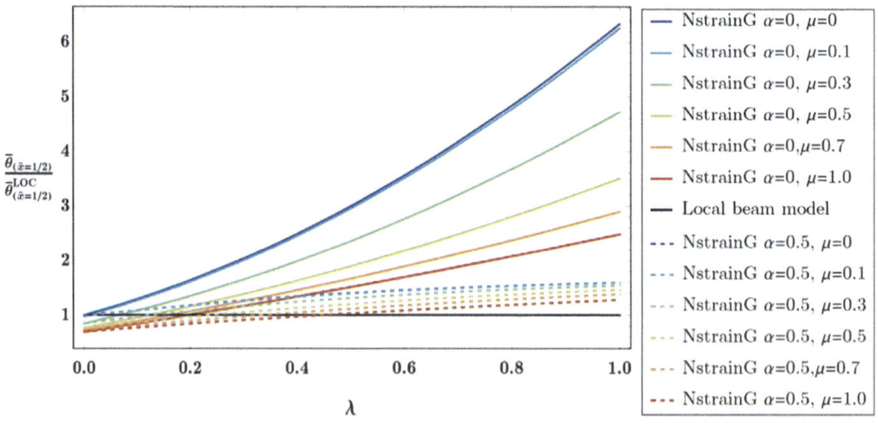

Fig. 1 Cantilever NstrainG nano-beams under uniform couples: normalized mid-span torsional rotation

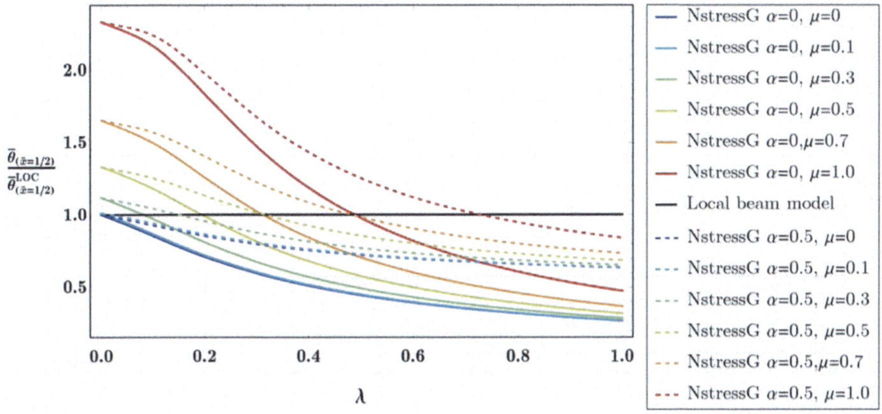

Fig. 2 Cantilever NstressG nano-beams under uniform couples: normalized mid-span torsional rotation

$\{0, 0.1, 0.3, 0.5, 0.7, 1.0\}$ and two values of the mixture parameter as $\alpha = 0$ and $\alpha = 0.5$ are prescribed. It is deduced from Figs. 1, 2, 3 and 4 that the size-dependent NstrainG model exhibits a softening behavior in terms of nonlocal characteristic parameter λ, that is a larger λ involves a larger torsional rotation for given gradient and mixture parameters. The torsional rotation of elastic nano-beams decreases as the gradient or the mixture parameters increase, and accordingly, NstrainG theory demonstrates a stiffening behavior in terms of gradient and mixture parameters for a given value of λ. Effects of characteristic parameters are more pronounced in NstrainG beams with fully-clamped ends. On the contrary, a softening response is demonstrated in the framework of NstressG for increasing gradient or mixture

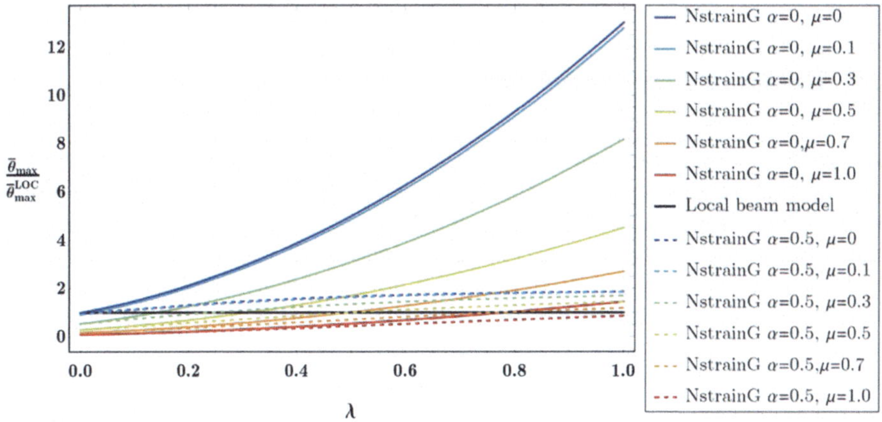

Fig. 3 Fully-clamped NstrainG nano-beams under uniform couples: normalized maximum torsional rotation

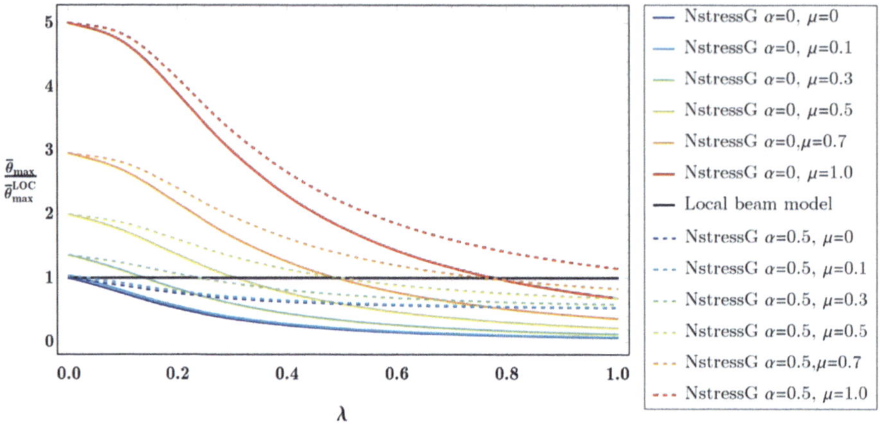

Fig. 4 Fully-clamped NstressG nano-beams under uniform couples: normalized maximum torsional rotation

parameters and a stiffening behavior is detected for increasing nonlocal characteristic parameter. In the framework of NstressG model, size-dependent elastostatic responses of fully-clamped beams are more affected by characteristic parameters.

As expected, the size-dependent elastic torsional rotation of nano-beams in accordance with either of the NstrainG or NstressG models coincides with the local response for vanishing small-scale characteristic and mixture parameters. Additionally, size-dependent effects of nonlocal and gradient parameters are less noticeable in the presence of non-vanishing mixture parameter $\alpha \neq 0$. Notably, NstrainG and NstressG theories demonstrate different softening and stiffening structural responses in terms of characteristic parameters due to profound differences in the fundamental assumptions of the models to capture scale phenomena.

4.2 Torsional Free Vibrations

In order to examine torsional free vibrations of nano-beams, the relevant elastodynamic problems associated with NstrainG and NstressG are formulated in terms of the torsional rotation field. The distributed couple is allowed to vanish, and consequently, twisting moment field T can be determined by prescribing the differential condition of equilibrium Eq. (6) to the constitutive differential law of NstrainG Eq. (14) or NstressG Eq. (22) as

$$\begin{aligned}\frac{1}{c^2}T_{\mathrm{NstrainG}}(x,t) &= J_\rho \partial_{tt}\chi(x,t)\\ &+ \frac{1}{c^2}J_G\chi(x,t) - J_G\left(\alpha + \frac{\ell^2}{c^2}\right)\partial_{xx}\chi(x,t)\\ \frac{1}{c^2}T_{\mathrm{NstressG}}(x,t) &= \left(\alpha + \frac{\ell^2}{c^2}\right)J_\rho \partial_{tt}\chi(x,t)\\ &+ \frac{1}{c^2}J_G\chi(x,t) - J_G\partial_{xx}\chi(x,t)\end{aligned} \tag{33}$$

The differential condition of dynamic equilibrium governing torsional vibrations of nano-beams can be expressed in terms of torsional rotation field by applying kinematic compatibility. Accordingly, the differential condition of dynamic equilibrium consistent with the NstrainG is

$$\begin{aligned}J_\rho \partial_{ttxx}\theta(x,t) + \frac{1}{c^2}J_G\partial_{xx}\theta(x,t) - J_G\left(\alpha + \frac{\ell^2}{c^2}\right)\partial_{xxxx}\theta(x,t)\\ = \frac{1}{c^2}J_\rho\partial_{tt}\theta(x,t)\end{aligned} \tag{34}$$

and NstressG writes as

$$\begin{aligned}\left(\alpha + \frac{\ell^2}{c^2}\right)J_\rho \partial_{ttxx}\theta(x,t) + \frac{1}{c^2}J_G\partial_{xx}\theta(x,t) - J_G\partial_{xxxx}\theta(x,t)\\ = \frac{1}{c^2}J_\rho\partial_{tt}\theta(x,t)\end{aligned} \tag{35}$$

equipped with the classical (standard) boundary conditions Eq. (7) and corresponding constitutive boundary conditions associated with NstrainG Eq. (15) or NstressG Eq. (23). A standard procedure of separating spatial and time variables is subsequently employed to study torsional free vibrations

$$\theta(x,t) = \Theta(x)\exp(i\omega t) \tag{36}$$

with $i = \sqrt{-1}$, Θ and ω denoting the spatial mode shapes and natural frequency of torsional vibrations. Imposing the separation of variables Eq. (36) on the differential conditions of dynamic equilibrium Eqs. (34)–(35), the differential condition of torsional coordinate functions for NstrainG is

$$-J_G\left(\alpha + \frac{\ell^2}{c^2}\right)\frac{d^4\Theta}{dx^4}(x) + \left(\frac{1}{c^2}J_G - J_\rho\omega^2\right)\frac{d^2\Theta}{dx^2}(x) + \frac{1}{c^2}J_\rho\omega^2\Theta(x) = 0 \tag{37}$$

and for NstressG is obtained as

$$-J_G\frac{d^4\Theta}{dx^4}(x) + \left(\frac{1}{c^2}J_G - \left(\alpha + \frac{\ell^2}{c^2}\right)J_\rho\omega^2\right)\frac{d^2\Theta}{dx^2}(x) + \frac{1}{c^2}J_\rho\omega^2\Theta(x) = 0 \tag{38}$$

The torsional coordinate function can be analytically detected as

$$\Theta(x) = \Upsilon_1 \exp(\beta_1 x) + \Upsilon_2 \exp(\beta_2 x) + \Upsilon_3 \exp(\beta_3 x) + \Upsilon_4 \exp(\beta_4 x) \quad (39)$$

where unknown integration constants Υ_k $(k = 1, 2, 3, 4)$ have yet to be determined along with β_k $(k = 1, 2, 3, 4)$ being the roots of the characteristic equation associated with the differential equations of Eq. (37) or Eq. (38).

For the NstrainG cantilever nano-beam, a homogeneous fourth-order algebraic system in terms of the unknown integration constants $\Upsilon_k(\, k = 1, 2, 3, 4)$ is established as a result of imposing (standard) classical BCs Eq. (31) along with CBCs Eq. (15) to the closed form solution of the torsional coordinate function Eq. (39). In the same way, homogeneous fourth-order algebraic systems can be determined for nano-beams associated with either of NstrainG or NstressG models. To detect the non-trivial solution of torsional free vibrations, the system of algebraic equations has to be singular. Accordingly, a highly nonlinear characteristic equation is obtained for nano-beams consistent with either of nonlocal gradient models that is numerically solved.

Fundamental torsional frequencies of cantilever and fully-clamped nano-beams associated with NstrainG and NstressG theories are numerically detected and demonstrated in Figs. 5, 6, 7 and 8. The detected torsional frequencies are also normalized employing their corresponding local natural frequencies $\bar{\omega}_{\mathrm{LOC}}$. The characteristic and mixture parameters are assumed to have the same ranging set as the elasto-static torsional response exhibited in Figs. 1 through 4.

It is inferred from the illustrative results associated with the NstrainG model that the nonlocal characteristic parameter λ has the effect of decreasing the fundamental torsional frequencies, that is a larger λ involves a smaller natural torsional frequency. The natural torsional frequencies consistent with the NstrainG model, therefore, demonstrate a softening structural response in terms of nonlocal characteristic parameter λ. Furthermore, the natural frequencies associated with the NstrainG model increase by increasing the gradient characteristic or the mixture parameter, and

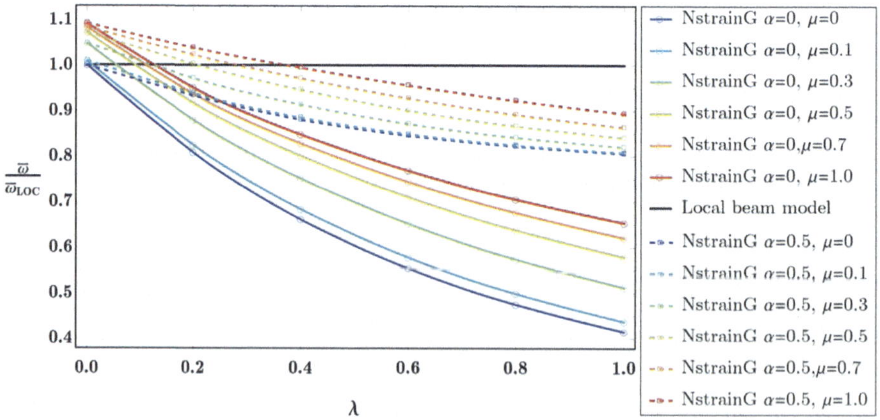

Fig. 5 Normalized torsional fundamental frequency of NstrainG cantilever nano-beams

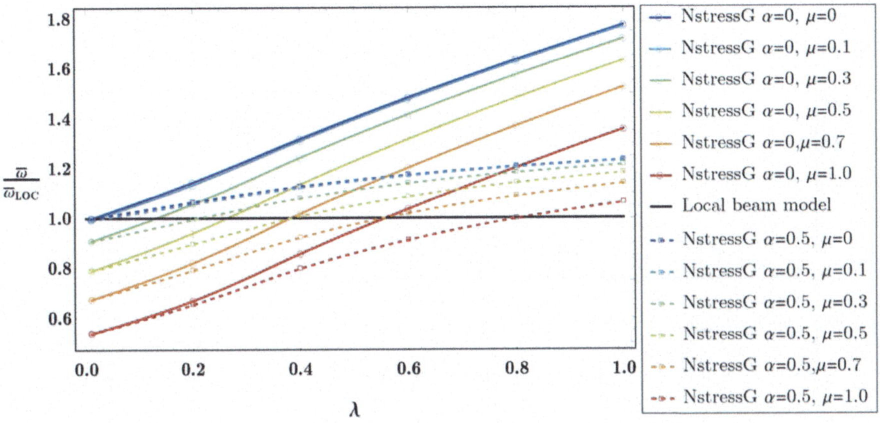

Fig. 6 Normalized torsional fundamental frequency of NstressG cantilever nano-beams

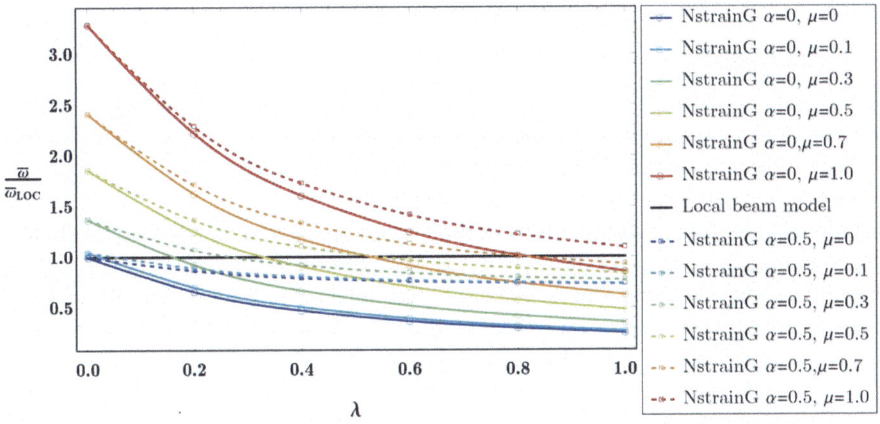

Fig. 7 Normalized torsional fundamental frequency of NstrainG fully-clamped nano-beams

accordingly, demonstrating a stiffening structural response in terms of gradient and mixture parameters. Conversely, a softening behavior is detected for torsional frequencies consistent with NstressG for increasing gradient or mixture parameters and a stiffening response is demonstrated for increasing nonlocal characteristic parameter. In both NstrainG and NstressG models, the effects of characteristic parameters are more noticeable in nano-beams with fully-clamped ends. For non-vanishing mixture parameter $\alpha \neq 0$, the fundamental torsional frequency of nano-beams is less affected by the nonlocal and gradient characteristic parameters. Fundamental torsional frequencies of the local elastic beam model can be recovered as the small-scale characteristic and mixture parameters approach zero. NstrainG and NstressG models are founded on different theoretical bases, and thus, exhibit different softening and stiffening structural responses in terms of characteristic parameters.

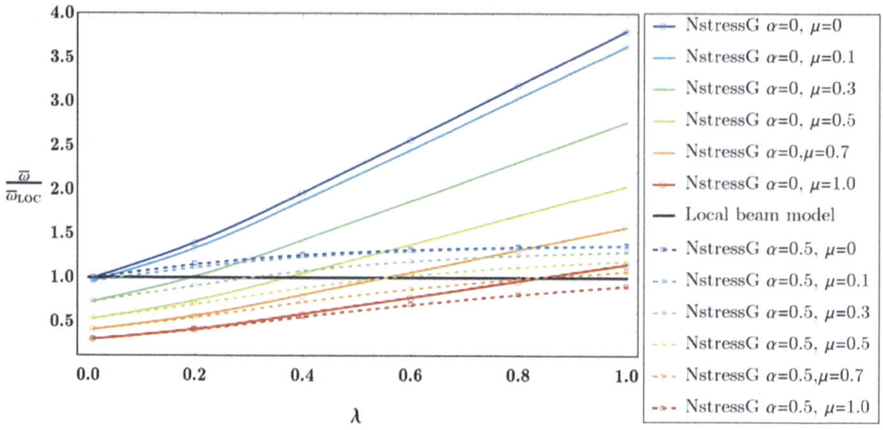

Fig. 8 Normalized torsional fundamental frequency of NstressG fully-clamped nano-beams

5 Conclusions

The nonlocal gradient elasticity theory of inflected nano-beams is extended in this chapter to the mechanics of elastic nano-beams under torsion. Size-dependent torsional response of the elastic beams is investigated by making recourse to a consistent variational constitutive formulation equipped with appropriately selected test fields. The convolution integrals of the constitutive law are transformed into equivalent differential conditions subject to non-standard boundary conditions of nonlocal type. A consistent unified theory of nonlocal gradient elasticity is established for the elastic torsion problem comprising both strain- and stress-driven nonlocal approaches. The well-established elasticity theories adopted in the mechanics of nano-structures including purely nonlocal stress-driven model, two-phase local/nonlocal strain- and stress-driven models as well as modified nonlocal strain gradient model are recovered as special cases. The novel NstrainG and NstressG theories of elasticity are applied to advantageously investigate the size-dependent torsional response of structural schemes of nano-technology applications. Elasto-static and -dynamic torsional responses of nano-beams are examined applying an efficient analytical solution procedure. The variational nonlocal gradient theory is demonstrated to lead to mathematically well-posed problems of mechanics of nano-structures, generally defined in bounded domains. The nonlocal strain- and stress-driven gradient models of elasticity can effectively simulate both stiffening and softening structural responses, and accordingly, provide an innovative viable approach for design and optimization of nano-engineered structures exploited in ground-breaking NEMS.

Acknowledgements The financial support of the Italian Ministry for University and Research (P.R.I.N. National Grant 2017, Project code 2017J4EAYB; "University of Naples Federico II" Research Unit) is gratefully acknowledged.

References

1. Mercan K, Emsen E, Civalek Ö (2019) Effect of silicon dioxide substrate on buckling behavior of Zinc Oxide nanotubes via size-dependent continuum theories. Compos Struct 218:130–141
2. Duan K, Li L, Wang F, Meng W, Hu Y, Wang X (2019) Importance of Interface in the Coarse-Grained Model of CNT/Epoxy Nanocomposites. Nano Mater 9:1479
3. Mercan K, Numanoglu HM, Akgöz B, Demir C, Civalek Ö (2017) Higher-order continuum theories for buckling response of silicon carbide nanowires (SiCNWs) on elastic matrix. Arch Appl Mech 87:1797–1814
4. Duan K, Zhang J, Li L, Hu Y, Zhu W, Wang X (2019) Diamond nanothreads as novel nanofillers for cross-linked epoxy nanocomposites. Compos Sci Technol 174:84–93
5. Pouresmaeeli S, Fazelzadeh SA, Ghavanloo E, Marzocca P (2018) Uncertainty propagation in vibrational characteristics of functionally graded carbon nanotube-reinforced composite shell panels. Int J Mech Sci 149:549–558
6. Ghavanloo E, Rafii-Tabar H, Fazelzadeh SA (2019) New insights on nonlocal spherical shell model and its application to free vibration of spherical fullerene molecules. Int J Mech Sci 161–162:105046
7. Barretta R, Faghidian SA, Marotti de Sciarra F (2019) A consistent variational formulation of Bishop nonlocal rods. Continuum Mech Thermodyn. https://doi.org/10.1007/s00161-019-00843-6
8. Hache F, Challamel N, Elishakoff I (2019) Asymptotic derivation of nonlocal beam models from two-dimensional nonlocal elasticity. Math Mech Solids 24:2425–2443
9. Numanoğlu HM, Civalek Ö (2019) On the torsional vibration of nanorods surrounded by elastic matrix via nonlocal FEM. Int J Mech Sci 161–162:105076
10. Hache F, Challamel N, Elishakoff I (2019) Asymptotic derivation of nonlocal plate models from three-dimensional stress gradient elasticity. Continuum Mech Thermodyn 31:47–70
11. Tashakorian M, Ghavanloo E, Fazelzadeh SA, Hodges DH (2018) Nonlocal fully intrinsic equations for free vibration of Euler-Bernoulli beams with constitutive boundary conditions. Acta Mech 229:3279–3292
12. Barretta R, Fazelzadeh SA, Feo L, Ghavanloo E, Luciano R (2018) Nonlocal inflected nano-beams: A stress-driven approach of bi-Helmholtz type. Compos Struct 200:239–245
13. Tuna M, Kirca M, Trovalusci P (2019) Deformation of atomic models and their equivalent continuum counterparts using Eringen's two-phase local/nonlocal model. Mech Res Commun 97:26–32
14. Zhu X, Li L (2017) Longitudinal and torsional vibrations of size-dependent rods via nonlocal integral elasticity. Int J Mech Sci 133:639–650
15. Fernández-Sáez J, Zaera R (2017) Vibrations of Bernoulli-Euler beams using the two-phase nonlocal elasticity theory. Int J Eng Sci 119:232–248
16. Barretta R, Faghidian SA, Marotti de Sciarra F (2019) Aifantis versus Lam strain gradient models of Bishop elastic rods. Acta Mech 230:2799–2812
17. Li L, Hu Y (2019) Torsional statics of two-dimensionally functionally graded microtubes. Mech Adv Mater Struct 26:430–442
18. Dilena M, Dell'Oste MF, Fernández-Sáez J, Morassi A, Zaera R (2019) Mass detection in nanobeams from bending resonant frequency shifts. Mech Syst Sig Process 116:261–276
19. Bagheri E, Asghari M, Danesh V (2019) Analytical study of micro-rotating disks with angular acceleration on the basis of the strain gradient elasticity. Acta Mech 230:3259–3278
20. Zaera R, Serrano Ó, Fernández-Sáez J (2020) Non-standard and constitutive boundary conditions in nonlocal strain gradient elasticity. Meccanica 55:469–479
21. Zaera R, Serrano Ó, Fernández-Sáez J (2019) On the consistency of the nonlocal strain gradient elasticity. Int J Eng Sci 138:65–81
22. Barretta R, Faghidian SA, Marotti de Sciarra F, Vaccaro MS (2020) Nonlocal strain gradient torsion of elastic beams: variational formulation and constitutive boundary conditions. Arch Appl Mech 90:691–706

23. Pinnola FP, Faghidian SA, Barretta R, Marotti de Sciarra F (2020) Variationally consistent dynamics of nonlocal gradient elastic beams. Int J Eng Sci 149:103220
24. Barretta R, Faghidian SA, Marotti de Sciarra F, Penna R, Pinnola FP (2020) On torsion of nonlocal Lam strain gradient FG elastic beams. Compos Struct 233:111550
25. Rafii-Tabar H, Ghavanloo E, Fazelzadeh SA (2016) Nonlocal continuum-based modeling of mechanical characteristics of nanoscopic structures. Phys Rep 638:1–97
26. Eringen A (1983) On differential equations of nonlocal elasticity and solutions of screw dislocation and surface waves. J Appl Phys 54:4703–4710
27. Ghavanloo E, Rafii-Tabar H, Fazelzadeh SA (2019) Computational continuum mechanics of nanoscopic structures: nonlocal elasticity approaches. Springer
28. Rafii-Tabar H (2008) Computational physics of carbon nanotubes. Cambridge University Press, Cambridge
29. Romano G, Luciano R, Barretta R, Diaco M (2018) Nonlocal integral elasticity in nanostructures, mixtures, boundary effects and limit behaviours. Continuum Mech Thermodyn 30:641–655
30. Romano G, Barretta R (2017) Nonlocal elasticity in nanobeams: the stress-driven integral model. Int J Eng Sci 115:14–27
31. Barretta R, Caporale A, Faghidian SA, Luciano R, Marotti de Sciarra F, Medaglia CM (2019) A stress-driven local-nonlocal mixture model for Timoshenko nano-beams. Compos Part B 164:590–598
32. Apuzzo A, Barretta R, Fabbrocino F, Faghidian SA, Luciano R, Marotti de Sciarra F (2019) Axial and torsional free vibrations of elastic nano-beams by stress-driven two-phase elasticity. J Appl Comput Mech 5:402–413
33. Aifantis EC (2011) On the gradient approach-relation to Eringen's nonlocal theory. Int J Eng Sci 49:1367–1377
34. Lim CW, Zhang G, Reddy JN (2015) A higher-order nonlocal elasticity and strain gradient theory and its applications in wave propagation. J Mech Phys Solids 78:298–313
35. Barretta R, Marotti de Sciarra F (2018) Constitutive boundary conditions for nonlocal strain gradient elastic nano-beams. Int J Eng Sci 130:187–198
36. Barretta R, Marotti de Sciarra F (2019) Variational nonlocal gradient elasticity for nano-beams. Int J Eng Sci 143:73–91
37. Romano G, Barretta A, Barretta R (2012) On torsion and shear of Saint-Venant beams. Eur J Mech A-Solid 35:47–60
38. Faghidian SA (2016) Unified formulation of the stress field of Saint-Venant's flexure problem for symmetric cross-sections. Int J Mech Sci 111–112:65–72

Reformulation of the Boundary Value Problems of Nonlocal Type Elasticity: Application to Beams

Xiao-Jian Xu

Abstract Nonlocal elasticity has been widely used in studying the mechanical behavior of nanostructures. However, when one deals with the boundary value problems (BVPs) of mechanical models subject to clamped-free boundary conditions, the paradox that the stiffening phenomena is observed. The purpose of this chapter is concerned with the reformulation of the newly BVPs of nanostructures. To this end, the weighted residual method is used. The asymptotic theory for nonlocal elasticity is established as well as the illustrating examples of the reformulation of the BVPs of several widely used beam models. In comparison with the corresponding nonlocal integral type, the present BVPs are relatively more convenient for the analysis.

1 Introduction

When the characteristic size of structures reduces to micro- to nano-scale, the obviously size-effects may be observed. As a result, several elastic theories have been established for capturing such size-effects. Among these theories, the nonlocal elasticity has now being a widely used theory, especially for capturing the materials/structures exhibiting softening phenomena [1]. However, when one uses this theory to develop elastic models of engineering structures, such as beams and plates, the paradox may be encountered, as evidence by several authors [2, 3]. In order to solve the paradox, several attempts have been conducted [4–9].

The first way is to utilize the corresponding nonlocal integral type, and use the mathematical transformation method. The central idea is to convert the origin nonlocal integral BVPs to the equivalent differential ones [7–11]. This method, however, may naturally increase the differential order(s), and the extra boundary conditions. As a result, the mathematical challenge may be encountered when one attempts to solve the corresponding equivalent higher-order BVPs.

X.-J. Xu (✉)
School of Highway Chang'an University, Xi'an 710064, People's Republic of China
e-mail: xuxiaojian@mail.nwpu.edu.cn

© The Author(s), under exclusive license to Springer Nature Switzerland AG 2021
E. Ghavanloo et al. (eds.), *Size-Dependent Continuum Mechanics Approaches*,
Springer Tracts in Mechanical Engineering,
https://doi.org/10.1007/978-3-030-63050-8_8

The second way, widely used decades ago, is to directly use the BVPs of nonlocal differential equations. However, this method has been found to be failed when one deals with the structures with clamped-free boundary conditions, as evidenced by the paradox reported in the literature [2, 3]. We now comes to a question, if is it possible to use the differential BVPs to capture the softening phenomena, and at the same time, to avoid the solution challenges encountered when one solves the BVPs of the corresponding integral ones? The purpose of this chapter is to give a positive answer to this problem.

2 Problem Formulation

In order to answer the above question in a theoretical aspect, it is neces-sary to firstly develop an asymptotic theory of nonlocal elasticity. To begin and provide a preliminary for readers, we now summarize the main concept of the nonlocal elasticity.

2.1 *Nonlocal Elasticity*

In accord with atomic theory of lattice dynamics and experimental observations on phonon dispersions, Eringen [12] developed a nonlocal elasticity theory by the use of an integral-type nonlocality, where the non-local stress σ_{ij}^{non} at point $\mathbf{x}$ is related to the associated local stress σ_{ij} by

$$\sigma_{ij}^{non}(\mathbf{x}) = \int_V \alpha(\mathbf{s})\,\sigma_{ij}(\mathbf{x}+\mathbf{s})\,\mathrm{d}V \tag{1}$$

with V is material domain, $\alpha(\mathbf{s})$ denotes the nonlocal weight function that is non-negative and increasing for decreasing values of $\mathbf{s}$.

However, for elastic bodies, the linear theory results into a set of integro-differential equations for the displacement field, which is generally difficult to solve. For infinite domain, Eringen provided the following differential constitutive equation as

$$\sigma_{ij}^{non} - \eta^2\sigma_{ij,kk}^{non} = \sigma_{ij} = C_{ijkl}\varepsilon_{kl} \tag{2}$$

where $\eta = e_0 a$ is the nonlocal parameter, C_{ijkl} is the elastic constant, ε_{kl} is strain related to the displacement u_i by

$$\varepsilon_{ij} = \frac{1}{2}\left(u_{i,j} + u_{j,i}\right) \tag{3}$$

In the absence of the body force, the equilibrium equation between the nonlocal stress and the inertia is given by

$$\sigma_{ij,j}^{non} = \rho \ddot{u}_i \tag{4}$$

where ρ is mass density of the body, and the dot over a symbol denotes partial differential with respect to time t.

Upon combining Eqs. (2) and (4), one can easily arrive at the following equilibrium equation

$$\sigma_{ij,j} = \rho \ddot{u}_i - \rho \eta^2 \ddot{u}_{i,jj} \tag{5}$$

For a homogeneous isotropic linearly elastic media, the local constitutive equation reads

$$\sigma_{ij} = \lambda u_{k,k} \delta_{ij} + \mu \left(u_{i,j} + u_{j,i} \right) \tag{6}$$

where λ and μ are the Lamé coefficients and δ_{ij} is the Kronecker delta.

Further after the substitution of Eq. (6) into Eq. (5) gives the following displacement equation

$$\mu \left(u_{i,jj} + u_{j,ij} \right) + \lambda u_{k,kj} \delta_{ij} = \rho \ddot{u}_i - \rho \eta^2 \ddot{u}_{i,jj} \tag{7}$$

The partial differential equation in Eq. (7) should be solved in combination with the boundary conditions. The higher-order boundary conditions involved are not properly given, until Polizzotto [13–15] who published a sequence of papers attempting to address the gradient-induced higher-order boundary conditions. Our next step is to present the corresponding boundary conditions for the nonlocal elasticity theory in the weak form.

As is well-known, the virtual displacement principle for the classical elasticity theory is widely used in the literature. Normally, this principle is used to generate boundary-value problems in weak forms. In this subsection, we extend this principle to determine the boundary conditions of the nonlocal elasticity.

We consider a linear elasticity continuum of occupying domain V and boundary surface A, which is subjected to the inertia within V and prescribed displacement u_i on the constrained boundary surface A_c. With a view toward determining the boundary conditions of the nonlocal elasticity theory from the weighted residual method, we firstly multiply a virtual displacement δu_i on both side of Eq. (5), and then integrate the resulting equation over the body domain V and time interval t_0 and t_1 to yield

$$\int_V \int_{t_0}^{t_1} \sigma_{ij,j} \delta u_i \, \mathrm{d}V \mathrm{d}t = \int_V \int_{t_0}^{t_1} \left(\rho u_{i,tt} - \rho \eta^2 u_{i,jjtt} \right) \delta u_i \, \mathrm{d}V \mathrm{d}t \tag{8}$$

which, after integration by parts, becomes

$$
\begin{aligned}
&-\int_V \int_{t_0}^{t_1} \left(\sigma_{ij}\delta\varepsilon_{ij} - \rho\dot{u}_i\delta\dot{u}_i - \rho\eta^2\dot{u}_{i,j}\delta\dot{u}_{i,j}\right) \mathrm{d}V\mathrm{d}t \\
&-\int_V \left(\rho\dot{u}_i\delta u_i + \rho\eta^2\dot{u}_{i,j}\delta u_i\right)\Big|_{t_0}^{t_1} \mathrm{d}V \\
&+\int_A \int_{t_0}^{t_1} \left(\sigma_{ij}n_j + \rho\eta^2\ddot{u}_{i,j}n_j\right)\delta u_i \,\mathrm{d}A\mathrm{d}t = 0
\end{aligned}
\tag{9}
$$

where n_j is the unit outward normal vector to A.

The important consequences of Eq. (9) immediately provide us the following initial/final conditions:

$$
\left.\begin{aligned}
u_i(\mathbf{x}, t_0) = \bar{u}_{i1}(\mathbf{x}), \quad \dot{u}_i(\mathbf{x}, t_0) + \eta^2\dot{u}_{i,j}(\mathbf{x}, t_0) = \bar{u}_{i2}(\mathbf{x}) \\
u_i(\mathbf{x}, t_1) = \bar{u}_{i3}(\mathbf{x}), \quad \dot{u}_i(\mathbf{x}, t_1) + \eta^2\dot{u}_{i,j}(\mathbf{x}, t_1) = \bar{u}_{i4}(\mathbf{x})
\end{aligned}\right\} \text{ in } V
\tag{10}
$$

and the higher-order boundary conditions:

$$
\sigma_{ij}n_j + \rho\eta^2\ddot{u}_{i,j}n_j = 0 \quad \text{or} \quad u_i = \bar{u}_i \quad \text{for arbitrary time } t
\tag{11}
$$

Noteworthy, we can decompose the traction into two parts such that the first part is linked to the local stress σ_{ij} by the Cauchy theorem, and the other part is due to the inertia of the elastic body. For convenience of the illustration, we omit the body force below. To fix ideas, we first consider a nonlocal elastic body featuring a nonlocal stress tensor σ_{ij}^{non}, the first variation of Hamilton-type functional is given by

$$
\delta H^{non} = \int_{t_0}^{t_1} \int_V \left(\rho\dot{u}_i\delta\dot{u}_i - \sigma_{ij}^{non}\delta\varepsilon_{ij}\right) \mathrm{d}V\mathrm{d}t = 0
\tag{12}
$$

which, by Eq. (3), results in

$$
\begin{aligned}
\delta H^{non} = &\int_{t_0}^{t_1} \int_V \left(\sigma_{ij,j}^{non} - \rho\ddot{u}_i\right)\delta u_i \,\mathrm{d}V\mathrm{d}t \\
&- \int_{t_0}^{t_1} \int_A \sigma_{ij}^{non} n_j \delta u_i \,\mathrm{d}A\mathrm{d}t + \int_V \left[\rho\dot{u}_i\delta u_i\right]\Big|_{t_0}^{t_1} \mathrm{d}V = 0
\end{aligned}
\tag{13}
$$

Note that the integrand of the last term of Eq. (13) vanishes for the elastic body whose configurations at $t = t_0$ and t_1 are prescribed. Then, for the arbitrariness of δu_i, we obtain the balance equation (4), and the boundary condition

$$
\sigma_{ij}^{non} n_j = 0
\tag{14}
$$

It is emphasized that Eq. (14) is equivalent to, by combining Eqs. (2) and (4), the boundary condition $(11)_1$.

Analogously, we can directly write down the following first variation of Hamilton-type functional:

$$\delta H = \int_{t_0}^{t_1} \int_V \left(\rho \dot{u}_i \delta \dot{u}_i + \rho \eta^2 \dot{u}_{i,j} \delta \dot{u}_{i,j} - \sigma_{ij} \delta \varepsilon_{ij}\right) \mathrm{d}V \mathrm{d}t = 0 \tag{15}$$

which, after integration by parts, reads

$$\begin{aligned} \delta H = & \int_{t_0}^{t_1} \int_V \left(\sigma_{ij,j} - \rho \ddot{u}_i + \rho \eta^2 \ddot{u}_{i,jj}\right) \delta u_i \, \mathrm{d}V \mathrm{d}t \\ & - \int_{t_0}^{t_1} \int_A \left(\sigma_{ij} n_j + \rho \eta^2 \ddot{u}_{i,j} n_j\right) \delta u_i \, \mathrm{d}A \mathrm{d}t \\ & + \int_V \left[\left(\rho \dot{u}_i - \rho \eta^2 \dot{u}_{i,jj}\right) \delta u_i\right] \Big|_{t_0}^{t_1} \mathrm{d}V \\ & + \int_A \left[\rho \eta^2 \dot{u}_{i,j} n_j \delta u_i\right] \Big|_{t_0}^{t_1} \mathrm{d}A = 0 \end{aligned} \tag{16}$$

Note again that the integrand of the last line of Eq. (13) vanishes for an elastic body whose configurations at $t = t_0$ and t_1 are prescribed. Then, for the arbitrariness of δu_i, we obtain the balance equation (5) and the weak-form boundary condition $(11)_1$.

2.2 *Asymptotic Theory for Nonlocal Elasticity*

It is shown in Sect. 2.1 that the solutions of the boundary-value problems in Eqs. (7) and (11) are a challenging task with respect to the classical problems. Therefore, developing an asymptotic theory that is related to the small length parameter is preferable. The objective of the present subsection is to establish an asymptotic theory in the nonlocal elasticity theory, and later to illustrate the results for one-dimensional problems. The exact solutions of the boundary-value problems will be ap-proximately replaced by a sequent of asymptotic ones truncated at different orders of the small length parameter.

We consider a small harmonic motion with frequency ω in which the displacement $u_i(\mathbf{x}, t)$ can be decomposed into

$$u_i(\mathbf{x}, t) = u_i(\mathbf{x}) \exp\left(\sqrt{-1}\omega t\right) \tag{17}$$

Substitution of Eq. (17) into the equilibrium equation in (5) yields

$$\sigma_{ij,j} = -\rho \omega^2 u_i + \rho \omega^2 \eta^2 u_{i,jj} \tag{18}$$

Analogously, recalling Eqs. (17) and (11), the boundary conditions read

$$\sigma_{ij} n_j - \rho \eta^2 \omega^2 u_{i,j} n_j = 0 \quad \text{or} \quad u_i = \bar{u}_i \tag{19}$$

where the stress σ_{ij} is given by Eq. (6).

Our next step is to establish an asymptotic theory of nonlocal elasticity theory on the basis of the boundary condition $(19)_1$. We here are concerned with the frequency of a nonlocal body with the emphasis of developing an asymptotic theory. The key feature of our asymptotic theory is to introduce a small parameter ϵ which is related to the small length parameter by $\epsilon = \eta^2$.

Equations (18) and (19) allow us to write the displacement field and fre-quency in the following form:

$$u_i(\mathbf{x}) = \sum_{m=0}^{\infty} \epsilon^m u_i^{(m)}(\mathbf{x}), \quad \omega = \sum_{m=0}^{\infty} \epsilon^m \omega_m \tag{20}$$

Inserting Eq. (20) into Eq. (18), and equating to zero the like powers of ϵ, we arrive at a hierarchy of equations to be solved order by order for $u_i^{(m)}$ and ω_m with the associated orders of boundary conditions.

At the leading order, we get an eigenvalue problem

$$o\left(\epsilon^0\right): \quad \sigma_{ij,j}^{(0)} + \rho\omega_0^2 u_i^{(0)} = 0 \tag{21}$$

subjected, remembering Eq. (11), to the leading order boundary condition:

$$\sigma_{ij}^{(0)} n_j = 0 \tag{22}$$

where $\sigma_{ij}^{(m)} = \lambda u_{k,k}^{(m)} \delta_{ij} + \mu \left[u_{i,j}^{(m)} + u_{j,i}^{(m)}\right]$. As expected, this leading order boundary-value problem is exactly the same as that for the classical ones. Therefore, we can solve this boundary-value problem for $u_i^{(0)}$ without any difficulty.

To this end, we can integrate over the domain V the product of Eq. (21) and $u_i^{(0)}$, and then, apply integration by parts, to obtain

$$\int_A \sigma_{ij}^{(0)} n_j u_i^{(0)} \, \mathrm{d}A = \int_V \left[\sigma_{ij}^{(0)} u_{i,j}^{(0)} - \rho\omega_0^2 u_i^{(0)2}\right] \mathrm{d}V \tag{23}$$

which, by Eq. (22), results in $\int_V \left[\sigma_{ij}^{(0)} u_{i,j}^{(0)} - \rho\omega_0^2 u_i^{(0)2}\right] \mathrm{d}V = 0$.Then, the frequency ω_0 can be easily obtained as

$$\omega_0^2 = \frac{\int_V \sigma_{ij}^{(0)} u_{i,j}^{(0)} \, \mathrm{d}V}{\int_V \rho u_i^{(0)2} \, \mathrm{d}V} \tag{24}$$

In order to obtain the next order asymptotic frequency parameter, we now proceed to the next order to obtain the frequency ω_1. At the next order, the equation is

$$o\left(\epsilon^1\right): \quad \sigma_{ij,j}^{(1)} + \rho\left[\omega_0^2 u_i^{(1)} + 2\omega_0\omega_1 u_i^{(0)} - \omega_0^2 u_{i,jj}^{(0)}\right] = 0 \tag{25}$$

subjected, by Eq. (11), to the next order boundary condition:

$$\sigma_{ij}^{(1)} n_j - \rho \omega_0^2 u_{i,j}^{(0)} n_j = 0 \tag{26}$$

To solve the next order boundary-value problem, we can multiply Eq. (25) by $u_i^{(0)}$ on the both hand side, and then, integrate over the domain V. The final result is

$$\int_V \left\{ \sigma_{ij,j}^{(1)} + \rho \left[\omega_0^2 u_i^{(1)} + 2\omega_0 \omega_1 u_i^{(0)} - \omega_0^2 u_{i,jj}^{(0)} \right] \right\} u_i^{(0)} \, \mathrm{d}V = 0 \tag{27}$$

Using the divergence theorem for the first and the fourth integrals, Eq. (27) becomes

$$\begin{aligned} &\int_V \left\{ \rho \left[\omega_0^2 u_i^{(1)} + 2\omega_0 \omega_1 u_i^{(0)} \right] \right\} u_i^{(0)} \, \mathrm{d}V \\ &\quad + \int_V \left[\rho \omega_0^2 u_{i,j}^{(0)2} - \sigma_{ij}^{(1)} u_{i,j}^{(0)} \right] \mathrm{d}V \\ &+ \int_A \left[\sigma_{ij}^{(1)} n_j - \rho \omega_0^2 u_{i,j}^{(0)} n_j \right] u_i^{(0)} \, \mathrm{d}A = 0 \end{aligned} \tag{28}$$

Then, we have the following expression by using the Eq. (26)

$$\begin{aligned} &\int_V \left\{ \rho \left[\omega_0^2 u_i^{(1)} + 2\omega_0 \omega_1 u_i^{(0)} \right] \right\} u_i^{(0)} \, \mathrm{d}V \\ &+ \int_V \left[\rho \omega_0^2 u_{i,j}^{(0)2} - \sigma_{ij}^{(1)} u_{i,j}^{(0)} \right] \mathrm{d}V = 0 \end{aligned} \tag{29}$$

We now subtract the integral of Eq. (21), multiplied by $u_i^{(1)}$, over the domain V, from Eq. (29), which results in

$$\begin{aligned} &\int_V \left[2\rho \omega_0 \omega_1 u_i^{(0)} \right] u_i^{(0)} \, \mathrm{d}V \\ &+ \int_V \left[\rho \omega_0^2 u_{i,j}^{(0)2} - \sigma_{ij}^{(1)} u_{i,j}^{(0)} - \sigma_{ij,j}^{(0)} u_i^{(1)} \right] \mathrm{d}V = 0 \end{aligned} \tag{30}$$

Using the divergence theorem for the last term of Eq. (30), and by Eq. (22) and Betti's reciprocal theorem, we have

$$\int_V \left[2\rho \omega_0 \omega_1 u_i^{(0)} u_i^{(0)} + \rho \omega_0^2 u_{i,j}^{(0)2} \right] \mathrm{d}V = 0 \tag{31}$$

As a result, the only unknown frequency ω_1 can be determined by solving Eq. (31) as

$$\omega_1 = -\frac{1}{2} \frac{\int_V \omega_0 u_{i,j}^{(0)2} \, \mathrm{d}V}{\int_V u_i^{(0)2} \, \mathrm{d}V} < 0 \tag{32}$$

Importantly, we observe that the frequency ω_0 and the displacement $u_i^{(0)}$ play an important role on the first order frequency parameter ω_1. Furthermore, it is also observed that is independent upon the other ordered displacement parameters.

The second order equation for $u_2^{(2)}$ is

$$o\left(\epsilon^2\right): \quad \sigma_{ij,j}^{(2)} + \rho\left(2\omega_0\omega_2 + \omega_1^2\right)u_i^{(0)} - 2\rho\omega_0\omega_1 u_{i,jj}^{(0)} - \rho\omega_0^2 u_{i,jj}^{(1)} + 2\rho\omega_0\omega_1 u_i^{(1)} + \rho\omega_0^2 u_i^{(2)} = 0 \tag{33}$$

subjected to the second order boundary condition:

$$\sigma_{ij}^{(2)} n_j - 2\rho\omega_0\omega_1 u_{i,j}^{(0)} - \rho\omega_0^2 u_{i,j}^{(1)} = 0 \tag{34}$$

Analogously, we integrate over the body domain V the product of Eq. (33) and $u_i^{(0)}$ minus the product of Eq. (21) and $u_i^{(2)}$. The final result, after using the divergence theorem for the terms with $\sigma_{ij,j}^{(2)}$, $u_{i,jj}^{(0)}$, $u_{i,jj}^{(1)}$ and $\sigma_{ij,j}^{(0)}$, is

$$\begin{aligned}
&\int_A \left[\sigma_{ij}^{(2)} u_i^{(0)} - \sigma_{ij}^{(0)} u_i^{(2)} - 2\rho\omega_0\omega_1 u_{i,j}^{(0)} u_i^{(0)} - \rho\omega_0^2\left(u_{i,j}^{(1)} u_i^{(0)} - u_{i,j}^{(0)} u_i^{(1)}\right)\right] n_j \,\mathrm{d}A \\
&+ \int_V \left(\sigma_{ij}^{(0)} u_{i,j}^{(2)} - \sigma_{ij}^{(2)} u_{i,j}^{(0)}\right) \mathrm{d}V + \int_V \left[\rho\left(2\omega_0\omega_2 + \omega_1^2\right) u_i^{(0)2} + 2\rho\omega_0\omega_1 u_{i,j}^{(0)2}\right] \mathrm{d}V \\
&+ \int_V \left[2\rho\omega_0\omega_1 u_i^{(0)} - \rho\omega_0^2 u_{i,jj}^{(0)}\right] u_i^{(1)} \,\mathrm{d}V = 0
\end{aligned} \tag{35}$$

The last line of Eq. (35) involves the knowledge of $u_i^{(1)}$, which motivates us to multiply $u_i^{(1)}$ in Eq. (25) and to integrate over the body domain V. After integrating by parts with respect to $\sigma_{ij,j}^{(1)}$, we have

$$\begin{aligned}
&\int_V \left[2\rho\omega_0\omega_1 u_i^{(0)} - \rho\omega_0^2 u_{i,jj}^{(0)}\right] u_i^{(1)} \,\mathrm{d}V = \\
&\int_V \left[\sigma_{ij}^{(1)} u_{i,j}^{(1)} - \rho\omega_0^2 u_i^{(1)2}\right] \mathrm{d}V - \int_A \sigma_{ij}^{(1)} u_i^{(1)} n_j \,\mathrm{d}A
\end{aligned} \tag{36}$$

Substituting Eq. (36) into Eq. (35), and after some re-arrangements, we arrive at

$$\begin{aligned}
&\int_A \left(\sigma_{ij}^{(2)} n_j - 2\rho\omega_0\omega_1 u_{i,j}^{(0)} - \rho\omega_0^2 u_{i,j}^{(1)}\right) u_i^{(0)} \,\mathrm{d}A \\
&+ \int_V \left[\sigma_{ij}^{(1)} u_{i,j}^{(1)} - \rho\omega_0^2 u_i^{(1)2} + \rho\left(2\omega_0\omega_2 + \omega_1^2\right) u_i^{(0)2} + 2\rho\omega_0\omega_1 u_{i,j}^{(0)2}\right] \mathrm{d}V \\
&- \int_A \sigma_{ij}^{(0)} n_j u_i^{(2)} + \left(\sigma_{ij}^{(1)} n_j - \rho\omega_0^2 u_{i,j}^{(0)} n_j\right) u_i^{(1)} \,\mathrm{d}A \\
&+ \int_V \left(\sigma_{ij}^{(0)} u_{i,j}^{(2)} - \sigma_{ij}^{(2)} u_{i,j}^{(0)}\right) \mathrm{d}V = 0
\end{aligned} \tag{37}$$

Inserting the ordered boundary conditions from Eqs. (22), (26) and (34) into the first line of Eq. (37) and recalling the Betti's reciprocal theorem, we can observe that the first two lines of Eq. (37) vanish. Therefore, it further enables us to obtain the second order frequency ω_2 as

$$\omega_2 = -\frac{\int_V \left[\sigma_{ij}^{(1)} u_{i,j}^{(1)} - \rho\omega_0^2 u_i^{(1)^2} + \rho\omega_1^2 u_i^{(0)^2} + 2\rho\omega_0\omega_1 u_{i,j}^{(0)^2}\right] \mathrm{d}V}{2\rho\omega_0 \int_V u_i^{(0)^2} \mathrm{d}V} \tag{38}$$

Noteworthy, we can see that the second order frequency ω_2 is closely related to the displacements $u_i^{(0)}$, $u_i^{(1)}$ and frequency parameters ω_0 and ω_1.

To summary, given a boundary-value problem in the nonlocal elasticity theory, we are able to solve the leading order equation (21) associated with boundary condition (22) to determine both the frequency parameter ω_0 and the displacement $u_i^{(0)}$. Then, substitution of $u_i^{(0)}$ into Eq. (32) yields the frequency parameter ω_1. Finally, solving Eq. (25) with boundary condition (26) enables us to obtain displacement $u_i^{(1)}$; the known values of the first two displacements and the frequency parameters admit us to determine the frequency parameter ω_2 from Eq. (38). Finally, it is emphasized that the above asymptotic procedure is formulated without considering any stress-strain relation that is independent on the specific constitutive material behaviors. As a result, the above relations are expected to apply in both the isotropic and anisotropic materials within the framework of the nonlocal type elasticity theory.

3 BVPs for Euler–Bernoulli Beam Model

The BVPs for the Euler–Bernoulli type beam models may be found in numerous works. The remarkable feature is that they transformed the original integral equation into the equivalent differential equation with introducing the extra constitutive boundary conditions [11, 16, 17]. Due to the mathematical challenge encountered, we here present the BVPs for the Euler–Bernoulli type beam models by using the weighted residual method.

3.1 Reformulation

Let u and w be axial and transverse displacements, and let l be the strain gradient parameter. We now consider a beam with length L subject to the axial force f and transverse distributed force q. The governing equations of motion of the Euler–Bernoulli beams within the framework of nonlocal strain gradient elastic theory are given by [18]

$$\left(1-l^2\frac{\partial^2}{\partial x^2}\right)\left(A_{xx}u''-B_{xx}w'''\right)-\left(1-\eta^2\frac{\partial^2}{\partial x^2}\right)\left(m_0\ddot{u}-m_1\ddot{w}'+f\right)=0$$
$$\left(1-l^2\frac{\partial^2}{\partial x^2}\right)\left(D_{xx}w''''-B_{xx}u'''\right)$$
$$+\left(1-\eta^2\frac{\partial^2}{\partial x^2}\right)\left(m_1\ddot{u}'-m_2\ddot{w}''+m_0\ddot{w}+q\right)=0 \tag{39}$$

where A_{xx}, B_{xx} and D_{xx} are the axial, axial-transverse coupling and bending stiffness of the beam, m_0, m_1 and m_2 are the mass inertia of the beam; the expressions can be found in Li et al. [18].

The weak form of Eq. (39) may be given by

$$\int_0^t\int_0^L\left[N_{cl,x}-l^2N_{cl,xxx}-B_{xx}w'''+B_{xx}w'''''+f_{,xx}-f\right]\delta u\,\mathrm{d}x\mathrm{d}t$$
$$+\int_0^t\int_0^L\left[m_0\left(\eta^2\ddot{u}_{,xx}-\ddot{u}\right)+m_1\ddot{w}'-m_1\eta^2\ddot{w}'''\right]\delta u\,\mathrm{d}x\mathrm{d}t$$
$$+\int_0^t\int_0^L\left[M_{cl,xx}-l^2M''''_{cl}-q+\eta^2q_{,xx}\right]\delta w\,\mathrm{d}x\mathrm{d}t$$
$$-\int_0^t\int_0^L\left[\left(1-\eta^2\frac{\partial^2}{\partial x^2}\right)\left(m_0\ddot{w}+m_1\ddot{u}'-m_2\ddot{w}_{,xx}\right)\right]\delta w\,\mathrm{d}x\mathrm{d}t=0 \tag{40}$$

where the classical stress resultants are introduced as

$$N_{cl}=A_{xx}u',\quad M_{cl}=B_{xx}u'-D_{xx}w'' \tag{41}$$

Then, integrating Eq. (40) by parts with respect to the higher-order derivatives, and after some mathematical manipulations, we have the following boundary conditions:

$$N_{xx}=N_{cl}-l^2N''_{cl}+m_0\eta^2\ddot{u}'-m_1\eta^2\ddot{w}''\quad\text{or}\quad u$$
$$N_{xx}^{(1)}=l^2N'_{cl}\quad\text{or}\quad u'$$
$$Q_{xz}=M'_{cl}-l^2M'''_{cl}+\left(m_2+m_0\eta^2\right)\ddot{w}'-m_2\eta^2\ddot{w}'''+m_1\eta^2\ddot{u}''\quad\text{or}\quad w$$
$$M=M_{cl}-l^2M''_{cl}-m_2\eta^2\ddot{w}''+m_1\eta^2\ddot{u}'\quad\text{or}\quad w'$$
$$M^{(1)}=-l^2M'_{cl}\quad\text{or}\quad w'' \tag{42}$$

It is noted from Eq. (42) that the higher-order static or kinematic boundary conditions are found. In practice, the procedure to choose these boundary conditions is determined by several possible selections; See for example Refs. [11, 19, 20].

The above boundary conditions can be degenerated as special cases to those of strain gradient Euler–Bernoulli models [21, 22] and those of nonlocal Euler–Bernoulli models [6]. In addition, if the strain gradient effect is omit (e.g., $l=0$), the higher order mass inertia is ignored, and the axial-transverse coupling effect is neglect, the boundary conditions of a homogenous beam may be reduced to the

Table 1 Various stress resultants used for the Euler–Bernoulli (EBT) and Timoshenko beam theory (TBT)

Beam	Resultant	Lu [2]	Reddy and Pang [23]	Present work
EBT	M	$-EIw'' + \rho A\eta^2\ddot{w}$	$-EIw'' + \rho A\eta^2\ddot{w}$	$-EIw''$
	Q	$-EIw''' + \rho A\eta^2\ddot{w}'$	$-EIw''' + \rho A\eta^2\ddot{w}'$	$-EIw''' + \rho A\eta^2\ddot{w}'$
TBT	M	–	$-EI\phi' + \rho A\eta^2\ddot{w}$	$-EI\phi' - \rho I\eta^2\ddot{\phi}'$
	Q_{xz}	–	$KGA(\phi - w') - \rho A\eta^2\ddot{w}'$	$KGA(\phi - w') - \rho A\eta^2\ddot{w}'$

nonlocal type

$$\begin{aligned} Q_{xz} &= -D_{xx}w''' + m_0\eta^2\ddot{w}' \quad \text{or} \quad w \\ M &= -D_{xx}w'' \quad \text{or} \quad w' \end{aligned} \tag{43}$$

The differences of the stress resultants for the Euler–Bernoulli beams can be found in Table 1.

3.2 Dynamic Behavior of Nonlocal Cantilevers: Euler–Bernoulli Beam

As illustrated in the introduction, the paradox may be found when one solve the BVPs for nonlocal cantilevers. We here attempt to use the present boundary conditions (see Eq. (43)) to solve the BVPs for nonlocal cantilevers. It is stated that we only use the weighted residual method to the BVPs. Without loss of the generality, we assume that the left end is fixed and the right end is free, and the natural frequency is ω.

Firstly, we obtain according to Eq. $(39)_2$ the equation of motion of the following form:

$$EIw'''' + \left(1 - \eta^2\frac{\partial^2}{\partial x^2}\right)\rho A\ddot{w} = 0 \tag{44}$$

and the boundary conditions of the form:

$$\begin{aligned} &w(0) = w'(0) = 0 \\ &Q_{xz}(L) = -D_{xx}w'''(L) + m_0\eta^2\ddot{w}'(L) = 0 \\ &M(L) = -D_{xx}w''(L) = 0 \end{aligned} \tag{45}$$

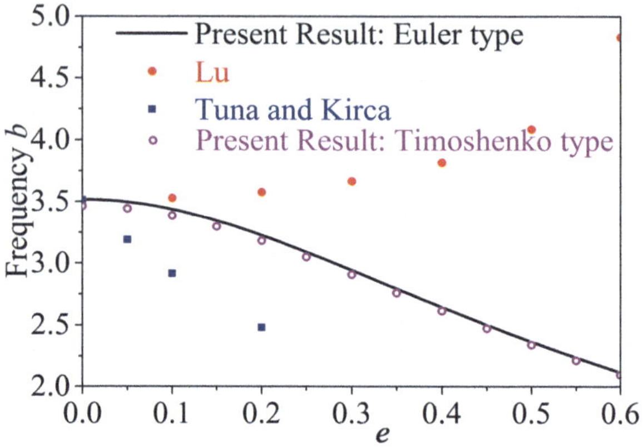

Fig. 1 The fundamental frequency versus the nonlocal parameter for Euler–Bernoulli beam models and Timoshenko beams. Lines: Euler type by Eq. (46); Red circles: Euler type data from Lu [2]; rectangular: Euler type data from Tuna and Kirca [8]; Open circles: Timoshenko type by Eq. (59)

The characteristic equation for BVPs of Eqs. (44) and (45) reads [6]

$$2b^2 + \left(b^4e^4 + 2b^2\right)\cos\alpha\cosh\beta - b^3e^2\sin\alpha\sinh\beta = 0 \tag{46}$$

where

$$\begin{pmatrix}\alpha\\ \beta\end{pmatrix} = \sqrt{\left(\pm b^2e^2 + b\sqrt{b^2e^4+4}\right)/2}, \quad e = \eta/L, \quad b = \omega\sqrt{\frac{\rho AL^4}{EI}} \tag{47}$$

By numerically solving Eq. (46), the fundamental frequency versus nonlocal parameter is depicted in Fig. 1. It is observed that the frequency decreases with increasing the nonlocal parameter, indicating that the present numerical result capture the anticipated softening phenomena. To highlight the present result, we also add the numerical results presented by Lu [2] and Tuna and Kirca [8]. It can be seen that the integral nonlocal elasticity can also capture the expected softening phenomena. Interestingly, the difference between the present result and that of Tuna and Kirca [8] is that the latter integral nonlocal elasticity predicts a more softening phenomena.

4 BVPs for Timoshenko Beam Model

The BVPs for the Euler–Bernoulli beam model have been formulated in Sect. 3. We here directly extend the Euler–Bernoulli model to the Timoshenko beam.

4.1 Reformulation

Let ϕ and k_s be the rotation and shear correction factor. The governing equations of motion of the Timoshenko beams within the framework of nonlocal strain gradient elastic theory are given by [18]

$$\left(1-l^2\frac{\partial^2}{\partial x^2}\right)\left(A_{xx}u''-B_{xx}\phi''\right)-\left(1-\eta^2\frac{\partial^2}{\partial x^2}\right)\left(m_0\ddot{u}-m_1\ddot{\phi}+f\right)=0$$
$$\left(1-l^2\frac{\partial^2}{\partial x^2}\right)\left(k_sA_{xz}w'-k_sA_{xz}\phi-B_{xx}u''+D_{xx}\phi''\right)$$
$$+\left(1-\eta^2\frac{\partial^2}{\partial x^2}\right)\left(m_1\ddot{u}-m_2\ddot{\phi}\right)=0$$
$$k_sA_{xz}\left(1-l^2\frac{\partial^2}{\partial x^2}\right)\left(w''-\phi'\right)-\left(1-\eta^2\frac{\partial^2}{\partial x^2}\right)\left(m_0\ddot{w}+q\right)=0 \tag{48}$$

It is noted that Eq. (48) is derived based on Reddy's beam model. Analogously, Eq. (48) may be written in the weak form as

$$\int_0^{t_0}\int_0^L\left(1-l^2\frac{\partial^2}{\partial x^2}\right)\left(A_{xx}u''-B_{xx}\phi''\right)\delta u\,\mathrm{d}x\mathrm{d}t$$
$$-\int_0^{t_0}\int_0^L\left(1-\eta^2\frac{\partial^2}{\partial x^2}\right)\left(m_0\ddot{u}-m_1\ddot{\phi}+f\right)\delta u\,\mathrm{d}x\mathrm{d}t$$
$$+\int_0^{t_0}\int_0^L\left(1-l^2\frac{\partial^2}{\partial x^2}\right)\left(k_sA_{xz}w'-k_sA_{xz}\phi-B_{xx}u''+D_{xx}\phi''\right)\delta\phi\,\mathrm{d}x\mathrm{d}t$$
$$+\int_0^{t_0}\int_0^L\left(1-\eta^2\frac{\partial^2}{\partial x^2}\right)\left(m_1\ddot{u}-m_2\ddot{\phi}\right)\delta\phi\,\mathrm{d}x\mathrm{d}t$$
$$+\int_0^{t_0}\int_0^L k_sA_{xz}\left(1-l^2\frac{\partial^2}{\partial x^2}\right)\left(w''-\phi'\right)\delta w\,\mathrm{d}x\mathrm{d}t$$
$$-\int_0^{t_0}\int_0^L\left(1-\eta^2\frac{\partial^2}{\partial x^2}\right)\left(m_0\ddot{w}+q\right)\delta w\,\mathrm{d}x\mathrm{d}t=0 \tag{49}$$

Then, integrating Eq. (49) by parts with respect to the higher-order derivatives, we have

$$\delta\int_0^t(T-U+V)\,\mathrm{d}t+\int_0^t\left([N_{xx}\delta u]_0^L+\left[N_{xx}^{(1)}\delta u'\right]_0^L+\left[Q_{xz}\delta w\right]_0^L\right)dt$$
$$+\int_0^t\left(-[M\delta\phi]_0^L+\left[Q_{xz}^{(1)}\delta w'\right]_0^L+\left[M^{(1)}\delta\phi'\right]_0^L\right)dt=0 \tag{50}$$

where

$$T=\int_0^L\left[\rho A\left(\dot{u}_1\delta\dot{u}_1+\eta^2\dot{u}_1'\delta\dot{u}_1'\right)+\rho A\left(\dot{u}_3\delta\dot{u}_3+\eta^2\dot{u}_3'\delta\dot{u}_3'\right)\right]\mathrm{d}x \tag{51}$$

$$U = \int_0^L \left[N_{cl}\delta u' + l^2 N'_{cl}\delta u'' - M_{cl}\delta\phi' - l^2 M'_{cl}\delta\phi''\right] \mathrm{d}x + \int_0^L \left[-l^2 Q'_{cl}\delta\phi' - Q_{cl}\delta\phi + Q_{cl}\delta w' + l^2 Q'_{cl}\delta w''\right] \mathrm{d}x \tag{52}$$

$$V = -\int_0^L \left[\left(1 - \eta^2\frac{\partial^2}{\partial x^2}\right)(f\delta u + q\delta w)\right] \mathrm{d}x \tag{53}$$

In Eqs. (49)–(51), T, U and V are the kinetic energy, strain energy and work done by the external forces, respectively. In Eqs. (50) and (52), we have defined the following classical stress resultants:

$$\begin{aligned} N_{cl} &= A_{xx}u' - B_{xx}\phi' \\ Q_{cl} &= k_s A_{xz}\left(w' - \phi\right) \\ M_{cl} &= B_{xx}u' - D_{xx}\phi' \end{aligned} \tag{54}$$

In addition, the other stress resultants are given by

$$\begin{aligned} N_{xx} &= N_{cl} - l^2 N''_{cl} + m_0\eta^2\ddot{u}' - m_1\eta^2\ddot{\phi}' \\ N_{xx}^{(1)} &= l^2 N'_{cl} \\ Q_{xz} &= Q_{cl} - l^2 Q''_{cl} + m_0\eta^2\ddot{w}' \\ M &= M_{cl} + l^2 Q'_{cl} - l^2 M''_{cl} - m_2\eta^2\ddot{\phi}' + m_1\eta^2\ddot{u}' \\ Q_{xz}^{(1)} &= l^2 Q'_{cl} \\ M^{(1)} &= -l^2 M'_{cl} \end{aligned} \tag{55}$$

Interestingly, it is found that when u and w are decoupled and the nonlocal parameter is not considered, the boundary conditions expressed by Eqs. (50) and (55) reduce to the results of Nojoumian and Salarieh [24]. On the other hand, if one only consider a nonlocal beam, Eqs. (50) and (55) reduces to Xu et al. [6] without including any non-classical boundary conditions. Noteworthy, the corresponding boundary conditions by Li et al. [18] can not be recovered to those given by Xu et al. [6]. Similarly, the above boundary conditions can be reduced as special cases to those of nonlocal Timoshenko models. How to choose the above higher-order boundary conditions, the interested readers may refer to Refs. [20, 25] where several possible boundary-value problems were extensively studied to assess the boundary effects.

In addition, if the strain gradient effect is omit (e.g., $l = 0$), the higher order mass inertia is ignored, and the axial-transverse coupling effect is neglect, the boundary conditions of a homogenous beam may be reduced to the nonlocal type:

$$\begin{aligned} Q_{xz} &= Q_{cl} + m_0\eta^2\ddot{w}' \quad \text{or} \quad w \\ M &= M_{cl} \quad \text{or} \quad \phi \end{aligned} \tag{56}$$

For clarity, the differences of the stress resultants for nonlocal Timoshenko beams are highlighted in Table 1.

4.2 Dynamic Behavior of Nonlocal Cantilevers: Timoshenko Type

We here attempt to solve the paradox encountered in the literature by modifying the boundary conditions (see Eqs. (50) and (55)) and solve the BVPs for nonlocal cantilevers using Timoshenko beam theory. Firstly, according to Eqs. $(48)_2$ and $(48)_3$, the equation of motion of nonlocal Timoshenko beams can be obtained as

$$k_s GA\left(w'' - \phi'\right) - P\left(1 - \eta^2 \frac{\partial^2}{\partial^2 x}\right) w'' - \rho A\left(1 - \eta^2 \frac{\partial^2}{\partial^2 x}\right) \ddot{w} = 0$$
$$EI\phi'' + k_s GA\left(w' - \phi\right) - \rho I\left(1 - \eta^2 \frac{\partial^2}{\partial^2 x}\right) \ddot{\phi} = 0 \tag{57}$$

Analogously, the corresponding boundary conditions of Eq. (57) are given by

$$\begin{aligned} &w(0) = \phi(0) = 0 \\ &Q_{xz}(L) = k_s GA\left[\phi(L) - w'(L)\right] - \rho A\eta^2 \ddot{w}'(L) = 0 \\ &M(L) = -EI\phi'(L) - \rho I\eta^2 \ddot{\phi}'(L) = 0 \end{aligned} \tag{58}$$

If one defines slender ratio effect $r = \sqrt{I/AL^2}$ and shear effect $s = \sqrt{EI/k_s GAL^2}$, the characteristic equation for BVPs of Eqs. (57) and (58) obtains from [6]

$$\begin{aligned} \Psi_\alpha \Psi_\beta \left(\beta^2 - \alpha^2\right) \sin\alpha \sinh\beta + \left(\alpha^2 \Psi_\alpha^2 + \beta^2 \Psi_\beta^2\right) \cos\alpha \cosh\beta \\ + 2\alpha\beta\Psi_\alpha\Psi_\beta = 0 \end{aligned} \tag{59}$$

where

$$\Psi_\alpha = \frac{\alpha^2 - b^2 e^2 s^2 \alpha^2 - b^2 s^2}{\alpha}, \quad \Psi_\beta = \frac{\beta^2 - b^2 e^2 s^2 \beta^2 + b^2 s^2}{\beta}$$
$$\begin{pmatrix} \alpha \\ \beta \end{pmatrix} = \left[\frac{\pm A_2 + \sqrt{A_2^2 - 4A_0 A_4}}{2A_4}\right]^{1/2}, \quad A_4 = 1 + b^2 e^2\left(e^2 r^2 s^2 b^2 - r^2 - s^2\right)$$
$$A_2 = b^2\left(e^2 + r^2 + s^2 - 2e^2 r^2 s^2 b^2\right), \quad A_0 = b^2\left(r^2 s^2 b^2 - 1\right) \tag{60}$$

Noteworthy, the other dimensionless parameters are the same with those of Euler–Bernoulli ones. Normally, the numerical results can be easily obtained by numerically solving Eq. (59). The fundamental frequency calculated by Eq. (59) as a function of

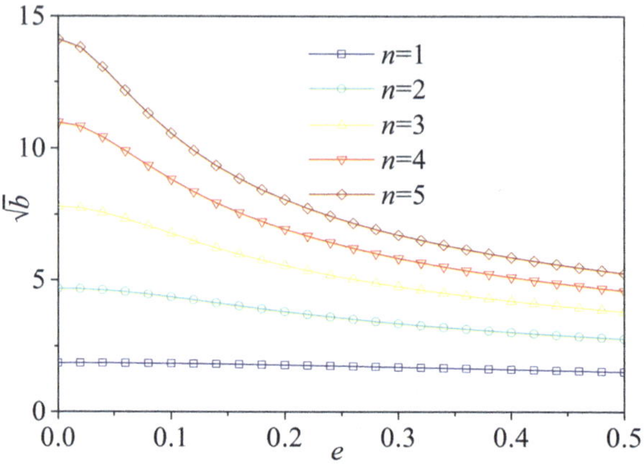

Fig. 2 The first five frequency parameters versus the nonlocal parameter for Timoshenko beams. Numerical results are obtained by solving Eq. (59)

the nonlocal parameter is also added in Fig. 1. Interestingly, the softening phenomena for nonlocal Timoshenko cantilevers is also captured by the present work. Moreover, it can also be seen from Fig. 1 that the frequencies calculated by the Euler–Bernoulli beam theory are larger than those by the corresponding Timoshenko beam theory. The differences may be more prominent for small values of nonlocal parameter. For the calculation of the frequency of the Timoshenko type, we take the following parameters for carbon nanotubes [3]: $E = 5.5$ TPa, $\nu = 0.19$, $D = 0.678$ nm, $h = 0.066$ nm and $k_s = 0.563$. D, h and k_s are the diameter, thickness and shear factor of the beam, respectively.

For a sufficiently thin walled carbon nanotube, the area and the second moment of inertia of the beam are approximately calculated based on $A = \pi Dh$ and $I = \pi D^3 h/8$, respectively. We also take the aspect ratio $L/D = 10$ until otherwise stated, where L is the length of the beam.

Figure 2 shows the first five frequency parameters as a function of the nonlocal parameter. It is observed that the frequency parameter decrease as the nonlocal parameter increases. It means that the nonlocal parameter may reveal a stiffness-softening effect for nonlocal beams. Moreover, it can also be seen that the softening effect may be remarkable for higher-order mode frequencies.

As is well-known, the Timoshenko beam model involves the effect of shear deformation and rotary inertia. In order to see these two effects, Fig. 3 displays the first four frequency parameters against the aspect ratio L/D for Timoshenko beams. As expected, the frequency parameters increase as aspect ratio increases. Moreover, the present Timoshenko beam model will reduce to the corresponding Euler–Bernoulli beam model when the aspect ratio is sufficiently large. On the other hand, the nonlocal parameter will play a prominent effect on the dynamic behavior of beams for higher-order mode numbers when one compares with these figures.

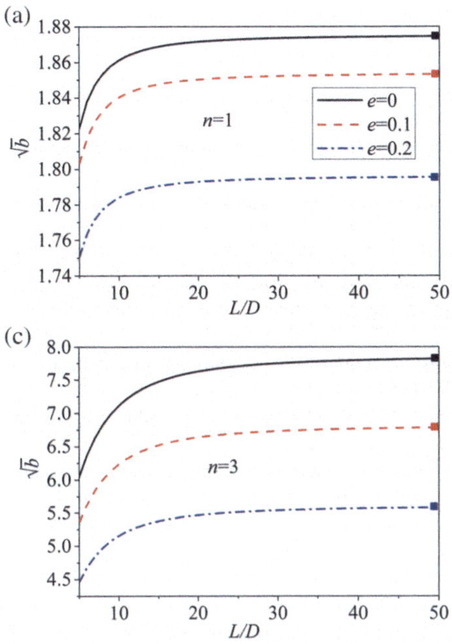

Fig. 3 The frequency parameters versus the aspect ratio L/D for Timoshenko beams: **a** $n = 1$, **b** $n = 2$, **c** $n = 3$, **d** $n = 4$. Numerical results are obtained by solving Eq. (59). Symbols are the corresponding Euler–Bernoulli results based on Eq. (46)

5 Conclusions

By using the weighted residual method, we established an asymptotic theory for the nonlocal elasticity, from which the frequencies for different order of magnitude were obtained as functions of the nonlocal parameter. Then, we used this method to reformulate the BVPs of Euler–Bernoulli and Timoshenko beam models within the framework of the newly developed nonlocal strain gradient elasticity theory. In case studies, we presented the closed-form solutions for characteristic frequency equations for both nonlocal Euler–Bernoulli beams and nonlocal Timoshenko beams based on the nonlocal elasticity. Our numerical results showed that the softening phenomena can be captured by the present formulations. In addition, our numerical results for differential type nonlocal elasticity showed a stiffening effect when they were compared with the corresponding nonlocal integral ones.

Acknowledgements This work was supported by Natural Science Basic Research Plan in Shaanxi Province of China (No. 2020JQ-337), and the Fundamental Research Funds for the Central Universities, CHD (No. 300102219315).

References

1. Thai H-T, Vo TP, Nguyen T-K, Kim S-E (2017) A review of continuum mechanics models for size-dependent analysis of beams and plates. Compos Struct 177:196–219
2. Lu P, Lee HP, Lu C, Zhang PQ (2006) Dynamic properties of flexural beams using a nonlocal elasticity model. J Appl Phys 99:073510
3. Wang CM, Zhang YY, He XQ (2007) Vibration of nonlocal Timoshenko beams. Nanotechnology 18:105401
4. Fernández-Sáez J, Zaera R, Loya JA, Reddy JN (2016) Bending of Euler-Bernoulli beams using Eringen's integral formulation: A paradox resolved. Int J Eng Sci 99:107–116
5. Challamel N, Wang CM (2008) The small length scale effect for a non-local cantilever beam: a paradox solved. Nanotechnology 19:345703
6. Xu X-J, Deng Z-C, Zhang K, Xu W (2016) Observations of the softening phenomena in the nonlocal cantilever beams. Compos Struct 145:43–57
7. Romano G, Barretta R, Diaco M, Marotti de Sciarra F (2017) Constitutive boundary conditions and paradoxes in nonlocal elastic nanobeams. Int J Mech Sci 121:151–156
8. Tuna M, Kirca M (2016) Exact solution of Eringen's nonlocal integral model for vibration and buckling of Euler-Bernoulli beam. Int J Eng Sci 107:54–67
9. Eptaimeros KG, Koutsoumaris CC, Tsamasphyros GJ (2016) Nonlocal integral approach to the dynamical response of nanobeams. Int J Mech Sci 115–116:68–80
10. Zhu X, Li L (2017) On longitudinal dynamics of nanorods. Int J Eng Sci 120:129–145
11. Apuzzo A, Barretta R, Luciano R, Marotti de Sciarra F, Penna R (2017) Free vibrations of Bernoulli-Euler nano-beams by the stress-driven nonlocal integral model. Compos Part B 123:105–111
12. Eringen AC (1983) On differential equations of nonlocal elasticity and solutions of screw dislocation and surface waves. J Appl Phys 54:4703–4710
13. Polizzotto C (2014) Stress gradient versus strain gradient constitutive models within elasticity. Int J Solids Struct 51:1809–1818
14. Polizzotto C (2015) A unifying variational framework for stress gradient and strain gradient elasticity theories. Eur J Mech- A/Solids 49:430–440
15. Polizzotto C (2016) Variational formulations and extra boundary conditions within stress gradient elasticity theory with extensions to beam and plate models. Int J Solids Struct 80:405–419
16. Apuzzo A, Barretta R, Faghidian SA, Luciano R, Marotti de Sciarra F (2019) Nonlocal strain gradient exact solutions for functionally graded inflected nano-beams. Compos. Part B 164:667–674
17. Zaera R, Serrano Ó, Fernández-Sáez J (2019) On the consistency of the nonlocal strain gradient elasticity. Int J Eng Sci 138:65–81
18. Li L, Li X, Hu Y (2016) Free vibration analysis of nonlocal strain gradient beams made of functionally graded material. Int J Eng Sci 102:77–92
19. Lam DCC, Yang F, Chong ACM, Wang J, Tong P (2003) Experiments and theory in strain gradient elasticity. J Mech Phys Solids 51:1477–1508
20. Xu X-J, Zheng M-L (2019) Analytical solutions for buckling of size-dependent Timoshenko beams. Appl Math Mech 40:953–976
21. Kahrobaiyan MH, Rahaeifard M, Ahmadian MT (2011) A nonlinear strain gradient beam formulation. Int J Eng Sci 49:1256–1267
22. Karparvarfard SMH, Asghari M, Vatankhah R (2015) A geometrically nonlinear beam model based on the second strain gradient theory. Int J Eng Sci 91:63–75
23. Reddy JN, Pang SD (2008) Nonlocal continuum theories of beams for the analysis of carbon nanotubes. J Appl Phys 103:023511
24. Nojoumian MA, Salarieh H (2016) Comment on "A micro scale Timoshenko beam model based on strain gradient elasticity theory". Eur J Mech-A/Solids 60:361–362
25. Xu X-J, Wang X-C, Zheng M-L, Ma Z (2017) Bending and buckling of nonlocal strain gradient elastic beams. Compos Struct 160:366–377

Application of Combined Nonlocal and Surface Elasticity Theories to Vibration Response of a Graded Nanobeam

Sami El-Borgi, Prakash Rajendran, and Mohamed Trabelssi

Abstract This chapter investigates the combined nonlocal and surface effects on the free and forced vibration response of a graded nanobeam resting on a nonlinear elastic foundation. Material gradation is assumed to be through the beam depth according to a power-law model. Instead of adopting the customary choice of the geometrical central axis, this study uses instead the physical neutral axis to eliminate the stretching-bending coupling effect due to the unsymmetrical material variation. It is also shown that the choice of the physical neutral axis leads to the elimination of the quadratic nonlinearity from the equation of motion. The Euler–Bernoulli beam theory along with the von Kármán geometric nonlinearity is formulated while accounting for Eringen's nonlocal elasticity differential model, determined neutral axis and surface effects. Under certain assumptions, a nonlinear fourth-order partial differential equation is derived and the Garlerkin technique with a single-mode approximation is used to obtain a second-order ordinary equation in the time domain with cubic nonlinearity. The Method of Multiple Scales (MMS) is used initially to derive the expression of the nonlinear natural frequency for the graded nanobeam accounting for both nonlocal and surface effects. Then, MMS is utilized to study the primary resonance of an externally forced nanobeam after obtaining its frequency

S. El-Borgi (✉)
Mechanical Engineering Program, Texas A&M University at Qatar, Engineering Building, Education City, P.O. Box 23874, Doha, Qatar
e-mail: sami.el_borgi@tamu.edu

P. Rajendran
Department of Mechanical Engineering, National Institute of Technology, Tiruchirappalli 620015, Tamil Nadu, India
e-mail: rprakash@nitt.edu

M. Trabelssi
Applied Mechanics and Systems Research Laboratory, Tunisia Polytechnic School, University of Carthage, B.P. 743, 2078 La Marsa, Tunisia
e-mail: mtrabel@g.clemson.edu; mohamed.trabelsi@ensit.u-tunis.tn

Department of Mechanical Engineering, Tunis Higher National Engineering School, University of Tunis, 1008 Tunis, Tunisia

© The Author(s), under exclusive license to Springer Nature Switzerland AG 2021

E. Ghavanloo et al. (eds.), *Size-Dependent Continuum Mechanics Approaches*, Springer Tracts in Mechanical Engineering,
https://doi.org/10.1007/978-3-030-63050-8_9

response curve. The primary objective of this work is to investigate the effects of the nonlocal and surface effects parameters, power-law index and boundary conditions on the free and forced vibration response of the nanobeam.

1 Introduction

Miniature beams are used in several electromechanical systems such as micro-actuators, micro-switches, bio-sensors, nano-wires, ultra-thin films, MEMS and NEMS as the basic structural elements [1–4]. Because the beam lengths are in the nanoscale dimension, the effect of material length scale has to be considered, which is not described by a classical continuum model. Several models have been developed over the past decade to incorporate for the size dependency such as Eringen's nonlocal elastic theory [5–7], Modified Coupled Stress theory [8, 9], strain gradient theory [10, 11] and a combination of nonlocal and strain gradient theory [12–14].

The nonlocal elasticity theory postulates that the stress in a continuum at a given location depends not only on the strain at that location but also on the strains at all points of the body. This dependency on the nonlocal strain is captured by a size effect parameter called the nonlocal parameter. The nonlocal elasticity theory was proposed by Eringen and co-workers [5–7]. Initially, the nonlocal elasticity was formulated using an integral form. Later on, a differential form was developed [6] based on a specific kernel function. This simplified the implementation of the theory in practical problems. However, Fernandez-Saez et al. [15] and Romano et al. [16], and the references therein, highlighted a paradox in the transformation from the integral form of the nonlocal model to the differential form for beam bending problems with an exponential nonlocal kernel. They showed that the transformation implied a relationship between the bending moment and the spatial derivative of the bending moment at the boundaries that must be satisfied. The bending moment obtained from the solution of the differential equation should be checked to ensure the obtained solution is also a solution to the integral form of the equation. This is readily done for problems with displacement type boundary conditions, since the bending moment will be the solution of a second order differential equation, and the constants of integration can be used to satisfy the bending moment boundary conditions. However, it should also be recognized that the nonlocal model in integral form is unable to model detailed local effects at boundaries, and hence there are always likely to be discrepancies between the actual and simulated bending moment at the boundary. Given that these discrepancies at the boundaries are always likely to be present whichever model used, here the differential form of the equations is used.

Functionally Graded Materials (FGMs) are novel heterogeneous composites in which the volume fraction of the constituent materials varies from one face to another. This enables gradual change in properties along a given direction. Furthermore, small changes in the composition make FGMs less prone to stress concentration and consequently to fracture [17]. As technology develops in the demand for nanostructures, FGMs are being used in several applications at the micro/nano scale in the form of micro-switches [18], shape memory alloy thin films [19], atomic force microscopes [20] and electrically actuated actuators [21].

Several researchers have studied the free and forced linear vibration response of a graded nanobeam using Eringen's nonlocal model. Free vibration of nonlocal nanobeam was modeled by Eltaher et al. [22] using the Finite Element Method (FEM). Uymaz [23] studied the free and forced vibration of a Functionally Graded (FG) nanobeam using Navier's method, while Rahmani and Pedram [24] used the same method to study the behavior of a graded Timoshenko beam. Nejad and Hadi [25] performed the free vibration analysis of a bi-directional FG Euler Bernoulli nanobeam using Hamilton's principle and Generalized differential quadrature method (GDQM).

The nonlinear vibration response of nanobeams has been investigated by several authors who included the von Kármán geometric nonlinearity in their formulation. Simsek [26] utilized He's variational method to model the nonlinear free vibration of a FG Euler-Bernoulli nanobeam. Niknam and Aghdam [27] used He's variational method for analyzing the large amplitude free vibration and buckling load of a FG Euler-Bernoulli nanobeam resting on a nonlinear elastic foundation. Nazemnezhad and Hosseini-Hashemi [28] studied the nonlinear free vibration of a functionally graded nanobeam while incorporating surface effects using the method of multiple scales. Finally, El-Borgi et al. [29] studied the free and forced vibration of a FG nanobeam resting on an elastic foundation using the Method of Multiple Scales (MMS), whereas Trablessi et al. [30] studied the same problem using the so-called Locally adaptive Differential Quadrature Method (LaDQM).

Recently researchers have been focusing on the investigation of surface effects in nanomaterials. These effects can be predominant in the behavior of nanostructures due to sudden increasing ratio of surface area to volume at nanoscale. In general, it is expected that surface stress or surface tension and surface elasticity properties can affect the size-dependent mechanical properties at nanoscale [31–35]. The analytical studies on the nonlinear free vibration of functionally graded nanobeams incorporating surface effects have been demonstrated in [36–38]. The effects of nonlocal parameter, residual surface stress and surface elasticity modulus on the nonlinear vibrational frequencies of Timoshenko beam have been studied in [39, 40]. Hosseini-Hashemi et al. [41] investigated the coupling between nonlocal and surface stress effects for the nonlinear vibration response of nanobeams.

In the above studies, the undeformed plane of the beam was assumed to coincide with the midplane of the section which in turn does not overlap with the neutral axis because of the material property gradation through the beam thickness. This assumption has led to an error in predicting the response of graded beams such as frequencies which can reach as much as 10% [42]. Therefore, instead of adopting the geometrical central axis, few investigators have recently used the physical neutral axis in their studies on graded beams [42–48]. This has also simplified their formulation by eliminating the stretching-bending coupling effect due to the unsymmetrical material variation.

Based on the above review and to the authors best knowledge, it can be concluded that few investigators have focused on studying the nonlinear free vibration of functionally graded nanobeams incorporating a combination of nonlocal and surface effects [33] or surface effects only [34–36]. In addition, the combination of nonlocal and surface effects as well as the physical neutral axis for studying the vibration response of graded beams has not been solved in the published literature to-date. Also, a limited number of researchers have used analytical methods such as MMS [28, 29], He's variational method [49, 50] and Navier's method [23, 24, 51] to study the free and forced vibration size-dependent response of homogenous and functionally graded nanobeams.

To fill the above gaps, this study is proposed to account for surface and nonlocal effects in addition to using the physical neutral axis to investigate the nonlinear, free and forced vibration response of a Euler-Bernoulli nanobeam using MMS. This chapter is arranged as follows. Following this introduction, Sects. 2 and 3 provide the governing equations for classical and nonlocal graded Euler-Bernoulli beam theory accounting for surface effects. The solutions for the free and forced vibrations obtained using MMS are presented in Sects. 4 and 5, respectively. Numerical results are provided and discussed in Sect. 6. Finally, concluding remarks are outlined in Sect. 7.

2 Classical Euler–Bernoulli Beam Theory with Surface Effects

This section uses the dynamic version of the principle of virtual displacements [52] to derive the equations of motion of the Euler–Bernoulli beam theory while considering the modified von Kármán nonlinearity as well as surface effects. This study focuses on a functionally graded straight nanobeam of length L, width b and height h, which is resting on a nonlinear elastic foundation and subjected to a distributed loading f_z as shown in Fig. 1. Due to the nature of graded material, the neutral axis of the beam does not coincide with the beam's mid-plane, thus two frames are introduced in this study:

- (x_m, y_m, z_m) is a frame located at the mid-plane of the beam as shown in Fig. 1, where x_m, y_m and z_m are coordinates taken along the length, width and thickness of the beam respectively.
- (x, y, z) is another reference frame located at the neutral axis of the beam, where $x = x_m$, $y = y_m$ and $z = z_m - h_0$, as depicted in Fig. 1. Here, h_0 is the distance from the midplane to the neutral axis and whose expression will be computed later.

All displacements (u, v, w) along (x, y, z) are functions of the x and z coordinates and time t. The bending is further assumed to be in the xz-plane, hence the displacement v is identically set to zero. The material properties such as elastic modulus, E, mass density, ρ are assumed to vary based on the following power law function:

$$P(z_m) = (P_u - P_L)\left(\frac{z_m}{h} + \frac{1}{2}\right)^n + P_L \tag{1}$$

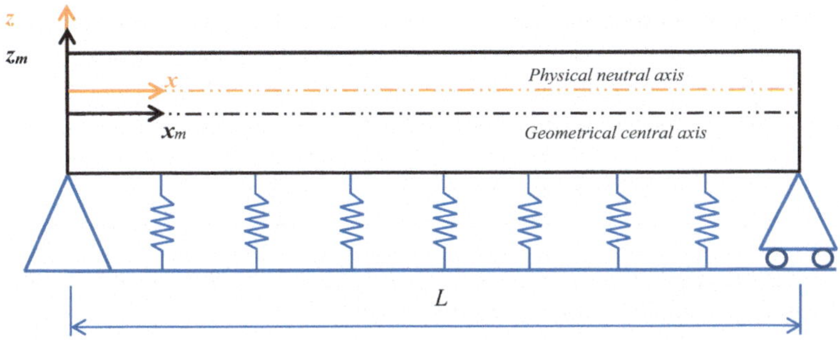

Fig. 1 Schematic of simply supported FG beam on nonlinear elastic foundation

where P_u and P_L are, respectively, the values of the material property P at the upper and lower surface of the nanobeam and n gives the variation profile of the material properties across the thickness of the nanobeam. The case of $n = 0$ corresponds to an isotropic beam with bulk properties of the upper surface. The displacement field according to the Euler-Bernoulli beam theory (EBT) can be expressed as follows:

$$\mathbf{u}(\mathbf{x}, \mathbf{t}) = [u(x, t) + z\theta_x]\,\hat{\mathbf{e}}_x + w(x, t)\hat{\mathbf{e}}_z, \quad \theta_x \equiv -\frac{\partial w}{\partial x} \tag{2}$$

where (u, w) are the axial and transverse displacements of the point $(x, 0)$ on the neutral axis of the beam, i.e., $z = 0$, and $(\hat{\mathbf{e}}_x, \hat{\mathbf{e}}_z)$ are unit vectors along the (x, z) coordinates. The Lagrangian strain tensor is of the form

$$\mathbf{E} \approx \boldsymbol{\varepsilon} = \left[\frac{\partial u}{\partial x} + z\frac{\partial \theta_x}{\partial x} + \frac{1}{2}\left(\frac{\partial w}{\partial x}\right)^2\right]\hat{\mathbf{e}}_x\hat{\mathbf{e}}_x \tag{3}$$

where the term $1/2\,(\partial w/\partial x)^2$ in the above equation represents the Von Kármán nonlinear strain. The following isotropic stress-strain relations is adopted:

$$\sigma_{xx}\,(x, z) = E\,(z)\,(\varepsilon_{xx}^{(0)}(x) + z\varepsilon_{xx}^{(1)}(x)) \tag{4}$$

where

$$\varepsilon_{xx}^{(0)} = \frac{\partial u}{\partial x} + \frac{1}{2}\left(\frac{\partial w}{\partial x}\right)^2, \quad \varepsilon_{xx}^{(1)} = \frac{\partial \theta_x}{\partial x} \tag{5}$$

and E is Young's modulus and the beam's constitutive relations are based on uniaxial stress strain relations. The surface elasticity effects in the beam can be modeled as follows [33]:

$$\sigma^s = E^s\,(z)\,\varepsilon_{xx} \tag{6}$$

where E^s is the surface elastic modulus through thickness direction of the beam.

The principle of virtual displacements for the problem at hand has the form

$$0 = \int_0^T \int_0^L \Bigg[m_0 \left(\frac{\partial u}{\partial t}\frac{\partial \delta u}{\partial t} + \frac{\partial w}{\partial t}\frac{\partial \delta w}{\partial t} \right) + m_2 \frac{\partial^2 w}{\partial x \partial t}\frac{\partial^2 \delta w}{\partial x \partial t} + cw\frac{\partial \delta w}{\partial t}$$
$$- M_{xx}^{(0)}\,\delta\varepsilon_{xx}^{(0)} - M_{xx}^{(1)}\,\delta\varepsilon_{xx}^{(1)} + f_z\,\delta w \Bigg] dx\,dt \tag{7}$$

Integration by parts the above equation and separation into transverse and axial displacements yield the equations of motion below

$$\frac{\partial M_{xx}^{(0)}}{\partial x} = m_0 \frac{\partial^2 u}{\partial t^2} \tag{8a}$$

$$\frac{\partial^2 M_{xx}^{(1)}}{\partial x^2} + \frac{\partial}{\partial x}\left[M_{xx}^{(0)} \frac{\partial w}{\partial x} \right] + f_z = m_0 \frac{\partial^2 w}{\partial t^2} - m_2 \frac{\partial^4 w}{\partial t^2 \partial x^2} + c\frac{\partial w}{\partial t} \tag{8b}$$

where ($M_{xx}^{(i)}, i = 0, 1$) are the stress resultants and (m_0, m_2) are the mass inertias which can be written as

$$M_{xx}^{(i)} = \int_A (z)^i \sigma_{xx} dA + \oint_\Gamma (z)^i \sigma^s ds, \ (i = 0, 1) \tag{9a}$$

$$m_0 = \int_A \rho\, dA, \quad m_2 = \int_A \rho z^2\, dA \tag{9b}$$

where $A = bh$ is the cross sectional area and Γ is the boundary of the cross sectional area. The geometric and force boundary conditions are given by

$$\text{Geometric:} \quad u, \qquad w, \qquad \theta_x$$
$$\text{Force:} \quad M_{xx}^{(0)}, \quad V_x \equiv M_{xx}^{(0)} \frac{\partial w}{\partial x} + \frac{\partial M_{xx}^{(1)}}{\partial x}, \quad M_{xx}^{(1)} \tag{10}$$

Using Eqs. (2)–(6), the local stress resultants ($M_{xx}^{(0)}$ and $M_{xx}^{(1)}$) can be expressed in terms of displacements

$$M_{xx}^{(0)} = \int_A E\,(z)\,(\varepsilon_{xx}^{(0)}(x) + z\varepsilon_{xx}^{(1)}(x)) dA + \oint_\Gamma E^s(z)(\varepsilon_{xx}^{(0)}(x) + z\varepsilon_{xx}^{(1)}(x)) ds$$
$$= \tilde{A}\left[\frac{\partial u}{\partial x} + \frac{1}{2}\left(\frac{\partial w}{\partial x}\right)^2 \right] - \tilde{B}\left(\frac{\partial^2 w}{\partial x^2}\right) \tag{11a}$$

$$M_{xx}^{(1)} = \int_A zE(z)(\varepsilon_{xx}^{(0)}(x) + z\varepsilon_{xx}^{(1)}(x))dA + \oint_\Gamma zE^s(z)(\varepsilon_{xx}^{(0)}(x) + z\varepsilon_{xx}^{(1)}(x))ds$$
$$= \tilde{B}\left[\frac{\partial u}{\partial x} + \frac{1}{2}\left(\frac{\partial w}{\partial x}\right)^2\right] - \tilde{D}\left(\frac{\partial^2 w}{\partial x^2}\right) \tag{11b}$$

where

$$\tilde{A} = \int_A E(z)\,dA + \oint_\Gamma E^s(z)ds,\quad \tilde{B} = \int_A E(z)z\,dA + \oint_\Gamma E^s(z)zds,$$
$$\tilde{D} = \int_A E(z)z^2\,dA + \oint_\Gamma E^s(z)z^2ds \tag{12}$$

The substitution of Eqs. (11a) and (11b) for $M_{xx}^{(0)}$ and $M_{xx}^{(1)}$ into Eqs. (8a) and (8b) yields the equations of motion and boundary conditions being represented using the displacement components u and w; similarly, the boundary conditions can also be expressed in terms of the displacements.

At this point, the position of the neutral axis is obtained by requiring that the total axial resultant is zero along this axis. Hence, Eq. (11a) becomes

$$M_{xx}^{(0)} = \tilde{A}\left[\frac{\partial u}{\partial x} + \frac{1}{2}\left(\frac{\partial w}{\partial x}\right)^2\right] - \tilde{B}\left(\frac{\partial^2 w}{\partial x^2}\right) = 0 \tag{13}$$

In determining the neutral axis position, the effect of axial displacement (u) is neglected and higher-order terms are assumed to be small (i.e., small deformation theory requires that $(\partial w/\partial x)^2$ is negligible). Hence, the above equation becomes

$$\tilde{B}\left(\frac{\partial^2 w}{\partial x^2}\right) = 0 \Leftrightarrow \tilde{B} = 0 \tag{14}$$

For the remainder of this chapter, $\tilde{B}$ will be set to zero but the main equations corresponding to a non-neutral axis $\tilde{B} \neq 0$ will be summarized in Appendix to show the effect of $\tilde{B}$ in the formulation. Simplifying Eq. (14) leads to

$$b\int_{\frac{-h}{2}-h_0}^{\frac{h}{2}-h_0} E(z)z\,dz + b\left[E_U^s\left(\frac{h}{2} - h_0\right) + E_L^s\left(-\frac{h}{2} - h_0\right)\right]$$
$$+ 2\int_{\frac{-h}{2}-h_0}^{\frac{h}{2}-h_0} E^s(z)z\,dz = 0 \tag{15}$$

where E_U^s and E_L^s are, respectively, the surface elastic moduli along the upper and lower faces of the beam. Making the following change of variable $z = z_m - h_0$, the above equation becomes

$$b\int_{\frac{-h}{2}}^{\frac{h}{2}} E(z_m)(z_m - h_0)\,dz_m + b\left[E_U^s\left(\tfrac{h}{2} - h_0\right) + E_L^s\left(-\tfrac{h}{2} - h_0\right)\right] + 2\int_{\frac{-h}{2}}^{\frac{h}{2}} E^s(z_m)(z_m - h_0)\,dz_m = 0 \tag{16}$$

The above equation can be further simplified to yield the expression of the neutral axis position

$$h_0 = \left[b\int_{\frac{-h}{2}}^{\frac{h}{2}} E(z_m)z_m\,dz_m + \frac{h}{2}b\left[E_U^s - E_L^s\right] + 2\int_{\frac{-h}{2}}^{\frac{h}{2}} E^s(z_m)z_m\,dz_m\right] / \left[b\int_{\frac{-h}{2}}^{\frac{h}{2}} E(z_m)\,dz_m + b\left[E_U^s + E_L^s\right] + 2\int_{\frac{-h}{2}}^{\frac{h}{2}} E^s(z_m)\,dz_m\right] \tag{17}$$

which can be simplified to

$$h_0 = \left[-h\left[\left(b\left(n^2+3n+2\right)+2\,hn\right)\left(E_L^s - E_U^s\right) + bhn(E_L - E_U)\right]\right] / \left[2(n+2)\left(bnE_L^s + bE_L^s + bnE_U^s + bE_U^s + 2\,hnE_L^s + 2\,hE_U^s + bhnE_L + bhE_U\right)\right] \tag{18}$$

The expressions of $\tilde{A}$ and $\tilde{D}$ given by Eq. (12) can be further simplified given the fact that the neutral axis position is determined

$$\begin{aligned}
\tilde{A} &= b\int_{-\frac{h}{2}}^{\frac{h}{2}} E(z_m)\,dz_m + 2\int_{-\frac{h}{2}}^{\frac{h}{2}} E^s(z_m)dz_m + b\left[E_U^s + E_L^s\right] \\
&= \frac{bh\left(nE_L + E_U\right)}{n+1} + b\left(E_L^s + E_U^s\right) + \frac{2\,h\left(nE_L^s + E_L^s\right)}{n+1}, \\
\tilde{D} &= b\int_{-\frac{h}{2}}^{\frac{h}{2}} E(z_m)\left(z_m - h_0\right)^2\,dz_m + 2\int_{-\frac{h}{2}}^{\frac{h}{2}} E^s(z_m)\left(z_m - h_0\right)^2\,dz_m \\
x &= \frac{1}{12}h\left(h^2 + 12\,h_0^2\right)\left(bE_L + 2E_L^s\right) + \frac{1}{4}b\left(\left(h + 2\,h_0\right)^2 E_L^s + \left(h - 2\,h_0\right)^2 E_U^s\right) \\
&\quad + \left(-\frac{h^3}{n+3} + \frac{\left(h + 2\,h_0\right)h^2}{n+2} - \frac{\left(h + 2\,h_0\right)^2 h}{4(n+1)}\right) \\
&\quad \times \left(b(E_L - E_U) + 2(E_L^s - E_U^s)\right)
\end{aligned} \tag{19}$$

Due to the Von Kármán non-linearity which appeared in Eq. (3) and thereafter defined, the equations of motion Eqs. (8a) and (8b) governing the axial displacement u and the transverse displacement w become coupled. A strategy is applied to eliminate the axial displacement u from the equations of motion and approximation of the von Kármán nonlinear term by a constant [53]. The strategy assumes the following: (1) the longitudinal frequency of a slender beam is much greater than the transverse frequency, hence, the inertia term $(m\ddot{u})$ can be neglected; (2) The nanobeam is fixed at its ends in the longitudinal direction, i.e. $u(0,t)=0$ and $u(L,t)=0$. The first assumption along with Eq. (8a) yields $M_{xx}^{(0)}$ as a constant. Thus, we obtain

$$\frac{\partial u}{\partial x}+\frac{1}{2}\left(\frac{\partial w}{\partial x}\right)^2=c_1(t) \tag{20}$$

Integrating Eq. (20) yields

$$u(x,t)=-\int_0^x \frac{1}{2}\left(\frac{\partial w(s,t)}{\partial s}\right)^2 ds+c_1(t)x+c_2(t) \tag{21}$$

where c_1 and c_2 are obtained from the boundary conditions from assumption 2.

$$c_1(t)=\frac{1}{L}\int_0^L \frac{1}{2}\left(\frac{\partial w}{\partial x}\right)^2 dx \quad \text{and} \quad c_2(t)=0 \tag{22}$$

Consequently, Eqs. (11a) and (11b) are reduced, respectively, to the following equations:

$$M_{xx}^{(0)}=\tilde{A}c_1(t) \tag{23a}$$

$$M_{xx}^{(1)}=-\tilde{D}\frac{\partial^2 w}{\partial x^2} \tag{23b}$$

where c_1 is given by Eq. (22). Substituting $M_{xx}^{(0)}$ and $M_{xx}^{(1)}$ from Eqs. (23a) and (23b) into the transverse equation of motion (8b) gives

$$m_0\ddot{w}-m_2\ddot{w}''+c\dot{w}+\tilde{D}w''''-\tilde{A}c_1w''=f_z \tag{24}$$

To make the above partial differential equation more compact, it was decided to use the dot $(\dot{\ })$ notation to denote differentiation with respect to t and and the prime $(')$ notation to indicate differentiation with respect to x. Both the distributed harmonic force and the reaction force of the elastic foundation constitute f_z in Eq. (24), which can be represented as [54]:

$$f_z=-k_Lw-k_{NL}w^3+k_sw''+F\cos(\theta t) \tag{25}$$

where k_L, k_{NL} and k_s are the linear, nonlinear and shear coefficients of the elastic foundation respectively and F is the amplitude of the driving harmonic force.

Substitution of f_z from Eq. (25) into Eq. (24) yields the following classical fourth-order partial differential equation governing motion of the beam in the transverse direction:

$$\begin{aligned} m_0\ddot{w}-m_2\ddot{w}''+c\dot{w}+\tilde{D}w''''-[\tilde{A}c_1+k_s]w'' \\ +k_Lw+k_{NL}w^3=F\cos(\theta t) \end{aligned} \tag{26}$$

The following non-dimensional variables are introduced to simplify the subsequent parametric analysis:

$$\hat{x} = \frac{x}{L}, \quad \hat{t} = \frac{t}{\tau}, \quad \hat{w} = \frac{w}{r} \tag{27}$$

where $r = \sqrt{\frac{I}{A}}$ is the radius of gyration of the cross section. The equation of motion in dimensionless form is derived using Eq. (26) and non-dimensional parameters defined in Eq. (27) as

$$\ddot{\hat{w}} - \hat{\alpha}\ddot{\hat{w}}'' + \hat{c}\dot{\hat{w}} + \hat{w}'''' - \hat{w}'' \int_0^1 [\kappa_0(\hat{w}')^2] d\hat{x} - k_s\hat{w}'' + \hat{k}_L\hat{w} + \hat{k}_{NL}\hat{w}^3 = \hat{F}\cos(\hat{\theta}\hat{t}) \tag{28}$$

where

$$\begin{gathered} \hat{\alpha} = m_2/(m_0L^2), \quad \tau = L^2\sqrt{m_0/\tilde{D}}, \quad \hat{c} = c\tau/m_0, \quad \hat{k}_s = k_s\tau^2/(m_0L^2), \\ \hat{k}_L = k_L\tau^2/m_0, \quad \hat{k}_{NL} = k_{NL}r^2\tau^2/m_0, \quad \hat{F} = F\tau^2/(m_0r), \\ \hat{\theta} = \theta\tau, \quad \kappa_0 = \tilde{A}r^2\tau^2/(2m_0L^4) \end{gathered} \tag{29}$$

3 Nonlocal Euler–Bernoulli Beam Theory with Surface Effects

According to Eringen [5, 6], in an elastic continuum, the state of stress $\boldsymbol{\sigma}$ at a point $\mathbf{x}$ in depends not just on the strain field $\boldsymbol{\varepsilon}$ at the point, but on the strains at all other points of the body. The atomic theory of lattice dynamics and experimental observations on phonon dispersion were used to show the above theory. Hence, the nonlocal stress tensor $\bar{\boldsymbol{\sigma}}$ at point $\mathbf{x}$ is expressed as

$$\bar{\boldsymbol{\sigma}} = \int_{\Omega} K(|\mathbf{x}' - \mathbf{x}|, \tau)\, \boldsymbol{\sigma}(\mathbf{x}')\, d\mathbf{x}' \tag{30}$$

where $\boldsymbol{\sigma}(\mathbf{x}')$ is the classical, macroscopic, second order Piola–Kirchoff stress tensor. The nonlocal modulus is represented using the kernel function $K(|\mathbf{x}' - \mathbf{x}|)$. The distance between points $\mathbf{x}$ and $\mathbf{x}'$ (in Euclidean norm) is given by $|\mathbf{x}' - \mathbf{x}|$ and τ is a material parameter that depends on internal and external characteristic lengths. In a Hookean solid, the macroscopic stress at a point $\boldsymbol{\sigma}$ at point $\mathbf{x}$ is related to the local strain $\boldsymbol{\varepsilon}$ by generalized Hooke's law

$$\boldsymbol{\sigma}(\mathbf{x}) = \mathbf{C}(\mathbf{x}) : \boldsymbol{\varepsilon}(\mathbf{x}) \tag{31}$$

where **C** is the fourth-order elasticity tensor and : denotes the 'double-dot product'. Due to the inconvenient form of the constitutive relation as an integral, the following differential equation model was preferred by Eringen [6]

$$\left(1-\mu_0^2\nabla^2\right)\bar{\boldsymbol{\sigma}}=\boldsymbol{\sigma},\quad \mu_0=\tau^2\ell^2=e_0^2a^2 \tag{32}$$

where ∇^2 is the Laplacian operator, e_0 is a material constant, a and ℓ are the internal and external characteristic lengths, respectively and μ_0 is the so-called nonlocal parameter. Furthermore, it is presumed that the size-dependency is present in the longitudinal direction only, for a beam shape. If x is chosen as the longitudinal direction of the nanobeam, Eq. (32) can be reduced to

$$\bar{\sigma}_{xx}-\mu_0\frac{d^2\bar{\sigma}_{xx}}{dx^2}=E\,(z)\,\varepsilon_{xx} \tag{33}$$

where $\bar{\sigma}_{xx}$ is the nonlocal stress.

The nonlocal stress resultants in terms of the nonlocal strain are obtained by integrating the above equation over the beam cross-sectional area as

$$\bar{M}_{xx}^{(0)}=\mu_0\frac{\partial^2\bar{M}_{xx}^{(0)}}{\partial x^2}+\tilde{A}\,\varepsilon_{xx}^{(0)} \tag{34a}$$

$$\bar{M}_{xx}^{(1)}=\mu_0\frac{\partial^2\bar{M}_{xx}^{(1)}}{\partial x^2}+\tilde{D}\,\varepsilon_{xx}^{(1)} \tag{34b}$$

Substituting Eqs. (2) and (5) into Eqs. (34a) and (34b), the above quantities can be written in terms of displacements as

$$\bar{M}_{xx}^{(0)}=\mu_0\frac{\partial^2\bar{M}_{xx}^{(0)}}{\partial x^2}+\tilde{A}\left[\frac{\partial u}{\partial x}+\frac{1}{2}\left(\frac{\partial w}{\partial x}\right)^2\right] \tag{35a}$$

$$\bar{M}_{xx}^{(1)}=\mu_0\frac{\partial^2\bar{M}_{xx}^{(1)}}{\partial x^2}-\tilde{D}\left(\frac{\partial^2 w}{\partial x^2}\right) \tag{35b}$$

The non-classical equations of motion are derived from the dynamic principle of virtual displacements (7) as

$$\frac{\partial\bar{M}_{xx}^{(0)}}{\partial x}=m_0\frac{\partial^2 u}{\partial t^2} \tag{36a}$$

$$\frac{\partial^2\bar{M}_{xx}^{(1)}}{\partial x^2}+\frac{\partial}{\partial x}\left[\bar{M}_{xx}^{(0)}\frac{\partial w}{\partial x}\right]+f_z=m_0\frac{\partial^2 w}{\partial t^2}-m_2\frac{\partial^4 w}{\partial t^2\partial x^2}+c\frac{\partial w}{\partial t} \tag{36b}$$

Differentiating Eq. (36a) once and substitution of the values of $\partial^2\bar{M}_{xx}^{(0)}/\partial x^2$ and $\partial^2\bar{M}_{xx}^{(1)}/\partial x^2$ from Eq. (36b) into Eqs. (35a) and (35b), we obtain

$$\bar{M}_{xx}^{(0)} = \tilde{A}\left[\frac{\partial u}{\partial x} + \frac{1}{2}\left(\frac{\partial w}{\partial x}\right)^2\right] + \mu_0\left(m_0\frac{\partial^3 u}{\partial x \partial t^2}\right) \tag{37a}$$

$$\bar{M}_{xx}^{(1)} = -\tilde{D}\left(\frac{\partial^2 w}{\partial x^2}\right) + \mu_0\left(m_0\frac{\partial^2 w}{\partial t^2} - m_2\frac{\partial^4 w}{\partial t^2 \partial x^2} + c\frac{\partial w}{\partial t} - \frac{\partial}{\partial x}\left[\bar{M}_{xx}^{(0)}\frac{\partial w}{\partial x}\right] - f_z\right) \tag{37b}$$

The assumptions used in Sect. 2 are employed to eliminate the axial displacement u in equations Eqs. (37a), (35b), and (37b) and derive expressions of the stress resultants in terms of the transverse displacement w. Thus,

$$\bar{M}_{xx}^{(0)} = \tilde{A}c_1(t) \tag{38a}$$

$$\bar{M}_{xx}^{(1)} = -\tilde{D}\frac{\partial^2 w}{\partial x^2} + \mu_0\left(m_0\frac{\partial^2 w}{\partial t^2} - m_2\frac{\partial^4 w}{\partial t^2 \partial x^2} + c\frac{\partial w}{\partial t} - \frac{\partial}{\partial x}\left[\bar{M}_{xx}^{(0)}\frac{\partial w}{\partial x}\right] - f_z\right) \tag{38b}$$

where $c_1(t)$ is defined in Eq. (22). After substituting $\bar{M}_{xx}^{(0)}$ from Eq. (38a) into Eq. (38b), $\bar{M}_{xx}^{(1)}$ can be further expressed as

$$\bar{M}_{xx}^{(1)} = \mu_0\left(m_0\frac{\partial^2 w}{\partial t^2} - m_2\frac{\partial^4 w}{\partial t^2 \partial x^2} + c\frac{\partial w}{\partial t} - \frac{\partial}{\partial x}(\tilde{A}c_1)\frac{\partial w}{\partial x} - f_z\right) \tag{39}$$

Substituting Eqs. (38a), (38b) and (39) into Eq. (36b) gives the nonlocal equation of motion of the nanobeam, which after substituting for the external load from Eq. (25), takes the following form:

$$\begin{aligned}
&m_0(\ddot{w} - \mu_0\ddot{w}'') - m_2(\ddot{w}'' - \mu_0\ddot{w}'''') + c(\dot{w} - \mu_0\dot{w}'') + \tilde{D}w'''' \\
&-(w'' - \mu_0 w'''')\int_0^1 [\kappa_0(w')^2]dx + k_L\left(w - \mu_0 w''\right) - k_s\left(w'' - \mu_0 w''''\right) \\
&+k_{NL}\left(w^3 - \mu_0\left(6w\left(w'\right)^2 + 3w^2 w''\right)\right) = F\cos(\theta t)
\end{aligned} \tag{40}$$

The non-dimensional parameters used in Eq. (27) are introduced here as well to simplify the parametric analysis. Thus, the normalized equation of motion becomes

$$\begin{aligned}
&\ddot{\hat{w}} - \hat{\mu}_0\ddot{\hat{w}}'' - \hat{\alpha}\left(\ddot{\hat{w}}'' - \hat{\mu}_0\ddot{\hat{w}}''''\right) + \hat{c}\left(\dot{\hat{w}} - \hat{\mu}_0\dot{\hat{w}}''\right) + \hat{w}'''' \\
&-\left(w'' - \hat{\mu}_0 w''''\right)\int_0^1 [\kappa_0(\hat{w}')^2]d\hat{x} + \hat{k}_L\left(\hat{w} - \hat{\mu}_0\hat{w}''\right) - \hat{k}_s\left(\hat{w}'' - \hat{\mu}_0\hat{w}''''\right) \\
&+\hat{k}_{NL}\left(\hat{w}^3 - \hat{\mu}_0\left(6\hat{w}\left(\hat{w}'\right)^2 + 3\hat{w}^2\hat{w}''\right)\right) = \hat{F}\cos(\hat{\theta}\hat{t})
\end{aligned} \tag{41}$$

where $\hat{\mu}_0 = \mu_0/L^2$ is the non-dimensional nonlocal parameter and the rest have been defined in Eq. (29). Setting $\hat{\mu}_0 = 0$, the classical equation of motion of the nanobeam (28) can be recovered.

4 Free Vibration Analysis

4.1 Problem Formulation

Two terms from Eq. (41) are dropped for studying a free vibration case, the damping term $\hat{c}$ and the external harmonic function $\hat{F}\cos(\hat{\theta}\hat{t})$. The resulting equation becomes

$$\ddot{\hat{\bar{w}}} - \hat{\mu}_0\ddot{\hat{\bar{w}}}'' - \hat{\alpha}\left(\ddot{\hat{\bar{w}}}'' - \hat{\mu}_0\ddot{\hat{\bar{w}}}''''\right) + \hat{c}\left(\dot{\hat{\bar{w}}} - \hat{\mu}_0\dot{\hat{\bar{w}}}''\right) + \hat{\bar{w}}'''' \\ -\left(\hat{\bar{w}}'' - \hat{\mu}_0\hat{\bar{w}}''''\right)\int_0^1[\kappa_0\left(\hat{\bar{w}}'\right)^2]\hat{d}x + \hat{k}_L\left(\hat{\bar{w}} - \hat{\mu}_0\hat{\bar{w}}''\right) \\ +\hat{k}_{NL}\left(\hat{\bar{w}}^3 - \hat{\mu}_0\left(6\hat{\bar{w}}\left(\hat{\bar{w}}'\right)^2 + 3\hat{\bar{w}}^2\hat{\bar{w}}''\right)\right) - \hat{k}_s\left(\hat{\bar{w}}'' - \mu_0\hat{\bar{w}}''''\right) = 0 \tag{42}$$

Equation (42) can be converted to a time-dependent ordinary differential equation by employing Galerkin Approximation, whereby the assumption $\hat{\bar{w}}(\hat{x}, \hat{t}) = \phi(\hat{x})q(\hat{t})$ is used. The term $\phi(\hat{x})$ is obtained from the linearized mode shapes of vibration of the beam as in Table 1 for different boundary conditions used in this study (S-S: simply supported at both ends; C-C: clamped at both ends).

To obtain the ordinary differential equation, the aforementioned assumption and linear mode shapes are substituted into Eq. (42). Finally, the equation is integrated from 0 to 1, giving:

$$\frac{d^2q}{d\hat{t}^2} + \beta_1\, q + \beta_3\, q^3 = 0 \tag{43a}$$

$$q(0) = A, \dot{q}(0) = 0 \tag{43b}$$

where the coefficients β_1 and β_3 are defined as

Table 1 Normalized mode shape functions of a uniform beam for different boundary conditions

B.C.	Mode shape, $\phi(\hat{x})$
S-S	$\sin(\pi\hat{x})$
C-C	$\cosh(q\hat{x}) - \cos(q\hat{x}) - \frac{\cosh(q)-\cos(q)}{\sinh(q)-\sin(q)}(\sinh(q\hat{x}) - \sin(q\hat{x}))$; $q = 4.7300$

$$\beta_1 = \frac{1}{\Delta}\int_0^1 \left\{\phi'''' + \hat{k}_L\left(\phi - \hat{\mu}_0\phi''\right) - \hat{k}_s(\phi'' - \hat{\mu}_0\phi'''')\right\}\phi\, d\hat{x} \tag{44a}$$

$$\beta_3 = \frac{1}{\Delta}\int_0^1 \hat{k}_{NL}\left(-3\mu_0\phi^2\phi'' - 6\mu_0\phi\left(\phi'\right)^2 + \phi^3\right)\phi\, d\hat{x} + \frac{1}{\Delta}\int_0^1 [\hat{\kappa}_0(\hat{\mu}_0\phi'''' - \phi'')\int_0^1 \left(\phi'\right)^2 d\hat{x}]\phi\, d\hat{x} \tag{44b}$$

$$\text{in which } \Delta = \int_0^1 \left\{\phi - \hat{\mu}_0\phi'' - \hat{\alpha}\left(\phi'' - \hat{\mu}_0\phi''''\right)\right\}\phi\, d\hat{x} \tag{44c}$$

and A is the amplitude. It is clear from Eq. (43a) that only cubic nonlinearity exists. In Eq. (43a), $\sqrt{\beta_1}$ is the non-dimensional linear natural frequency.

4.2 Analysis Using the Method of Multiple Scales

Using a non-dimensionalization strategy, the non-dimensional linear frequency is set to unity which enables comparison of results with other authors

$$q = A\bar{q}, \hat{t} = \frac{\bar{t}}{\sqrt{\beta_1}} \tag{45}$$

Equation (45) is used in Eqs. (43a) and (43b), the coefficient of $\hat{q}$ is made unity and the hat symbols over q and t are dropped. This gives the following equations

$$\frac{d^2\bar{q}}{d\bar{t}^2} + \bar{q} + \alpha_2\bar{q}^3 = 0 \tag{46a}$$

$$\bar{q}(0) = 1, \dot{\bar{q}}(0) = 0 \tag{46b}$$

where $\alpha_2 = \beta_3 A^2/\beta_1$.

The solution for the above equation is obtained through the method of multiple scales [55]. This method represents the solution in terms of multiple independent variables (time scales) in place of a single variable, through an *n-th* order uniform expansion. The solution is expressed using the expansion

$$\bar{q}(\bar{t};\varepsilon) = \varepsilon\bar{q}_1(T_0, T_1, T_2) + \varepsilon^2\bar{q}_2(T_0, T_1, T_2) + ... \tag{47}$$

where ε is a small parameter measuring the amplitude of oscillation. The fast time scale $T_0 = t$ is defined as the one on which main oscillatory behavior occurs, whereas the slow timescale $T_n = \varepsilon^n t, n \geq 1$ is defined as the one for which the amplitude and phase modulation take place. The time differentiation operation is expanded as

$$\frac{d}{d\bar{t}} = D_0 + \varepsilon D_1 + \varepsilon^2 D_2 + \varepsilon^3 D_3 + ... \tag{48}$$

$$\frac{d^2}{d\bar{t}^2} = D_0^2 + 2\varepsilon D_0 D_1 + \varepsilon^2(D_1^2 + 2D_0 D_2) + \varepsilon^3(2D_1 D_2 + 2D_0 D_3) + ... \tag{49}$$

where $D_i = \partial/\partial T_i$. One then substitutes these equations into the time-dependent ODE to obtain a system of linear differential equations based on the range of expansion of terms.

Equation (42) is solved using a third-order uniform expansion. Hence, three timescales are used, namely, T_0, T_1 and T_2 and the solution is of the form

$$\bar{q}(\bar{t}) = \varepsilon\bar{q}_1 + \varepsilon^2\bar{q}_2 + \varepsilon^3\bar{q}_3 \tag{50}$$

The following system of equations are obtained by substituting Eq. (50) into Eq. (46a). Each equation is representative of a different order or power of ε

$$\varepsilon^1: \; D_0^2\bar{q}_1 + \bar{q}_1 = 0 \tag{51a}$$
$$\varepsilon^2: \; D_0^2\bar{q}_2 + \bar{q}_2 = -2D_0 D_1\bar{q}_1 \tag{51b}$$
$$\varepsilon^3: \; D_0^2\bar{q}_3 + \bar{q}_3 = -(D_1^2 + 2D_2 D_0)q_1 - \alpha_2\bar{q}_1^3 - 2D_0 D_1\bar{q}_2 \tag{51c}$$

The solution for Eq. (51a) may be conveniently be written in the following form

$$\bar{q}_1 = A(T_1, T_2)e^{iT_0} + \overline{A}(T_1, T_2)e^{-iT_0} \tag{52}$$

where i is $\sqrt{-1}$, A is an unknown complex function and $\overline{A}$ is its complex conjugate. Substitution of Eq. (52) into Eq. (51b) yields

$$D_0^2\bar{q}_2 + \bar{q}_2 = -2iD_1Ae^{iT_0} + cc \tag{53}$$

where cc is the complex conjugate of the preceding terms on the right hand side. In order to remove secular terms from the above expression, this requires selecting the coefficients of e^{iT_0} and setting them to zero

$$\begin{aligned} -2iD_1A &= 0 \;\Rightarrow D_1A = 0 \\ -2iD_1\overline{A} &= 0 \;\Rightarrow D_1\overline{A} = 0 \end{aligned} \tag{54}$$

This suggests that A is independent of the time scale T_1 and is entirely a function of T_2. Substituting this simplification into Eq. (53) and solving the resulting differential equation yields the following solution:

$$\bar{q}_2 = 0 \tag{55}$$

The solutions q_1 from Eq. (52) and q_2 from Eq. (55) are then substituted into Eq. (51c) while also taking into account the simplification obtained from Eq. (54) leading to

$$D_0^2\bar{q}_3 + \bar{q}_3 = \left(-2i D_2 A - 3\alpha_2 A^2\overline{A}\right) e^{iT_0} - \alpha_2 A^3 e^{3iT_0} + cc \tag{56}$$

Similarly as before, secular terms are eliminated from the above equation by setting to zero the coefficient that multiplies e^{iT_0}. Therefore,

$$-2i D_2 A - 3\alpha_2 A^2\overline{A} = 0 \tag{57}$$

The solution of the above equation may be written conveniently in polar form as follows [55]:

$$A = \frac{1}{2} a e^{i\psi} \tag{58}$$

where a and ψ are real functions of T_2. Substituting Eq. (58) into Eq. (57) and separating the real and imaginary parts, one obtains the amplitude and phase equations respectively as follows:

$$\frac{da}{dT_2} = 0 \tag{59a}$$

$$a\frac{d\psi}{dT_2} - \frac{3}{8}\alpha_2 a^3 = 0 \tag{59b}$$

This indicates that $a(T_2)$ is a constant. Also the ODE for ψ can be solved as

$$a(T_2) = constant \tag{60a}$$

$$\psi = \frac{3}{8}\alpha_2 a^2 T_2 + \psi_0 \tag{60b}$$

where ψ_0 is a constant and $T_2 = \varepsilon^2 \bar{t}$. Substituting a and ψ from Eq. (60) into Eq. (58), it can be concluded that

$$A(T_2) = \frac{1}{2} a e^{i\left(\frac{3}{8}\alpha_2 a^2 \varepsilon^2 \bar{t} + \psi_0\right)} \tag{61}$$

Substituting for $\bar{q}_1$ and $\bar{q}_2$ from equations Eqs. (52) and (55) into Eq. (50) and dropping ε^3 then the expression of $\bar{q}$ becomes

$$\bar{q} = \varepsilon\bar{q}_1 = \varepsilon a \cos\left(\Omega\bar{t} + \psi_0\right) \tag{62}$$

where Ω is the nonlinear nonlocal frequency which can be written as [55]

$$\Omega = 1 + \frac{3}{8}\alpha_2 a^2 \varepsilon^2 \tag{63}$$

For this solution to satisfy the initial conditions given by Eq. (46b), a trigonometric manipulation is introduced whereby the error associated with a third-order expansion is accounted for as follows [28]:

$$\bar{q} = \varepsilon a \cos\left(\Omega \bar{t} + \psi_0\right) + \varepsilon^2 b \cos(\bar{t} + \rho) \tag{64}$$

The values of constants in Eq. (60) are attained by substituting the initial conditions in Eq. (46b) into Eq. (64) as

$$\psi_0 = 0, \quad a = \frac{1}{\varepsilon}, \quad \rho = 0, \quad b = 0 \tag{65}$$

A simplified version of the nonlinear nonlocal frequency can be obtained by substituting the variables in Eq. (65) into Eq. (63)

$$\Omega = 1 + \frac{3}{8}\alpha_2 \tag{66}$$

Further, Eq. (66) is multiplied by $\sqrt{\beta_1}$ to give a non-dimensional form for the nonlinear nonlocal frequency which will be employed in the validation and discussion of results (Sect. 6).

5 Forced Vibration Analysis

5.1 Problem Formulation

In the study of forced vibration, the entire equation of motion given by Eq. (41) is considered as it includes the damping coefficient as well as the external harmonic forcing function that excites the nanobeam. A Galerkin discretization with a one-mode approximation is employed to obtain an ordinary differential equation (ODE) from the partial differential equation (41). The entire equation is then multiplied by the mode shape function $\phi(\hat{x})$ as indicated in Table 1 and integrated from 0 to 1. As a result, the resulting ODE is obtained

$$\frac{d^2 q}{d\hat{t}^2} + 2\beta_d \frac{dq}{d\hat{t}} + \omega^2 q + \beta_3 q^3 = F_0 \cos(\hat{\theta}\hat{t}) \tag{67}$$

where

$$\beta_d = \frac{1}{2\Delta}\int_0^1 \left(\phi - \hat{\mu}_0\phi''\right)\hat{c}\phi \, d\hat{x} \tag{68}$$

$$F_0 = \frac{1}{\Delta}\int_0^1 \phi \hat{F} d\hat{x} \tag{69}$$

and $\omega^2 = \beta_1$, β_3 and Δ have been defined in equations Eqs. (44a)–(44c).

5.2 *Analysis Using Method of Multiple Scales*

For the primary resonance, the linear frequency of the system ω and frequency of excitation, $\hat{\theta}$ are assumed to be nearly identical, i.e. $\hat{\theta} \approx \omega$. By assuming a solution of third order expansion, the damping and forcing terms are magnified to ensure they appear in the third order expansion.

$$F_0 = \varepsilon^3 F_0, \quad \beta_d = \varepsilon^2 \beta_d \tag{70}$$

The magnified coefficients in Eq. (70) and the solution of initial expansion in Eq. (50) are subsequently substituted in Eq. (67), to obtain the following set of linear equations:

$$\varepsilon^1: \quad D_0^2 q_1 + \omega^2 q_1 = 0 \tag{71a}$$

$$\varepsilon^2: \quad D_0^2 q_2 + \omega^2 q_2 = -2D_0 D_1 q_1 \tag{71b}$$

$$\varepsilon^3: \quad D_0^2 q_3 + \omega^2 q_3 = -(D_1^2 + 2D_0 D_2)q_1 - \beta_d D_0 q_1 - \beta_3 q_1^3 - 2D_0 D_1 q_2 + F_0\left(\frac{1}{2}e^{i\hat{\theta}T_0} + \frac{1}{2}e^{-i\hat{\theta}T_0}\right) \tag{71c}$$

The solution for Eq. (71a) is assumed as

$$q_1 = A(T_1, T_2)e^{i\omega T_0} + \overline{A}(T_1, T_2)e^{-i\omega T_0} \tag{72}$$

where A is an unknown complex function and $\overline{A}$ is its complex conjugate.

Substitution of Eq. (72) into Eq. (71b) yields

$$D_0^2 q_2 + \omega^2 q_2 = -2iD_1 A\omega e^{i\omega T_0} + 2iD_1\overline{A}\omega e^{-i\omega T_0} \tag{73}$$

In order to remove secular terms from the above expression, this requires selecting the coefficients of $e^{i\omega T_0}$ and setting them to zero

$$-2iD_1A\omega = 0 \Rightarrow D_1A = 0 \tag{74a}$$

$$-2iD_1\overline{A}\omega = 0 \Rightarrow D_1\overline{A} = 0 \tag{74b}$$

This suggests that A is independent of the time scale T_1 and is entirely a function of T_2. Substituting this simplification into Eq. (73) and solving the partial differential equation, yields the solution of Eq. (71b) as:

$$q_2 = 0 \tag{75}$$

The solutions q_1 from Eq. (72) and q_2 from Eq. (75) are then substituted into Eq. (73) while also taking into account the simplification obtained from Eq. (74)

$$D_0^2q_3 + q_3 = \left(-2iD_2A\omega - 3\beta_3A^2\overline{A} - i\beta_dA\omega\right)e^{i\omega T_0} \\ - \beta_3A^3e^{3iT_0} + \frac{1}{2}F_0e^{i\hat{\theta}T_0} + cc \tag{76}$$

A detuning parameter σ is introduced which shows the nearness of $\hat{\theta}$, the excitation frequency to the linear natural frequency ω

$$\hat{\theta} = \omega + \varepsilon^2\sigma \tag{77}$$

Secular terms are eliminated from q_3 in Eq. (76) by collecting the coefficients of $e^{i\omega T_0}$. Keeping in mind that $\varepsilon^2 = \frac{T_2}{T_0}$, one obtains

$$-2iD_2A\omega - 3\beta_3A^2\overline{A} - i\beta_dA\omega + \frac{1}{2}F_0e^{iT_2\sigma} = 0 \tag{78}$$

which gives us the second expression for the secular term. A is now expressed in its polar form by

$$A = \frac{1}{2}ae^{i\psi} \tag{79}$$

where a and ψ are real functions of T_2. Substituting Eq. (79) into Eq. (78) and separating the real and imaginary parts, one obtains the amplitude and phase equations respectively

$$-\omega\frac{da}{dT_2} - \beta_d a\omega + \frac{1}{2}F_0\sin(-\psi + T_2\sigma) = 0 \tag{80a}$$

$$a\omega\frac{d\psi}{dT_2} - \frac{3}{8}\beta_3a^3 + \frac{1}{2}F_0\cos(-\psi + T_2\sigma) = 0 \tag{80b}$$

The amplitude and phase equations can be transformed into an autonomous system (i.e. one in which T_2 does not appear explicitly) by setting:

$$\gamma = -\psi + \sigma T_2 \tag{81}$$

And can be rewritten as

$$-\omega\frac{da}{dT_2} - \beta_d a\omega + \frac{1}{2}F_0 \sin(\gamma) = 0 \tag{82a}$$

$$-a\omega\frac{d\gamma}{dT_2} + \omega a\sigma - \frac{3}{8}\beta_3 a^3 + \frac{1}{2}F_0 \cos(\gamma) = 0 \tag{82b}$$

Steady-state motion occurs when $a' = \gamma' = 0$ [55]. This corresponds to the singular points of Eq. (82), i.e, they correspond to the solutions of

$$\begin{aligned} \sin(\gamma) &= \frac{2\beta_d a\omega}{F_0} \\ \cos(\gamma) &= \frac{1}{4}\left(\frac{3\beta_3 a^3 - 8\omega a\sigma}{F_0}\right) \end{aligned} \tag{83}$$

Squaring and adding these equations, the frequency-response equation can be determined for the case of primary resonance

$$a^2\left[\left(\frac{3}{8}\frac{\beta_3 a^2}{\omega} - \sigma\right)^2 + \beta_d^2\right] = \left(\frac{1}{2}\frac{F_0}{\omega}\right)^2 \tag{84}$$

MMS is well proven in developing frequency-response equations in the analysis of forced vibrations with cubic nonlinearity [55] and therefore it has been employed in this study.

6 Numerical Results and Discussion

6.1 Validation Studies

The first validation consists of computing the nonlocal linear frequency for different values of the nonlocal parameter $\hat{\mu}_0$ and comparing the results with those obtained by Kasirajan et al. [33], Reddy [56] and Thai [57]. The beam aspect ratio is $L/h = 100$ and the material properties are $E = 70$ GPa and the density $\rho = 2700$ kg/m^3. The length of the beam is $L = 10$ nm and breadth and height of the beam are considered to be $b = h = 0.01L$. No surface effects are considered in this first validation study. The values of the non-dimensional natural frequency for different cases of the nonlocal parameter $\hat{\mu}_0$ are reported in Table 2.

Table 2 Variation of the non-dimensional natural frequency ratio obtained from linear analysis for the simply supported (S-S) case with various values of the nonlocal parameter $\hat{\mu}_0$ $(L/h = 100)$

$\hat{\mu}_0$	Present work	Reference [33]	Reference [56]	Reference [57]
0	9.8691	9.8677	9.8683	9.8679
0.01	9.4154	9.4141	9.4147	9.4143
0.02	9.0190	9.0179	9.0183	9.0180
0.03	8.6688	8.6676	8.6682	8.6678
0.04	8.3565	8.3553	8.3558	8.3555

Table 3 Variation of frequency parameter $(\omega_{nonlinear}/\omega_{linear})$ with various amplitudes for the C-C case

A	Present work	Reference [58]	Reference [59]
1	1.0217	1.0222	1.0222
2	1.0868	1.0858	1.0858
3	1.1953	1.1833	1.1833

It can be seen that increasing the nonlocal parameter results in a decrease of the natural frequency indicating a softening of the nanobeam and that the obtained results are in agreement with other authors.

The second validation study is the nonlinear frequency parameter defined as the ratio of the nonlinear natural frequency to the linear natural frequency for various amplitude ratios obtained from a nonlinear analysis. The geometrical and material properties are same as those considered in the first validation study. The mode shape of C-C case is normalized such that the maximum amplitude is equal to 1. Results of the present study are reported in Table 3 and compared with those obtained from references [58] and [59]. The obtained results are in good agreement with the published data and indicate an increase in the frequency parameter as the amplitude increases.

The third validation study is conducted to study the effect of the nonlocal frequency ratio, which is defined as the ratio of nonlocal nonlinear natural frequency to classical nonlinear natural frequency and is given by

$$\text{Nonlocal frequency ratio} = \frac{\text{Nonlocal nonlinear natural frequency}}{\text{Classical nonlinear natural frequency}} \tag{85}$$

The obtained results are reported in Table 4 for the S-S case along with results published by [60] and [61] for different values of the amplitude A and the gradation index n. The length of the beam is $L = 10nm$ and the breadth and height of the beam are taken to be $b = h = 0.1L$. The material properties of the beam are taken from [60]. No elastic foundation effects are considered in this validation study. It is observed that nonlocal frequency ratio increases substantially when the amplitude and beam length increases, but it decreases slightly as the power index increases. Again, excellent agreement was obtained between the obtained and published results.

Table 4 Variation of nonlocal nonlinear frequency ratio for the S-S case, with μ_0 = 2 nm and no elastic foundation ($\hat{k}_L = 0, \hat{k}_{NL} = 0, \hat{k}_s = 0$)

A	n	L(nm)	$\hat{\mu}_0$	Present work	Reference [60]	Reference [61]
0	3	10	0.02	0.9139	0.9139	0.9139
		20	0.005	0.9762	0.9762	0.9762
1	0	10	0.02	0.9293	0.9293	0.9280
		20	0.005	0.9803	0.9803	0.9800
	3	10	0.02	0.9291	0.9221	0.9139
		20	0.005	0.9803	0.9785	0.9762
2	0	10	0.02	0.9631	0.9631	0.9517
		20	0.005	0.9893	0.9893	0.9865
	3	10	0.02	0.9626	0.9413	0.9526
		20	0.005	0.9892	0.9840	0.9867

In the last three validations, the midplane axis and the physical neutral axis of the beam are the same. The fourth validation study is conducted to verify the accuracy of the non-dimensional linear natural frequency obtained from the assumption that the undeformed plane coincides with the physical neutral axis rather than the midplane axis. The frequency was computed for material homogeneity index $n = 0$ and 4 and for different values of elastic modulus ratios E_U/E_L. The material properties of the beam are selected to be the same as those used in the work by Eltaher et al. [42]: $E_L = 210$ GPa and density $\rho_L = 7800$ kg/m^3. The length L, breadth b and height h of the beam are chosen as 100 nm, 10 nm and 1 nm, respectively. The obtained results are tabulated in Table 5 along with those published by Eltaher et al. [42] and are found to be in good agreement. In this table, the third column corresponds to the eigenvalue λ normalized by Eltaher et al. [42] while the fourth column reports the frequency $\hat{\omega}$ computed using the normalization used in this study (ie, Eqs. (27) and (29)). The relationship between λ and $\hat{\omega}$ is given in the caption of Table 5. The obtained and published results indicate that the frequency (or eigenvalue) increases substantially as the elastic modulus ratio increases and the h_0/h remains zero for all $n = 0$ cases indicating that the neutral and midplane axes coincide in this case. On the other hand, a slight increase in the frequency for $n = 4$ case is witnessed along with an increase of h_0/h indicating a further shifting of the neutral axis from the midplane axis.

Table 5 Effect of elastic modulus ratios on the first non-dimensional natural frequency obtained from the linear analysis for the simply supported (S-S) beam with shifted neutral axis ($\hat{\mu}_0 = 0$, $L/h = 100$, $\rho_U/\rho_L = 1$ and $\lambda = \sqrt{\frac{\hat{\omega}}{\tau}}L\sqrt[4]{\frac{\rho_L A}{E_L I}}$)

E_U/E_L	$n = 0$	Present			Reference [42]		
		λ	$\hat{\omega}$	h_0/h	λ	$\hat{\omega}$	h_0/h
1		3.1415	9.8691	0	3.1417	9.8702	0
2		3.7359	9.8691	0	3.7361	9.8701	0
4		4.4427	9.8691	0	4.4430	9.8701	0
8		5.2833	9.8691	0	5.2836	9.8699	0
10		5.5865	9.8691	0	5.5867	9.8698	0
E_U/E_L	$n = 4$	λ	$\hat{\omega}$	h_0/h	λ	$\hat{\omega}$	h_0/h
1		3.1415	9.8691	0	3.1417	9.8702	0
2		3.3348	9.8691	0.0555	3.3350	9.8699	0.0556
4		3.5566	9.8691	0.1250	3.5570	9.8711	0.1250
8		3.7867	9.8690	0.1944	3.7869	9.8698	0.1944
10		3.8627	9.8689	0.2142	3.8629	9.8699	0.2143

6.2 Parametric Study

6.2.1 Free Vibration Response

The free vibration response is quantified using three measures, namely, (a) frequency ratio; (b) surface frequency ratio and (c) non-dimensional frequency. The frequency and surface frequency ratios are respectively, defined as follows:

$$\text{Frequency ratio} = \frac{\text{Nonlinear natural frequency with nonlocal effect}}{\text{Nonlinear natural frequency without nonlocal effect}} \tag{86a}$$

$$\text{Surface Frequency ratio} = \frac{\text{Nonlinear natural frequency with surface effect}}{\text{Nonlinear natural frequency without surface effect}} \tag{86b}$$

Parametric study, it was decided to vary the nonlocal parameter $\hat{\mu}_0$ from 0 to 0.04 and to fix the elastic modulus ratio $E_U/E_L = 10$ as the critical cases selected to study the effect of the frequency ratio over the aspect ratio of the beam as shown in Fig. 2a–b for S-S and C-C cases respectively. It can be seen that increasing $\hat{\mu}_0$ tends to decreases the hardening effect for both S-S and C-C graded beams. For the S-S case, the aspect ratio does not seem to affect the frequency ratio for a given value of $\hat{\mu}_0$. However, for the C-C case, a slight increase in the frequency ratio is observed up to a value of L/h equal to about 20 and then remains constant.

The second part of this section provides the effect of the surface modulus ratio E_U^s/E_L^s on the variation of the surface frequency ratio with respect to various ampli-

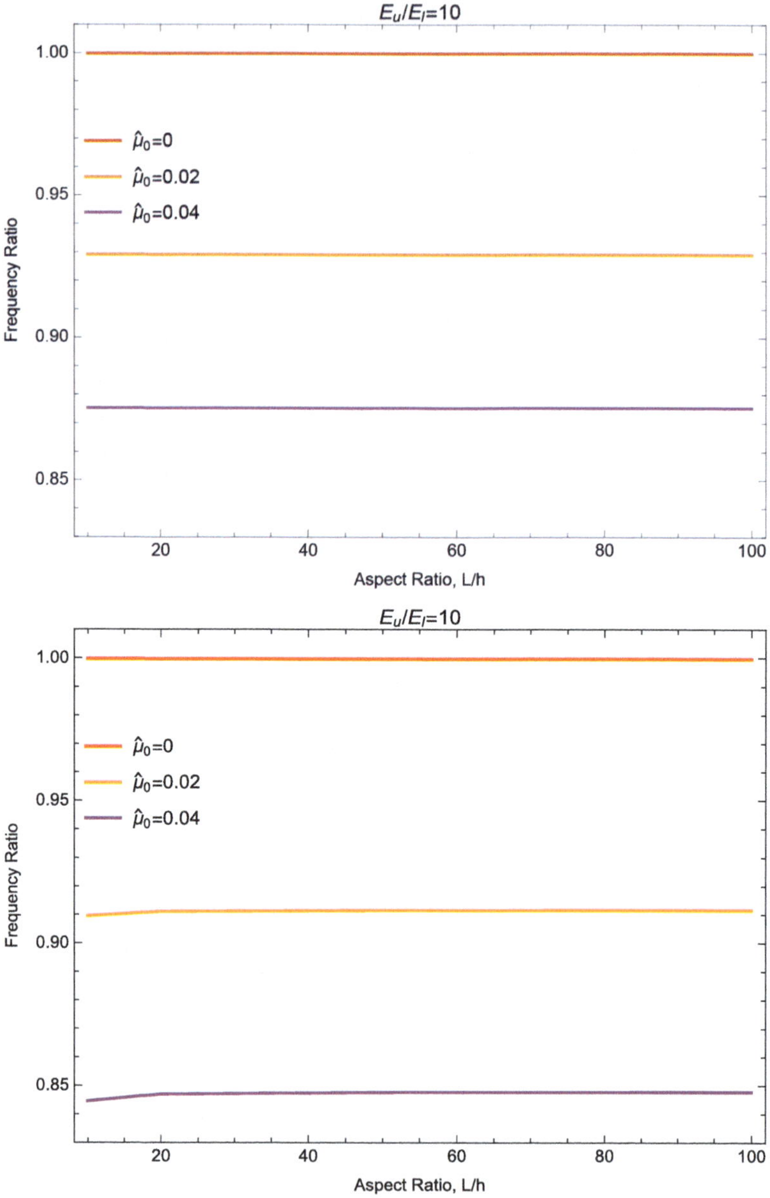

Fig. 2 Frequency ratio versus aspect ratio L/H for **a** S-S beam **b** C-C beam ($\rho_U/\rho_L = 1$, $E_s = 0$, $E_U/E_L = 10$, $n = 0$, $A = 1$, $L = 10$ nm, $\hat{k}_L = 0$, $\hat{k}_{NL} = 0$, $\hat{k}_s = 0$)

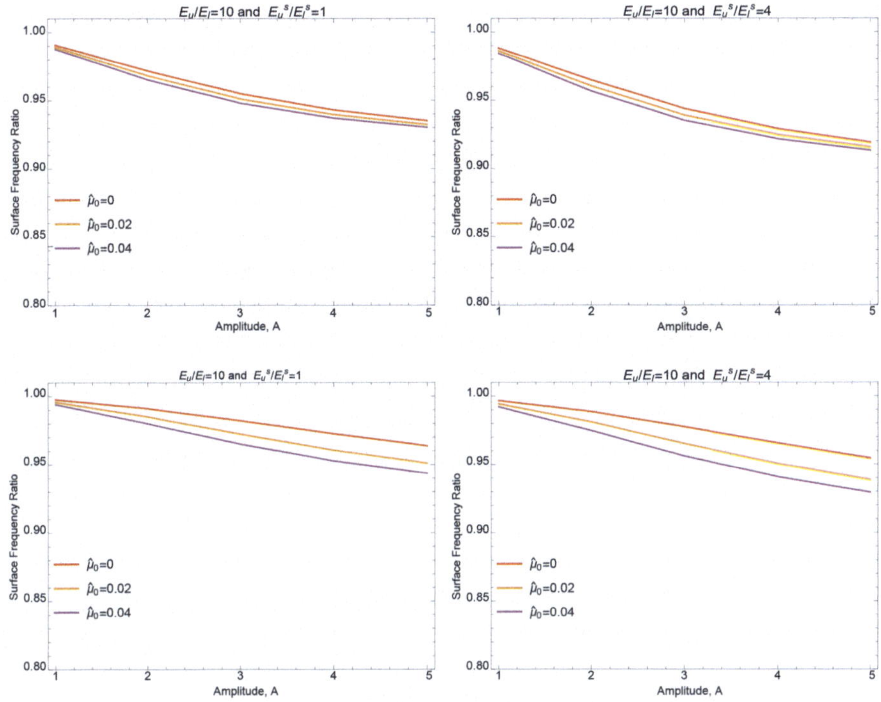

Fig. 3 Surface frequency ratio versus amplitude for **a–b** S-S beam **c–d** C-C beam ($\rho_U/\rho_L = 1$, $n = 1$, $A = 1$, $L/h = 10$, $L = 10$ nm, $\hat{k}_L = 0$, $\hat{k}_{NL} = 0$, $\hat{k}_s = 0$)

tudes for both S-S and C-C cases. The critical values of the nonlocal parameter and elastic modulus ratio are chosen as similar to the frequency ratio analysis mentioned in the previous paragraph. Figure 3a–d show the trend of decrease in the surface frequency ratio with an increase of amplitudes for both S-S and C-C graded beams, respectively. It is clear that softening effect is observed when increasing the nonlocal parameter for the S-S and C-C cases. On the other hand, increasing the amplitude causes softening effect for both S-S and C-C cases. Overall, the effect of nonlocal parameter dominates the surface elasticity effects.

In the last part of this section, the effects of various parameters such as nonlocal, material inhomogeneity, elastic modulus ratio, surface elastic modulus ratio and stiffness coefficients of the elastic foundation on the non-dimensional natural frequency of nanobeam with physical neutral axis for different boundary conditions are examined using MMS method. The parametric analysis is performed by modelling the graded nanobeam with a square cross-section ($b = h = 0.1L$) and the length of the beam is 10 nm. In the graded nanobeam, the bottom surface ($z = -h/2$) is rich with Aluminum properties and grading varies through thickness based on the power law as given in Eq. (1). The material properties at the bottom surface are $E = 70$ GPa, density $\rho = 2700$ kg/m^3 and $E_s = 5.1882$ N/m. The mode shape functions from Table 2 are used to obtain the results.

Table 6 The effect of elastic modulus ratios on the non-dimensional frequency (Ω) with physical neutral axis including surface elasticity effects for a nonlocal S-S graded nanobeam ($\rho_U/\rho_L = 1$, $A = 1$, $\hat{k}_L = 0$, $\hat{k}_{NL} = 0$, $\hat{k}_s = 0$)

n	$\hat{\mu}_0$	E_U^s/E_L^s	$E_U/E_L = 1$		$E_U/E_L = 4$		$E_U/E_L = 10$	
			$\hat{\omega}_{NL}$	h_0/h	$\hat{\omega}_{NL}$	h_0/h	$\hat{\omega}_{NL}$	h_0/h
1	0	1	10.5793	0.0000	10.7373	0.0893	10.9041	0.1294
		2	10.5409	0.0342	10.7211	0.1016	10.8939	0.1345
		4	10.5139	0.0851	10.7011	0.1228	10.8772	0.1439
		8	10.5224	0.1482	10.6875	0.1554	10.8549	0.1604
		10	10.5362	0.1691	10.6877	0.1681	10.8477	0.1675
	0.04	1	9.2085	0.0000	9.3965	0.0893	9.5949	0.1293
		2	9.1634	0.0342	9.3778	0.1016	9.5830	0.1344
		4	9.1325	0.0851	9.3548	0.1228	9.5637	0.1439
		8	9.1449	0.1481	9.3403	0.1554	9.5382	0.1603
		10	9.1622	0.1690	9.3411	0.1681	9.5300	0.1675
4	0	1	10.5793	0.0000	10.5891	0.1054	10.7524	0.1937
		2	10.5288	0.0335	10.5771	0.1234	10.7526	0.2021
		4	10.4833	0.0875	10.5673	0.1543	10.7572	0.2173
		8	10.4732	0.1624	10.5780	0.2015	10.7774	0.2429
		10	10.4840	0.1894	10.5914	0.2199	10.7912	0.2536
	0.04	1	9.2085	0.0000	9.2219	0.1054	9.4190	0.1937
		2	9.1491	0.0335	9.2084	0.1234	9.4198	0.2021
		4	9.0964	0.0875	9.1982	0.1543	9.4263	0.2173
		8	9.0876	0.1624	9.2135	0.2015	9.4520	0.2429
		10	9.1018	0.1894	9.2305	0.2199	9.4691	0.2536

It is noticed from Table 6 that the trend of the nonlinear frequency decreases and increases on increasing the values of surface modulus ratio for the initial elastic modulus cases. However, for higher elastic modulus ratio, only a decreasing behavior of frequency is observed. Thus, the surface modulus effects are dominated in high elastic modulus ratio cases. A substantial decrease in a nonlinear frequency is observed when increasing the nonlocal parameter. However, the effect of material inhomogeneity index is insignificant. On the other hand, it can be seen that the value of h_0/h increases while increasing all the parameters which indicates that the physical neutral axis of the graded beam shifts from the geometrical central axis. It is also observed from Table 7 that the effect of elastic foundation significantly increases the natural frequency, but a more pronounced decrease in a nonlinear frequency is predicted on increasing the nonlocal parameter and maintains consistency in the results as compared to the Table 6. For the C-C case, it is seen from Table 8 a substantial increase in the frequency due to the rigidity of this boundary condition. Also, a decreasing and increasing trend of frequency is observed as the surface modulus ratio increases, for the higher elastic modulus ratio and the rest of the frequency behavior is similar to

Table 7 The effect of elastic modulus ratios on the non-dimensional frequency (Ω) with physical neutral axis including surface elasticity without elastic foundation effects for a nonlocal S-S graded nanobeam ($\rho_U/\rho_L = 1$, $A = 1$, $\hat{k}_L = 100$, $\hat{k}_{NL} = 50$, $\hat{k}_s = 15$)

n	$\hat{\mu}_0$	E_U^s/E_L^s	$E_U/E_L = 1$		$E_U/E_L = 4$		$E_U/E_L = 10$	
			$\hat{\omega}_{NL}$	h_0/h	$\hat{\omega}_{NL}$	h_0/h	$\hat{\omega}_{NL}$	h_0/h
1	0	1	19.6622	0.0000	19.7406	0.0894	19.8232	0.1294
		2	19.6410	0.0342	19.7304	0.1017	19.8168	0.1345
		4	19.6225	0.0851	19.7165	0.1229	19.8062	0.1439
		8	19.6169	0.1482	19.7031	0.1554	19.7909	0.1604
		10	19.6197	0.1691	19.7003	0.1682	19.7854	0.1675
	0.04	1	18.9571	0.0000	19.0394	0.0894	19.1262	0.1293
		2	18.9351	0.0342	19.0290	0.1017	19.1196	0.1344
		4	18.9162	0.0851	19.0148	0.1229	19.0932	0.1439
		8	18.9115	0.1481	19.0015	0.1554	19.0932	0.1603
		10	18.9149	0.1690	18.9990	0.1682	19.0877	0.1675
4	0	1	19.6622	0.0000	19.6597	0.1054	19.7284	0.1937
		2	19.6346	0.0335	19.6506	0.1234	19.7262	0.2021
		4	19.6059	0.0875	19.6395	0.1543	19.7243	0.2173
		8	19.5878	0.1624	19.6336	0.2015	19.7270	0.2429
		10	19.5870	0.1894	19.6355	0.2199	19.7307	0.2536
	0.04	1	18.9571	0.0000	18.9554	0.1054	19.0289	0.1937
		2	18.9284	0.0335	18.9461	0.1234	19.0269	0.2021
		4	18.8990	0.8758	18.9352	0.1543	19.0254	0.2173
		8	18.8815	0.1624	18.9304	0.2015	19.0291	0.2429
		10	18.8814	0.1894	18.9328	0.2199	19.0333	0.2536

that of S-S case. From Table 9, it is observed that the behavior of frequency decreases as the values of surface modulus and nonlocal parameter increases which reflect the softening effects in the graded beam. However, the frequency increases as the elastic modulus increases which indicates the hardening effect of the graded beam.

6.2.2 Forced Vibration Response

In the parametric study related to forced vibration, the effect of nonlocal parameter, material inhomogeneity, linear, nonlinear and shear coefficients of the elastic foundation in addition to surface elasticity effects on the frequency response of graded beam are investigated. Here the damping coefficient is assumed as $\hat{c} = 0.1$ and the external harmonic force amplitude is chosen as $\hat{F}_0 = 1$ for the S-S case. However, for the C-C case, the harmonic force is set as $\hat{F}_0 = 2$ to demonstrate a clearer nonlinear frequency response.

Table 8 The effect of elastic modulus ratios on the non-dimensional frequency (Ω) with physical neutral axis including surface elasticity without elastic foundation effects for a nonlocal C-C graded nanobeam ($\rho_U/\rho_L = 1$, $A = 1$, $\hat{k}_L = 0$, $\hat{k}_{NL} = 0$, $\hat{k}_s = 0$)

n	$\hat{\mu}_0$	E_U^s/E_L^s	$E_U/E_L = 1$		$E_U/E_L = 4$		$E_U/E_L = 10$	
			$\hat{\omega}_{NL}$	h_0/h	$\hat{\omega}_{NL}$	h_0/h	$\hat{\omega}_{NL}$	h_0/h
1	0	1	22.6663	0.0000	22.7434	0.0894	22.8243	0.1294
		2	22.6442	0.0342	22.7320	0.1016	22.8173	0.1344
		4	22.6229	0.0851	22.7159	0.1228	22.8053	0.1439
		8	22.6114	0.1481	22.6986	0.1554	22.7877	0.1603
		10	22.6116	0.1690	22.6942	0.1681	22.7812	0.1675
	0.04	1	19.0252	0.0000	19.1976	0.0893	19.3790	0.1293
		2	18.9789	0.0342	19.1755	0.1016	19.3652	0.1344
		4	18.9388	0.0851	19.1453	0.1228	19.3420	0.1439
		8	18.9280	0.1481	19.1168	0.1554	19.3090	0.1603
		10	18.9347	0.1690	19.1111	0.1681	19.2972	0.1675
4	0	1	22.6663	0.0000	22.6594	0.1054	22.7191	0.1937
		2	22.6377	0.3352	22.6484	0.1234	22.7157	0.2021
		4	22.6058	0.0875	22.6337	0.1543	22.7112	0.2173
		8	22.5798	0.1624	22.6211	0.2015	22.7093	0.2429
		10	22.5753	0.1894	22.6200	0.2199	22.7110	0.2536
	0.04	1	19.0252	0.0000	19.0208	0.1054	19.1731	0.1937
		2	18.9649	0.0335	19.0011	0.1234	19.1687	0.2021
		4	18.9027	0.0875	18.9775	0.1543	19.1650	0.2173
		8	18.8647	0.1624	18.9663	0.2015	19.1719	0.2429
		10	18.8639	0.1894	18.9710	0.2199	19.1804	0.2536

Figure 4 depict the effect of material inhomogeneity n on the frequency response of the beam without and with surface effects for (a)–(b) S-S beam (c)–(d) C-C beam respectively. For high elastic modulus ratio $E_U/E_L = 10$, it is observed that increasing the power law index results in a decrease in hardening effect of the beam as can be seen from Fig. 4a. However, a little discrepancy is noticed as the power law index varies from 1 to 4 due to the surface elasticity effect (see Fig. 4b). It is understood from Fig. 4c–d that the C-C beam produces narrow frequency responses and surface elasticity effects are barely noticed when increasing power law index.

Figure 5a–b illustrate the effect of nonlocal parameter on the frequency response of S-S case. A decreasing trend in the amplitude of frequency response is observed as the nonlocal parameter increases thereby displaying a hardening effect. However, hardly no changes due to surface elasticity effects are noticed. On the other hand, despite the high amplitude response of the C-C beam, less hardening effect is observed as shown in Fig. 5c–d.

Table 9 The effect of elastic modulus ratios on the non-dimensional frequency (Ω) with physical neutral axis including surface elasticity with elastic foundation effects (Ω) for a nonlocal C-C graded nanobeam (ρ_U/ρ_L = 1, A = 1, $\hat{k}_L = 100$, $\hat{k}_{NL} = 50$, $\hat{k}_s = 15$)

n	$\hat{\mu}_0$	E_U^s/E_L^s	$E_U/E_L = 1$		$E_U/E_L = 4$		$E_U/E_L = 10$	
			$\hat{\omega}_{NL}$	h_0/h	$\hat{\omega}_{NL}$	h_0/h	$\hat{\omega}_{NL}$	h_0/h
1	0	1	28.6908	0.0000	28.7472	0.0894	28.8062	0.1294
		2	28.6724	0.0342	28.7366	0.1017	28.7997	0.1345
		4	28.6514	0.0851	28.7206	0.1229	28.7885	0.1439
		8	28.6328	0.1482	28.7010	0.1554	28.7712	0.1604
		10	28.6287	0.1691	28.6948	0.1682	28.7645	0.1675
	0.04	1	28.4434	0.0000	28.5445	0.0894	28.6503	0.1294
		2	28.4107	0.0342	28.5259	0.1017	28.6389	0.1345
		4	28.3739	0.0851	28.4979	0.1229	28.6191	0.1439
		8	28.3421	0.1482	28.4637	0.1554	28.5888	0.1604
		10	28.3354	0.1691	28.4531	0.1682	28.5771	0.1675
4	0	1	28.6908	0.0000	28.6781	0.1054	28.7088	0.1937
		2	28.6672	0.3352	28.6667	0.1234	28.7039	0.2021
		4	28.6375	0.0875	28.6494	0.1543	28.6963	0.2173
		8	28.6048	0.1624	28.6286	0.2015	28.6873	0.2429
		10	28.5951	0.1894	28.6227	0.2199	28.6852	0.2536
	0.04	1	28.4434	0.0000	28.4218	0.1054	28.4789	0.1937
		2	28.4014	0.0335	28.4018	0.1234	28.4704	0.2021
		4	28.3491	0.08758	28.3718	0.1543	28.4574	0.2173
		8	28.2926	0.1624	28.3364	0.2015	28.4425	0.2429
		10	28.2763	0.1894	28.3267	0.2199	28.4393	0.2536

Figure 6a–d show the effect of the foundation linear coefficient $\hat{k}_L$ on the frequency response without and with surface effects for (a)–(b) S-S beam (c)–(d) C-C beam respectively. According to Fig. 6a, it can be concluded that upon increasing the linear coefficient leads to a decrease in the hardening effect of the graded beam. This effect can be understood from the expression β_1 which contains the square of the linear frequency and also depends on linear and shear coefficients. Figure 6c shows narrow curves and the linear coefficient effect is insignificant in the C-C case. However, there are no changes observed due to surface elasticity effects as can seen from Fig. 6b, d. Figure 7a–d depict the effect of the foundation nonlinear coefficient $\hat{k}_{NL}$ on the frequency response without and with surface effects for (a)–(b) S-S beam and (c)–(d) C-C beam respectively. It can be observed from Fig. 7a that increasing the nonlinear stiffness coefficient substantially improves the hardening effect in the S-S beam while a little hardening is noticed in the C-C beam as illustrated in Fig. 7c. This effect is due to the constant β_3 which is dependent on the nonlinear coefficient. Figure 8a–d illustrate the effect of the foundation shear coefficient $\hat{k}_S$ on the frequency response without and with surface effects for (a)–(b) S-S beam (c)–(d) C-C beam

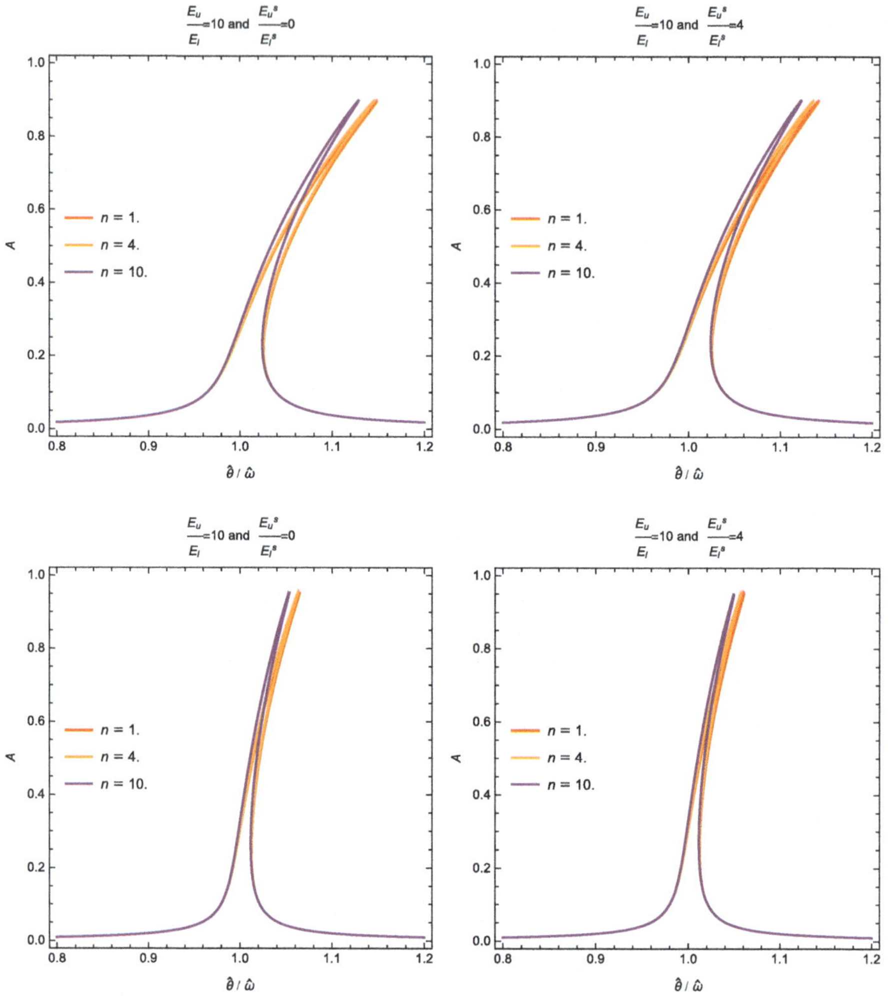

Fig. 4 Effect of material inhomogeneity n on frequency response of FG beam without and with surface effects for **a–b** S-S beam **c–d** C-C beam respectively ($\rho_U/\rho_L = 1$, $\hat{\mu}_0 = 0.02$, $\hat{k}_L = 10$, $\hat{k}_{NL} = 50$, $\hat{k}_s = 5$)

respectively. It can be concluded that increasing the shear coefficient provides a softening effect in the S-S beam, as shown in Fig.8a. However, the effect of the shear coefficient is less dominant in the C-C beam as clearly evident from Fig. 8c. Finally, the overall effect of shear coefficient is greater than that of the linear coefficient on the frequency response. It is observed that the nonlocal effect has more impact than the surface elasticity effect and power law index on the free and vibration studies.

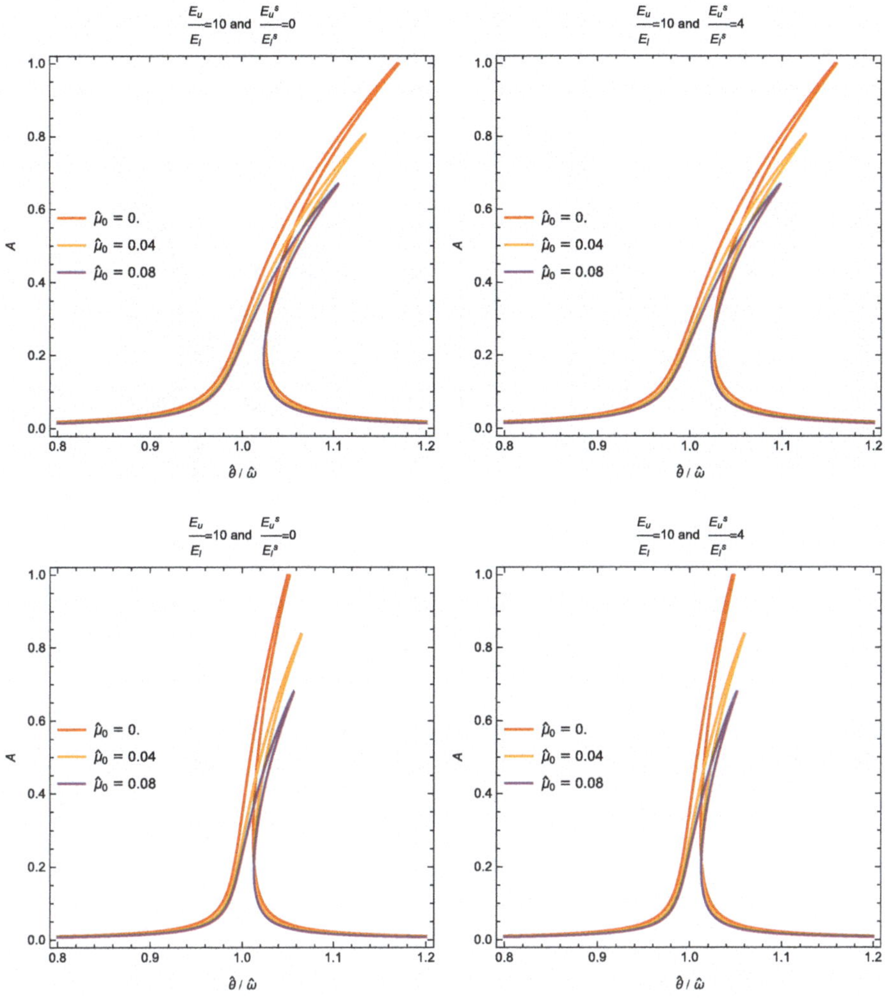

Fig. 5 Effect of nonlocal parameter $\hat{\mu}_0$ on frequency response of FG beam without and with surface effects for **a–b** S-S beam **c–d** C-C beam respectively ($\rho_U/\rho_L = 1$, $n = 2$, $\hat{k}_L = 10$, $\hat{k}_{NL} = 50$, $\hat{k}_s = 5$)

7 Conclusions

In this study the effects of surface elasticity on the nonlocal nonlinear, free and forced response of a graded Euler-Bernoulli beam resting on a nonlinear elastic foundation were investigated. The geometric nonlinearity of the beam was modeled using von Kármán nonlinearity and the length-scale effect was modeled using Eringen's nonlocal elasticity model. Instead of adopting the usual choice of the geometrical central axis, this study uses instead the physical neutral axis to eliminate the

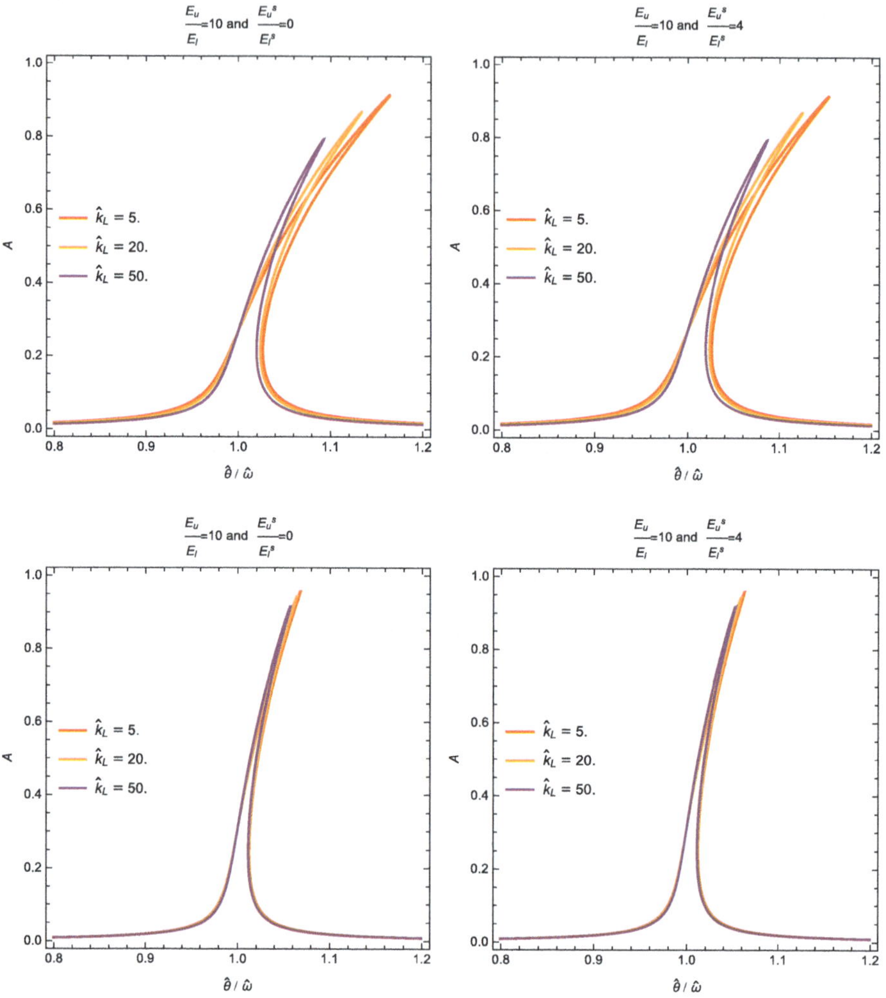

Fig. 6 Effect of linear coefficient of elastic foundation $\hat{k}_L$ on frequency response of FG beam without and with surface effects for **a–b** S-S beam **c–d** C-C beam respectively ($\rho_U/\rho_L = 1, n = 2,$ $\hat{\mu}_0 = 0.02$, $\hat{k}_{NL} = 50$, $\hat{k}_s = 5$)

stretching-bending coupling effect due to the unsymmetrical material variation which was shown to eliminate the quadratic nonlinearity from the equation of motion. These equations were obtained using the principle of virtual displacements while accounting for surface effects. A first-order Galerkin approximation was utilized to obtain the second-order nonlinear ordinary differential equation which was solved using the MMS to obtain the nonlinear natural frequency and the frequency response of the graded nanobeam. The results obtained from the present study were validated with preliminary cases in the existing literature and close agreements were obtained.

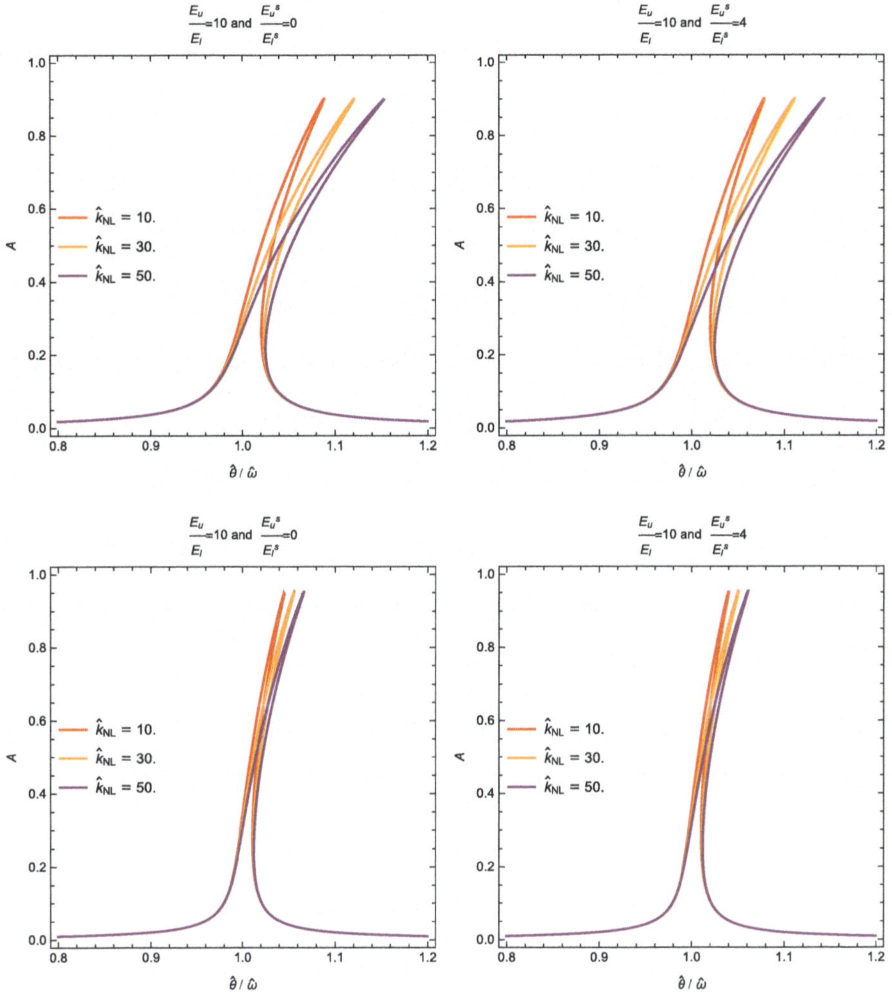

Fig. 7 Effect of nonlinear coefficient of elastic foundation $\hat{k}_{NL}$ on frequency response of FG beam without and with surface effects for **a–b** S-S beam **c–d** C-C beam respectively ($\rho_U/\rho_L = 1, n = 2$, $\hat{\mu}_0 = 0.02$, $\hat{k}_L = 10$, $\hat{k}_s = 5$)

A parametric study was conducted to investigate the effect of the inhomogeneity index, the nonlocal parameter, the surface modulus ratio, the elastic modulus ratio, the foundation coefficients of elastic and different boundary conditions on the free and forced vibration responses of the graded nanobeam. The essential conclusions of this study can be summarized as follows:

- The nonlocal effect tends to decrease the natural frequency and this decreasing effect is more visible in the C-C nanobeam than the S-S nanobeam. However, the material length scale effect is insignificant on the frequency ratio.

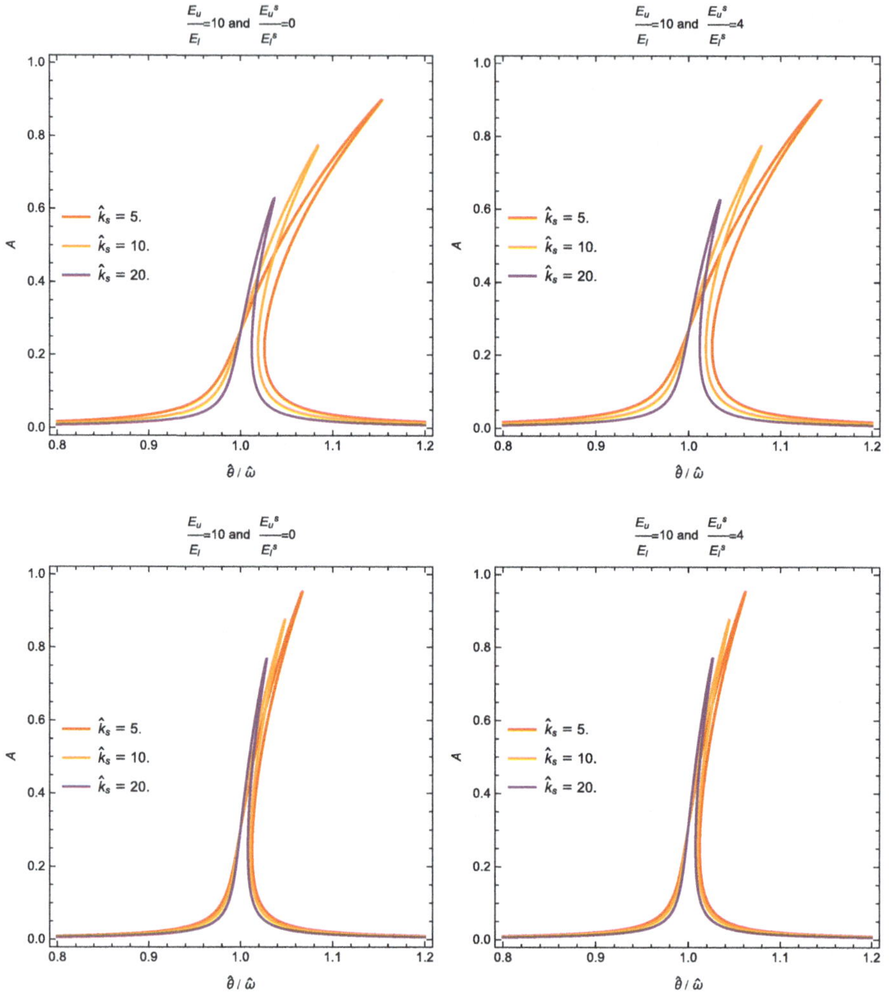

Fig. 8 Effect of shear coefficient of elastic foundation $\hat{k}_S$ on frequency response of FG beam without and with surface effects for **a–b** S-S beam **c–d** C-C beam respectively ($\rho_U/\rho_L = 1, n = 2$, $\hat{\mu}_0 = 0.02$, $\hat{k}_L = 10$, $\hat{k}_{NL} = 50$)

- Inclusion of surface elasticity also reduces the natural frequency on increasing the vibration amplitude in the graded beam. Yet, the surface elasticity effect has less impact than the nonlocal effect.
- A decrease in a natural frequency is observed when the nonlocal parameter increases and it contributes more than the power law index, surface modulus, elastic modulus. However, the nonlinear frequency is significantly influenced by the coefficients of elastic foundation.

- An increase in the power law index provides a softening effect on the system, whereas introduction of nonlocal parameter increases a hardening effect on the frequency response of the graded beam.
- The increase in the nonlinear coefficient of the elastic foundation provides a hardening effect in the frequency response of the graded nanobeam, however the increase in the linear and shear coefficients yield the opposite effect.
- The influence of the surface elasticity has only a subtle effect on the frequency response of the S-S and C-C graded nanobeams.
- The authors also showed that the choice of the physical neutral axis leads to the elimination of the quadratic nonlinearity from the equation of motion.

Acknowledgements The authors gratefully acknowledge the help of Mrs Hedia Layouni El-Borgi in typesetting the Latex document. The first author is grateful to the support of Texas A&M University at Qatar.

Appendix

Midplane axis equation of motion to be compared with Eq. (28) (surface effects only):

$$\ddot{\hat{w}} - \hat{\alpha}\ddot{\hat{w}}'' + \hat{c}\dot{\hat{w}} + \hat{w}'''' - \hat{w}'' \int_0^1 \left[\kappa_0 \left(\hat{w}'\right)^2 - \kappa_1 \hat{w}''\right] d\hat{x} - k_s w'' + \hat{k}_L \hat{w} + \hat{k}_{NL} \hat{w}^3 = \hat{F}\cos(\hat{\theta}\hat{t}) \tag{87}$$

Midplane axis equation of motion to be compared with Eq. (41) **(both nonlocal and surface effects):**

$$\ddot{\hat{w}} - \hat{\mu}_0 \ddot{\hat{w}}'' - \hat{\alpha}\left(\ddot{\hat{w}}'' - \hat{\mu}_0 \ddot{\hat{w}}''''\right) + \hat{c}\left(\dot{\hat{w}} - \hat{\mu}_0 \dot{\hat{w}}''\right) + \hat{w}'''' - \left(\hat{w}'' - \hat{\mu}_0 \hat{w}''''\right) \int_0^1 \left[\kappa_0 \left(\hat{w}'\right)^2 - \kappa_1 \hat{w}''\right] d\hat{x} + \hat{k}_L \left(\hat{w} - \hat{\mu}_0 \hat{w}''\right) + \hat{k}_{NL}\left(\hat{w}^3 - \hat{\mu}_0 \left(6\hat{w}\left(\hat{w}'\right)^2 + 3\hat{w}^2 \hat{w}''\right)\right) - \hat{k}_s \left(w'' - \mu_0 w''''\right) = \hat{F}\cos(\hat{\theta}\hat{t}) \tag{88}$$

Midplane axis discretized forced vibration equation of motion to be compared with Eq. (68) **(both nonlocal and surface effects):**

$$1\frac{d^2 q}{d\hat{t}^2} + 2\beta_d \frac{dq}{d\hat{t}} + \beta_1 q + \beta_2 q^2 + \beta_3 q^3 = \hat{F}\cos(\hat{\theta}\hat{t}) \tag{89a}$$

$$q(0) = A, \quad \dot{q}(0) = 0 \tag{89b}$$

where the coefficients β_i $(i = 1, 2, 3)$ are defined as

$$\beta_1 = \frac{1}{\Delta}\int_0^1 \left\{\phi^{(4)} + \hat{k}_L\left(\phi - \hat{\mu}_0\phi''\right) - \hat{k}_S\left(\phi'' - \hat{\mu}_0\phi^{(4)}\right)\right\}\phi\, dx \tag{90a}$$

$$\beta_2 = \frac{1}{\Delta}\int_0^1 \left\{\left(\phi'' - \hat{\mu}_0\phi^{(4)}\right)\int_0^1 -\kappa_1\phi''\, dx\right\}\phi\, d\hat{x} \tag{90b}$$

$$\beta_3 = \frac{1}{\Delta}\int_0^1 \hat{k}_{NL}\left(-3\hat{\mu}_0\phi\phi'' - 6\hat{\mu}_0\left(\phi'\right)^2 + \phi^2\right)\phi\, d\hat{x} + \frac{1}{\Delta}\int_0^1 [\kappa_0(\hat{\mu}_0\phi'''' - \phi'')\int_0^1 \left(\phi'\right)^2\, d\hat{x}]\phi\, d\hat{x} \tag{90c}$$

$$\beta_d = \frac{1}{2\Delta}\int_0^1 \left(\phi - \hat{\mu}_0\phi''\right)\hat{c}\phi\, d\hat{x} \tag{90d}$$

$$F_0 = \frac{1}{\Delta}\int_0^1 \phi\hat{F}d\hat{x} \tag{90e}$$

in which $\Delta = \int_0^1 \left\{\phi - \hat{\mu}_0\phi'' - \hat{\alpha}\left(\phi'' - \hat{\mu}_0\phi''''\right)\right\}\phi\, d\hat{x}$, $\kappa_0 = \tilde{A}r^2\tau^2/2m_0L^4$, $\kappa_1 = \tilde{B}r\tau^2/m_0L^4$.

References

1. Hung ES, Senturia SD (1999) Extending the travel range of analog-tuned electrostatic actuators. J Microelectromech Sys 8:497–505
2. Li L, Bhushan B, Takashima K, Baek CW, Kim YK (2003) Mechanical characterization of micro/nanoscale structures for MEMS/NEMS applications using nanoindentation techniques. Ultramicroscopy 97:481–494
3. Moser Y, Gijs MAM (2007) Miniaturized flexible temperature sensor. J Microelectromech Sys 16:1349–1354
4. Pei J, Tian F, Thundat T (2004) Glucose biosensor based on the microcantilever. Anal Chem 76:292–297
5. Eringen AC, Edelen DGB (1972) On nonlocal elasticity. Int J Eng Sci 10:233–248
6. Eringen AC (1983) On differential equations of nonlocal elasticity and solutions of screw dislocation and surface waves. J Appl Phys 54:4703–4710
7. Eringen AC (2002) Nonlocal continuum field theories. Springer
8. Mindlin RD (1963) Influence of couple-stresses on stress concentrations. Exp Mech 3:1–7
9. Toupin RA (1964) Theories of elasticity with couple-stress. Arch Rat Mech Anal 17:85–112
10. Mindlin RD (1965) Second gradient of strain and surface-tension in linear elasticity. Int J Solids Struct 1:417–438
11. Lam DCC, Yang F, Chong ACM, Wang J, Tong P (2003) Experiments and theory in strain gradient elasticity. J Mech Phys Solids 51:1477–1508
12. Aifantis EC (2011) On the gradient approach-relation to Eringen's nonlocal theory. Int J Eng Sci 49:1367–1377

13. Narendar S, Gopalakrishnan S (2010) Ultrasonic wave characteristics of nanorods via nonlocal strain gradient models. J Appl Phys 107:084312
14. Lim CW, Zhang G, Reddy JN (2015) A higher-order nonlocal elasticity and strain gradient theory and its applications in wave propagation. J Mech Phys Solids 78:298–313
15. Fernández-Sáez J, Zaera R, Loya JA, Reddy JN (2016) Bending of Euler-Bernoulli beams using Eringen's integral formulation: a paradox resolved. Int J Eng Sci 99:107–116
16. Romano G, Barretta R, Diaco M, Marotti de Sciarra F (2017) Constitutive boundary conditions and paradoxes in nonlocal elastic nanobeams. Int J Mech Sci 121:151–156
17. Erdogan F (1995) Fracture mechanics of functionally graded materials. Compos Eng 5:753–770
18. Shariat BS, Liu Y, Rio G (2013) Modelling and experimental investigation of geometrically graded shape memory alloys. Smart Mater Struct 22:025030
19. Lü CF, Lim CW, Chen WQ (2009) Size-dependent elastic behavior of FGM ultra-thin films based on generalized refined theory. Int J Solids Struct 46:1176–1185
20. Kahrobaiyan MH, Asghari M, Rahaeifard M, Ahmadian MT (2010) Investigation of the size-dependent dynamic characteristics of atomic force microscope microcantilevers based on the modified couple stress theory. Int J Eng Sci 48:1985–1994
21. Zhang J, Fu Y (2012) Pull-in analysis of electrically actuated viscoelastic microbeams based on a modified couple stress theory. Meccanica 47:1649–1658
22. Eltaher MA, Emam SA, Mahmoud FF (2012) Free vibration analysis of functionally graded size-dependent nanobeams. Appl Math Comput 218:7406–7420
23. Uymaz B (2013) Forced vibration analysis of functionally graded beams using nonlocal elasticity. Compos Struct 105:227–239
24. Rahmani O, Pedram O (2014) Analysis and modeling the size effect on vibration of functionally graded nanobeams based on nonlocal Timoshenko beam theory. Int J Eng Sci 77:55–70
25. Nejad MZ, Hadi A (2016) Non-local analysis of free vibration of bi-directional functionally graded euler-bernoulli nano-beams. Int J Eng Sci 105:1–11
26. Şimşek M (2014) Large amplitude free vibration of nanobeams with various boundary conditions based on the nonlocal elasticity theory. Compos Part B 56:621–628
27. Niknam H, Aghdam MM (2015) A semi analytical approach for large amplitude free vibration and buckling of nonlocal FG beams resting on elastic foundation. Compos Struct 119:452–462
28. Nazemnezhad R, Hosseini-Hashemi S (2014) Nonlocal nonlinear free vibration of functionally graded nanobeams. Compos Struct 110:192–199
29. El-Borgi S, Fernandes R, Reddy JN (2015) Non-local free and forced vibrations of graded nanobeams resting on a non-linear elastic foundation. Int J Nonlin Mech 77:348–363
30. Trabelssi M, El-Borgi S, Ke LL, Reddy JN (2017) Nonlocal free vibration of graded nanobeams resting on a nonlinear elastic foundation using DQM and LaDQM. Compos Struct 176:736–747
31. Jiang LY, Yan Z (2010) Timoshenko beam model for static bending of nanowires with surface effects. Physica E 42:2274–2279
32. Challamel N, Elishakoff I (2012) Surface stress effects may induces softening: Euler-Bernoulli and Timoshenko buckling solutions. Physica E 44:1862–1867
33. Kasirajan P, Amirtham R, Reddy JN (2015) Surface and non-local effects for non-linear analysis of Timoshenko beams. Int J Nonlin Mech 76:100–111
34. Shanab RA, Attia MA, Mohamed SA (2017) Nonlinear analysis of functionally graded nanoscale beams incorporating the surface energy and microstructure effects. Int J Mech Sci 131:908–923
35. Yang L, Fan T, Yang L, Han X, Chen Z (2017) Bending of functionally graded nanobeams incorporating surface effects based on Timoshenko beam model. Theor Appl Mech Lett 7:152–158
36. Hosseini-Hashemi S, Nazemnezhad R (2013) An analytical study on the nonlinear free vibration of functionally graded nanobeams incorporating surface effects. Compos Part B 52:199–206
37. Lewandowski R (1987) Application of the ritz method to the analysis of non-linear free vibrations of beams. J Sound Vib 114:91–101
38. Zare M, Hashemi SH, Nazemnezhad R (2013) An exact analytica lapproach for free vibration of Mindlin rectangular nano-plates via nonlocal elasticity. Compos Struct 100:290–299

39. Lei XW, Natsuki T, Shi JX, Ni QQ (2012) Surface effects on the vibrational frequency of double-walled carbon nanotubes using the nonlocal Timoshenko beam model. Compos Part B 43:64–69
40. Lee HL, Chang WJ (2010) Surface and small-scale effects on vibration analysis of a non-uniform nanocantilever beam. Physica E 43:466–469
41. Hosseini-Hashemi S, Nazemnezhad R, Rokni H (2015) Nonlocal nonlinear free vibration of nanobeams with surface effects. Eur J Mech A/Solids 52:44–53
42. Eltaher MA, Alshorbagy AE, Mahmoud FF (2013) Determination of neutral axis position and its effect on natural frequencies of functionally graded macro/nano beams. Compos Struct 99:193–201
43. Simsek M (2016) Nonlinear free vibration of a functionally graded nanobeam using nonlocal strain gradient theory and a novel hamiltonian approach. Int J Eng Sci 105:12–27
44. Al-Basyouni KS, Tounsi A, Mahmoud SR (2015) Size dependent bending and vibration analysis of functionally graded micro beams based on modified couple stress theory and neutral surface position. Compos Struct 125:621–630
45. Ahouel M, Houari MSA, Bedia EA, Tounsi A (2016) Size-dependent mechanical behavior of functionally graded trigonometric shear deformable nanobeams including neutral surface position concept. Steel Compos Struct 20:963–981
46. Barretta R, Feo L, Luciano R, Marotti de Sciarra F, Penna R (2016) Functionally graded timoshenko nanobeams, a novel nonlocal gradient formulation. Compos Part B 100:208–219
47. Ebrahimi F, Barati MR (2017) Hygrothermal effects on vibration characteristics of viscoelastic FG nanobeams based on nonlocal strain gradient theory. Compos Struct 159:433–444
48. Wang CM, Ke LL, Chowdhury ANR, Yang J, Kitipornchai S, Fernando D (2017) Critical examination of midplane and neutral plane formulations for vibration analysis of FGM beams. Eng Struct 130:275–281
49. Şimşek M (2014) Nonlinear static and free vibration analysis of microbeams based on the nonlinear elastic foundation using modified couple stress theory and he's variational method. Compos Struct 112:264–272
50. Şimşek M (2015) Size dependent nonlinear free vibration of an axially functionally graded (AFG) microbeam using he's variational method. Compos Struct 131:207–214
51. Şimşek M, Yurtcu HH (2013) Analytical solutions for bending and buckling of functionally graded nanobeams based on the nonlocal Timoshenko beam theory. Compos Struct 97:378–386
52. Reddy JN (2013) An introduction to continuum mechanics. Cambridge University Press
53. Nayfeh AH, Pai PF (2008) Linear and non-linear structural mechanics. John Wiley and Sons, New-York
54. Fallah A, Aghdam MM (2011) Nonlinear free vibration and post-buckling analysis of functionally graded beams on nonlinear elastic foundation. Eur J Mech A/Solids 30:571–583
55. Nayfeh AH, Mook DT (1995) Nonlinear oscillations. John Wiley & Sons
56. Reddy JN (2007) Nonlocal theories for bending, buckling and vibration of beams. Int J Eng Sci 45:288–307
57. Thai H (2012) A nonlocal beam theory for bending, buckling, and vibration of nanobeams. Int J Eng Sci 52:56–64
58. Malekzadeh P, Shojaee M (2013) Surface and nonlocal effects on the nonlinear free vibration of non-uniform nanobeams. Compos Part B 52:84–92
59. Ribeiro P (2001) Hierarchical finite element analyses of geometrically non-linear vibration of beams and plane frames. J Sound Vib 246:225–244
60. Nazemnezhad R, Hosseini-Hashemi S (2014) Nonlocal nonlinear free vibration of functionally graded nanobeams. Compos Struct 110:192–199
61. Niknam H, Aghdam MM (2015) A semi analytical approach for large amplitude free vibration and buckling of nonlocal FG beams resting on elastic foundation. Compos Struct 119:452–462

Finite Element Nonlocal Integral Elasticity Approach

Maysam Naghinejad, Hamid Reza Ovesy, Mohsen Taghizadeh, and Seyyed Amir Mahdi Ghannadpour

Abstract In the current chapter, a finite element theory has been developed based on the nonlocal integral elasticity using Hamilton's principle. Formulations have been derived using Euler-Bernoulli beam theory and classical plate theory in order to study the bending, buckling and vibration behavior of nanostructures. The current method is capable of modelling complex geometries and boundary conditions. Besides, effects of nonlocal parameter, geometrical parameters, boundary conditions and viscoelastic parameter on the mechanical behavior of nano-scaled beams and plates have been studied.

1 Introduction

There are several fields, which cannot be examined thoroughly using classical theories. Solid fracture, stress field on the dislocation core and crack tip, singularities at points where loads are applied, sharp corners and discontinuities in the body, elastic short-wavelength behavior prediction, and viscosity increase of fluid flows in microscopic channels are some weaknesses of the classical theories. Besides, polaritons, gyrotropic effects and superconductivity cannot be investigated using the classical field theories. Also, classical continuum field theories are unable to predict the correct behavior of materials in micro and nano-scale. So, for studying the mechanical

M. Naghinejad · H. R. Ovesy (✉)
Aerospace Engineering Department, Amirkabir University of Technology, Tehran, Iran
e-mail: ovesy@aut.ac.ir

M. Naghinejad
e-mail: m.naghinejad@aut.ac.ir

M. Taghizadeh
Mechanical Engineering Department, Hakim Sabzevari University, Sabzevar, Iran
e-mail: m.taghizadeh@hsu.ac.ir

S. A. M. Ghannadpour
New Technologies and Engineering Department, Shahid Beheshti University, Tehran, Iran
e-mail: a_ghannadpour@sbu.ac.ir

© The Author(s), under exclusive license to Springer Nature Switzerland AG 2021
E. Ghavanloo et al. (eds.), *Size-Dependent Continuum Mechanics Approaches*,
Springer Tracts in Mechanical Engineering,
https://doi.org/10.1007/978-3-030-63050-8_10

behavior of materials in small-scale and phenomena that inherently happens in small-scale, atomic or nonlocal theories, which can consider the long-range interatomic force effects, should be employed. Atomic theories have relatively higher computational cost than nonlocal theories.

In summary, in nonlocal theories, the behavior of the material at one point is a function of the state of all points. One of the most popular non-local theories for studying nano-scale structures is Eringen's nonlocal theory [1]. In Eringen's nonlocal theory, stress at a point is a function of strains at all points of the material [1]. Under certain conditions, the nonlocal differential form could be extracted from the more general nonlocal integral theory. Despite the relative simplicity of nonlocal differential elasticity theory, it has some restrictions, e.g., some individual kernels must be considered for nonlocal differential elasticity to be derived, applying the natural boundary conditions for nano-plates are somehow ambiguous, investigating the complex geometries are intricate. Therefore, it is more reasonable to use the nonlocal integral theory for the accuracy of some applications and problems. Some researchers have conducted some investigations using the nonlocal integral elasticity theory. Finite element nonlocal integral elasticity theory, which first have been prepared by Polizzotto [2], is one of the best-known methods, that have advantages of both the nonlocal integral theory and finite element method. Also, Marotti de Sciarra [3, 4] has proved the complete family of variational principles, developed a consistent nonlocal finite element procedure based on a suitable definition of weight function and provided some numerical applications for a bar in tension. The nonlocal FE method has been then further developed by Pisano et al. [5] (for analyzing a nano-plates under tension), Taghizadeh et al. [6, 7] (for analyzing the bending and buckling of nano-beams and nano-plates), Koutsoumaris et al. [8–10] (for investigating the bending and dynamic response of nano-beams considering the modified kernel type), Norouzzadeh and Ansari [11] (for Bending of nanobeams), Tuna and Kirca [12] (for analyzing the bending, buckling and vibration of nano-beams) and Naghinejad and Ovesy [13–16] (for considering the viscoelastic effects and studying the vibrations and buckling of nano-beams and nano-plates).

For accuracy of analyzing the nanostructures in some applications, the viscoelastic properties have been taken into account, in some studies. For example, the proper functioning of oscillators depends on considering their damping characteristics [17]. Also, by assuming the viscoelastic characteristics in the nano-scale mass sensors, which works by measuring the shift in the vibration frequency, more accurate detections can be achievable. Also, considering these properties leads to the excellent image quality of atomic force microscopes (AFM) [18]. Knowing the importance of considering viscoelasticity in some applications of nanostructures, recently some researchers have combined the nonlocal theory with viscoelastic properties. For instance, Ghavanloo and Fazalzadeh [19] have studied the flexural vibration of viscoelastic carbon nanotubes conveying fluid and embedded in viscous fluid. They considered the nonlocal Timoshenko beam model and used Hamilton's principle to derive the formulations. Lei et al. [20, 21] have investigated the free vibration of nano-beams based on nonlocal differential elasticity theory by considering viscoelastic properties. They obtained complex frequencies by the transfer function method. Pouresmaeeli et al. [22] have investigated the vibration behavior of viscoelastic orthotropic nanoplates using the nonlocal differential theory and the Kelvin-Voigt viscoelastic model. They have studied the effects of nonlocal parameter, structural damping, stiffness and damping coefficient of the foundation on the vibration frequency. Free vibration of multi-nanoplate system embedded in viscoelastic medium have been studied by Karlicic et al. [23] using nonlocal differential theory. Governing equations have been derived using D'Alamberts principle and Kirchhoff-Love plate theory. Naghinejad and Ovesy [15] have developed a non-local integral finite element method to investigate the free vibration of viscoelastic Euler-Bernoulli nano-scaled beam. The formulations have been obtained using Hamilton's principle, and effects of different parameters and boundary conditions on the free vibration behavior have been discussed. They have also studied the viscoelastic buckling of nano-scaled beams using the nonlocal differential theory and the Kelvin-Voigt viscoelastic model [14].

In the current chapter, firstly, the constitutive equations of nonlocal elasticity and viscoelasticity theories are derived, then the aforementioned finite element nonlocal integral elasticity theory is developed using Hamilton's principle and the formulations are explained. Besides, applying the boundary conditions for beams and plates in the current method and the governing formulations of bending, buckling, and vibration are discussed. Finally, bending, buckling, and vibration of nano-scaled beams and plates are investigated through numerical examples, and the effects of different parameters on the mechanical behavior are studied.

2 Nonlocal Integral Theory

2.1 Elastic Constitutive Equations

In classical elasticity theory, the stress-strain relations are stated only for a single material point. Whereas, weighted averages of a material state-variable over a region should be taken into account for defining the constitutive law at a point, in an integral-type nonlocal theory. Eringen and Edelen [24, 25] developed the theories of nonlocal elasticity in which the nonlocal characteristics have been attached to many fields, including mass, entropy, internal energy, and body forces. Because of their ambiguity, yet to be verified and be used in real problems, simplifications have been considered later. After that, equilibrium and kinematic equations have been used as a standard form, and only the nonlocal form of the constitutive equations have been considered [26, 27]. Thus, in mentioned non-local theory, stress at any point of the material is a function of all strains at the vicinity. For a linear isotropic elastic continuum, the constitutive equation is given by Eq. (1)

$$\mathbf{t}(\mathbf{x}) = \int_V \alpha\left(\left|\mathbf{x}' - \mathbf{x}\right|,\ \tau\right) \boldsymbol{\sigma}\left(\mathbf{x}'\right) dV, \quad \forall \mathbf{x} \in V \tag{1}$$

in which $\mathbf{t}$, α and $\boldsymbol{\sigma}$ are nonlocal stress tensor, nonlocal kernel function, and local stress tensor respectively. The local stress tensor is written as $\boldsymbol{\sigma}(\mathbf{x}) = \mathbf{D} : \boldsymbol{\varepsilon}(\mathbf{x})$ in which $\mathbf{D}$ is the fourth-order elastic moduli tensor, and ε is a strain tensor. The nonlocal kernel is a function of distance, between any point in the domain and reference point, and $\tau = e_0 a/l$ in which a and l are the internal (e.g. lattice parameter or granular distance) and external (e.g. wavelength or crack length) characteristic lengths. e_0 is a constant for any material which can be calibrated by comparing the results with different methods such as molecular dynamics or lattice dynamics according to various parameters, e.g. geometry, mode shapes and boundary conditions [13]. As a result, the nonlocal stress is expressed as a weighted value of the strain field:

$$\mathbf{t}(\mathbf{x}) = \int_V \alpha\left(\left|\mathbf{x}' - \mathbf{x}\right|, \tau\right) \mathbf{D} : \boldsymbol{\varepsilon}\left(\mathbf{x}'\right) dV', \quad \forall \mathbf{x} \in V \tag{2}$$

It is known that, if the nonlocal kernel was taken as the Dirac delta function, the nonlocal constitutive equation reduces to that of the local theory. It is worth noting that the nonlocal constitutive equation can be considered as two-phase [2, 5]. The first phase corresponds to the local part (local fraction ζ_1) and the second phase to the nonlocal part (non-local fraction ζ_2). It is evident that $\zeta_1 + \zeta_2 = 1$ and $\zeta_1, \ \zeta_2 > 0$. ζ_1 and ζ_2 can be obtained using different methods. An approach to calculate values of ζ_1 and ζ_2 can be determining the contribution of reference point and all points in the region (except the reference point) on the stiffness [13, 15, 28, 29]. For example, in the current study, for determining the values of ζ_1 and ζ_2 on each element, nonlocal kernel is integrated on the whole domain (the value is called I_1). Then kernel is integrated on the element containing the reference point (namely I_2). Obviously the remaining portion is named $I_3 = I_1 - I_2$. So, the ratio of I_2 to I_1 corresponds to ζ_1 and the ratio of I_3 to I_1 corresponds to ζ_2. The two-phase model mathematically handles the constitutive equation. In particular, the two-phase model transforms the first kind Fredholm integral equation into a second kind one. In applications in a finite domain, the first kind Fredholm integral equation leads to an ill-posed problem, which is difficult to deal with [30]. Assuming the kernel function as follows:

$$\alpha\left(\left|\mathbf{x}' - \mathbf{x}\right|, \tau\right) = \zeta_1 \delta\left(\mathbf{x}' - \mathbf{x}\right) + \zeta_2 \bar{\alpha}\left(\left|\mathbf{x}' - \mathbf{x}\right|, \tau\right) \tag{3}$$

in which, $\bar{\alpha}$ is the typical nonlocal kernel function, Eq. (2) could be written as the two-phase nonlocal constitutive equation:

$$\mathbf{t}(\mathbf{x}) = \zeta_1 \boldsymbol{\sigma}(\mathbf{x}) + \zeta_2 \int_V \bar{\alpha}\left(\left|\mathbf{x}' - \mathbf{x}\right|, \tau\right) \mathbf{D} : \boldsymbol{\varepsilon}\left(\mathbf{x}'\right) dV \quad \forall \mathbf{x} \in V \tag{4}$$

The better performance of the latter form in some applications has led to the usage of this form over the original form in some recent articles [13, 15, 16, 31].

2.2 Viscoelastic Constitutive Equations

For a nonlocal viscoelastic material, Eq. (1) can be written as follows [15, 20]

$$\mathbf{t}^{ve}(\mathbf{x}) = \int_V \alpha\left(\left|\mathbf{x}' - \mathbf{x}\right|, \tau\right) \boldsymbol{\sigma}^{ve}\left(\mathbf{x}'\right) dV \quad \forall \mathbf{x} \in V \tag{5}$$

in which, $\mathbf{t}^{ve}$ and $\boldsymbol{\sigma}^{ve}$ are the nonlocal and local viscoelastic stress tensors, respectively [15, 20]. It is known that the viscoelastic constitutive equation for a linear homogeneous solid is as follows.

$$\boldsymbol{\sigma}^{ve} = \mathbf{G}(t)\,\boldsymbol{\varepsilon}(0) + \int_0^t \mathbf{G}(t-T)\frac{\partial \boldsymbol{\varepsilon}(T)}{\partial T} dT \tag{6}$$

where $\mathbf{G}$ is the stress relaxation tensor and $\boldsymbol{\varepsilon}$ is the strain tensor. Substituting Eq. (6) into Eq. (5) gives

$$\mathbf{t}^{ve}(\mathbf{x}) = \int_V \alpha\left(\left|\mathbf{x}' - \mathbf{x}\right|, \tau\right) \mathbf{G}(t)\,\boldsymbol{\varepsilon}\left(\mathbf{x}', 0\right) dV$$

$$+ \int_V \alpha\left(\left|\mathbf{x}' - \mathbf{x}\right|, \tau\right)\left(\int_0^t \mathbf{G}(t-T)\frac{\partial \boldsymbol{\varepsilon}\left(\mathbf{x}', T\right)}{\partial T} dT\right) dV \quad \forall \mathbf{x} \in V \tag{7}$$

Equation (7) is the constitutive equation of a nano-material considering nonlocal integral theory and viscoelastic properties. The relaxation tensor of a general viscoelastic model (see Fig. 1) could be stated as

$$\mathbf{G}(t) = \mathbf{G}_0 - \sum_{n=1}^{N} \mathbf{G}_n\left(1 - e^{-t\mathbf{T}_n{}^{-1}}\right) = \mathbf{G}_\infty + \sum_{n=1}^{N} \mathbf{G}_n e^{-t\mathbf{T}_n{}^{-1}} \tag{8}$$

in which, $\mathbf{G}_0 = \mathbf{G}_\infty + \Sigma\mathbf{G}_n$ is the initial relaxation tensor and $\mathbf{T}_n = \mathbf{G}_n{}^{-1}\boldsymbol{\eta}_n$ are the relaxation times (in which $\mathbf{T}_n$ and $\boldsymbol{\eta}_n$ are diagonal). Using the Boltzmann superposition principle and substituting Eq. (8) into Eq. (7), the nonlocal viscoelastic constitutive equation is obtained as (due to the fact that, as time passes the values of $t\mathbf{T}_n{}^{-1}$ and strain increases, and the value of $\mathbf{T}_n$ is relatively small, we can neglect $\sum \mathbf{G}_n e^{-t\mathbf{T}_n^{-1}}\boldsymbol{\varepsilon}(0)$ compared to other terms)

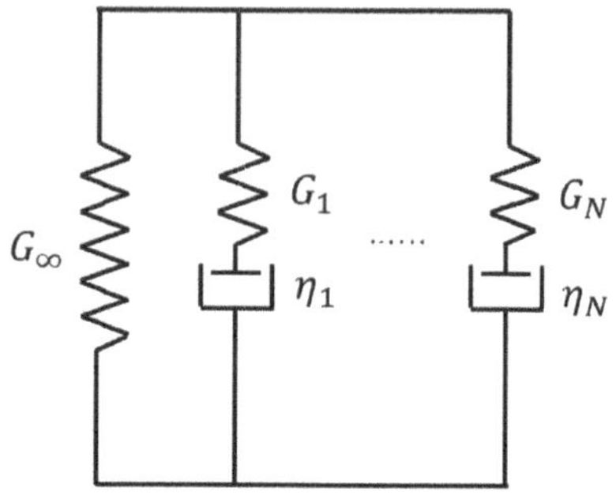

Fig. 1 General viscoelastic model schematic according to relaxation modulus

$$\mathbf{t}^{ve}(\mathbf{x}) = \int_V \alpha\left(\left|\mathbf{x}' - \mathbf{x}\right|,\ \tau\right)\mathbf{G}_\infty \boldsymbol{\varepsilon}\left(\mathbf{x}',\ t\right)dV + \int_V \alpha\left(\left|\mathbf{x}' - \mathbf{x}\right|,\ \tau\right)\left(\int_0^t \sum_{n=1}^N \mathbf{G}_n e^{-(t-T)\mathbf{T}_n{}^{-1}} \frac{\partial \boldsymbol{\varepsilon}\left(\mathbf{x}',\ T\right)}{\partial T} dT\right)dV \tag{9}$$

Different viscoelastic models are assumed by considering different values for parameters of Eq. (9). For example, by taking $N = 1$ and $\mathbf{G}_1 \to \infty$ Kelvin-Voigt model is given, and by considering $N = 1$, the three-parameter standard viscoelastic model can be yielded [20]. By considering the above parameters the Kelvin-Voigt and the three-parameter standard model are respectively obtained as follows. For the Kelvin-Voigt model the derivation process is as follows

$$\mathbf{t}^{ve}(\mathbf{x}) = \int_V \alpha\left(\left|\mathbf{x}' - \mathbf{x}\right|,\ \tau\right)\mathbf{G}_\infty \boldsymbol{\varepsilon}\left(\mathbf{x}',\ t\right)dV + \int_V \alpha\left(\left|\mathbf{x}' - \mathbf{x}\right|,\ \tau\right)\left(\mathbf{G}_1 e^{-(t-T)\mathbf{T}_1{}^{-1}} \frac{\partial \boldsymbol{\varepsilon}\left(\mathbf{x}',\ T\right)}{\partial T} dT\right)dV \tag{10}$$

By carrying out the integration, considering the Kelvin-Voigt assumptions,

$$\mathbf{t}^{ve}(\mathbf{x}) = \int_V \alpha\left(\left|\mathbf{x}' - \mathbf{x}\right|,\ \tau\right)\left(\mathbf{G}_\infty \boldsymbol{\varepsilon}\left(\mathbf{x}',\ t\right) + \frac{\partial \boldsymbol{\varepsilon}\left(\mathbf{x}',\ t\right)}{\partial t}\mathbf{G}_1 \mathbf{T}_1\right)dV \tag{11}$$

So,

$$\mathbf{t}^{ve}(\mathbf{x}) = \int_V \alpha\left(\left|\mathbf{x}' - \mathbf{x}\right|,\ \tau\right)\left(\mathbf{G}_\infty \boldsymbol{\varepsilon}\left(\mathbf{x}',\ t\right) + \frac{\partial \boldsymbol{\varepsilon}\left(\mathbf{x}',\ t\right)}{\partial t}\boldsymbol{\eta}_1\right)dV \tag{12}$$

and by introducing $\mathbf{T}_d = \mathbf{G}_\infty{}^{-1}\boldsymbol{\eta}_1$, finally we have

$$\mathbf{t}^{ve}(\mathbf{x}) = \int_V \alpha\left(\left|\mathbf{x}' - \mathbf{x}\right|,\ \tau\right)\mathbf{G}_\infty\left(\boldsymbol{\varepsilon}\left(\mathbf{x}',\ t\right) + \mathbf{T}_d\frac{\partial\boldsymbol{\varepsilon}\left(\mathbf{x}', t\right)}{\partial t}\right)dV \tag{13}$$

It is noted that the classical form of the above constitutive equation is as follows:

$$\boldsymbol{\sigma}^{ve} = \mathbf{G}_\infty\left(\boldsymbol{\varepsilon} + \mathbf{T}_d\frac{\partial\boldsymbol{\varepsilon}}{\partial t}\right) \tag{14}$$

Also considering the three-parameter viscoelastic model, Eq. (9) can be simply written as

$$\begin{aligned}\mathbf{t}^{ve}(\mathbf{x}) &= \int_V \alpha\left(\left|\mathbf{x}' - \mathbf{x}\right|,\ \tau\right)\mathbf{G}_\infty\boldsymbol{\varepsilon}\left(\mathbf{x}',\ t\right)dV\\ &+\int_V \alpha\left(\left|\mathbf{x}' - \mathbf{x}\right|,\ \tau\right)\left(\mathbf{G}_1 e^{-(t-T)\mathbf{T}_1{}^{-1}}\frac{\partial\boldsymbol{\varepsilon}\left(\mathbf{x}',\ T\right)}{\partial T}dT\right)dV\end{aligned} \tag{15}$$

For a two-phase nonlocal theory, using Eqs. (4) and (9) the constitutive equation will be obtained as

$$\begin{aligned}\mathbf{t}^{ve}(\mathbf{x}) &= \zeta_1\left(\mathbf{G}_\infty\boldsymbol{\varepsilon}\left(\mathbf{x},\ t\right) + \int_0^t\sum_{n=1}^{N}\mathbf{G}_n e^{-(t-T)\mathbf{T}_n^{-1}}\frac{\partial\boldsymbol{\varepsilon}\left(\mathbf{x},\ T\right)}{\partial T}dT\right)\\ &+\zeta_2\int_V \bar{\alpha}\left(\left|\mathbf{x}' - \mathbf{x}\right|,\ \tau\right)\mathbf{G}_\infty\boldsymbol{\varepsilon}\left(\mathbf{x}',\ t\right)dV\\ &+\zeta_2\int_V \bar{\alpha}\left(\left|\mathbf{x}' - \mathbf{x}\right|,\ \tau\right)\left(\int_0^t\sum_{n=1}^{N}\mathbf{G}_n e^{-(t-T)\mathbf{T}_n^{-1}}\frac{\partial\boldsymbol{\varepsilon}\left(\mathbf{x}',\ T\right)}{\partial T}dT\right)dV\end{aligned} \tag{16}$$

2.3 Notes on the Kernel Type

Looking at the constitutive Eq. (1), the three dimensional nonlocal kernel $\bar{\alpha}\left(\left|\mathbf{x}' - \mathbf{x}\right|, \tau\right)$ has the (length)$^{-3}$ dimension. Besides, it is known that $\bar{\alpha}$ is a function of the characteristic length ratio (a/l). Also, the nonlocal kernel has some notable specifications. For example, as it has been expressed by Eringen [1]:

(i) Nonlocal kernels maximum value happens at the reference point (i.e., $\mathbf{x}' = \mathbf{x}$).
(ii) $\bar{\alpha}$ converts to Dirac-delta whenever $\tau \to 0$, i.e., the classical theory must be extracted when $\tau \to 0$.
(iii) $\bar{\alpha}$ can be determined for a certain material by curve-fitting the plane waves dispersion-curves with those of experiments or atomic lattice dynamics.

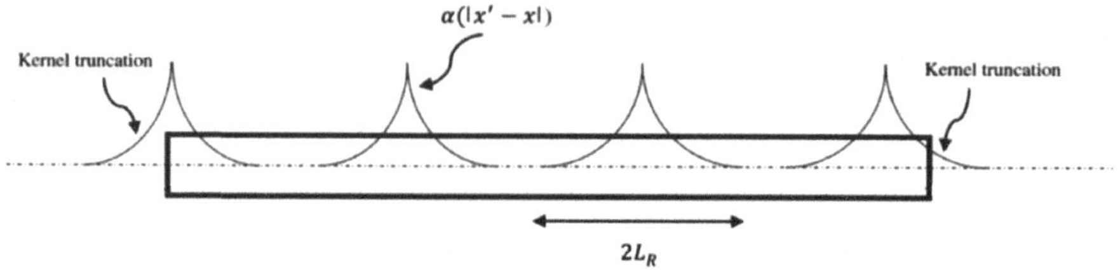

Fig. 2 Modifications of kernel function near the boundaries

It is noted that because of defining the kernel function on the infinite domain, for using it in the finite domain it must be normalized and, in the current approach, truncated at the boundaries [5, 7, 13] (e.g., see Fig. 2). The normalization process is carried out since the nonlocal kernel should satisfy the following condition

$$\int_{V_\infty} \alpha\left(\left|\mathbf{x}' - \mathbf{x}\right|, \tau\right) dV = 1 \tag{17}$$

where V_∞ is an infinite domain, in which the working domain is embedded [2, 5]. The condition (17) also guarantees that for an infinite domain, when the strain is uniform, the related stress will also be uniform [2]. For this purpose in the current study the kernel function, which may violate the condition (17) considering a finite length or truncating near the boundaries, is normalized by dividing it by the normalization parameter. The normalization parameter is obtained by integrating the kernel value in the finite domain of the problem.

Some examples of kernel function are [1]:

- One-dimensional form:

$$\alpha\left(\left|x' - x\right|, \frac{e_0 a}{l}\right) = \begin{cases} \frac{1}{e_0 a}\left(1 - \frac{\left|x' - x\right|}{e_0 a}\right) & \left|x' - x\right| \leq e_0 a \\ 0 & \left|x' - x\right| > e_0 a \end{cases} \tag{18}$$

$$\alpha\left(\left|x' - x\right|, \frac{e_0 a}{l}\right) = \frac{1}{2e_0 a} \exp\left(-\left|x' - x\right| / e_0 a\right) \tag{19}$$

$$\alpha\left(\left|x' - x\right|, \frac{e_0 a}{l}\right) = \frac{1}{\sqrt{\pi}\, e_0 a l} \exp\left(-\left(x' - x\right)^2 / e_0 a l\right) \tag{20}$$

- Two-dimensional form:

$$\alpha\left(\left|\mathbf{x}' - \mathbf{x}\right|, \frac{e_0 a}{l}\right) = \frac{1}{2\pi \left(e_0 a\right)^2} K_0(\left|\mathbf{x}' - \mathbf{x}\right| / e_0 a) \tag{21}$$

in which K_0 is the modified Bessel function of the second kind.

$$\alpha\left(\left|\mathbf{x}'-\mathbf{x}\right|,\frac{e_0a}{l}\right)=\frac{1}{\pi e_0al}\exp\left(-\left(\mathbf{x}'-\mathbf{x}\right).\left(\mathbf{x}'-\mathbf{x}\right)/e_0al\right) \tag{22}$$

- Three-dimensional form

$$\alpha\left(\left|\mathbf{x}'-\mathbf{x}\right|,t\right)=\frac{1}{8(\pi t)^{\frac{3}{2}}}\exp\left(-\frac{\left(\mathbf{x}'-\mathbf{x}\right).\left(\mathbf{x}'-\mathbf{x}\right)}{4t}\right) \tag{23}$$

where $t=e_0al/4$.

$$\begin{aligned}\alpha\left(\left|\mathbf{x}'-\mathbf{x}\right|,\frac{e_0a}{l}\right)&=\frac{1}{4\pi\left(e_0a\right)^2}\\&\times\frac{1}{\sqrt{\left(\mathbf{x}'-\mathbf{x}\right).\left(\mathbf{x}'-\mathbf{x}\right)}}\exp\left(-\frac{\sqrt{\left(\mathbf{x}'-\mathbf{x}\right).\left(\mathbf{x}'-\mathbf{x}\right)}}{e_0a}\right)\end{aligned} \tag{24}$$

In addition to the above solution to deal with the normalization problem, Koutsoumaris et al. [9, 30] have used the so-called modified kernel which has been defined by Bazant and Jirasek [32]. This modified kernel preserves symmetry with respect to **x**, satisfies the normalization condition at each point of a finite domain V and satisfies all the properties of a nonlocal kernel [9]. In addition it is seen that the modified kernel recalls the locality only when the normalization condition is violated [9] (i.e. near the boundaries).

$$\begin{aligned}\alpha_{mod}\left(\left|\mathbf{x}-\mathbf{x}'\right|,\ \tau\right)=&\left(1-\int_V\alpha\left(\left|\mathbf{x}'-\mathbf{x}\right|,\tau\right)dV\right)\delta\left(\left|\mathbf{x}-\mathbf{x}'\right|\right)\\&+\alpha\left(\left|\mathbf{x}-\mathbf{x}'\right|,\tau\right)\end{aligned} \tag{25}$$

3 Nonlocal Integral Finite Element Method

3.1 Variational Equations

In this section, the finite element nonlocal integral method is developed considering viscoelastic properties as it has been proposed by Polizzotto [2] for elastic materials. The finite element formulations are derived using Hamilton's principle based on nonlocal integral theory. These formulations are then extended in the following sections to study the bending, buckling, and vibration of nanostructures. Considering the two-phase non-local theory using Eqs. (4) and (5), and taking into account the inertia effects, the total potential energy can be written as follows

$$\Pi = \frac{1}{2}\zeta_1 \int_V \boldsymbol{\varepsilon}(\mathbf{x}) : \boldsymbol{\sigma}^{ve}(\mathbf{x})\, dV + \frac{1}{2}\zeta_2 \int_V \int_{V'} \alpha\left(\left|\mathbf{x}' - \mathbf{x}\right|,\ \tau\right) \boldsymbol{\varepsilon}(\mathbf{x}) : \boldsymbol{\sigma}^{ve}\left(\mathbf{x}'\right) dV' dV - W_{inrt} - W_{ext} - W_g \tag{26}$$

where W_{inrt} is the work done by inertia forces, W_{ext} is the work done by external forces, and W_g shows the energy due to geometric stiffness. Considering the Kelvin-Voigt model (Eqs. (13)), (26) can be written as [15]

$$\begin{aligned}\Pi = &\frac{1}{2}\zeta_1 \int_V \boldsymbol{\varepsilon}(\mathbf{x}) : D : \left(\boldsymbol{\varepsilon}(\mathbf{x}, t) + \mathbf{T}_d \frac{\partial \varepsilon(\mathbf{x}, t)}{\partial t}\right) dV \\ &+ \frac{1}{2}\zeta_2 \int_V \int_{V'} \alpha\left(\left|\mathbf{x}' - \mathbf{x}\right|,\ \tau\right) \boldsymbol{\varepsilon}(\mathbf{x}) : \mathbf{D} : \boldsymbol{\varepsilon}\left(\mathbf{x}'\right) dV' dV \\ &+ \frac{1}{2}\zeta_2 \int_V \int_{V'} \alpha\left(\left|\mathbf{x}' - \mathbf{x}\right|,\ \tau\right) \boldsymbol{\varepsilon}(\mathbf{x}) : \mathbf{D} : \left(\mathbf{T}_d \frac{\partial \boldsymbol{\varepsilon}\left(\mathbf{x}', t\right)}{\partial t}\right) dV' dV \\ &- W_{inrt} - W_{ext} - W_g\end{aligned} \tag{27}$$

in which $\mathbf{D} = \mathbf{G}_\infty$ is the fourth rank elastic moduli tensor. It is noted that the above statement can be extended for other viscoelastic models by using the explained procedure. Now, applying the variations, the total potential energy (Eq. (27)) is minimized as

$$\begin{aligned}\delta\Pi = &\zeta_1 \int_V \delta\boldsymbol{\varepsilon}(\mathbf{x}) : \mathbf{D} : \left(\boldsymbol{\varepsilon}(\mathbf{x}, t) + \mathbf{T}_d \frac{\partial \varepsilon(\mathbf{x}, t)}{\partial t}\right) dV \\ &+ \zeta_2 \int_V \int_{V'} \alpha\left(\left|\mathbf{x}' - \mathbf{x}\right|,\ \tau\right) \delta\boldsymbol{\varepsilon}(\mathbf{x}) : \mathbf{D} : \boldsymbol{\varepsilon}\left(\mathbf{x}'\right) dV' dV \\ &+ \zeta_2 \int_V \int_{V'} \alpha\left(\left|\mathbf{x}' - \mathbf{x}\right|,\ \tau\right) \delta\boldsymbol{\varepsilon}(\mathbf{x}) : \mathbf{D} : \left(\mathbf{T}_d \frac{\partial \boldsymbol{\varepsilon}\left(\mathbf{x}', t\right)}{\partial t}\right) dV' dV \\ &- \delta W_{inrt} - \delta W_{ext} - \delta W_g = 0\end{aligned} \tag{28}$$

By substituting the corresponding terms of W_{inrt}, W_{ext} and W_g, we have

$$\begin{aligned}\delta\Pi = &\zeta_1 \int_V \delta\boldsymbol{\varepsilon}(\mathbf{x}) : \mathbf{D} : \left(\boldsymbol{\varepsilon}(\mathbf{x}, t) + \mathbf{T}_d \frac{\partial \boldsymbol{\varepsilon}(\mathbf{x}, t)}{\partial t}\right) dV \\ &+ \zeta_2 \int_V \int_{V'} \alpha\left(\left|\mathbf{x}' - \mathbf{x}\right|,\ \tau\right) \delta\boldsymbol{\varepsilon}(\mathbf{x}) : \mathbf{D} : \boldsymbol{\varepsilon}\left(\mathbf{x}'\right) dV' dV\end{aligned}$$

$$+ \zeta_2 \int\limits_V \int\limits_{V'} \alpha\left(\left|\mathbf{x}' - \mathbf{x}\right|, \tau\right) \delta\boldsymbol{\varepsilon}(\mathbf{x}) : \mathbf{D} : \left(\mathbf{T}_d \frac{\partial \boldsymbol{\varepsilon}\left(\mathbf{x}', t\right)}{\partial t}\right) dV' dV$$

$$- \int\limits_V \delta\mathbf{u} \cdot (-\rho\ddot{\mathbf{u}})\, dV - \int\limits_V \bar{\mathbf{b}} \cdot \delta\mathbf{u}\, dV$$

$$- \int\limits_{S_t} \bar{\mathbf{t}} \cdot \delta\mathbf{u}\, dS - \int\limits_V \boldsymbol{\sigma}_{0m} \delta\boldsymbol{\varepsilon}_{nl}(\mathbf{x})\, dV = 0 \tag{29}$$

where $\boldsymbol{\varepsilon}_{nl}$ is the nonlinear strain tensor, $\boldsymbol{\sigma}_{0m}$ is the external compressive stress, ρ is the mass density, $\bar{\mathbf{t}}$ and $\bar{\mathbf{b}}$ are respectively the surface force on the surface S_t and body force in volume V and $\mathbf{u}$ is the displacement field. It is seen that Eq. (29) is of a general form that can be used for analyzing the various mechanical behavior of nanostructures. For example, it can be used for extracting the formulations for studying the bending, buckling, and vibration.

3.2 *Finite Element Formulations*

In this section, the foundation of finite element formulation is to be established. For developing the finite element nonlocal integral theory Eq. (29) should be discretized. So, domain V will be partitioned to N subdomains, and displacement field $\mathbf{u}(\mathbf{x})$ and strain tensor $\boldsymbol{\varepsilon}(\mathbf{x})$ of the n-th element can be written as

$$\mathbf{u}(\mathbf{x}) = \mathbf{N}_n(\mathbf{x})\, \mathbf{d}_n, \qquad n = 1, \ldots, N \tag{30}$$

$$\boldsymbol{\varepsilon}(\mathbf{x}) = \mathbf{B}_n(\mathbf{x})\, \mathbf{d}_n, \qquad n = 1, \ldots, N \tag{31}$$

Matrices $\mathbf{N}_n(\mathbf{x})$, $\mathbf{B}_n(\mathbf{x})$ and $\mathbf{d}_n(\mathbf{x})$ include the shape functions, corresponding partial derivatives, and node degrees of freedom, respectively. Applying the discretizing process and using Eqs. (30) and (31), then Eq. (29) can be written as

$$\delta\Pi = \zeta_1 \sum_{n=1}^{N} \delta\mathbf{d}_n^T \left(\int\limits_{V_n} \mathbf{B}_n^T : \mathbf{D} : \mathbf{B}_n dV \right) \mathbf{d}_n$$

$$+ \zeta_1 \sum_{n=1}^{N} \delta\mathbf{d}_n^T \left(\int\limits_{V_n} \mathbf{T}_d \mathbf{B}_n^T : \mathbf{D} : \mathbf{B}_n dV \right) \dot{\mathbf{d}}_n$$

$$+\zeta_2\sum_{n=1}^{N}\sum_{m=1}^{N}\delta\mathbf{d}_n^T\left(\int_{V_n}\int_{V_m}\alpha\left(\left|\mathbf{x}'-\mathbf{x}\right|,\tau\right)\mathbf{B}_n^T:\mathbf{D}:\mathbf{B}_m^{'}dV'dV\right)\mathbf{d}_m$$

$$+\zeta_2\sum_{n=1}^{N}\sum_{m=1}^{N}\delta\mathbf{d}_n^T\left(\int_{V_n}\int_{V_m}\mathbf{T_d}\alpha\left(\left|\mathbf{x}'-\mathbf{x}\right|,\tau\right)\mathbf{B}_n^T:\mathbf{D}:\mathbf{B}_m^{'}dV'dV\right)\dot{\mathbf{d}}_m$$

$$-\sum_{n=1}^{N}\left(\delta\mathbf{d}_n^T\int_{V_n}\mathbf{N}_n^T.\bar{\mathbf{b}}\,dV\right)-\sum_{n=1}^{N}\left(\delta\mathbf{d}_n^T\int_{S_{\mathbf{t}n}}\mathbf{N}_n^T.\bar{\mathbf{t}}\,dS\right)$$

$$-\sum_{n=1}^{N}\left(\delta\mathbf{d}_n^T\left(\int_{V_n}\mathbf{N}_n^T.\left(-\rho\mathbf{N}_n\right)dV\right)\ddot{\mathbf{d}}_n\right)$$

$$-\sum_{n=1}^{N}\left(\delta\mathbf{d}_n^T\left(\int_{V_n}\mathbf{B}_n^{gT}:\boldsymbol{\sigma}_0:\mathbf{B}_n^g dV\right)\mathbf{d}_n\right)=0 \tag{32}$$

It is noted that matrix $\mathbf{B}_n^g$ is related to nonlinear strains and $\dot{\mathbf{d}}_n=\partial\mathbf{d}_n/\partial t$. Moreover, as it has been defined ζ_1 and ζ_2 correspond to the local and nonlocal fractions $(\zeta_1+\zeta_2=1,\ \zeta_1,\ \zeta_2>0)$. Using the Boolean matrix $\mathbf{Q}_n$ the degrees of freedom of the n-th element $(\mathbf{d}_n)$ is connected to the structural DOF matrix $(\mathbf{U})$

$$\mathbf{d}_n=\mathbf{Q}_n\mathbf{U} \tag{33}$$

Substituting Eq. (33) into Eq. (32) gives

$$\delta\Pi=\zeta_1\sum_{n=1}^{N}\delta\mathbf{U}^T\mathbf{Q}_n^T\left(\int_{V_n}\mathbf{B}_n^T:\mathbf{D}:\mathbf{B}_n dV\right)\mathbf{Q}_n\mathbf{U}$$

$$+\zeta_1\sum_{n=1}^{N}\delta\mathbf{U}^T\mathbf{Q}_n^T\left(\int_{V_n}\mathbf{T}_d\mathbf{B}_n^T:\mathbf{D}:\mathbf{B}_n dV\right)\mathbf{Q}_n\dot{\mathbf{U}}$$

$$+\zeta_2\sum_{n=1}^{N}\sum_{m=1}^{N}\delta\mathbf{U}^T\mathbf{Q}_n^T\left(\int_{V_n}\int_{V_m}\alpha\left(\left|\mathbf{x}'-\mathbf{x}\right|,\tau\right)\mathbf{B}_n^T:\mathbf{D}:\mathbf{B}_m^{'}dV'dV\right)\mathbf{Q}_m\mathbf{U}$$

$$+\zeta_2\sum_{n=1}^{N}\sum_{m=1}^{N}\delta\mathbf{U}^T\mathbf{Q}_n^T\left(\int_{V_n}\int_{V_m}\mathbf{T}_d\alpha\left(\left|\mathbf{x}'-\mathbf{x}\right|,\tau\right)\mathbf{B}_n^T:\mathbf{D}:\mathbf{B}_m^{'}dV'dV\right)\mathbf{Q}_m\dot{\mathbf{U}}$$

$$-\sum_{n=1}^{N}\left(\delta\mathbf{U}^T\mathbf{Q}_n^T\int_{V_n}\mathbf{N}_n^T.b\,dV\right)-\sum_{n=1}^{N}\left(\delta\mathbf{U}^T\mathbf{Q}_n^T\int_{S_{tn}}\mathbf{N}_n^T.t\,dS\right)$$

$$-\sum_{n=1}^{N}\left(\delta\mathbf{U}^{T}\mathbf{Q}_{n}^{T}\left(\int_{V_n}\mathbf{N}_{n}^{T}.(-\rho\mathbf{N}_{n})\,dV\right)\mathbf{Q}_{n}\ddot{\mathbf{U}}\right)$$

$$-\sum_{n=1}^{N}\left(\delta\mathbf{U}^{T}\mathbf{Q}_{n}^{T}\left(\int_{V_n}\mathbf{B}^{g}{}_{n}^{T}:\boldsymbol{\sigma}_{0}:\mathbf{B}_{n}^{g}dV\right)\mathbf{Q}_{n}\mathbf{U}\right)=0 \quad (34)$$

Equation (34) is the governing finite element equation based on the nonlocal integral theory considering the Kelvin-Voigt viscoelastic model. Mechanical behavior of nanostructures (e.g., bending, vibration, and buckling) can be analyzed by considering the related terms of the current equation.

3.3 Element Types

Noting that the proposed method has a finite element base, various types of elements can be used for analyzing different problems. For example, to analyze the simple tension problem of a plate (for axial behavior), elements of the C^0 continuity class will suffice, however, for analyzing the bending, buckling or vibration assuming the classical beam theory elements of the C^1 continuity class is usually needed. C^0 continuity means that the displacements between the elements are continuous, but their first derivatives are not (Lagrange elements). However, in C^1 continuity, both the displacements and their first derivatives are continuous (Hermite elements). In this section, some of the sample elements that have relatively more applications are introduced.

Hermite beam element (Fig. 3) is used later in this chapter for analyzing the mechanical behavior of Euler-Bernoulli beam elements. These elements include two nodes with two degrees of freedom for each node, i.e. displacement and rotation. The displacement matrix and shape functions of Hermite beam elements in the local coordinate system are considered as

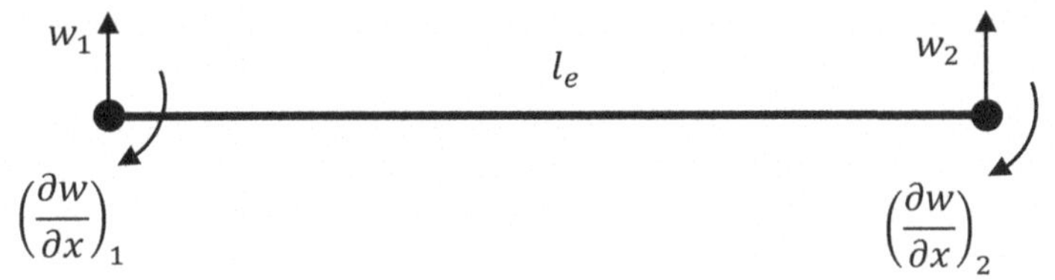

Fig. 3 Hermite beam element

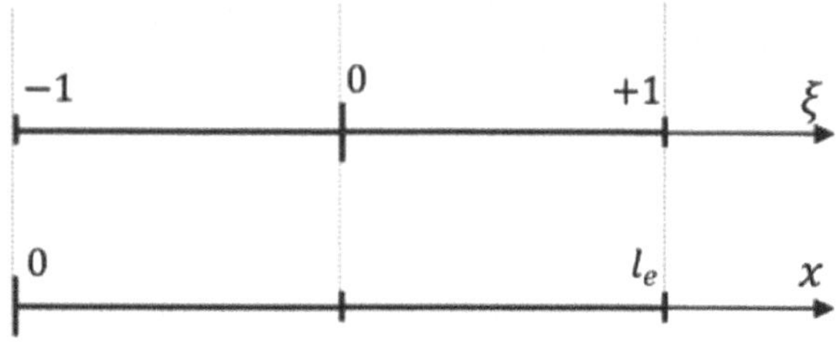

Fig. 4 Local and general coordinates of the beam

$$\mathbf{d}_n = \begin{Bmatrix} w_1 \\ \left(\frac{\partial w}{\partial x}\right)_1 \\ w_2 \\ \left(\frac{\partial w}{\partial x}\right)_2 \end{Bmatrix} \tag{35}$$

$$\mathbf{N}_n^T(\xi) = \frac{1}{8}\begin{Bmatrix} 2(1-\xi)^2(2+\xi) \\ l_e(1-\xi)^2(1+\xi) \\ 2(1+\xi)^2(2-\xi) \\ l_e(1+\xi)^2(\xi-1) \end{Bmatrix} \tag{36}$$

The general coordinates can be converted to local ones using (see Fig. 4)

$$\xi = \frac{2x}{l_e} - 1; \quad 0 \le x \le l_e, \quad -1 \le \xi \le 1 \tag{37}$$

in which l_e is the element length.

Conforming elements (displacements are always continuous between adjacent elements) are mostly based on higher degree polynomials and need relatively high computational cost for producing the corresponding matrices. Moreover, the nonlocal integral finite element method has more computational cost in comparison to the local method. Therefore, it is more efficient to use elements with polynomials of lower degree (e.g. non-conforming elements) for modelling the nonlocal plate problems. If the assumed elements pass the "patch test", the results will converge [33]. The original patch test, developed by Irons et al. [34] is a check that determines whether a patch of elements subject to a constant strain reproduced the constitutive behavior of the material and resulted in correct stresses when it became infinitesimally small [35]. In other words, patch test is a sufficient requirement for convergence.

The Adini-Clough quadrilateral element (Fig. 5) is a non-conforming element, which despite being unable to pass the patch test, is known to give good convergent results for bending and buckling of thin plates, i.e. it is a generalized conforming element [36]. These elements consist of four nodes with three degrees of freedom for each. Matrix $\mathbf{d}_n$ for the n-th element is expressed as

Fig. 5 Adini-Clough quadrilateral plate element

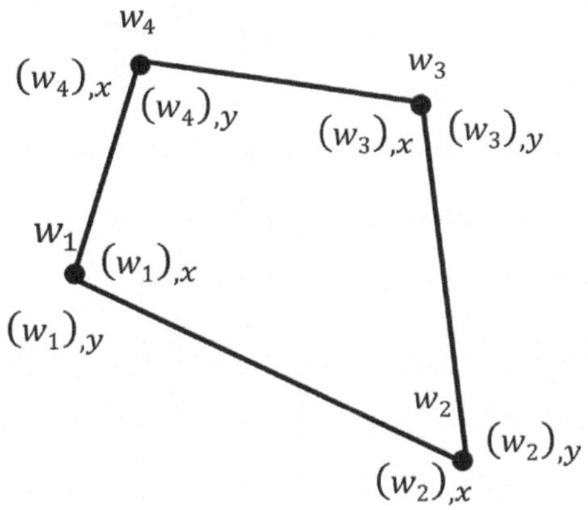

$$
\begin{aligned}
\mathbf{d}_n = {} & [w_1\ w_{1,x}\ w_{1,y}\ w_2\ w_{2,x}\ w_{2,y}\ \ldots \\
& \ldots\ w_3\ w_{3,x}\ w_{3,y}\ w_4\ w_{4,x}\ w_{4,y}]^T
\end{aligned} \tag{38}
$$

Also, the shape functions of the Adini-Clough elements are as $(-1 < \xi < 1,\ -1 < \eta < 1)$

$$
\mathbf{N}_n^T(\xi,\eta) = \frac{1}{8}\left\{\begin{array}{c}
e(a-\xi-\eta) \\
e\left(1-\xi^2\right) \\
e\left(1-\eta^2\right) \\
f(a+\xi-\eta) \\
-f\left(1-\xi^2\right) \\
f\left(1-\eta^2\right) \\
g(a+\xi+\eta) \\
-g\left(1-\xi^2\right) \\
-g\left(1-\eta^2\right) \\
h(a-\xi+\eta) \\
h\left(1-\xi^2\right) \\
-h\left(1-\eta^2\right)
\end{array}\right\} \tag{39}
$$

where

$$
\begin{aligned}
& a = 2-\xi^2-\eta^2, \quad e = (1-\xi)(1-\eta), \quad f = (1+\xi)(1-\eta), \\
& g = (1+\xi)(1+\eta), \quad h = (1-\xi)(1+\eta)
\end{aligned} \tag{40}
$$

The 8-node C^0-quadratic isoparametric Serendipity elements with 2 degrees of freedom per node can be used for analyzing the beam in 3-dimensions [6] or tension of a plate [5]. $\mathbf{d}_n$ and $\mathbf{N}_n$ matrices are as

$$\mathbf{d}_n = [u_1\ v_1\ u_2\ v_2\ u_3\ v_3\ u_4\ v_4 \ldots \\ \ldots u_5\ v_5\ u_6\ v_6\ u_7\ v_7\ u_8\ v_8]^T \tag{41}$$

$$\mathbf{N}_n^{T^{u\ (or)\ v}}(\xi,\eta) = \frac{1}{4}\begin{Bmatrix} -(\xi+\eta+1)(\xi-1)(\eta-1) \\ 2(\xi-1)(\xi+1)(\eta-1) \\ (-\xi+\eta+1)(\xi+1)(\eta-1) \\ -2(\eta-1)(\xi+1)(\eta+1) \\ (\xi+\eta-1)(\xi+1)(\eta+1) \\ -2(\xi-1)(\xi+1)(\eta+1) \\ -(-\xi+\eta-1)(\xi-1)(\eta+1) \\ 2(\xi-1)(\eta+1)(\eta-1) \end{Bmatrix} \tag{42}$$

where $\mathbf{N}_n^{T^{u\ (or)\ v}}$ includes the shape functions of the element corresponding to both u and v degrees of freedom.

3.4 Notes on the Boundary Conditions

In this section, by considering a sample case, it is attempted to demonstrate the incapability of nonlocal differential elasticity theory for efficiently applying the natural boundary conditions on a plate. Consider a plate of length and width l_x and l_y, respectively. The boundary conditions of the mentioned plate are simply-supported on all four edges. Using the nonlocal differential constitutive equation, the following set of equations can be obtained [37]

$$M_{xx} - l^2\tau^2\nabla^2 M_{xx} = -D_{11}\frac{\partial^2 w}{\partial x^2} - D_{12}\frac{\partial^2 w}{\partial y^2} \tag{43a}$$

$$M_{yy} - l^2\tau^2\nabla^2 M_{yy} = -D_{12}\frac{\partial^2 w}{\partial x^2} - D_{22}\frac{\partial^2 w}{\partial y^2} \tag{43b}$$

$$M_{xy} - l^2\tau^2\nabla^2 M_{xy} = -2D_{66}\frac{\partial^2 w}{\partial x \partial y} \tag{43c}$$

Assuming the bottom left corner of the plate as the center of the coordinate system, the boundary condition at $x = 0$ would be like $M_{xx} = 0$. So the Eq. (43) becomes

$$\frac{\partial^2 w}{\partial x^2} = \frac{l^2\tau^2}{D_{11}}\frac{\partial^2 M_{xx}}{\partial x^2} - \frac{D_{12}}{D_{11}}\frac{\partial^2 w}{\partial y^2} \tag{44}$$

It is noted that the equilibrium equations for a plate are as [37]

$$\frac{\partial^2 M_{xx}}{\partial x^2} + 2\frac{\partial^2 M_{xy}}{\partial x \partial y} + \frac{\partial^2 M_{yy}}{\partial y^2} + \frac{\partial}{\partial x}\left(N_0^x \frac{\partial w}{\partial x}\right) + \frac{\partial}{\partial y}\left(N_0^y \frac{\partial w}{\partial y}\right) + \frac{\partial}{\partial x}\left(N_0^{xy} \frac{\partial w}{\partial y}\right) + \frac{\partial}{\partial y}\left(N_0^{xy} \frac{\partial w}{\partial x}\right) - m_0 \frac{\partial^2 w}{\partial t^2} + q = 0 \tag{45a}$$

$$Q_x - \frac{\partial M_{xx}}{\partial x} - \frac{\partial M_{xy}}{\partial y} - N_0^x \frac{\partial w}{\partial x} - N_0^{xy} \frac{\partial w}{\partial y} = 0 \tag{45b}$$

$$Q_y - \frac{\partial M_{xy}}{\partial x} - \frac{\partial M_{yy}}{\partial y} - N_0^y \frac{\partial w}{\partial y} - N_0^{xy} \frac{\partial w}{\partial x} = 0 \tag{45c}$$

Now the Navier solution can only be applied when $\partial^2 w/\partial n^2$ is zero across the boundary of the plate. It means that, for considering the Navier solution, according to Eq. (44) $(l^2\tau^2/D_{11})[(\partial^2 M_{xx}/\partial x^2) - (D_{12}/D_{11})(\partial^2 w/\partial y^2)]$ should be zero at $x = 0$, but it is not. Even for a free boundary condition, the mentioned problem arises for a plate. It is where the finite element nonlocal integral method comes in and makes things easier! No such complexity is seen in the FEM based method, and various kinds of rather complex boundary conditions can be dealt with.

4 Nano-Scaled Beams

The finite element formulations for bending, vibration, and buckling of nano-scaled beams are prepared in the following sections. It is noted that various beam theories can be included in the formulation, but for the sake of brevity only Euler-Bernoulli beam theory is assumed here.

4.1 Applying Boundary Conditions for Nano-Scaled Beams

The boundary conditions can be applied rather quickly in the currently proposed method. For instance, some of the common boundary conditions are as follow (see Fig. 6)

Both-ends simply supported beam:

$$\begin{aligned} w(x) &= 0 \quad \text{at} \quad x = 0 \\ w(x) &= 0 \quad \text{at} \quad x = l \end{aligned} \tag{46}$$

Both-ends clamped beam:

$$\begin{aligned} w(x) = w_{,x}(x) &= 0 \quad \text{at} \quad x = 0 \\ w(x) = w_{,x}(x) &= 0 \quad \text{at} \quad x = l \end{aligned} \tag{47}$$

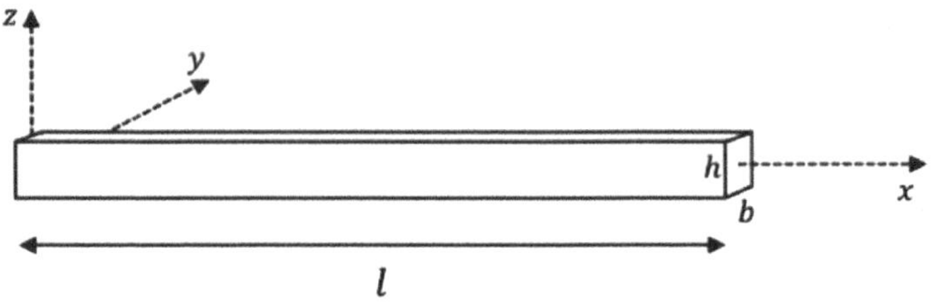

Fig. 6 The geometry of a beam

Clamped-free beam:

$$w(x) = w_{,x}(x) = 0 \quad \text{at} \quad x = 0 \tag{48}$$

Simply supported-clamped beam:

$$\begin{aligned} w(x) &= 0 \quad \text{at} \quad x = 0 \\ w(x) &= w_{,x}(x) = 0 \quad \text{at} \quad x = l \end{aligned} \tag{49}$$

4.2 Bending of Elastic Nano-Scaled Beams

The bending formulations are given for an Euler-Bernoulli nano-scaled beam. It is noted that this method can be extended for other beam theories. Considering the classical beam theory, the displacement field is given as

$$u(x, z) = u_0(x) - z\frac{\partial w_0}{\partial x} \tag{50}$$

$$w(x, z) = w_0(x) \tag{51}$$

in which u and w are longitudinal and transverse displacements of the beam, respectively. Also, u_0 and w_0 are the displacement components at the mid-axis. Also, neglecting the higher-order terms, the strain can be expressed as

$$\varepsilon = \varepsilon_x = \frac{\partial u}{\partial x} = \frac{\partial u_0}{\partial x} - z\frac{\partial^2 w_0}{\partial x^2} \tag{52}$$

For an Euler-Bernoulli beam, if the pure bending condition is assumed, Eq. (52) might be written as

$$\varepsilon = \varepsilon_x = -z\frac{\partial^2 w_0}{\partial x^2} = \mathbf{B}_n\mathbf{d}_n = z\mathbf{B}_n^b\mathbf{d}_n \tag{53}$$

where $\mathbf{B}_n^b$ consists of the partial derivatives of the shape functions of the subject beam. Substituting Eq. (53) into Eq. (34) and only considering the terms corresponding to elastic bending, the local and nonlocal stiffness matrices and external force matrix can be obtained as

$$\mathbf{K}_n^L = \zeta_1\left(I_{yy}\int_{-1}^{1}\mathbf{B}_n^{b^T}(\xi)\,E\mathbf{B}_n^b(\xi)\det(\mathbf{J}(\xi))\,d\xi\right) \tag{54}$$

$$\mathbf{K}_{nm}^{NL} = \zeta_2[bh\,I_{yy}\int_{-1}^{1}\int_{-1}^{1}\alpha\left(\left|x'(\xi) - x(\xi)\right|, \tau\right)\,\mathbf{B}_n^{b^T}(\xi)\,E\mathbf{B}_m^b(\xi')$$
$$\times\,\det(\mathbf{J}(\xi'))\det(\mathbf{J}(\xi))\,d\xi' d\xi] \tag{55}$$

$$\mathbf{F}_n = bh\int_{-1}^{1}\mathbf{N}_n^T.\bar{\mathbf{b}}\det(\mathbf{J}(\xi))\,d\xi + b\int_{-1}^{1}\mathbf{N}_n^T.\bar{\mathbf{t}}\det(\mathbf{J}(\xi))\,d\xi \tag{56}$$

where $\mathbf{K}_{nm}^{NL}$ shows the nonlocal effect of the m-th element on the n-th one. $\mathbf{J}$ is the Jacobian matrix, and I_{yy} is the second moment of area. Also, total matrices can be written as

$$\mathbf{K}_{total}^L = \sum_{n=1}^{N}\mathbf{Q}_n^T\left(\mathbf{K}_n^L\right)\mathbf{Q}_n \tag{57}$$

$$\mathbf{K}_{total}^{NL} = \sum_{n=1}^{N}\sum_{m=1}^{N}\mathbf{Q}_n^T\left(\mathbf{K}_{nm}^{NL}\right)\mathbf{Q}_m \tag{58}$$

$$\mathbf{F}_{total} = \sum_{n=1}^{N}\mathbf{Q}_n^T\left(\mathbf{F}_n\right)\mathbf{Q}_n \tag{59}$$

It is noted that for carrying out the corresponding integration, numerical integration schemes can be used. Substituting Eqs. (54)–(58) into Eq. (34) and assuming bending related terms, the corresponding governing bending equation of the elastic beam is developed as follows

$$\left(\mathbf{K}_{total}^{L} + \mathbf{K}_{total}^{NL}\right)\mathbf{U} = \mathbf{F}_{total} \tag{60}$$

By solving Eq. (60), displacement matrix **U** can be calculated as

$$\mathbf{U} = \left(\mathbf{K}_{total}^{L} + \mathbf{K}_{total}^{NL}\right)^{-1}\mathbf{F}_{total} \tag{61}$$

4.3 Vibration of Nano-Scaled Beams

For analyzing the vibration of the beam, mass (inertia effects) and stiffness of the beam should be considered. It is noted that for a viscoelastic beam, the damping matrix should also be taken into account. So, Eq. (34) can be written as

$$\begin{aligned}
\delta\Pi &= \zeta_1 \sum_{n=1}^{N} \delta\mathbf{U}^T \mathbf{Q}_n^T \left(\int_{V_n} \mathbf{B}_n^T : \mathbf{D} : \mathbf{B}_n dV \right) \mathbf{Q}_n \mathbf{U} \\
&+ \zeta_1 \sum_{n=1}^{N} \delta\mathbf{U}^T \mathbf{Q}_n^T \left(\int_{V_n} \mathbf{T}_d \mathbf{B}_n^T : \mathbf{D} : \mathbf{B}_n dV \right) \mathbf{Q}_n \dot{\mathbf{U}} \\
&+ \zeta_2 \sum_{n=1}^{N} \sum_{m=1}^{N} \delta\mathbf{U}^T \mathbf{Q}_n^T \left(\int_{V_n} \int_{V_m} \alpha\left(\left|\mathbf{x}' - \mathbf{x}\right|, \tau\right) \mathbf{B}_n^T : \mathbf{D} : \mathbf{B}_m^{'} dV' dV \right) \mathbf{Q}_m \mathbf{U} \\
&+ \zeta_2 \sum_{n=1}^{N} \sum_{m=1}^{N} \delta\mathbf{U}^T \mathbf{Q}_n^T \left(\int_{V_n} \int_{V_m} \mathbf{T}_d \alpha\left(\left|\mathbf{x}' - \mathbf{x}\right|, \tau\right) \mathbf{B}_n^T : \mathbf{D} : \mathbf{B}_m^{'} dV' dV \right) \mathbf{Q}_m \dot{\mathbf{U}} \\
&- \sum_{n=1}^{N} \left(\delta\mathbf{U}^T \mathbf{Q}_n^T \int_{V_n} \mathbf{N}_n^T . b \, dV \right) - \sum_{n=1}^{N} \left(\delta\mathbf{U}^T \mathbf{Q}_n^T \int_{S_{tn}} \mathbf{N}_n^T . t \, dS \right) \\
&- \sum_{n=1}^{N} \left(\delta\mathbf{U}^T \mathbf{Q}_n^T \left(\int_{V_n} \mathbf{N}_n^T \cdot (-\rho \mathbf{N}_n) \, dV \right) \mathbf{Q}_n \ddot{\mathbf{U}} \right) = 0
\end{aligned} \tag{62}$$

Local and nonlocal stiffness matrices are expressed as Eqs. (54)–(58). Mass matrix ($\mathbf{M}_n$) and damping matrices ($\mathbf{C}_n^L$ and $\mathbf{C}_{nm}^{NL}$) might be extracted for elements by substituting Eqs. (50)–(53) into Eq. (62).

$$\mathbf{M}_n = \left(bh \int_{-1}^{1} \mathbf{N}_n^T (\xi) \, \rho \mathbf{N}_n (\xi) \det (\mathbf{J} (\xi)) \, d\xi \right) \tag{63}$$

$$\mathbf{C}_n^L = \zeta_1 \left(I_{yy} \int_{-1}^{1} \mathbf{B}^{b^T}_n(\xi)\, E T_d \mathbf{B}_n^b(\xi) \det(\mathbf{J}(\xi))\, d\xi \right) \tag{64}$$

$$\mathbf{C}_{nm}^{NL} = \zeta_2 [bh\, I_{yy} \int_{-1}^{1}\int_{-1}^{1} \alpha\left(\left|x'(\xi) - x(\xi)\right|, \tau\right) \mathbf{B}_n^{b^T}(\xi)\, E T_d \mathbf{B}_m^b(\xi') \times \det\left(\mathbf{J}(\xi')\right) \det(\mathbf{J}(\xi))\, d\xi' d\xi] \tag{65}$$

Besides, the total mass and damping matrices are as follows

$$\mathbf{M}_{total} = \sum_{n=1}^{N} \mathbf{Q}_n^T\ (\mathbf{M}_n)\ \mathbf{Q}_n \tag{66}$$

$$\mathbf{C}_{total}^L = \sum_{n=1}^{N} \mathbf{Q}_n^T\ (\mathbf{C}_n^L)\ \mathbf{Q}_n \tag{67}$$

$$\mathbf{C}_{total}^{NL} = \sum_{n=1}^{N}\sum_{m=1}^{N} \mathbf{Q}_n^T\ (\mathbf{C}_{nm}^{NL})\ \mathbf{Q}_m \tag{68}$$

Substituting Eqs. (54)–(58) and (63)–(68) into Eq. (62), the following equation is obtained for viscoelastic vibration of nonlocal Euler-Bernoulli nano-scaled beam.

$$\mathbf{M}_{total}\ddot{\mathbf{U}} + \left(\mathbf{C}_{total}^L + \mathbf{C}_{total}^{NL}\right)\dot{\mathbf{U}} + \left(\mathbf{K}_{total}^L + \mathbf{K}_{total}^{NL}\right)\mathbf{U} = 0 \tag{69}$$

Free vibration frequencies and mode shapes are the eigenvalues and eigenfunctions of Eq. (69), respectively. For solving the eigenvalue problem of Eq. (69), the following parameters are first defined.

$$\mathbf{V} = \begin{bmatrix} \dot{\mathbf{U}} \\ \mathbf{U} \end{bmatrix} \tag{70}$$

$$\mathbf{M}^* = \begin{bmatrix} \mathbf{M}_{total} & \mathbf{0} \\ \mathbf{0} & \mathbf{I} \end{bmatrix} \tag{71}$$

$$\mathbf{C}^{*}=\begin{bmatrix}\mathbf{C}_{total}^{L}+\mathbf{C}_{total}^{NL} & \mathbf{K}_{total}^{L}+\mathbf{K}_{total}^{NL}\\ -\mathbf{I} & \mathbf{0}\end{bmatrix} \tag{72}$$

Assuming Eqs. (70)–(72), (69) becomes

$$\mathbf{M}^{*}\dot{\mathbf{V}}+\mathbf{C}^{*}\mathbf{V}=\mathbf{0} \tag{73}$$

For calculating the complex eigenvalues of the free vibration, the solution of the following form is assumed

$$\mathbf{V}=\boldsymbol{\phi}(x)\; e^{i\omega t} \tag{74}$$

Now substituting Eq. (74) in Eq. (73), the following relation has resulted.

$$\left(i\omega\mathbf{M}^{*}+\mathbf{C}^{*}\right)\boldsymbol{\phi}=0 \tag{75}$$

For obtaining the eigenvalues of Eq. (75), the determinant of the coefficient matrix should be set to zero as

$$\left|i\omega\mathbf{M}^{*}+\mathbf{C}^{*}\right|=0 \tag{76}$$

By solving Eq. (76), the values of ω are calculated and by substituting them into Eq. (75) mode shapes can be obtained. It is noted that for an elastic beam, Eq. (69) becomes

$$\mathbf{M}_{total}\ddot{\mathbf{U}}+\left(\mathbf{K}_{total}^{L}+\mathbf{K}_{total}^{NL}\right)\mathbf{U}=0 \tag{77}$$

By considering matrix $\mathbf{U}$ as

$$\mathbf{U}=\boldsymbol{\phi}(x)\; e^{i\omega t} \tag{78}$$

Equation (77) can be written as

$$\left(\left(\mathbf{K}_{total}^{L}+\mathbf{K}_{total}^{NL}\right)-\omega^{2}\mathbf{M}_{total}\right)\boldsymbol{\phi}(x)=0 \tag{79}$$

By equating the determinant of coefficient matrix of Eq. (79)to zero, natural frequencies and mode shapes can be calculated.

$$\left|\left(\mathbf{K}_{total}^{L}+\mathbf{K}_{total}^{NL}\right)-\omega^{2}\mathbf{M}_{total}\right|=0 \tag{80}$$

4.4 Buckling of Nano-Scaled Beams

For analyzing the viscoelastic buckling of structures (relaxation or creep model), an imperfection (initial displacement) is needed to be assumed. So, for deriving more general relations, imperfection effects are taken into account. Retaining the buckling related terms and the imperfection effects, Eq. (34) takes the form

$$
\begin{aligned}
\delta\Pi &= \zeta_1 \sum_{n=1}^{N} \delta\mathbf{U}^T\mathbf{Q}_n^T \left(\int_{V_n} \mathbf{B}_n^T : \mathbf{D} : \mathbf{B}_n dV \right) \mathbf{Q}_n\mathbf{U} \\
&+\zeta_1 \sum_{n=1}^{N} \delta\mathbf{U}^T\mathbf{Q}_n^T \left(\int_{V_n} \mathbf{T}_d\mathbf{B}_n^T : \mathbf{D} : \mathbf{B}_n dV \right) \mathbf{Q}_n\dot{\mathbf{U}} \\
&-\zeta_1 \sum_{n=1}^{N} \left(\delta\mathbf{U}^T\mathbf{Q}_n^T \left(\int_{V_n} \mathbf{B}^{0T}_n : \mathbf{D} : \mathbf{B}_n^0 dV \right) \mathbf{Q}_n\mathbf{U}_0 \right) \\
&+\zeta_2 \sum_{n=1}^{N}\sum_{m=1}^{N} \delta\mathbf{U}^T\mathbf{Q}_n^T \left(\int_{V_n}\int_{V_m} \alpha\left(\left|\mathbf{x}' - \mathbf{x}\right|, \tau\right) \mathbf{B}_n^T : \mathbf{D} : \mathbf{B}_m' dV'dV \right) \mathbf{Q}_m\mathbf{U} \\
&+\zeta_2 \sum_{n=1}^{N}\sum_{m=1}^{N} \delta\mathbf{U}^T\mathbf{Q}_n^T \left(\int_{V_n}\int_{V_m} \mathbf{T}_d\alpha\left(\left|\mathbf{x}' - \mathbf{x}\right|, \tau\right) \mathbf{B}_n^T : \mathbf{D} : \mathbf{B}_m' dV'dV \right) \mathbf{Q}_m\dot{\mathbf{U}} \\
&-\zeta_2 \sum_{n=1}^{N}\sum_{m=1}^{N} \left(\delta\mathbf{U}^T\mathbf{Q}_n^T \left(\int_{V_n}\int_{V_m} \alpha\left(\left|\mathbf{x}' - \mathbf{x}\right|, \tau\right) \mathbf{B}^{0T}{}_n : \mathbf{D} : \mathbf{B}_m^{0\prime} dV'dV \right) \mathbf{Q}_m\mathbf{U}_0 \right) \\
&-\sum_{n=1}^{N} \left(\delta\mathbf{U}^T\mathbf{Q}_n^T \left(\int_{V_n} \mathbf{B}^{gT}_n : \sigma_0 : \mathbf{B}_n^g dV \right) \mathbf{Q}_n\mathbf{U} \right) = 0
\end{aligned}
\tag{81}
$$

In addition to Eqs. (54)–(58) and (64), (65), (67) and (68), initial stiffness matrix and geometric stiffness matrix are defined as follows

$$
\mathbf{K}_n^{L^0} = \zeta_1 \left(I_{yy} \int_{-1}^{1} \mathbf{B}^{0bT}_n(\xi)\, E\mathbf{B}^{0b}_n(\xi) \det\mathbf{J}(\xi)\, d\xi \right) \tag{82}
$$

$$
\begin{aligned}
\mathbf{K}_{nm}^{NL^0} &= \zeta_2 [bh\, I_{yy} \int_{-1}^{1}\int_{-1}^{1} \alpha\left(\left|x'(\xi) - x(\xi)\right|, \tau\right)\, \mathbf{B}^{0bT}{}_n(\xi)\, E\mathbf{B}^{0b}{}_m(\xi') \\
&\times \det\left(\mathbf{J}(\xi')\right) \det\left(\mathbf{J}(\xi)\right) d\xi' d\xi
\end{aligned}
\tag{83}
$$

$$\mathbf{K}_n^g = \left(bh \int_{-1}^{1} \mathbf{B}_n^{gT}(\xi)\,\sigma_0 \mathbf{B}_n^g(\xi)\det(\mathbf{J}(\xi))\,d\xi \right) \tag{84}$$

It is known that, for an Euler-Bernoulli beam, the nonlinear strain can be assumed as

$$\varepsilon_{nl} = \frac{1}{2}\left(\frac{\partial w}{\partial x}\right)^2 \tag{85}$$

B_n^g is consistent with the nonlinear strain form. Taking into account the defined parameters, the governing viscoelastic buckling equation becomes

$$\begin{aligned} &\left(\mathbf{C}_{total}^{L} + \mathbf{C}_{total}^{NL}\right)\dot{\mathbf{U}} + \left(\mathbf{K}_{total}^{L} + \mathbf{K}_{total}^{NL}\right)\mathbf{U} \\ &- \left(\mathbf{K}_{total}^{L^0} + \mathbf{K}_{total}^{NL^0}\right)\mathbf{U}_0 - \mathbf{K}_{total}^{g}\mathbf{U} = 0 \end{aligned} \tag{86}$$

By numerically solving Eq. (86), the viscoelastic buckling solutions can be acquired. By viscoelastic buckling solution, we mean the condition in which the compressive load is kept constant and as time passes, the displacements increase. There comes a time in which the displacements exceed the assumed buckling condition. This time is called the viscoelastic buckling time. Also, by keeping out the time-related terms and imperfections in Eq. (86) the eigenvalue problem for the elastic buckling of perfect beam would result.

$$\left(\mathbf{K}_{total}^{L} + \mathbf{K}_{total}^{NL}\right)\mathbf{U} - \mathbf{K}_{total}^{g}\mathbf{U} = 0 \tag{87}$$

5 Nano-Scaled Plates

The appropriate formulations are to be introduced for analyzing the bending, vibration, and buckling of nano-scaled plates. As it has been assumed with respect to the beams, in the plate case also the classical plate theory is only taken into account, and other theories can be added to the formulation by following a similar procedure.

5.1 Applying Boundary Conditions for Nano-Scaled Plates

One of the main advantages of finite element integral elasticity approach is the straightforwardness of applying the boundary conditions for a plate (Fig. 7). It is noted that the boundary conditions are applied as they are imposed in conventional finite element methods of classical elasticity. For example, the following commonly used boundary conditions are outlined.

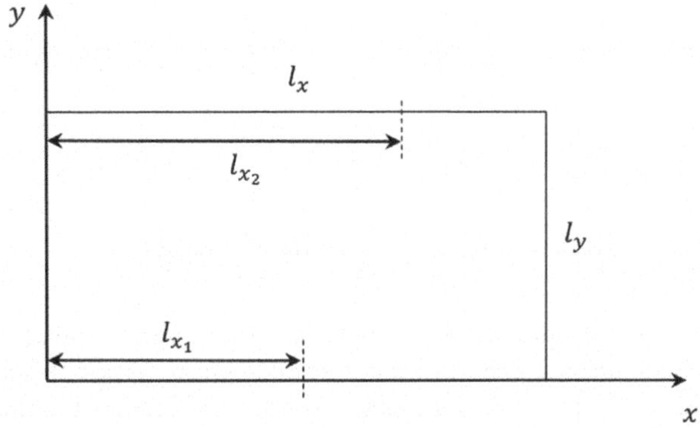

Fig. 7 Plate dimensions

All-sides simply supported plate:

$$\begin{aligned} w(x,y) &= 0 \quad \text{at} \quad x = 0, l_x \\ w(x,y) &= 0 \quad \text{at} \quad x = 0, l_y \end{aligned} \tag{88}$$

All-sides clamped plate:

$$\begin{aligned} w\,(x,y) = w_{,x}\,(x,y) &= 0 \quad \text{at} \quad x = 0, l_x \\ w\,(x,y) = w_{,x}\,(x,y) &= 0 \quad \text{at} \quad x = 0, l_y \end{aligned} \tag{89}$$

Two-sides clamped, two-sides free (opposite sides):

$$w\,(x,y) = w_{,x}\,(x,y) = 0 \quad \text{at} \quad x = 0, l_x \tag{90}$$

Two-sides simply supported, two-sides clamped (opposite sides):

$$\begin{aligned} w\,(x,y) &== 0 \quad \text{at} \quad x = 0, l_x \\ w\,(x,y) = w_{,x}\,(x,y) &= 0 \quad \text{at} \quad x = 0, l_y \end{aligned} \tag{91}$$

Interestingly enough, the partial boundary conditions can also be applied fairly easily and with little effort using the current method, for instance, some partial boundary conditions are expressed below.

Two-sides clamped, the other two sides each partially simply supported:

$$\begin{aligned} w\,(x,y) = w_{,x}\,(x,y) = 0 \quad &\text{at} \quad x = 0, l_x \\ w\,(x,y) = 0 \quad &\text{at} \quad y = 0,\ x = 0 - l_{x1} \\ w\,(x,y) = 0 \quad &\text{at} \quad y = l_y,\ x = 0 - l_{x2} \end{aligned} \tag{92}$$

Two-sides simply supported, the other two sides each partially clamped:

$$\begin{aligned} w(x,y) &= 0 \quad \text{at} \quad x = 0, l_x \\ w(x,y) = w_{,x}(x,y) &= 0 \quad \text{at} \quad y = 0,\ x = 0 - l_{x1} \\ w(x,y) = w_{,x}(x,y) &= 0 \quad \text{at} \quad y = l_y,\ x = 0 - l_{x2} \end{aligned} \tag{93}$$

5.2 Bending of Nano-Scaled Plates

Bending formulations considering the nonlocal integral theory and classical plate theory are presented in the current section. Through the classical plate theory, the following relation is assumed for the strain field of a plate.

$$\boldsymbol{\varepsilon}(\mathbf{x}) = \begin{Bmatrix} \varepsilon_x \\ \varepsilon_y \\ \gamma_{xy} \end{Bmatrix} = \begin{Bmatrix} u_{0,x} \\ v_{0,y} \\ u_{0,x} + v_{0,y} \end{Bmatrix} - z \begin{Bmatrix} w_{,xx} \\ w_{,yy} \\ 2w_{,xy} \end{Bmatrix} \tag{94}$$

in which, ε_x, ε_y, and γ_{xy} are the longitudinal strain, transverse strain, and in-plane shear strain, respectively. Also, u_0, v_0, and w are the displacements of mid-plane in the x-, y-, and z-directions. By considering the pure bending condition, Eq. (94) becomes

$$\boldsymbol{\varepsilon}(\mathbf{x}) = \begin{Bmatrix} \varepsilon_x \\ \varepsilon_y \\ \gamma_{xy} \end{Bmatrix} = -z \begin{Bmatrix} w_{,xx} \\ w_{,yy} \\ 2w_{,xy} \end{Bmatrix} = \mathbf{B}_n(\mathbf{x})\,\mathbf{d}_n = z\mathbf{B}_n^p(\mathbf{x})\,\mathbf{d}_n \tag{95}$$

Substituting Eq. (95) into Eq. (34), using Eq. (33) and considering the bending-related terms, for an elastic case, Eq. (34) becomes

$$\begin{aligned} \delta\Pi &= \zeta_1 \sum_{n=1}^{N} \delta\mathbf{U}^T\mathbf{Q}_n^T \left(\int_{V_n} \mathbf{B}_n^T : \mathbf{D} : \mathbf{B}_n dV \right) \mathbf{Q}_n\mathbf{U} \\ &+ \zeta_2 \sum_{n=1}^{N}\sum_{m=1}^{N} \delta\mathbf{U}^T\mathbf{Q}_n^T \left(\int_{V_n}\int_{V_m} \alpha\left(\left|\mathbf{x}' - \mathbf{x}\right|, \tau\right) \mathbf{B}_n^T : \mathbf{D} : \mathbf{B}_m' dV' dV \right) \mathbf{Q}_m\mathbf{U} \\ &- \sum_{n=1}^{N} \left(\delta\mathbf{U}^T\mathbf{Q}_n^T \int_{V_n} \mathbf{N}_n^T.\bar{\mathbf{b}}\, dV \right) - \sum_{n=1}^{N} \left(\delta\mathbf{U}^T\mathbf{Q}_n^T \int_{S_{tn}} \mathbf{N}_n^T.\bar{\mathbf{t}}\, dS \right) = 0 \end{aligned} \tag{96}$$

Substituting the last term of Eq. (95) into Eq. (96) the following relations are defined for local and nonlocal stiffness matrices of the n-th element.

$$\mathbf{K}_n^L = \zeta_1 \left(\int_{A_{elem}} \int_h \mathbf{B}_n^{pT} : \mathbf{D} : \mathbf{B}_n^p \det(\mathbf{J})\, z^2 dz\, dA_{elem} \right) \tag{97}$$

$$\mathbf{K}_{nm}^{NL} = \zeta_2 [\int_{A_{elem}} \int_h \int_{A'_{elem}} \int_{h'} \alpha \left(\left| \mathbf{x}'(\xi,\eta) - \mathbf{x}(\xi,\eta) \right|, \tau \right) z'^2\, \mathbf{B}_n^{pT} : \mathbf{D} : \mathbf{B}_m^{p'} \times \det(\mathbf{J}) \det\left(\mathbf{J}'\right) dz'\, dA'_{elem} dz\, dA_{elem}] \tag{98}$$

The total stiffness matrices are also given as Eqs. (57) and (58). Also, the bending governing equations (similar to Eq. (60)) can be solved by the same procedure as that explained in Sect. 4.2.

5.3 Vibration of Nano-Scaled Plates

For analyzing the vibration of nano-scaled plates, the general form of Eq. (62) could be used. For a nano-scaled plate, the local and nonlocal stiffness matrices can be expressed as Eqs. (97) and (98). Also, noting Eq. (62) the element mass matrix and damping matrices can be obtained as

$$\mathbf{M}_n = \int_{A_{elem}} \int_h \mathbf{N}_n^T . (\rho \mathbf{N}_n) \det(\mathbf{J})\, dz dA_{elem} \tag{99}$$

$$\mathbf{C}_n^L = \zeta_1 \left(\int_{A_{elem}} \int_h \mathbf{T}_d \mathbf{B}_n^{pT} : \mathbf{D} : \mathbf{B}_n^p \det(\mathbf{J})\, z^2 dz\, dA_{elem} \right) \tag{100}$$

$$\mathbf{C}_{nm}^{NL} = \zeta_2 [\int_{A_{elem}} \int_h \int_{A'_{elem}} \int_{h'} \alpha \left(\left| \mathbf{x}'(\xi,\eta) - \mathbf{x}(\xi,\eta) \right|, \tau \right) \mathbf{T}_d\, z'^2 \mathbf{B}_n^{pT} : \mathbf{D} : \mathbf{B}_m^{p'} \times \det(\mathbf{J}) \det\left(\mathbf{J}'\right) dz'\, dA'_{elem} dz\, dA_{elem}] \tag{101}$$

The total structural matrices can also be calculated by Eqs. (66)–(68). Also, the eigenvalues and eigenfunctions of vibration might also be given by applying the same procedure as that explained in Sect. 4.3.

5.4 Buckling of Nano-Scaled Plates

Buckling analysis of nano-scaled plates can also be carried out by considering the buckling form of general variational Eq. (34) as Eq. (81). The nonlinear strain for a plate is assumed as

$$\varepsilon_{nl} = \frac{1}{2}\begin{Bmatrix} w_{,x}^2 \\ w_{,y}^2 \\ 2w_{,x}w_{,y} \end{Bmatrix} \tag{102}$$

It is seen that the stiffness matrices are given by Eqs. (97) and (98), and geometric stiffness matrix and initial stiffness matrices are given as follow

$$\mathbf{K}_n^{L^0} = \zeta_1\left(\int_{A_{elem}}\int_h \mathbf{B}^{0p}{}_n^T : \mathbf{D} : \mathbf{B}^{0p}_n \det(\mathbf{J})\, z^2 dz\, dA_{elem}\right) \tag{103}$$

$$K_{nm}^{NL^0} = \zeta_2[\int_{A_{elem}}\int_h\int_{A'_{elem}}\int_{h'} \alpha\left(\left|\mathbf{x}'(\xi,\eta) - \mathbf{x}(\xi,\eta)\right|, \tau\right) z'^2\, \mathbf{B}^{0p}{}_n^T : \mathbf{D} : \mathbf{B}^{0p'}_m$$
$$\times \det(\mathbf{J}) \det\left(\mathbf{J}'\right) dz'\, dA'_{elem} dz\, dA_{elem}] \tag{104}$$

$$\mathbf{K}_n^g = \int_{A_{elem}}\int_h \mathbf{B}^{g}{}_n^T : \boldsymbol{\sigma}_0 : \mathbf{B}_n^g dz\, dA_{elem} \tag{105}$$

in which, $\boldsymbol{\sigma}_0$ is the initial stress which has the following form

$$\boldsymbol{\sigma}_0 = \begin{bmatrix} \sigma_{xx}^0 & \sigma_{xy}^0 \\ \sigma_{yx}^0 & \sigma_{yy}^0 \end{bmatrix} \tag{106}$$

Considering the mentioned equations, Eqs. (86) and/or (87) are used for nano-scaled plates to solve the viscoelastic/elastic buckling problems.

6 Numerical Examples and Discussions

In this section, the bending, buckling, and vibration of nano-scaled beams and plates are studied using the presented finite element nonlocal integral method. Due to a better correlation between current results with those of nonlocal differential elasticity, the Bessel kernel is used in the following analysis. The assumed kernel is [6, 13]

$$\bar{\alpha}\left(\left|\mathbf{x}-\mathbf{x}'\right|,\ \tau\right)=2\pi(e_0a)^2K_0\left(\frac{\left|\mathbf{x}-\mathbf{x}'\right|}{e_0a}\right) \tag{107}$$

6.1 *Elastic Beam and Plate Bending*

In this section, the bending of nano-beams and nano-plates are analyzed using some examples. It is noted that in the current section, the constitutive equation (2) (one-phase nonlocal integral elasticity theory) has been adopted for implementing the finite element nonlocal integral elasticity method.

6.1.1 Bending of Nano-Scaled Beams

The bending of nonlocal Euler-Bernoulli nano-scaled beams is investigated. Results are obtained using the approach explained in Sect. 4.2. The Hermite element (see Sect. 3.3) is assumed in the analysis, and for the numerical integration, the Gauss-Legendre quadrature rule is used by considering three Gauss points.

Figure 8 shows the non-dimensional maximum deflection (w/w_L) for two-sides simply supported beam, under $\bar{k}=1$ and $\bar{q}=1$ loading. $\bar{k}=Fl^2/EI$ is the non-dimensional central point load parameter, and $\bar{q}=ql^3/EI$ is the non-dimensional

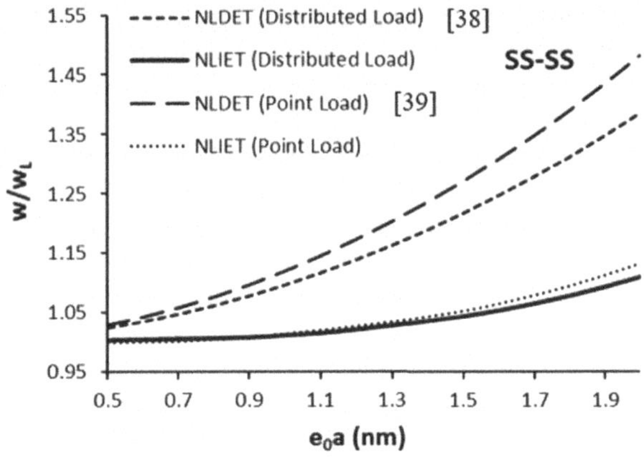

Fig. 8 Variations of non-dimensional deflection with the nonlocal parameter for both-sides simply supported Euler-Bernoulli nano-scale beam [6] (l = 10 nm and h = 0.1 nm)

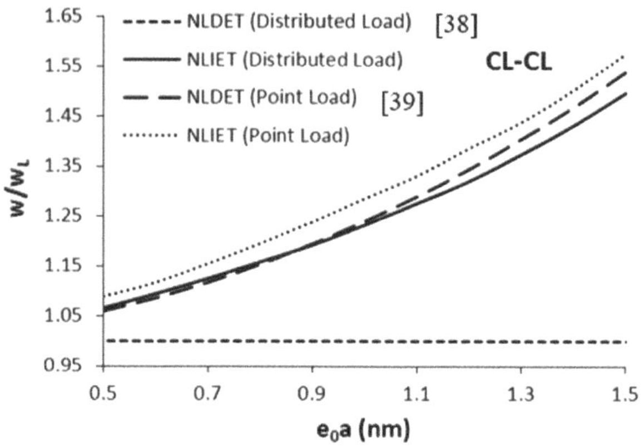

Fig. 9 Variations of non-dimensional deflection with the nonlocal parameter for both-sides clamped Euler-Bernoulli nano-scale beam [6] ($l = 10$ nm and $h = 0.1$ nm)

uniformly-distributed load parameter. The results of the finite element nonlocal integral elasticity (nonlocal integral elasticity theory NLIET) are compared with those of Wang et al. [38] (for distributed load) and Wang and Liew [39] (for point load) based on the nonlocal differential elasticity theory (NLDET). It is observed that for a beam with simply supported boundary conditions subjected to a point load, the nonlocal differential elasticity theory could not predict the bending deflection properly. However, both the differential and integral theories show a rise in deflection by increasing the nonlocal parameter for uniformly-distributed load case.

Figure 9 shows the variation of maximum deflection with the nonlocal parameter for two-sides clamped boundary condition ($\bar{k} = 1$, $\bar{q} = 1$). It is seen that in this case the nonlocal differential elasticity has not captured the effect of nonlocality by considering distributed load. However, the nonlocal integral theory shows an increase in deflection by considering the nonlocal parameter, for both loading conditions.

6.1.2 Bending of Nano-Scaled Plates

Figure 10 show the variations of deflection for the midpoint of a nano-scaled simply supported plate under the uniformly-distributed loading $q_0 = 1$ nN/nm. The assumed plate have properties of $l = 10$ nm, $E = 30 \times 10^3$ nN/nm^2, and $\nu = 0.3$. The Adini-clough element type is used for modeling the plate and for numerical integration, the Gauss-Legendre quadrature method assuming three integration points in each direction is considered. Also, the results of the current study have been compared with those of Aghababaei and Reddy [40] based on the nonlocal differential elasticity theory. As it is seen, by increasing the nonlocal parameter, both the integral and differential theories predict the decrease in the stiffness of the plate. However, the increase in deflection is more pronounced for the nonlocal integral elasticity method. It might have occurred because the nonlocal differential elasticity theory has been extracted from the general integral form and cannot satisfy the force boundary conditions properly [6].

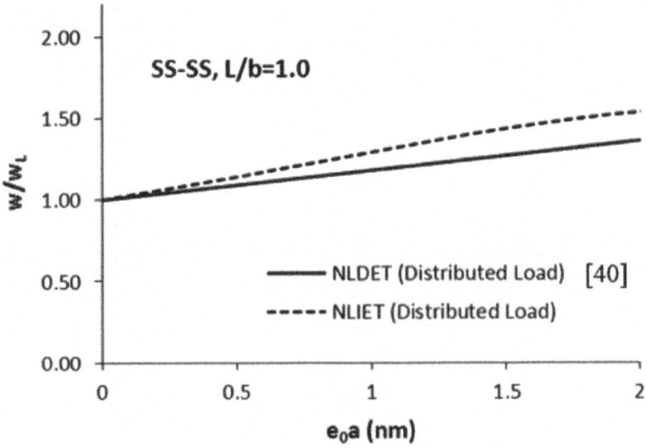

Fig. 10 Variations of non-dimensional deflection with the nonlocal parameter for all-sides simply supported nano-scale plate ($l/b = 1$) under distributed load

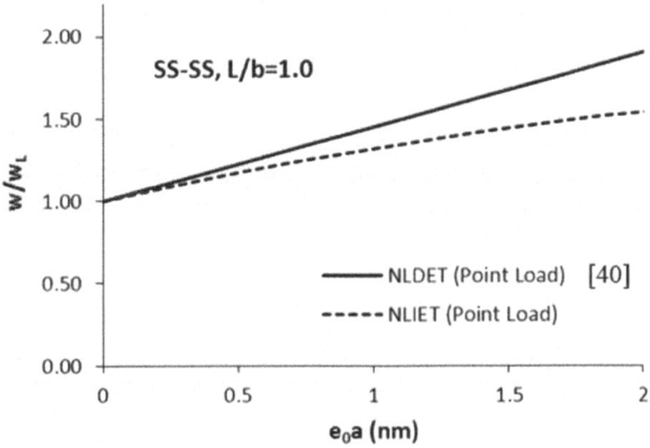

Fig. 11 Variations of non-dimensional deflection with the nonlocal parameter for all-sides simply supported nano-scale plate under point load

Figure 11 shows the variation of midpoint deflection of the all-sides simply supported nano-scaled plate with nonlocal parameter under point load $F_0 = 1$ nN. It is seen that the increase in deflection is more pronounced for nonlocal differential elasticity case in comparison with the nonlocal integral elasticity.

6.2 Elastic Beam and Plate Vibration

Free vibration of nano-beams and nano-plates are studied using the finite element nonlocal integral elasticity method and considering the two-phase nonlocal constitutive equation (Eq. (4)). The convergence study is carried out for obtaining the optimum number of elements. The results are then compared with those available in the literature, and finally, the effects of various parameters are investigated on free vibration behavior. For the subject beam $E = 1$ TPa and $\rho = 2000$ kg/m^3, and for the plate $E = 1$ TPa, $\nu = 0.16$ and $\rho = 2250$ kg/m^3.

Table 1 Convergence study of natural frequency ($\bar{\omega} = l\sqrt{\omega/C_0}$, $C_0 = \sqrt{EI/\rho A}$) by nonlocal Euler-Bernoulli beam theory [13]

Number of elements	Simply supported beam				Clamped beam			
	$\tau = 0$	$\tau = 0.01$	$\tau = 0.1$	$\tau = 0.2$	$\tau = 0$	$\tau = 0.01$	$\tau = 0.1$	$\tau = 0.2$
10	3.1417	3.1897	3.1772	3.1642	4.7301	4.8198	4.7956	4.7642
20	3.1416	3.1388	3.1351	3.0801	4.7300	4.7258	4.7139	4.6050
50	3.1416	3.1103	3.0482	2.9695	4.7300	4.6826	4.5685	4.3934
100	3.1416	3.0421	2.9561	2.8154	4.7300	4.5786	4.4126	4.2259
200	3.1416	2.9532	2.8738	2.7956	4.7300	4.4789	4.2661	4.0938
300	3.1416	2.9315	2.8339	2.7857	4.7300	4.4026	4.2053	4.0350
400	3.1416	2.9259	2.8332	2.7851	4.7300	4.4002	4.2018	4.0303

6.2.1 Free Vibration of Nano-Scaled Beams

The free vibration of nano-scaled beams is investigated through the assumption of Euler-Bernoulli beam theory. Thus, the procedure of Sect. 4.3 has been adopted for elastic beams, and Hermite beam elements have been considered (see Sect. 3.3). For the numerical integration, three integration points are used for Gauss-Legendre quadrature method. Dimensions of the beam are as $l = 10$ nm, $h = 1$ nm and $b = 0.5$ nm.

Table 1 shows the convergence study for free vibration of nano-scaled Euler-Bernoulli beam considering two types of boundary conditions. It is seen that the convergence rate for the local case ($\tau = 0$) is faster than the nonlocal case. This behavior can be explained by further investigation of the nonlocal kernel function characteristics. For a given reference point, the value of kernel function is maximum, i.e.g.oes to infinity, on that point and it decreases by moving further from it. So, for accurately capturing the nonlocal effects near the reference point, several elements are needed. Besides, by decreasing the size of elements the accuracy of numerical integrations increases.

Table 2 shows the comparison of the natural frequencies based on the nonlocal integral theory with those of differential elasticity theory which have been reported by Lu et al. [41], Reddy [42] and Ghannadpour [43]. It is seen that the nonlocal differential theory cannot predict the softening effect of the nonlocal parameter for the cantilever beam (considering the fundamental natural frequency). As it has been said, the current discrepancy might have been caused by the fact that the nonlocal differential theory is extracted from the integral one under certain assumptions and in the region far from the boundaries. However, there has been a generally good agreement between the results.

Figure 12 shows the variations of the fundamental natural frequency with the nonlocal parameter considering different length to thickness ratio for cantilever boundary conditions. It is seen that, by increasing the nonlocal parameter value, natural fre-

Table 2 Comparison of non-dimensional natural frequencies ($\bar{\omega} = l\sqrt{\omega/C_0}$, $C_0 = \sqrt{EI/\rho A}$) for nonlocal Euler-Bernoulli beam

Ref.	τ	Simply supported beam	Clamped beam	Cantilever beam
Present	0	3.14	4.73	1.88
	0.1	2.83	4.2	1.67
	0.2	2.78	4.03	1.63
	0.5	2.56	3.65	1.49
	0.7	2.49	3.55	1.45
Lu et al. [41]	0	3.14	4.73	1.88
	0.1	3.07	4.59	1.88
	0.2	2.89	4.28	1.89
	0.5	2.30	3.31	2.02
	0.7	2.02	2.89	–
Reddy [42]	0	3.14	–	–
	0.1	3.00	–	–
	0.2	2.87	–	–
Ghannadpour et al. [43]	0	3.14	4.73	1.88
	0.5	2.30	3.31	2.02
	0.7	2.02	2.89	–

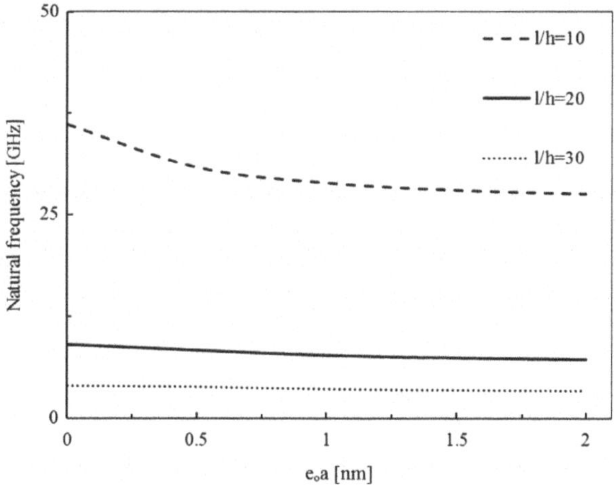

Fig. 12 Fundamental frequency variations with the nonlocal parameter considering different length to thickness ratios for cantilever nano-scaled beam [13]

quency decreases. Also, it is observed that for larger values of l/h, the sensitivity of frequency to e_0a decreases. This might be due to the fact that, the effects of nonlocality are more pronounced near the boundaries, so for shorter beams these effects can increase.

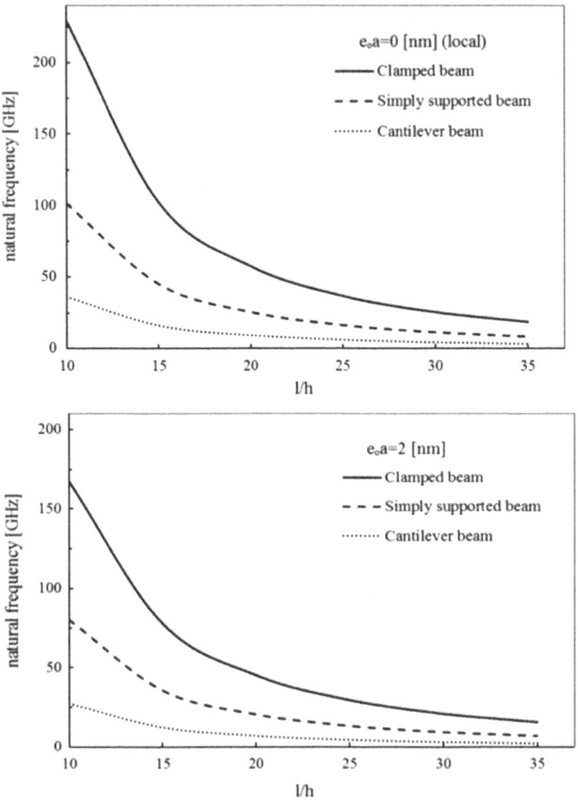

Fig. 13 Variations of the fundamental natural frequency with length to thickness ratio for nano-scaled beams [13]

Figure 13 shows the effect of length to thickness ratio on the natural frequency considering different nonlocal parameters and various boundary conditions. It is seen that regardless of local/nonlocal effects, the increase in l/h has led to the reduction of natural frequencies. The highest reduction belongs to both-sides clamped boundary condition and the lowest to cantilever.

6.2.2 Free Vibration of Nano-Scaled Plates

The classical plate theory is used to investigate the free vibration of nano-scaled plates [16]. The procedure of Sect. 5.3, in conjunction with Adini-Clough element types (3.3), have been adopted. Three points in each direction have been assumed for numerical integration. It is noted that the length and thickness of the square plate are 10 nm and 0.34 nm, respectively.

Convergence study of a square plate considering $e_0a = 1$ nm and all-sides simply supported boundary condition are shown in Table 3. Equal size square elements have been used for mesh allocation. It is seen that the results converge for 45×45 elements.

Table 3 Convergence study of free vibration considering a simply-supported square nano-scaled plate ($e_0a = 1$ nm)

Number of square elements	Fundamental natural frequency [GHz]
3 × 3	71.5887
7 × 7	59.9211
11 × 11	55.953
15 × 15	54.1543
19 × 19	53.1874
23 × 23	52.6066
27 × 27	52.2302
31 × 31	51.9721
33 × 33	51.8128
37 × 37	51.789
41 × 41	51.788
45 × 45	51.788

Table 4 Comparison of the results for free vibration of a square nano-scaled plate ($\omega_{ND} = \omega l^2\sqrt{\rho h/D}$)

Boundary conditions	Present ($e_0a = 1$ nm)	Reference [44]	Reference [45]	Reference [46]	Reference [46] MD (zigzag)	Reference [46] MD (armchair)
Simply supported	15.52	18.02	18.01	18.88	7.62	17.84
Clamped	28.16	–	–	33.78	34.36	34.84

Table 4 shows the comparison of non-dimensional fundamental natural frequency based on finite element integral nonlocal elasticity with those of nonlocal differential theory. Results of Pradhan and Phadikar [44] are obtained by using the Navier solution and nonlocal classical plate theory. Murmu and Pradhan [45] have also used nonlocal classical plate theory whereas Ansari et al. [46] have implemented nonlocal first-order shear deformation theory. Besides, the results of molecular dynamics have been included in the table. As it is predictable, considering the non-locality leads to some discrepancies between the results which are relatively more pronounced for all-sides simply supported boundary condition. By comparing the results of current study with those of molecular dynamics, it can be concluded that for the subject case the value of nonlocal parameter lies between 0 and 1 nm.

Figure 14 shows the effect of nonlocal parameter on the natural frequency of square nano-scaled plate considering different length to thickness ratios. As observed, by increasing the nonlocal parameter natural frequency decrease. This softening effect is more pronounced for the lower values of the plate length to thickness ratio, i.e., the variation of natural frequency with l/h is less noticeable for larger plates. This might be due to the importance of nonlocal effects near the boundaries, that is to say,

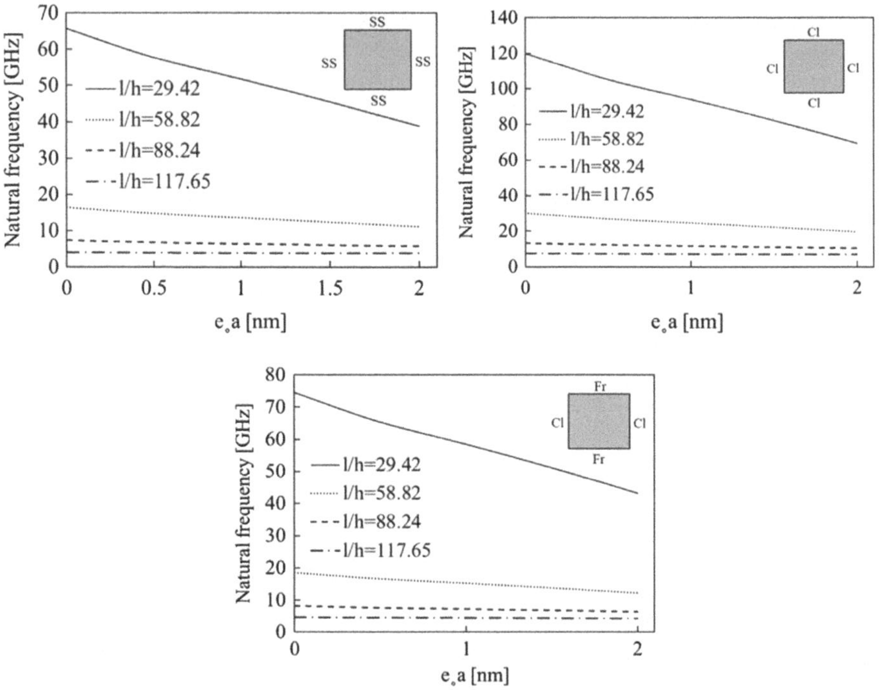

Fig. 14 Effect of the nonlocal parameter on the natural frequency of square plate with all-sides simply supported, all-sides clamped and cantilever boundary condition ($h = 0.34$ nm) [16]

for the smaller plates, the boundary conditions play an important role in nonlocal characteristics of the structure. It is noted that by increasing the length of the structure the e_0a/l value diminishes for a given value of e_0a.

Figure 15 shows the variations of the natural frequency with the length of the square plate for different boundary conditions considering $e_0a = 1$ nm. It is seen that, by increasing the length the natural frequency decreases. Also, decrease in natural frequency is more steep for all-sides clamped boundary condition in comparison with all-sides simply supported and clamped free conditions.

Figure 16 shows the effect of aspect ratio on free vibration of rectangular nano-scaled plate considering all-sides simply supported boundary condition. As it is seen, by increasing the aspect ratio, the natural frequencies decrease.

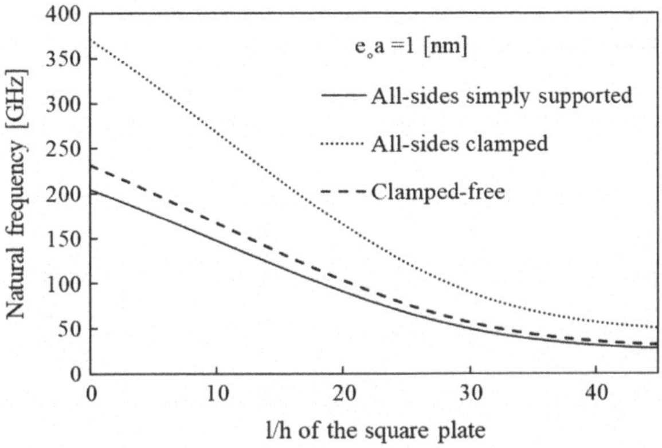

Fig. 15 Effect of the length of the square plate on the natural frequency for $e_0a = 1$ nm and $h = 0.34$ nm

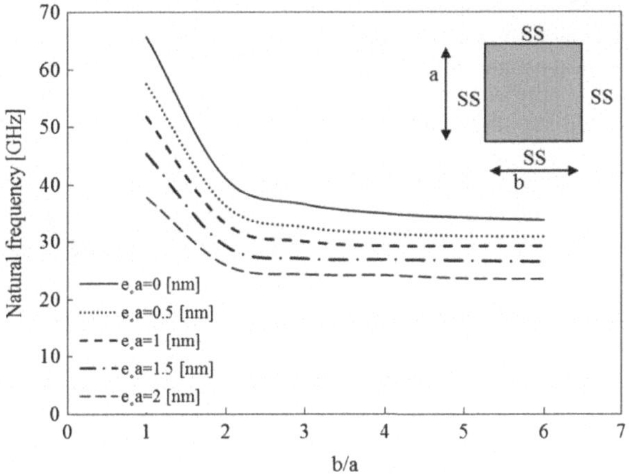

Fig. 16 Effect of the aspect ratio of the rectangular plate on the natural frequency considering all-sides simply supported boundary condition [16]

6.3 *Elastic Beam and Plate Buckling*

The buckling of nano-beams and nano-plates are investigated employing the finite element nonlocal integral elasticity method and considering the single-phase nonlocal constitutive equation (Eq. (2)).

6.3.1 Buckling of Nano-Scaled Beams

The procedure of Sect. 4.4 has been employed for the buckling analysis of Euler-Bernoulli nano-scaled beams. Hermite beam elements (see Sect. 3.3) are used and the beam properties are $E = 1$ TPa, $l = 20$ nm, $b = t = 1$ nm. Table 5 shows the convergence study for a two-sides clamped beam considering $e_0a = 1$ nm and 2 nm.

Figures 17 and 18 show the variations of buckling load ratio with the nonlocal parameter for both-sides simply supported and clamped boundary conditions, respectively. Also, the results of the current study have been compared with those

Table 5 Convergence study for buckling of nonlocal Euler-Bernoulli beam considering two-sides clamped boundary condition

Number of elements	$e_0a = 1$ nm	$e_0a = 2$ nm
10	12.3789	7.4291
20	9.2301	6.6915
30	8.4526	6.5072
50	7.9771	6.3892
100	7.7294	6.3213
200	7.6488	6.2952
300	7.6296	6.2879
400	7.6214	6.2845

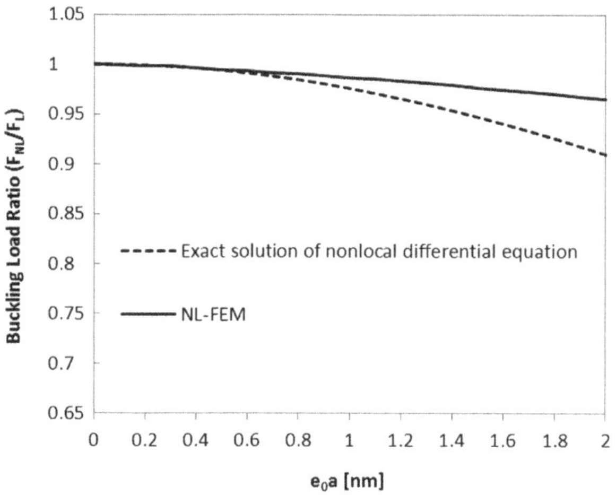

Fig. 17 Variations of buckling load ratio with the nonlocal parameter for both sides simply supported Euler-Bernoulli beam [7]

based on the nonlocal differential elasticity [7] considering the Timoshenko beam theory. It is seen that, by increasing the nonlocal parameter buckling load decreases.

Figures 19 and 20 show the variations of buckling load with length to thickness ratio for different boundary conditions considering $e_0a = 1$ nm. It is observed that, by increasing the length to thickness ratio, the discrepancy between the results of the current method with those of local elasticity and differential elasticity theory decreases. Also, it is seen that the effect of the nonlocal parameter is more pronounced for shorter beams.

6.3.2 Buckling of Nano-Scaled Plates

The buckling of classical plates are analyzed using the finite element nonlocal integral method, and the results are compared with those of nonlocal differential elasticity theory. The Adini-Clough element has been used for the analysis (see Sect. 3.3).

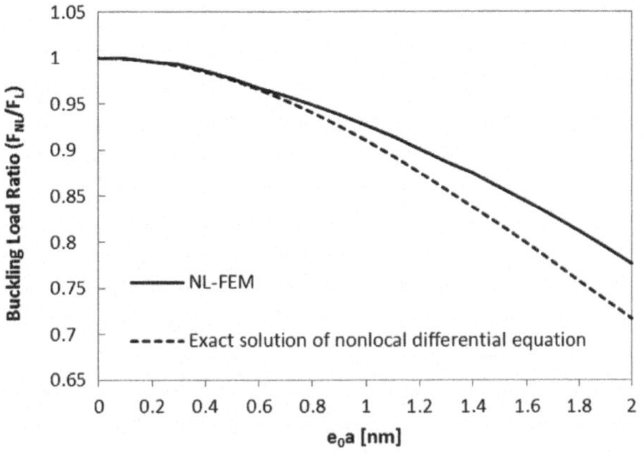

Fig. 18 Variations of buckling load ratio with the nonlocal parameter for both sides clamped Euler-Bernoulli beam [7]

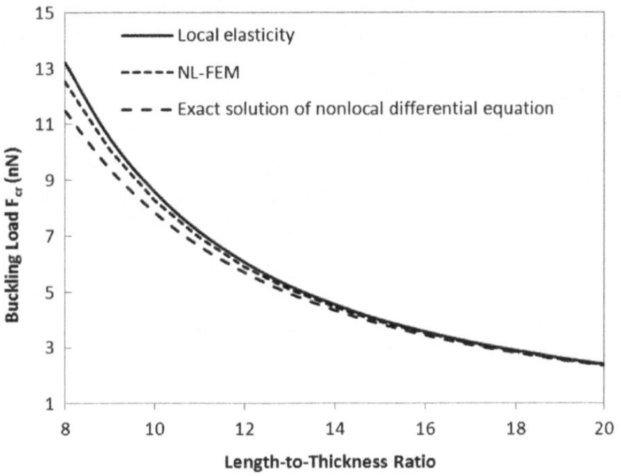

Fig. 19 Variations of buckling load with length to thickness ratio for $e_0a = 1$ nm considering both sides simply supported beam [7]

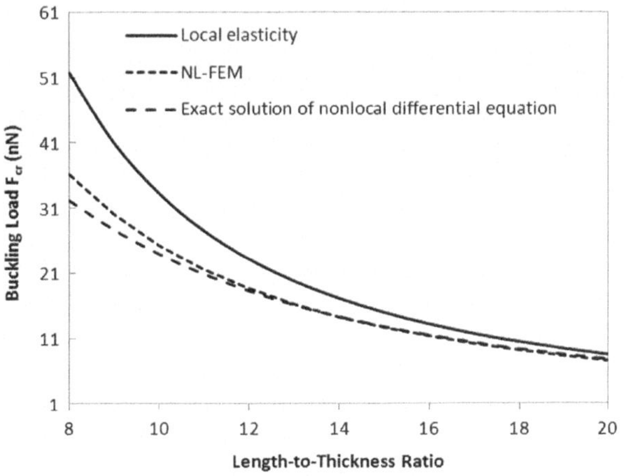

Fig. 20 Variations of buckling load with length to thickness ratio for $e_0a = 1$ nm considering both sides clamped beam [7]

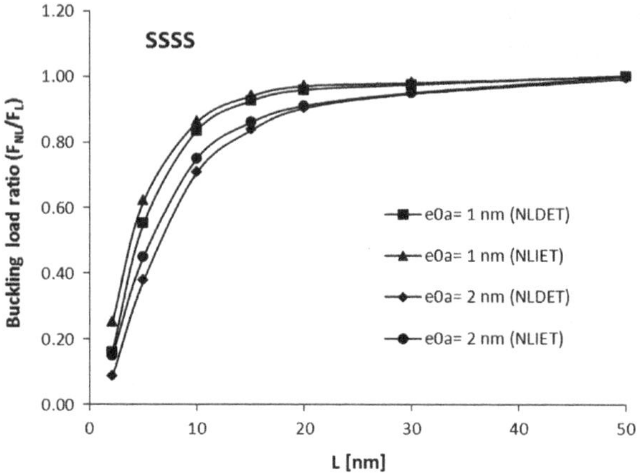

Fig. 21 Variations of non-dimensional buckling load with the length of the square plate considering all-sides simply supported boundary conditions

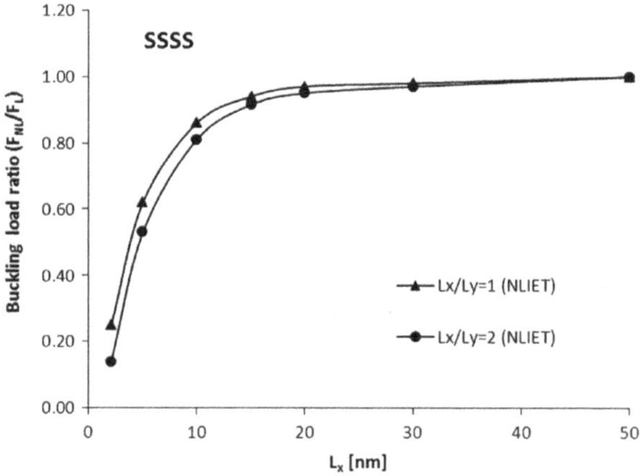

Fig. 22 Variations of non-dimensional buckling load with the length of the plate for different values of aspect ratio considering all-sides simply supported boundary condition

Figure 21 shows the variations of buckling load ratio with the length of square plate for all-sides simply supported boundary condition considering $e_0a = 1$ nm and $e_0a = 2$ nm. The results of the current study have been compared with those obtained by Levy method based on nonlocal differential elasticity theory [47]. It is seen that by increasing the length of the plate, the nonlocal effects decrease. These effects in smaller dimensions result in more reduction of buckling loads. In other hands, by increasing the e_0a discrepancy between the results of the current study with those of nonlocal differential elasticity increase. This might be due to the fact that by increasing the value of e_0a, the effects of boundary conditions can play a more important part.

Figure 22 shows the variations of buckling load ratio with the length of the plate for $l_x/l_y = 1$ and $l_x/l_y = 2$ considering $e_0a = 1$ nm and all-sides simply supported boundary conditions.

Figure 23 shows the effect of nonlocal parameter on the buckling ratio of nano-scaled plate considering all sides simply supported, and all sides clamped boundary

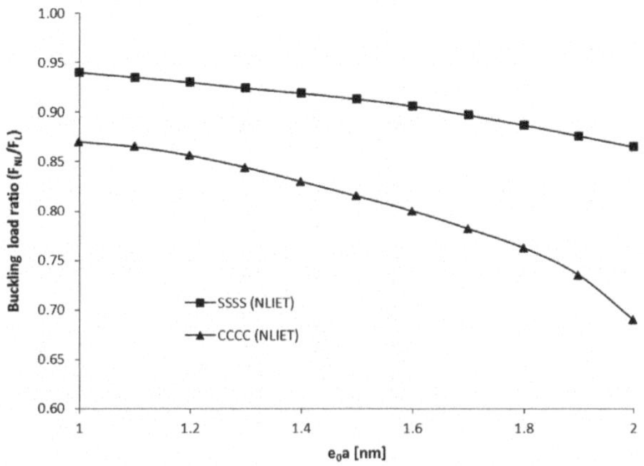

Fig. 23 Effect of the nonlocal parameter on the buckling load ratio for all-sides clamped and all-sides simply supported nano-scaled plate ($l_x = l_y = 1.5$ nm)

conditions. It is seen that for all sides clamped boundary condition, the reduction in buckling load is more pronounced in comparison with the all sides simply supported boundary conditions. It is due to the fact that, stronger boundary conditions lead to more pronounced nonlocal effects near the boundaries.

6.4 *Viscoelastic Free Vibration*

The free vibration of nano-scaled beams and plates are studied using the finite element nonlocal integral elasticity approach considering the Kelvin-Voigt viscoelastic model. Properties of the beam are $E = 1$ TPa and $\rho = 2000$ kg/m^3, and for the plate $E = 1$ TPa, $\nu = 0.16$ and $\rho = 2250$ kg/m^3.

6.4.1 Free Vibration of Viscoelastic Nano-Scaled Beams

For obtaining the results of the viscoelastic free vibration of nano-scaled beams, the procedure explained in Sect. 4.3 have been adopted. Hermite type elements have been used for meshing the beam (see Sect. 3.3). The complex eigenvalues of the current study based on the finite element nonlocal integral method are compared with those of Lu et al. [41] and Lei et al. [20] based on the nonlocal differential elasticity theory and those obtained by Abaqus/CAE commercial software (local viscoelasticity) in Table 6 [15]. The results have been extracted for different values of nonlocal and viscoelastic parameters considering various boundary conditions. The complex eigenvalues obtained by the finite element integral nonlocal method is lower than those of the nonlocal differential theory. Also, both the real and imaginary parts of the eigenvalue decrease by increasing the nonlocal parameter for both methods, except in the case of cantilever boundary condition for the nonlocal differential theory.

Table 6 Comparison of the complex eigenvalues ($\bar{\omega} = l\sqrt{\omega/C_0}$, $C_0 = \sqrt{EI/\rho A}$) for free vibration of nonlocal viscoelastic Euler-Bernoulli beam [15]

References	τ	Undamped elastic beam ($T_d = 0$ [ns])			Kelvin-Voigt viscoelastic beam ($T_d = 10^{-4}$ [ns])		
		Clamped	Simply supported	Cantilever	Clamped	Simply supported	Cantilever
Present	0	4.73	3.14	1.88	4.72 + 1.27i	3.14 + 0.59i	1.87 + 0.21i
	0.1	4.2	2.83	1.67	4.19 + 1i	2.82 + 0.46i	1.66 + 0.16i
	0.2	4.03	2.8	1.63	4.03 + 0.65i	2.79 + 0.31i	1.62 + 0.11i
Lei et al. [20]	0	4.96	3.14	1.88	4.95 + 1.4i	3.14 + 0.59i	1.87 + 0.21i
	0.1	4.72	3.07	1.88	4.71 + 1.26i	3.07 + 0.56i	1.88 + 0.21i
	0.2	4.23	2.89	1.89	4.22 + 1.02i	2.89 + 0.5i	1.89 + 0.21i
Lu et al. [41]	0	4.73	3.14	1.88	–	–	–
	0.1	4.59	3.07	1.88	–	–	–
	0.2	4.28	2.89	1.89	–	–	–
ABAQUS	0	4.16	3.06	1.85	4.16 + 1.12i	3.06 + 0.57i	1.84 + 0.21i

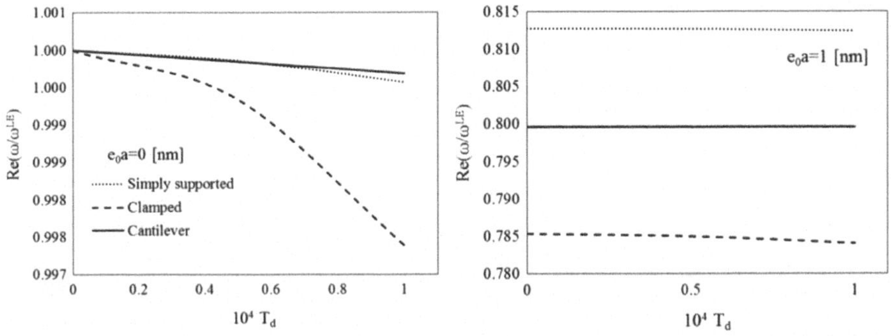

Fig. 24 Variations of the real part of the frequency with viscoelastic parameter for Euler-Bernoulli nano-scaled beam for $e_0a = 0$ nm and $e_0a = 1$ nm [15]

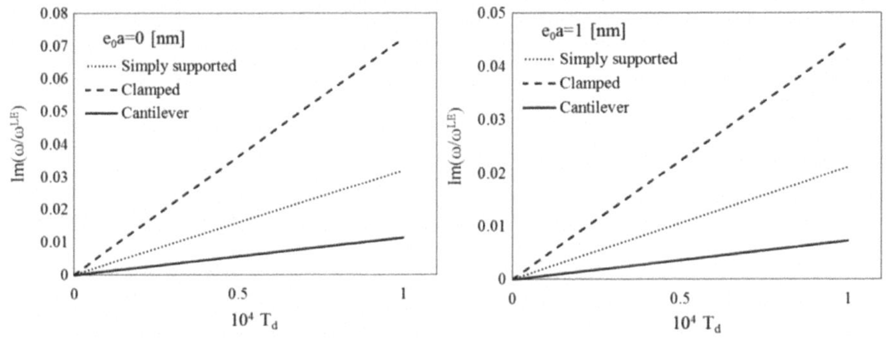

Fig. 25 Variations of the imaginary part of the frequency with viscoelastic parameter for Euler-Bernoulli nano-scaled beam for $e_0a = 0$ nm and $e_0a = 1$ nm [15]

Figures 24 and 25 show the variations of real and imaginary parts of non-dimensional eigenvalues, (non-dimensionalized by local elastic natural frequency ω_{LE}) with the viscoelastic parameter considering different values of nonlocal parameter and various boundary conditions. It is seen that by increasing the viscoelastic parameter the real part of complex eigenvalues decreases and the imaginary part increases. Also, it is observed that for the both-sides clamped boundary condition, in comparison with other boundary conditions, the effect of the viscoelastic parameter is more noticeable. However, by increasing the nonlocal parameter the change in eigenvalues due to viscoelastic parameter decreases.

Figure 26 shows the effect of nonlocal parameter on the real and imaginary parts of eigenvalues for viscoelastic nonlocal Euler-Bernoulli beam considering both sides simply supported boundary condition and various viscoelastic parameters and beam lengths. It is seen that by increasing the value both the real and imaginary parts of frequency decrease. For shorter beams, the variations of the eigenvalues are relatively more pronounced, because by decreasing the length of the beam, the effects

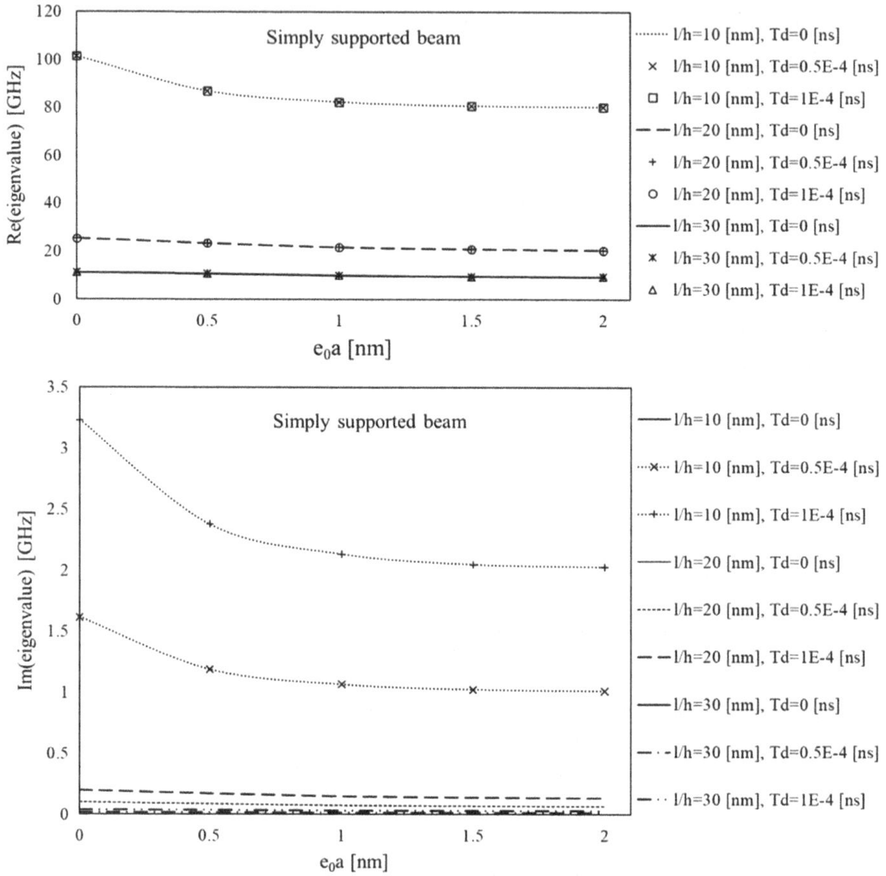

Fig. 26 Effects of the nonlocal parameter on the complex eigenvalues for both sides simply supported Euler-Bernoulli beam [15]

of boundary conditions become more important and nonlocality effects are stronger near the boundaries. Besides, for shorter beams, the effects of viscoelastic parameter on the imaginary part become more noticeable.

6.4.2 Free Vibration of Viscoelastic Nano-Scaled Plates

For analyzing the free vibration behavior of viscoelastic square nano-scaled plates, the procedure explained in Sect. 5.3 is followed. The plate has been meshed by Adini-Clough type elements (see Sect.3.3), and it has length and thickness of 10 and 0.34 nm, respectively. Also, in the current section terms of diagonal matrix $\mathbf{T}_d$ (viscoelastic parameters) are assumed to be equal and shown by T_d.

Table 7 Comparison between the fundamental natural frequency (GHz) of nano-scaled plates based on nonlocal integral theory and nonlocal differential theory considering all sides simply supported boundary condition

References	τ	Undamped elastic plate ($T_d = 0$ [ns])	Kelvin-Voigt viscoelastic plate ($T_d = 0.5 \times 10^{-4}$ [ns])	Kelvin-Voigt viscoelastic plate ($T_d = 10^{-4}$ [ns])
Pang et al. [48]	0.0	65.7909	65.7873 + 0.6799i	65.7768 + 1.3598i
	0.5	64.2253	64.2220 + 0.6479i	64.2122 + 1.2959i
	1.0	60.1240	60.1213 + 0.5678i	60.1132 + 1.1356i
	1.5	54.7472	54.7452 + 0.4708i	54.7391 + 0.9416i
	2.0	49.1803	49.1789 + 0.3799i	49.1745 + 0.7599i
Present study	0.0	65.7909	65.7873 + 0.6799i	65.7768 + 1.3598i
	0.5	57.5403	57.5380 + 0.5201i	57.5309 + 1.0401i
	1.0	51.788	51.7845 + 0.4243i	51.7810 + 0.8486i
	1.5	45.5904	45.5892 + 0.3265i	45.5857 + 0.6530i
	2.0	38.1002	38.0995 + 0.2280i	38.0974 + 0.4560i

Comparison between the fundamental eigenvalues based on the nonlocal integral theory with those of nonlocal differential theory [48] considering all sides simply supported boundary condition is shown in Table 7. It is seen that for a local case ($\tau = 0$), the agreement between the results is excellent. By increasing the nonlocal parameter, the results start to differentiate.

Figure 27 shows the effect of nonlocal parameter on the free vibration behavior of viscoelastic classical nano-scaled plate for different viscoelastic and length parameters considering all sides simply supported boundary condition. It is seen that by increasing the value of $e_0 a$, both the real and imaginary parts of eigenvalues decrease and this effect is more pronounced for smaller plates. Also, the decrease in the imaginary part due to increase in nonlocal parameter is more pronounced for larger values of T_d.

Figure 28 shows the variations of real and imaginary parts of non-dimensional eigenvalues (non-dimensionalized by the natural frequency of local elastic plate) with the viscoelastic parameter considering different boundary conditions for $e_0 a = 1$ nm. It is observed that by increasing the value of T_d, the real part of eigenvalues decrease and imaginary part increase. Besides, this variation is more pronounced for all-sides clamped boundary condition.

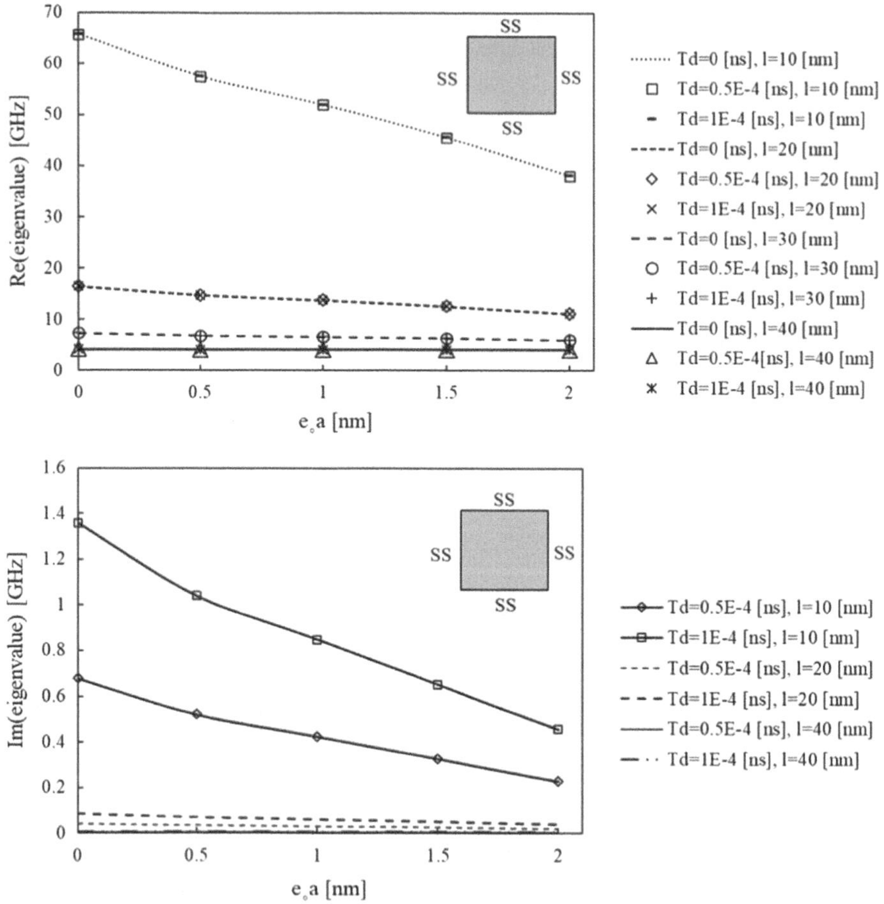

Fig. 27 Variations of the complex eigenvalues with nonlocal parameter for all-sides simply supported plate considering different viscoelastic parameters

7 Conclusions

In this chapter, a finite element method based on the nonlocal integral theory has been developed to study the mechanical behavior of nano-scaled beams and plates. The constitutive equation has been obtained considering the nonlocal integral theory and viscoelastic properties. Then, Hamilton's principle has been adopted for deriving the formulations and preparing the finite element foundation. Despite the simplicity of non-local differential theory in some aspects, it has been extracted from the more general nonlocal integral theory under certain conditions, thus it can only manage some limited problems (e.g. simple geometries, boundary conditions, and certain kernel types). However, the method presented in the current chapter can be used

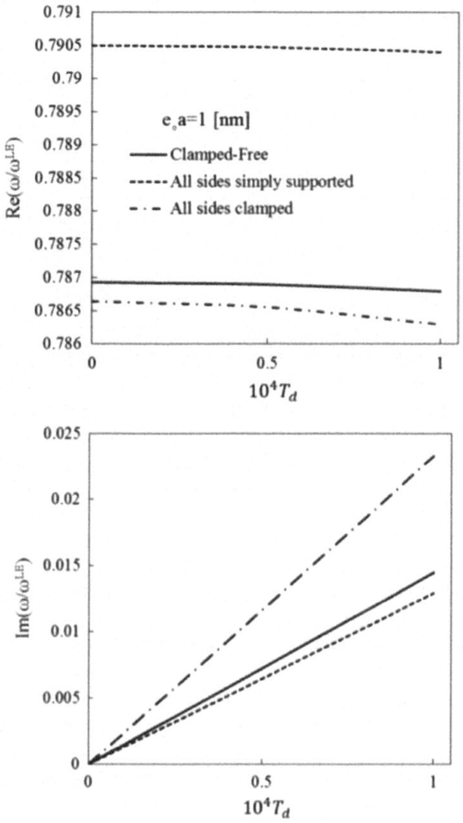

Fig. 28 Variations of the complex eigenvalues for a nano-scaled plate with the viscoelastic parameter for different boundary conditions

for modeling a broad range of problems, including complex geometries, various boundary conditions, and different kernel types. For instance, by using the finite element nonlocal theory the force boundary conditions can be analyzed properly for nano-scaled plates. In addition, the paradox, which has been seen in some cases considering the nonlocal differential theory, does not arise in the current method.

In previous sections, the formulations for studying bending, buckling, and vibration of nano-scaled beams and plates have been presented, and some examples have been discussed for understanding the effects of non-local parameters, geometrical parameters, boundary conditions and viscoelastic effects on the corresponding mechanical behavior.

Acknowledgements Maysam Naghinejad is grateful to Iran National Elites Foundation for the post-doctorate fellowship awarded to him.

References

1. Eringen AC (1983) On differential equations of nonlocal elasticity and solutions of screw dislocation and surface waves. J Appl Phys 54:4703–4710
2. Polizzotto C (2001) Nonlocal elasticity and related variational principles. Int J Solids Struct 38:7359–7380
3. Marotti de Sciarra F (2008) Variational formulations and a consistent finite-element procedure for a class of nonlocal elastic continua. Int J Solids Struct 45:4184–4202
4. Marotti de Sciarra F (2009) On non-local and non-homogeneous elastic continua. Int J Solids Struct 46:651–676
5. Pisano AA, Sofi A, Fuschi P (2009) Nonlocal integral elasticity: 2D finite element based solutions. Int J Solids Struct 46:3836–3849
6. Taghizadeh M, Ovesy HR, Ghannadpour SAM (2015) Nonlocal integral elasticity analysis of beam bending by using finite element method. Struct Eng Mech 54:755–769
7. Taghizadeh M, Ovesy HR, Ghannadpour SAM (2016) Beam buckling analysis by nonlocal integral elasticity. Int J Struct Stab Dyn 16:1–19
8. Eptaimeros KG, Koutsoumaris CC, Tsamasphyros GJ (2016) Nonlocal integral approach to the dynamical response of nanobeams. Int J Mech Sci 115–116:68–80
9. Koutsoumaris CC, Eptaimeros KG, Tsamasphyros GJ (2017) A different approach to Eringen's nonlocal integral stress model with applications for beams. Int J Solids Struct 112:222–238
10. Eptaimeros KG, Koutsoumaris CC, Dernikas IT, Zisis T (2018) Dynamical response of an embedded nanobeam by using nonlocal integral stress models. Compos. Part B Eng 150: 255–268
11. Norouzzadeh A, Ansari R (2017) Finite element analysis of nano-scale Timoshenko beams using the integral model of nonlocal elasticity. Physica E 88:194–200
12. Tuna M, Kirca M (2017) Bending, buckling and free vibration analysis of Euler-Bernoulli nanobeams using Eringen's nonlocal integral model via finite element method. Compos Struct 179:269–284
13. Naghinejad M, Ovesy HR (2018) Free vibration characteristics of nanoscaled beams based on nonlocal integral elasticity theory. J Vib Control 24:3974–3988
14. Naghinejad M, Ovesy HR (2018) Buckling analysis of Kelvin-Voigt viscoelastic nonlocal Euler-Bernoulli nano-beams. In: Proceedings of eighth international conference thin-walled structure Lisbon, Portugal, 2018
15. Naghinejad M, Ovesy HR (2019) Viscoelastic free vibration behavior of nano-scaled beams via finite element nonlocal integral elasticity approach. J Vib Control 25:445–459
16. Ovesy HR, Naghinejad M (2019) Nano-scaled plate free vibration analysis by nonlocal integral elasticity theory. AUT J Mech Eng 3:77–81
17. Calleja M, Kosaka PM (2012) Á. San Paulo, J. Tamayo, Challenges for nanomechanical sensors in biological detection. Nanoscale 4:4925
18. Payton OD, Picco L, Miles MJ, Homer ME, Champneys R (2012) Modelling oscillatory flexure modes of an atomic force microscope cantilever in contact mode whilst imaging at high speed. Nanotechnology 23:265702
19. Ghavanloo E, Fazelzadeh SA (2011) Flow-thermoelastic vibration and instability analysis of viscoelastic carbon nanotubes embedded in viscous fluid. Physica E 44:17–24
20. Lei Y, Murmu T, Adhikari S, Friswell MI (2013) Dynamic characteristics of damped viscoelastic nonlocal Euler-Bernoulli beams. Eur J Mech A/Solids 42:125–136
21. Lei Y, Adhikari S, Friswell MI (2013) Vibration of nonlocal Kelvin-Voigt viscoelastic damped Timoshenko beams. Int J Eng Sci 66–67:1–13
22. Pouresmaeeli S, Ghavanloo E, Fazelzadeh SA (2013) Vibration analysis of viscoelastic orthotropic nanoplates resting on viscoelastic medium. Compos Struct 96:405–410
23. Karličić D, Kozić P, Pavlović R (2014) Free transverse vibration of nonlocal viscoelastic orthotropic multi-nanoplate system (MNPS) embedded in a viscoelastic medium. Compos Struct 115:89–99

24. Eringen AC, Edelen DGB (1972) On nonlocal elasticity. Int J Eng Sci 10:233–248
25. Edelen DGB, Green AE, Laws N (1971) Nonlocal continuum mechanics. Arch Ration Mech Anal 43:36–44
26. Eringen AC, Kim BS (1974) Stress concentration at the tip of crack. Mech Res Commun 1:233–237
27. Eringen AC, Speziale CG, Kim BS (1977) Crack-tip problem in non-local elasticity. J Mech Phys Solids 25:339–355
28. Kröner E (ed) (1968) Mechanics of Generalized Continua. Springer
29. Eringen AC (1987) Theory of nonlocal elasticity and some applications. Res Mech 21:313–342
30. Koutsoumaris CC, Eptaimeros KG (2018) A research into bi-Helmholtz type of nonlocal elasticity and a direct approach to Eringen's nonlocal integral model in a finite body. Acta Mech 229:3629–3649
31. Khaniki HB, Hosseini-Hashemi S, Khaniki HB (2018) Dynamic analysis of nano-beams embedded in a varying nonlinear elastic environment using Eringen's two-phase local/nonlocal model. Eur Phys J Plus 133:283
32. Bažant ZP, Jirásek M (2002) Nonlocal integral formulations of plasticity and damage: survey of progress. J Eng Mech 128:1119–1149
33. Zienkiewicz OC, Taylor RL (2000) The finite element method, vol 2. Butterworth-Heinemann
34. Irons BM, Razzaque A (1972) Experience with the patch test for convergence of finite elements. In: Aziz AK (ed) The mathematical foundations of the finite element method with applications to partial differential equations. Academic Press
35. Zienkiewicz OC, Taylor RL (2000) The finite element method, vol 1. Butterworth-Heinemann
36. Long YQ, Cen S, Long ZF (2009) Advanced finite element method in structural engineering. Springer
37. Phadikar JK, Pradhan SC (2010) Variational formulation and finite element analysis for nonlocal elastic nanobeams and nanoplates. Comput Mater Sci 49:492–499
38. Wang CM, Kitipornchai S, Lim CW, Eisenberger M (2008) Beam bending solutions based on nonlocal Timoshenko beam theory. J Eng Mech 134:475–481
39. Wang Q, Liew KM (2007) Application of nonlocal continuum mechanics to static analysis of micro- and nano-structures. Phys Lett A 363:236–242
40. Aghababaei R, Reddy JN (2009) Nonlocal third-order shear deformation plate theory with application to bending and vibration of plates. J Sound Vib 326:277–289
41. Lu P, Lee HP, Lu C, Zhang PQ (2006) Dynamic properties of flexural beams using a nonlocal elasticity model. J Appl Phys 99:073510-1–9
42. Reddy JN (2007) Nonlocal theories for bending, buckling and vibration of beams. Int J Eng Sci 45:288–307
43. Ghannadpour SAM, Mohammadi B, Fazilati J (2013) Bending, buckling and vibration problems of nonlocal Euler beams using Ritz method. Compos Struct 96:584–589
44. Pradhan SC, Phadikar JK (2009) Nonlocal elasticity theory for vibration of nanoplates. J Sound Vib 325:206–223
45. Murmu T, Pradhan SC (2009) Vibration analysis of nano-single-layered graphene sheets embedded in elastic medium based on nonlocal elasticity theory. J Appl Phys 105:064319
46. Ansari R, Sahmani S, Arash B (2010) Nonlocal plate model for free vibrations of single-layered graphene sheets. Phys Lett A 375:53–62
47. Aksencer T, Aydogdu M (2011) Levy type solution method for vibration and buckling of nanoplates using nonlocal elasticity theory. Physica E 43:954–959
48. Pang M, Li ZL, Zhang YQ (2018) Size-dependent transverse vibration of viscoelastic nanoplates including high-order surface stress effect. Phys B 545:94–98

'Explicit' and 'Implicit' Non-local Continuum Descriptions: Plate with Circular Hole

Meral Tuna, Lorenzo Leonetti, Patrizia Trovalusci, and Mesut Kirca

Abstract Classical theory of elasticity fails to reflect the true behaviour of solids with internal material organization when internal and external length scales are of comparable orders. This drawback leads to emergence of non-classical continuum theories which are offered to be distinguished as 'implicit'/'weak' and 'explicit'/'strong' non-local models according to different interpretations of the incorporation of characteristic length scales. As an extension of recent works of authors, the presented chapter is focused on the correspondence between 'implicit' type Cosserat (micropolar) and 'explicit' type Eringen's two-phase local/non-local models, in terms of characteristic quantities. To this end, an example problem of practical importance; a plate with a circular hole, is studied by employing standard displacement based finite element method. The non-locality of Eringen's model is optimized regarding stress concentration factors reported for infinite Cosserat plates. The analysis of Eringen's model is repeated by adopting both Euclidean and geodetical distance during incorporation of long range interactions. According to the results, stress fields of explicitly and implicitly non-local models seem to be in a very good agreement considering plates with large scale ratios, as the missing neighbour relations appeared at domain boundaries of Eringen's model are not effective at the vicinity of the hole. Yet, obtaining different 'explicit' material parameters for each sample problem reveals that it is unlikely to have a unified relationship between characteristic quantities of Cosserat's and Eringen's models.

M. Tuna
Yasar University, Izmir, Turkey
e-mail: meral.tunaeroglu@yasar.edu.tr

L. Leonetti
University of Calabria, Rende, Italy
e-mail: lorenzo.leonetti@unical.it

P. Trovalusci (✉)
Sapienza University of Rome, Rome, Italy
e-mail: patrizia.trovalusci@uniroma1.it

M. Kirca (✉)
Istanbul Technical University, Istanbul, Turkey
e-mail: kircam@itu.edu.tr

© The Author(s), under exclusive license to Springer Nature Switzerland AG 2021

E. Ghavanloo et al. (eds.), *Size-Dependent Continuum Mechanics Approaches*,
Springer Tracts in Mechanical Engineering,
https://doi.org/10.1007/978-3-030-63050-8_11

1 Introduction

It is well-known that when internal (atomic/granular distance, size of heterogeneities, etc.) and external (sample length, wavelength, etc.) scales become comparable (e.g. $L/l \sim 1$), the discrete and heterogeneous nature of underlying material organization starts to play a fundamental role on its gross mechanical behaviour. Although, such structures can be reduced to Cauchy continua in certain limits ($L/l \gg 1$), they cannot be described within the scope of classical elasticity [1, 2]. Hence, in such cases, either discrete modelling techniques [3–6] or non-classical continuum models that account for size effects [7–13] must be resorted depending on the specificity of the problem.

In discrete modelling, the structure is assumed as an assembly of explicitly modelled rigid bodies connected by inter-particle potentials. Despite their capabilities of representing the actual behaviour of the media, direct discrete modelling techniques (e.g. limit analysis, molecular dynamics simulations, etc.) are not very practical due to their computational expense, which rapidly increases with total number of degrees of freedom (DOFs). Therefore, when a coarse modelling is to be preferred, it is useful to resort equivalent continua formulations to satisfactorily replicate the overall mechanical behaviour of discrete assembly.[1] Indeed, discrete to continuum approaches dates back to 19th century to the works of French scientists Navier [15], Cauchy [16], Poisson [17, 18], who tried to derive the constitutive equations of a classical continua starting from the description of lattice systems, later abandoned in the favour of energetic-continuum approach proposed by Green [19, 20]. Among them, contributions by Voigt [21, 22] and [23] who introduced particle rotations, couple interactions, and multi-body connections should be quoted.[2] On the track of Voigt's approach, systematic treatise of the Cosserat brothers [27] on the deformable bodies with both the translational and rotational degrees of freedom appeared to be the basis of historically first enriched non-classical continuum model followed by emergence of many other non-local theories.

Although any continua with presence of internal lengths and spatial dispersion properties are referred as non-local [28, 29], the theories offered to represent their behaviour are distinguished as 'implicit'/'weak' or 'explicit'/'strong' models depending on the mathematical interpretation of non-locality [10, 12, 14, 28]. In the so-called implicit non-local models (e.g. micromorphic theory, microstretch theory, micropolar theory, couple stress theory), also known as multifield continua, the body is treated as a collection of deformable or rigid particles that are endowed with additional degrees of freedom [12, 30–32]. As a result of newly introduced kinematical descriptors, which are used for representation of material microstructure, advanced strain and stress measures are emerged. Since size effects are incorporated through those non-standard primal fields, the theories belong to family of 'implicit' non-local models possess a 'weak' non-local character. On the other hand, in 'explicit'/'strong' non-local models, the primal kinematic and work-conjugated dynamic descriptor of classical theory of elasticity are preserved, while equations contain different operators

[1]These approaches are explained in detail in Ref. [14].

[2]A general survey about the original molecular models can be found in Refs. [24–26].

in spatial field [13]. As non-local parameter that includes information about material's internal length is directly linked to the bulk properties, this kind of theories possess a 'strong' non-local character. Among many different models of both classes, the focus of the present chapter will be on implicit type Cosserat [9, 12, 33, 34] and explicit type Eringen's non-local [13, 35] models.

Micropolar theory, which was first developed by Cosserat brothers [27], and further improved by many others as Nowacki [33] and Eringen [12, 36], corresponds to an assembly of rigid particles that are subjected to independent displacements and rotations, and interact through forces and couples. The theory has been widely employed in describing heterogeneous materials with microstructures, in particular block assemblies; granular rock, masonry, particle composites, etc. [1, 37, 38]. Due to the presence of relative rotation between micro (rotation of individual material points) and macro (local rigid) rotations, it is specifically suitable to investigate the behaviour of anisotropic/orthotropic media [1, 39–41]. The theory can be reduced to the couple stress in case micro rotations are forced to follow the macro rotation, while via further statical restriction on couples, the classical theory of elasticity should be recovered [37].

On the other hand, Eringen's non-local theory, which was first introduced in [42–45] and further improved by Eringen and Edelen [35], and Eringen [13, 46–51], is concerned with the physics of material bodies whose behaviour at a material point is influenced by the state of all points of the body. The theory covers long-range interactions by relating stress at a point to the strain of entire domain through an attenuation-type kernel function. Hence, it has been widely employed in investigation of the mechanical behaviour of nano/micro sized structures such as; carbon nanotubes, graphene sheets, atomic chains/arrays, etc. [52, 53].

Although constitutive relation in Eringen's theory is originally introduced in an *integral form* [13, 35], for specific kernel functions, it is later transformed to a *differential form* leading to easier handling of the problem mathematically [50]. Shortly after, by mixture of local and non-local constitutive relations through a weight regulating coefficient, another version of constitutive equation, also known as *two-phase local/non-local form* is proposed [51, 54].

All these forms of Eringen's non-local model with their enhanced versions have been constructed on mathematical and physical grounds, and employed by many researchers over the years [55–78][3]; yet, they may also have some drawbacks, especially in computational terms. For instance, differential form of constitutive equation has been shown to fail in representing actual mechanical response of bounded structures for some boundary conditions, such as cantilevered beams [79–85] while the original integral form may lead to ill-posed mathematical description that induces existence and uniqueness issues for some boundary conditions. [71, 86–88]. However, it has also been shown that the effect of ill-posedness is limited, even vanishes practically for physically reasonable (small) values of non-local parameter (κ/L).

[3]It should be noted that beyond this limited number of cited research, and researchers, there is an excessive literature.

This may also be pointed out by comparison of solutions of pure non-local model to those of two-phase model provided by [71] which is not ill-posed [88].

Although the application fields of Cosserat (micropolar) and Eringen's theories seem to be quite different due to distinct kinematic and dynamic descriptors they possess, comparison between various non-classical theories has the utmost significance in increasing the accuracy and precision of the modelling and design studies. Following the previous works of authors [89, 90], the possible correspondences and differences between 'implicit' type Cosserat and 'explicit' type Eringen's non-local models are investigated through a different example problem of pratical importance; a plate with a circular hole subjected to tensile loads. The results are obtained by employing standard displacement based finite element method (FEM) within the linearised kinematical framework. To study scale effects on solutions, the dimension of circular hole is fixed while the edge length of plates vary. The material parameters related to size effects are assumed to be known for 'implicit' model, while the non-locality of 'explicit' model is optimized with using an evolutionary optimization algorithm so that stress concentration factor (SCF) obtained for infinite plates will be coincide for both non-local models regarding the current problem under investigation. As tailoring non-locality through fraction coefficient is quite convenient computationally, two-phase local/non-local form of Eringen's constitutive relation is employed in the present work, while a bi-exponential type kernel function is preferred due to its wide range application in literature and its ease on implementation to FE algorithm. Moreover, since in the vicinity of voids (hole, crack, etc.), the geodetical distance, which is defined as the length of shortest path connecting two points without intersecting the boundaries [55], does not equal to Euclidean distance, the analysis of Eringen's model is performed adopting both distances. At last, the stress fields of both 'implicit' and 'explicit' type non-local models are compared with discussing possible improvements.

2 Materials and Methods

In this section, the continuum theories presented in the study are briefly explained for isotropic, homogeneous and linear elastic bodies, and corresponding displacement-based finite element (FE) formulations are derived regarding two-dimensional plane-strain problems within the linearised kinematical framework. A special attention has devoted on corresponding integration scheme, which is particularly challenging for Eringen's theory of elasticity due to its convolution type constitutive equation. For Eringen's model, FE formulation is implemented with developing an in-house Wolfram Mathematica code, while Cosserat model is implemented within the environment of the software COMSOL Multiphysics. The superscripts E and M that refer to Eringen's and Cosserat models, respectively, are used to distinguish the parameters appearing in both theories, and possessing different interpretations.

In two-dimensional framework the body under investigation can be assumed as a set of material points in 2D Euclidean space, occupying a domain $\Omega \in \{x, y\}$ with a uniform thickening of h, enclosed with a boundary, Γ which is subjected to traction vector, $\bar{\mathbf{t}}$, and couple traction vector, $\bar{\mathbf{m}}$ (considered for Cosserat media only).

The FE formulation is derived using the principle of minimum total potential energy, where for the static analysis, total potential energy functional, Π, which is written in terms of total elastic strain energy, U, and external work potential, W, as in the form; $\Pi = U + W$ must be minimum for equilibrium.

$$\frac{\partial \Pi\,[\mathbf{d}]}{\partial \mathbf{d}_i} = 0, \quad i = 1, 2, \ldots, N_{total} \tag{1}$$

where

$$\Pi\,[\mathbf{d}] = \sum_{i=1}^{N_{total}} \Pi_i \tag{2}$$

with N_{total} being the total number of elements, and $\mathbf{d}_i$ and $\mathbf{d}$ respectively referring to nodal unknowns vector of an element i, and the whole model. One can see that, as a result of 'weak' non-local character of constitutive equation for micropolar model, total potential energy of an element i depends only on its own displacement/rotation field:

$$\Pi_i^M = \Pi_i^M\left[\mathbf{d}_i^M\right], \quad i = 1, 2, \ldots, N_{total} \tag{3}$$

which resulting Eq. (1) to be simplified into following form,

$$\frac{\partial \Pi_1^M}{\partial \mathbf{d}_1^M} = \frac{\partial \Pi_2^M}{\partial \mathbf{d}_2^M} = \ldots = \frac{\partial \Pi_{N_{total}}^M}{\partial \mathbf{d}_{N_{total}}^M} = 0 \tag{4}$$

with the remaining terms (i.e. $\partial \Pi_i^M \big/ \partial \mathbf{d}_j^M$ for $i \neq j$) vanish.

On the other hand, due to 'strong' non-local character of Eringen's model, which accounts for long range interactions, total potential energy of an element i depends not only on its displacement field, but also to displacement fields of all other elements.

$$\Pi_i^E = \Pi_i^E\left[\mathbf{d}_1^E, \mathbf{d}_2^E, \ldots, \mathbf{d}_{N_{total}}^E\right], \quad i = 1, 2, \ldots, N_{total} \tag{5}$$

Hence all derivatives in Eq. (1) are recovered:

$$\frac{\partial \Pi_i^E}{\partial \mathbf{d}_1^E} = \frac{\partial \Pi_i^E}{\partial \mathbf{d}_2^E} = \ldots = \frac{\partial \Pi_i^E}{\partial \mathbf{d}_{N_{total}}^E} = 0, \quad i = 1, 2, \ldots, N_{total} \tag{6}$$

2.1 Micropolar Model

Micropolar theory belongs to a class of generalized continua, that, using the definition Kunin [28], Maugin [10], Eringen [12], Trovalusci [14], called as 'implicit' non-local model. As the material particles that constitute the continuum are described in terms of both their positions and orientations, in 2D media each particle possess two in-plane displacement components along x and y directions (u_x, u_y), and one out-of-plane micro-rotation component along z direction (ϕ_z) [9, 12, 33], which leads to following linearised kinematic relations:

$$\varepsilon_{ij}^M = u_{i,j} + e_{ijk}\phi_k, \quad \chi_{kj} = \phi_{k,j} \tag{7}$$

with $i, j = x, y$ and $k = z$. Here, comma symbol in the subscript stands for partial derivative operations, and ε_{ij} and χ_{kj} denote the components of strain and curvature tensors, respectively, while e_{ijk} is the usual third order permutation tensor. In case the micro-rotations are constrained to follow the local rigid (macro) rotations, the classical kinematic relations, i.e., $\varepsilon_{ij} = (u_{i,j} + u_{j,i})/2$, are recovered and the theory becomes couple stress [30, 37, 91].

From balance considerations, each component of surface traction, t_i^M, and surface couple-traction, m_k, are described in terms of non-symmetric stress tensor, σ_{ij}^M, couple stress tensor, μ_{kj}, and unit outward normal vector, n_j, as follows:

$$t_i^M = \sigma_{ij}^M n_j, \quad m_k = \mu_{kj} n_j \tag{8}$$

with subscripts referring to their components. If body forces and body couples are neglected, the equilibrium equations take the following form.

$$\sigma_{ij,j}^M = 0, \quad \mu_{kj,j} - e_{ijk}\sigma_{ij}^M = 0 \tag{9}$$

Considering linear elasticity, the stress-strain relation of an isotropic micropolar continua can be represented as:

$$\begin{aligned} \sigma_{ij}^M &= \lambda\varepsilon_{kk}^M\delta_{ij} + (\mu + \chi)\varepsilon_{ij}^M + \mu\varepsilon_{ji}^M, \\ \mu_{kj} &= \alpha\chi_{ii}\delta_{kj} + \beta\chi_{jk} + \gamma\chi_{kj} \end{aligned} \tag{10}$$

which requires, six independent elastic material constants for the complete description. Here λ and μ refer to Lame's constants, while α, β, γ and χ are constants related to micropolar theory. Note that, although for first Lame's constant the well-known relationship; $\lambda = E\nu/((1+\nu)(1-2\nu))$ is valid, second Lamé's constants, μ, is not equal to shear modulus, G, as in the classical convention, but holds following relation:

$$\mu = G - \frac{\chi}{2} \tag{11}$$

while Poisson's ratio, ν, is described as follows.

$$\nu = \frac{\lambda}{2(\lambda + G)} = \frac{\lambda}{2\lambda + 2\mu + \chi} \tag{12}$$

Equation (10) can be rewritten in matrix form as following

$$\boldsymbol{\sigma}^M = \mathbf{D}_\varepsilon^M \boldsymbol{\varepsilon}^M, \quad \boldsymbol{\mu} = \mathbf{D}_\chi^M \boldsymbol{\chi} \tag{13}$$

with elasticity matrices being

$$\mathbf{D}_\varepsilon^M = \begin{bmatrix} \lambda + 2G & \lambda & 0 & 0 \\ \lambda & \lambda + 2G & 0 & 0 \\ 0 & 0 & G + \chi/2 & G - \chi/2 \\ 0 & 0 & G - \chi/2 & G + \chi/2 \end{bmatrix}, \quad \mathbf{D}_\chi^M = \begin{bmatrix} \gamma & 0 \\ 0 & \gamma \end{bmatrix} \tag{14}$$

Since size effects and relative rotations are incorporated through an internal characteristic length, l_c, and a coupling number, N, with following relations hold [92]:

$$l_c^2 = \frac{\gamma}{2(2\mu + \chi)}, \quad N^2 = \frac{\chi}{2(\mu + \chi)} \tag{15}$$

constants appearing in $\mathbf{D}_\varepsilon^M$ do not retain the memory of internal lengths, while bending moduli $\mathbf{D}_\chi^M$ is responsible for scale effects. In case, l_c and N are taken small enough, the Cosserat effects become negligible, and the body behaves as Cauchy continua only if the material under investigation belongs at least to orthotetragonal symmetry class as in the presented study [1, 37].

2.1.1 Finite Element Formulation

For FE modelling, the field variables within an element e ($\mathbf{u}_e$, ϕ_e) are approximated considering a natural coordinate system, ζ, η, and using related interpolation function matrices:

$$\mathbf{u}_e = \mathbf{N}_u^M \mathbf{d}_{e\varepsilon}^M, \; \phi_e = \mathbf{N}_\phi \mathbf{d}_{e\phi}^M \tag{16}$$

Here, $\mathbf{N}_u^M$ is employed for interpolation of nodal in-plane displacements and consists of quadratic interpolation functions, while $\mathbf{N}_\phi$ is employed for nodal out-of-plane micro-rotations, and includes linear type shape functions

$$\mathbf{N}_u^M = \begin{bmatrix} N_u^1 & 0 & \cdots & N_u^9 & 0 \\ 0 & N_u^1 & \cdots & 0 & N_u^9 \end{bmatrix}, \quad \mathbf{N}_\phi = \begin{bmatrix} N_\phi^1 \ldots N_\phi^4 \end{bmatrix} \tag{17}$$

with considering following representation of nodal unknowns vectors for nine-node Lagrange element.

$$\mathbf{d}_{e\varepsilon}^M = \left\{ \tilde{u}_x^1 \; \tilde{u}_y^1 \ldots \tilde{u}_x^9 \; \tilde{u}_y^9 \right\}_e^T, \quad \mathbf{d}_{e\phi}^M = \left\{ \tilde{\phi}_z^1 \cdots \tilde{\phi}_z^4 \right\}_e^T \tag{18}$$

Fig. 1 For micropolar model, **a** 9-node Lagrange, **b** 4-node quadrilateral element in natural coordinate system

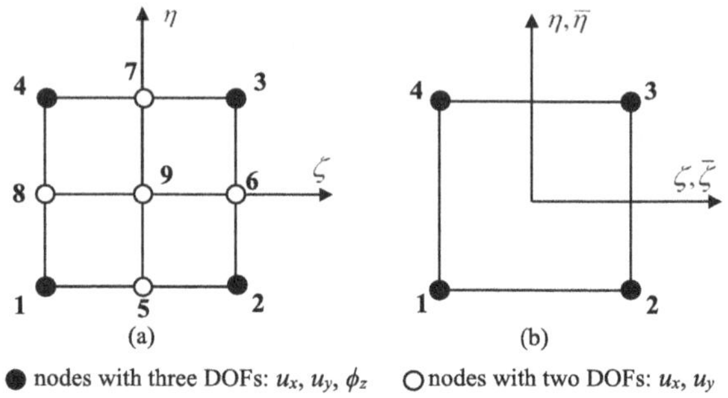

The over tilde symbol is used to indicate the nodal values and the superscripts refer to node numbers in an element.

As clearly seen from Fig. 1a and Eqs. (17) and (18), all the nodes of the nine-node Lagrange element possess displacement-type DOFs, whereas micro-rotation DOFs are attached only to the four corner nodes.

Important

Considering nine-node Lagrange element, the use of a lower interpolation order for micro-rotations (as depicted in Fig. 1a) allows an optimal compromise between numerical accuracy and computational efficiency to be achieved. In other words, the adoption of the same interpolation functions for both displacements and micro-rotations usually leads to negligible accuracy improvements against a notable increase in the number of DOFs [93].

In case a four-node linear element is considered for discretization, then both displacement and rotation DOFs are attached to same (corner) nodes as illustrated in Fig. 1b while $\mathbf{N}_u$ includes linear interpolation functions.

$$\mathbf{N}_u = \begin{bmatrix} N_u^1 & 0 & \cdots & N_u^4 & 0 \\ 0 & N_u^1 & \cdots & 0 & N_u^4 \end{bmatrix} \tag{19}$$

According strain and curvature fields that are ordered in vectors;

$$\boldsymbol{\varepsilon}_e^M = \left[\varepsilon_{xx}^M \; \varepsilon_{yy}^M \; \varepsilon_{xy}^M \; \varepsilon_{yx}^M\right]_e^T, \quad \boldsymbol{\chi}_e = \left[\chi_{zx} \; \chi_{zy}\right]_e^T \tag{20}$$

are obtained as follows for nine-node Lagrange element

$$\begin{aligned} \boldsymbol{\varepsilon}_e^M &= \mathbf{L}^M \mathbf{N}_u^M \mathbf{d}_{e\varepsilon}^M + \mathbf{M}\mathbf{N}_\phi \mathbf{d}_{e\phi}^M = \left[\mathbf{L}^M \mathbf{N}_u^M \; \mathbf{M}\mathbf{N}_\phi\right] \begin{Bmatrix} \mathbf{d}_{e\varepsilon}^M \\ \mathbf{d}_{e\phi}^M \end{Bmatrix} = \mathbf{B}_{e\varepsilon}^M \mathbf{d}_e^M, \\ \boldsymbol{\chi}_e &= \nabla\left(\mathbf{N}_\phi \mathbf{d}_{e\phi}^M\right) = \left[0 \; \nabla\mathbf{N}_\phi\right] \begin{Bmatrix} \mathbf{d}_{e\varepsilon}^M \\ \mathbf{d}_{e\phi}^M \end{Bmatrix} = \mathbf{B}_{e\chi}^M \mathbf{d}_e^M \end{aligned} \tag{21}$$

with $\mathbf{d}_e^M = \left\{ \tilde{u}_x^1 \; \tilde{u}_y^1 \; \ldots \; \tilde{u}_x^9 \; \tilde{u}_y^9 \; \tilde{\phi}_z^1 \; \ldots \; \tilde{\phi}_z^4 \right\}_e^T$, and $\mathbf{L}^M$, $\mathbf{M}$, ∇ respectively being the differential matrix operator, permutation vector and gradient operator:

$$\mathbf{L}^M = \begin{bmatrix} \frac{\partial}{\partial x} & 0 \\ 0 & \frac{\partial}{\partial y} \\ \frac{\partial}{\partial y} & 0 \\ 0 & \frac{\partial}{\partial x} \end{bmatrix}, \quad \mathbf{M} = \begin{bmatrix} 0 \\ 0 \\ +1 \\ -1 \end{bmatrix}, \quad \nabla = \begin{bmatrix} \frac{\partial}{\partial x} \\ \frac{\partial}{\partial y} \end{bmatrix} \tag{22}$$

Hence for a FE model, Eq. (13) takes the following form

$$\boldsymbol{\sigma}_e^M = \mathbf{D}_{e\varepsilon}^M \mathbf{B}_{e\varepsilon}^M \mathbf{d}_e^M, \quad \boldsymbol{\mu}_e = \mathbf{D}_{e\chi}^M \mathbf{B}_{e\chi}^M \mathbf{d}_e^M \tag{23}$$

where $\boldsymbol{\sigma}_e^M$ and $\boldsymbol{\mu}_e$ vectors are described as;

$$\boldsymbol{\sigma}_e^M = \left[\sigma_{xx}^M \; \sigma_{yy}^M \; \sigma_{xy}^M \; \sigma_{yx}^M \right]_e^T, \quad \boldsymbol{\mu}_e = \left[\mu_{zx} \; \mu_{zy} \right]_e^T \tag{24}$$

Note that, the derivations in $\mathbf{B}_{e\varepsilon}^M$ and $\mathbf{B}_{e\chi}^M$ matrices should be performed using chain rule and inverse Jacobian matrix of corresponding element e, $\mathbf{J}_e^{-1}$, as the interpolation function matrices are represented in terms of natural coordinate system.

Finally, formulation of mth element is derived as below according to principle of minimum total potential energy given in Eq. (4).

$$\begin{aligned} \mathbf{f}_m^M &= \underbrace{\left(h \int_{-1}^{1} \int_{-1}^{1} \left(\mathbf{B}_{m\varepsilon}^M\right)^{\mathrm{T}} \mathbf{D}_{m\varepsilon}^M \, \mathbf{B}_{m\varepsilon}^M \; \det |\mathbf{J}_m| \; \mathrm{d}\zeta \, \mathrm{d}\eta \right)}_{\mathbf{k}_{m\varepsilon}^M} \mathbf{d}_m^M \\ &+ \underbrace{\left(h \int_{-1}^{1} \int_{-1}^{1} \left(\mathbf{B}_{m\chi}^M\right)^{\mathrm{T}} \mathbf{D}_{m\chi}^M \, \mathbf{B}_{m\chi}^M \; \det |\mathbf{J}_m| \; \mathrm{d}\zeta \, \mathrm{d}\eta \right)}_{\mathbf{k}_{m\chi}^M} \mathbf{d}_m^M \\ \mathbf{f}_m^M &= \left(\mathbf{k}_{m\varepsilon}^M + \mathbf{k}_{m\chi}^M\right) \mathbf{d}_m^M \end{aligned} \tag{25}$$

By performing proper assemblage operations, the well-known linear equation system is achieved:

$$\mathbf{K}^M \, \mathbf{d}^M = \mathbf{f}^M \tag{26}$$

where $\mathbf{K}^M$ refers to global stiffness matrix and $\mathbf{f}^M$ is the global nodal force vector of discretized 2D micropolar media.

Attention

The integration operations in Eq. (25) can be performed either analytically or using a numerical integration scheme. Here Gauss Quadrature (GQ) method is employed. For first order elements with linear shape functions, 2 × 2, for second order elements with quadratic shape function, 3 × 3 Gauss sampling points are adequate to capture the exact value of corresponding integral. Hence, stiffness matrix of nine-node Lagrange element illustrated in Fig. 1a can be calculated as;

$$
\begin{aligned}
\mathbf{k}_m^M &= h \sum_{j=1}^{N_{GQ_3}} \sum_{i=1}^{N_{GQ_3}} w_i w_j \left(\mathbf{B}_{m\varepsilon}^M (\zeta_i, \eta_j)\right)^{\mathrm{T}} \mathbf{D}_{m\varepsilon}^M \mathbf{B}_{m\varepsilon}^M(\zeta_i, \eta_j) \det \left|\mathbf{J}_m(\zeta_i, \eta_j)\right| \\
&+ h \sum_{s=1}^{N_{GQ_2}} \sum_{r=1}^{N_{GQ_2}} w_r w_s \left(\mathbf{B}_{m\chi}^M (\zeta_r, \eta_s)\right)^{\mathrm{T}} \mathbf{D}_{m\chi}^M \mathbf{B}_{m\chi}^M (\zeta_r, \eta_s) \det \left|\mathbf{J}_m (\zeta_r, \eta_s)\right|
\end{aligned}
\tag{27}
$$

where N_{GQ_2} and N_{GQ_3} refer to number of Gauss sampling points (GP) in an element, and equal to 2 and 3, respectively. Since DOFs related to micro-rotations are attached only to corner nodes with a curvature field approximated using linear shape functions, it is convenient to calculate the integration of $\mathbf{k}_{m\chi}^M$ using 2 × 2 GP, while $\mathbf{k}_{m\varepsilon}^M$ requires 3 × 3 GQ due to existence of quadratic shape functions in $\mathbf{B}_{m\varepsilon}^M$. Here $\zeta_i, \eta_j, \zeta_r, \eta_s$ stand for natural coordinates with w_i, w_j, w_r, w_s being the Gauss weights of the corresponding GP.

2.2 *Eringen's Non-local Model*

Eringen's theory of elasticity is considered as an 'explicit' non-local model [10, 12, 14, 28], where stress is calculated by integrating the multiplication of strain field with an attenuation-type kernel function, which contains information about underlying discrete material organization, over the entire domain [13]. This theory is built upon eight fundamental axioms of constitutive equations which makes it one of the best candidates to explain the true nature of the matter among others, at least, at continuum level. This is the main reason why Eringen's theory will be utilized in this chapter. In this theory, primal fields of classical elasticity are conversed and following kinematical relations are obtained within the linearised framework:

$$
\varepsilon_{ij}^E = \frac{1}{2}\left(u_{i,j} + u_{j,i}\right) \tag{28}
$$

with $i, j = x, y$ for a 2D media. Correspondingly, each point contains two DOFs: in-plane displacement components u_x and u_y along x and y directions. Here ε_{ij}^E refers to components of the symmetric strain tensor.

For the continuum to be in balance, interactions between material points, that are characterized through traction forces only, $\mathbf{t}^E$, can be described in terms of symmetric stress tensor, $\boldsymbol{\sigma}^E$, and unit normal vector, $\mathbf{n}$, as follows:

$$t_i^E = \sigma_{ij}^E n_j \tag{29}$$

Since the axiom of objectivity provides the vanishing of non-local body force residuals, while strain-rate dependent non-local energy residual dies out under the quasi-static condition [13, 59, 94], in the absence of body forces, following equilibrium equation is carried out, similar to classical (local) elasticity theory.

$$\sigma_{ij,j}^E = 0 \tag{30}$$

Stress-strain relation of two phase local/non-local model is then written as follows for a linear elastic and isotropic 2D body [51, 54]:

$$\boldsymbol{\sigma}^E = \xi \mathbf{D}^E \boldsymbol{\varepsilon}^E + (1-\xi) \int_A \tau\left(r(\bar{\mathbf{x}}, \mathbf{x}), \kappa\right) \mathbf{D}^E \boldsymbol{\varepsilon}^E\left(\bar{\mathbf{x}}\right) dA(\bar{\mathbf{x}}) \tag{31}$$

with elasticity matrix being

$$\mathbf{D}^E = \begin{bmatrix} \lambda + 2G & \lambda & 0 \\ \lambda & \lambda + 2G & 0 \\ 0 & 0 & G \end{bmatrix} \tag{32}$$

Here $\mathbf{x}$ (or $\bar{\mathbf{x}}$) refers to $\mathbf{x}_i = \{x_i, y_i\}$ which represents the coordinates of any point i. As one can see, two additional parameters, ξ and κ, both of which attribute to non-local character of the structure, are introduced with Eringen's two phase local/non-local theory. Here ξ refers to fraction coefficient and controls the weight of local and non-local parts in the constitutive equation. As decreasing values imply more pronounced non-locality, assuming ξ equal to unity yields full local model while 0 corresponds to full non-local case, in which problems regarding existence and uniqueness of solution may arise. Meanwhile, non-local parameter κ, includes information about material's internal structure and defines the intensity of the kernel function (τ), according to which, the link between source point $\mathbf{x}$ and remaining points $\bar{\mathbf{x}}$ are identified. Although for a 2D domain, many different types of kernel function (e.g. error function, bell shape, conical shape, bi-exponential) exist [50, 55, 95], all of them have to fulfil the following requirements [50]:

1. Kernel function should be a positive decaying function with a maximum value at $\bar{\mathbf{x}} = \mathbf{x}$.
2. Considering regions sufficiently away from boundaries, kernel function should return to Dirac delta function for $\kappa \to 0$, in which case the classical theory of elasticity must be recovered.

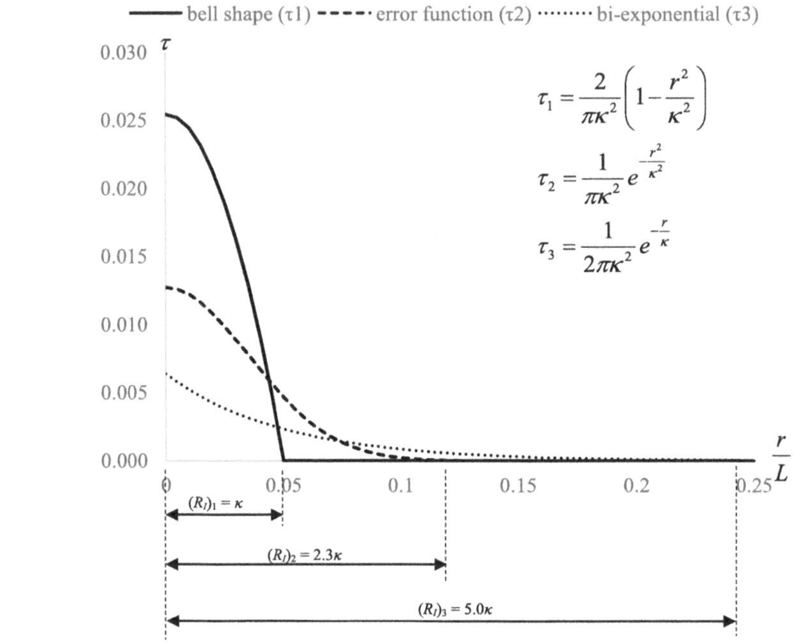

Fig. 2 Plots of attenuation functions

3. Eringen's non-local theory should approximate to lattice dynamics when $\kappa/L \to 1$.
4. Kernel function should satisfy the following normalization condition on V_∞.

$$\int\limits_{V_\infty} \tau\left(r(\bar{\mathbf{x}}, \mathbf{x}), \kappa\right) dV = 1 \tag{33}$$

Due to the decaying nature of kernel function, the effects of long range interactions practically vanish beyond a certain limit. This limit is called as *influence zone*, and defined with a radius (R_I) since the projection of any isotropic kernel function on 2D domain is circular. In Fig. 2, the cross section views of different attenuation functions are plotted for a non-local parameter $\kappa = 0.05L$, where R_I is determined as; $\tau|_{r\to R_I} = 0.0065\,\tau|_{r\to 0}$. Here L refers to external length of the 2D structure.

In accordance with literature [66], a bi-exponential type kernel function (τ_3 in Fig. 2) is used for calculations due to following reasons. First, the implementation of partitioned type kernel functions (e.g. bell shape, canonical shape, etc.) would require an additional loop in the algorithm, and increases the computational burden. Second, kernel functions with relatively narrow influence zone have a rapid decaying character, which drastically increases the number of Gauss sampling points for a good approximation of integration operations. Although the proper choice of kernel function is discussed in terms of computational expense only, it actually must be selected based on experimental data.

Important

Distance measure r appeared in kernel functions is generally defined as the length of a straight line that connects source point, $\mathbf{x}$, and all other points, $\bar{\mathbf{x}}$ such that; $r = |\bar{\mathbf{x}} - \mathbf{x}|$, while Polizzotto [55] proposed a more refined model that considers the existence of cracks and/or holes in the body. Accordingly, it was suggested that, the distance r of any pair $(\mathbf{x}, \bar{\mathbf{x}}) \in V$ should depend on geodetic path, $r(\mathbf{x}, \bar{\mathbf{x}})$, not to Euclidean distance, $|\bar{\mathbf{x}} - \mathbf{x}|$. Here geodetic distance is submitted as the minimum length that connects two points without intersecting the boundary surface of the body. Hence, it is clear that, for domains without voids the geodetical distance equals to Euclidean one ($r(\mathbf{x}, \bar{\mathbf{x}}) = |\bar{\mathbf{x}} - \mathbf{x}|$), while $r(\mathbf{x}, \bar{\mathbf{x}}) \geq |\bar{\mathbf{x}} - \mathbf{x}|$ relation holds for any domain.[4]

Considering our example problem possessing a hole (see Fig. 3b), the effect of geodetical distance on stress field should be studied. To this end, the analysis of Eringen's model is performed adopting both Euclidean and geodetical distances. Here, the geodetical path between two points is determined by following the steps listed below:

1. Tangents of corresponding points are obtained.
2. Among each pair of tangent lines, the ones that are closer to the other point are detected.
3. The intersection points of corresponding tangent lines are found.
4. The arc in between interaction points are identified.

According distances illustrated at Fig. 3b are calculated as following:

$$\begin{aligned} &\text{Euclidean}: \ |\mathbf{x}_{\mathrm{A}} - \mathbf{x}_{\mathrm{D}}| = \sqrt{(x_{\mathrm{A}} - x_{\mathrm{D}})^2 + (y_{\mathrm{A}} - y_{\mathrm{D}})^2} \\ &\text{Geodetical}: r(\mathbf{x}_{\mathrm{A}}, \mathbf{x}_{\mathrm{D}}) = |\mathbf{x}_{\mathrm{A}} - \mathbf{x}_{\mathrm{B}}| + a\theta_{\mathrm{BC}} + |\mathbf{x}_{\mathrm{C}} - \mathbf{x}_{\mathrm{D}}| \end{aligned} \tag{34}$$

where a refers to radius of hole and θ_{BC} denotes the angle between intersection points (B and C) of each tangent line.

2.2.1 Finite Element Formulation

For FE modelling of 2D non-local media, the domain is discretized using standard four-node linear elements where displacement field within an element e ($\mathbf{u}_e$) is approximated using nodal displacement vector, $\mathbf{d}_e^E$, and interpolation matrix, $\mathbf{N}_u$, (see Eq. (19)) such that; $\mathbf{u}_e = \mathbf{N}_u \mathbf{d}_e^E$.

[4] It is important to note that, although the geodetical distance between two points is unique, the geodetical path may sometimes be nonunique [55].

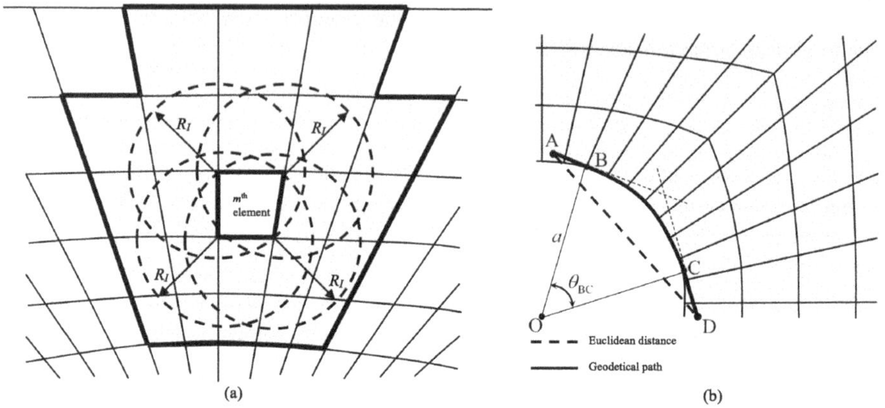

Fig. 3 **a** Illustration of influence zone of mth element (Elements that fall within the zone are shown in grey). **b** Geodetical path and Euclidean distance between source and neighbour points

Important

Even though second-order elements clearly suppress the first-order ones in problems with high gradients (e.g. stress concentration), it is not preferred herein as the integration of multiplication of quadratic shape functions and exponential type attenuation function would drastically increase the computational burden.

Corresponding strain field within an element e, i.e., $\boldsymbol{\varepsilon}_e^E = \left[\varepsilon_{xx}^E \; \varepsilon_{yy}^E \; 2\varepsilon_{xy}^E\right]_e^T$, is then defined as:

$$\boldsymbol{\varepsilon}_e^E = \mathbf{L}^E \mathbf{N}_u \mathbf{d}_e^E = \mathbf{B}_e^E \mathbf{d}_e^E$$

$$\mathbf{L}^E = \begin{bmatrix} \frac{\partial}{\partial x} & 0 \\ 0 & \frac{\partial}{\partial y} \\ \frac{\partial}{\partial x} & \frac{\partial}{\partial y} \end{bmatrix} \tag{35}$$

$$\mathbf{d}_e^E = \left\{\tilde{u}_x^1 \; \tilde{u}_y^1 \; \tilde{u}_x^2 \; \tilde{u}_y^2 \; \tilde{u}_x^3 \; \tilde{u}_y^3 \; \tilde{u}_x^4 \; \tilde{u}_y^4\right\}_e^T$$

with $\mathbf{L}^E$ being the differential matrix operator. As previously mentioned, the derivations in $\mathbf{B}_e^E$ is performed using chain rule and inverse Jacobian matrix of corresponding element e, $\mathbf{J}_e^{-1}$, as the interpolation functions are represented in terms of natural coordinate system.

The constitutive relation given in Eq. (31) is then transformed to following for an element e:

$$\boldsymbol{\sigma}_e^E = \xi_e \mathbf{D}_e^E \mathbf{B}_e^E \mathbf{d}_e^E + (1-\xi_e) \sum_{i=1}^{N_{total}} \int_{-1}^{1} \int_{-1}^{1} \frac{e^{-\frac{r}{\kappa}}}{2\pi\kappa^2} \mathbf{D}_i^E \bar{\mathbf{B}}_i^E \mathbf{d}_i^E \det\left|\bar{\mathbf{J}}_i\right| \mathrm{d}\bar{\zeta}\,\mathrm{d}\bar{\eta} \tag{36}$$

while stress tensor, $\boldsymbol{\sigma}_e^E$, is described as below:

$$\boldsymbol{\sigma}_e^E = \left[\sigma_{xx}^E \; \sigma_{yy}^E \; \sigma_{xy}^E\right]_e^T \tag{37}$$

Here the over bar denotes that, the related matrix is written in terms of $\bar{\zeta}, \bar{\eta}$ as a requirement of Eringen's constitutive relation, hence, should not be confused with overbars in prescribed surface traction and surface couple-traction.

According to principle of minimum total potential energy given in Eq. (6), formulation of mth element is derived as below.

$$\begin{aligned} \mathbf{f}_m^E &= \xi_m \mathbf{k}_m^E \mathbf{d}_m^E + (1-\xi_m)\left(\mathbf{k}_{mm}^E \mathbf{d}_m^E + \sum_{n=1,\, n\neq m}^{N_{total}} \mathbf{k}_{mn}^E \mathbf{d}_n^E\right) \\ &+ (1-\xi_n) \sum_{n=1,\, n\neq m}^{N_{total}} \left(\mathbf{k}_{nm}^E\right)^T \mathbf{d}_n^E \end{aligned} \tag{38}$$

where

$$\begin{aligned} \mathbf{k}_m^E &= h \int_{-1}^{1}\int_{-1}^{1} \left(\mathbf{B}_m^E\right)^T \mathbf{D}_m^E \mathbf{B}_m^E \det|\mathbf{J}_m| \,\mathrm{d}\zeta\,\mathrm{d}\eta \\ \mathbf{k}_{mm}^E &= h \int_{-1}^{1}\int_{-1}^{1}\int_{-1}^{1}\int_{-1}^{1} \frac{e^{-\frac{r}{\kappa}}}{2\pi\kappa^2} \left(\mathbf{B}_m^E\right)^T \mathbf{D}_m^E \bar{\mathbf{B}}_m^E \det\left|\bar{\mathbf{J}}_m\right| \det|\mathbf{J}_m| \,\mathrm{d}\bar{\zeta}\,\mathrm{d}\bar{\eta}\mathrm{d}\zeta\,\mathrm{d}\eta \\ \mathbf{k}_{mn}^E &= \frac{h}{2} \int_{-1}^{1}\int_{-1}^{1}\int_{-1}^{1}\int_{-1}^{1} \frac{e^{-\frac{r}{\kappa}}}{2\pi\kappa^2} \left(\mathbf{B}_m^E\right)^T \mathbf{D}_n^E \bar{\mathbf{B}}_n^E \det\left|\bar{\mathbf{J}}_n\right| \det|\mathbf{J}_m| \,\mathrm{d}\bar{\zeta}\,\mathrm{d}\bar{\eta}\mathrm{d}\zeta\,\mathrm{d}\eta \end{aligned} \tag{39}$$

The term with coefficient ξ represents the local part of two-phase model, while the terms with coefficient $1-\xi$ correspond to the non-local part.[5] In the non-local part, the first term stands for the contribution of the mth element to its own energy, while second and third terms account for the influence exerted on the mth element by the remaining elements, and the influence exerted by the mth element to other elements, respectively. Second and third terms in the non-local part, which expand element stiffness matrix to overall dimension and called as cross-stiffness matrices, are equal to each other only for homogenous material properties such that; $\mathbf{k}_{mn}^E = \left(\mathbf{k}_{nm}^E\right)^T$ for $\mathbf{D}_n^E = \mathbf{D}_m^E$.

[5]One can see that, Eqs. (38) and (39) have the same nature with the corresponding ones reported in [60].

To reduce the computational burden, only the elements inside the influence zone (see Fig. 3a) are assumed to contribute to cross-stiffness matrices $\mathbf{k}_{mn}^E$ and $\mathbf{k}_{nm}^E$, while the effect of others practically vanish due to the decaying nature of attenuation function, as previously described.

Here we designate the influence zone of an element by setting the corner nodes as centre points of the corresponding circle. Even if a very limited part of neighbouring elements reside to that zone, its contribution is included into the calculations. With this way, the zone has been extended as much as possible without compromising from computational efficiency. Please note that, after this point, the N_{total} expression in Eqs. (36) and (38) is replaced with total number of elements reside in the influence zone of corresponding element, N_{total_m}. The above mentioned simplification inherently leads to some sort of banding of the global stiffness matrix, $\mathbf{K}^E$ during assemblage operations.

Finally, linear equation system is obtained:

$$\mathbf{K}^E \, \mathbf{d}^E = \mathbf{f}^E \tag{40}$$

with $\mathbf{f}^E$ being the global nodal force vector of discretized 2D Eringen media.

Attention

The integration operations given in Eq. (39) are performed using GQ method. Although for local part ($\mathbf{k}_m^E$), 2×2 Gauss sampling points (GP) are sufficient regarding 4-node elements, for non-local part, the number of GP varies depending on the ratio between non-local parameter and element length as well as calculated part of stiffness matrix. As the κ / l_e ratio decreases, more Gauss points are needed in order to capture the trend of exponential function in integration, while cross stiffness matrices ($\mathbf{k}_{mn}^E$, $\mathbf{k}_{nm}^E$) requires less GP than $\mathbf{k}_{mm}^E$. Accordingly, the number of GP is arranged such that the relative difference between stiffness matrices obtained via GQ method and other numerical integration schemes embedded in commercial programs (e.g. Mathematica) remains below a prescribed limit (2%).
Below calculation of $\mathbf{k}_{mn}^E$ matrix of mth element is given.

$$\mathbf{k}_{mn}^E = \frac{h}{2} \sum_{s}^{N_{GQ_b}} \sum_{r}^{N_{GQ_b}} \sum_{i}^{N_{GQ_a}} \sum_{j}^{N_{GQ_a}} w_s w_r w_i w_j \frac{e^{-\frac{r}{\kappa}}}{2\pi\kappa^2} \left(\mathbf{B}_m^E \left(\zeta_r, \eta_s\right)\right)^T \mathbf{D}_m^E \bar{\mathbf{B}}_n^E \left(\bar{\zeta}_i, \bar{\eta}_j\right) \det \left|\bar{\mathbf{J}}_n \left(\bar{\zeta}_i, \bar{\eta}_j\right)\right| \det \left|\mathbf{J}_m \left(\zeta_r, \eta_s\right)\right| \tag{41}$$

As previously mentioned, N_{GQ_b} and N_{GQ_a} refer to number of GP in mth and nth elements, respectively. Accordingly, $\bar{\zeta}_i, \bar{\eta}_j$ stand for natural coordinates of nth element with w_i, w_j being the Gauss weights of the corresponding GP. A similar description holds for mth element, and $\zeta_r, \eta_s, w_r, w_s$.
If r is considered as Euclidean distance following relation holds

$$r = \sqrt{\left(X_m \left(\zeta_r, \eta_s\right) - \bar{X}_n \left(\bar{\zeta}_i, \bar{\eta}_j\right)\right)^2 + \left(Y_m \left(\zeta_r, \eta_s\right) - \bar{Y}_n \left(\bar{\zeta}_i, \bar{\eta}_j\right)\right)^2} \tag{42}$$

where $\mathbf{X}_{m(n)} = \{X_{m(n)}(\zeta, \eta), Y_{m(n)}(\zeta, \eta)\}$ defines coordinate mapping between physical and natural coordinate system of mth element. However, in case the Euclidean path between GPs of mth and nth elements transits through the hole as illustrated in Fig. 3b, geodetical distance defined in Eq. $(34)_2$ should be used to account for long-range interactions.

3 Differential Evolution Method

Differential Evolution method (DEM) was proposed in 1997 by Storn and Price [96] as a heuristic numerical optimization approach. It ensures good convergence properties with requiring only a few control parameters. Each iteration of the algorithm consists of three steps; mutation, crossover and selection. Below, these steps are briefly explained for a function with N_0 parameters to be optimized.[6]

First of all, an initial vector population, $\mathbf{p}$, is created. This population can be chosen either randomly or uniformly, and must cover the entire parameter space.

$$\mathbf{p}_i^J = \left[p_{i,1}^J \; p_{i,2}^J \cdots p_{i,N_0}^J \right]^T, \quad i = 1, 2, \ldots, N_P \tag{43}$$

Here superscript J stands for generation number while N_P refers to number of vectors at each population, and is one of the control parameters. According to Storn and Price [96] a reasonable choice for N_P is between $5 \times N_0$ and $10 \times N_0$ which cannot be altered during the optimization process.[7]

After formation of initial vector population, the generation of mutant vectors, $\mathbf{m}$, takes place. To this end, the weighted difference between randomly selected two target vectors are added to a third one, and as a result, a mutant vector is created for each target vector:

$$\mathbf{m}_i^{J+1} = \mathbf{p}_{r_1}^J + F\left(\mathbf{p}_{r_2}^J - \mathbf{p}_{r_3}^J\right), \quad i = 1, 2, \ldots, N_P \tag{44}$$

with $r_1, r_2, r_3 \in \{1, 2, \ldots, N_P\}$ and $r_1 \neq r_2 \neq r_3 \neq i$ being hold. F is another control parameter, and regulates the amplification of differential variation. Its value varies between 0 and 2, while 0.5 is suggested as a good initial choice. If the population converges too soon F should be increased with keeping in mind that the values greater than 1 are occasionally effective.

In order to increase the diversity of parameter vectors, the crossover step is introduced. The crossover (trial) vector;

$$\mathbf{c}_i^{J+1} = \left[c_{i,1}^{J+1} \; c_{i,2}^{J+1} \ldots c_{i,N_0}^{J+1} \right]^T, \quad i = 1, 2, \ldots, N_P \tag{45}$$

[6]For more information about the algorithm, the readers are referred to [96].

[7]N_P must be taken at least 4 to fulfil the requirement in mutation step.

is determined by comparing each term of target and mutation vectors according to following criteria:

$$c_{i,j}^{J+1} = \begin{Bmatrix} m_{i,j}^{J+1} \rightarrow \text{rand}\,(j) \leq CR \\ p_{i,j}^{J} \rightarrow \text{rand}\,(j) > CR \end{Bmatrix}, \quad j = 1, 2, \ldots, N_0 \tag{46}$$

where rand(j) indicates the jth evaluation of a uniform random number generator that has an outcome between 0 and 1, while CR corresponds to third control parameter and $\in [0, 1]$. Accordingly large values of CR accelerates the convergence, yet 0.1 is suggested as a good first choice by [96].

In the last step, the trial vector, $\mathbf{c}_i^{J+1}$, is compared with the target vector, $\mathbf{p}_i^J$, to decide whether or not it is suited to become a member of generation $J + 1$. This is achieved by comparing the objective functions, OF, which is aimed to be minimized.

$$\mathbf{p}_i^{J+1} = \begin{Bmatrix} \mathbf{c}_i^{J+1} \rightarrow \text{OF}\left(\mathbf{c}_i^{J+1}\right) \leq \text{OF}\left(\mathbf{p}_i^{J}\right) \\ \mathbf{p}_i^{J} \rightarrow \text{OF}\left(\mathbf{c}_i^{J+1}\right) > \text{OF}\left(\mathbf{p}_i^{J}\right) \end{Bmatrix}, \quad i = 1, 2, \ldots, N_P \tag{47}$$

The convergence is assumed to be fulfilled either the total number of stall iterations exceeds 50, or the objective function is minimized up to 0.001%.

4 Numerical Simulations

This section is concerned with the comparison of 'implicit' Cosserat and 'explicit' two-phase local/non-local Eringen's models through numerical examination of linear elastic plate with a circular hole, under uniform tensile stress (Fig. 4).

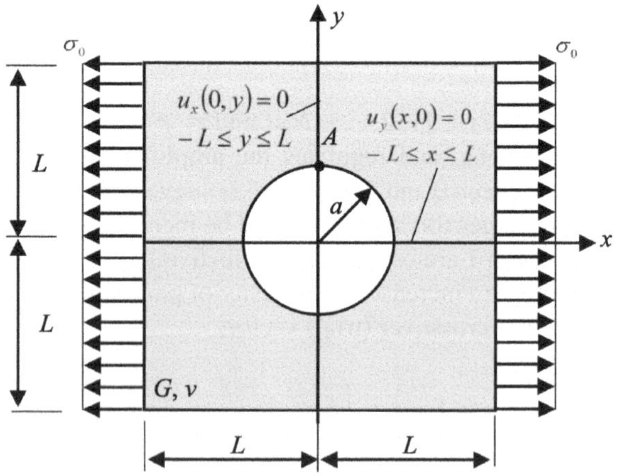

Fig. 4 Geometry, boundary and loading conditions of example problem

As loading condition, a constant stress of $\sigma_0 = 100$ (MPa) is applied to left and right vertical boundary edges, while boundary conditions are imposed by constraining the horizontal and vertical displacements of vertical and horizontal axes located in the middle, respectively. All the analyses are performed regarding the full domain since in the context of Eringen's non-local theory, imposing symmetry is much more complicated than only considering the symmetric part of the model as a result of the missing neighbour elements that should have contributed to cross stiffness matrices [65, 97].

In accordance with authors' previous publication [89], the simulations are repeated considering two different mesh configurations of different refinement (see Fig. 5), both characterized by three different scale ratios: $L/a = 3.0$ (Model 1), $L/a = 10.0$ (Model 2) and $L/a = 20.0$ (Model 3). The radius of hole, a, equals to 0.05 (m), while the width of the plate (structural/macro length), L, is modified in size to consider both finite and infinite domains. To study the effect of non-locality on solutions and to examine the correspondences and differences between 'implicit' and 'explicit' non-local continuum models, the analysis are repeated by modelling the plate as Cauchy, Cosserat and Eringen continua. For simulations, the non-locality of Eringen's model is optimized by tailoring the fraction coefficient ξ, while non-local parameter κ/a

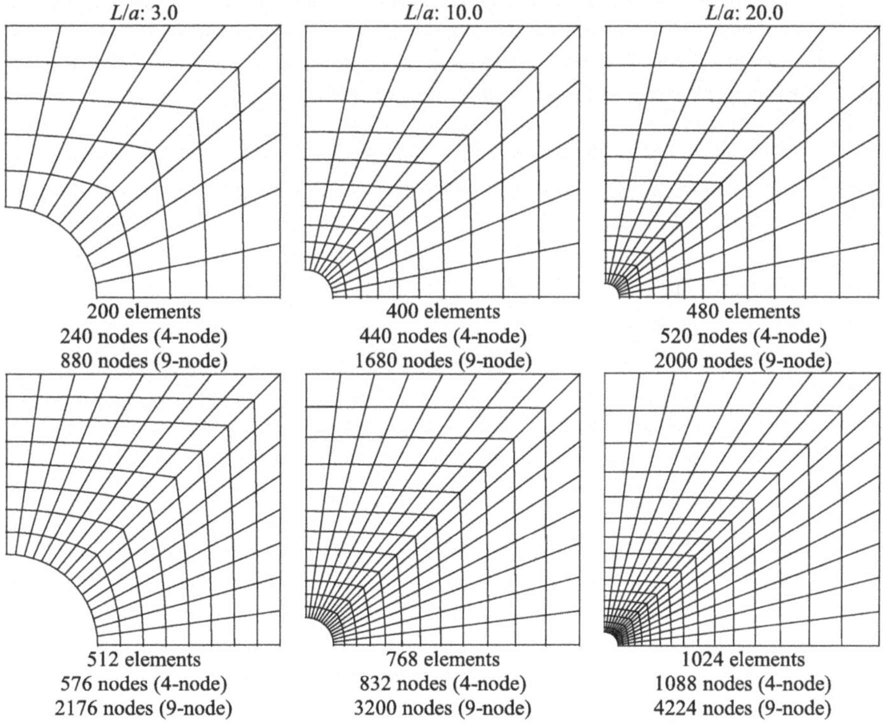

Fig. 5 Mesh configuration (top: coarse, bottom: fine) of models with a circular hole

equals to 0.2 in accordance with [66]. For this value of κ/a, the radius of influence zone (R_I) is taken as 0.05 (m). For optimization, Differential Evolution Method [96] is used with the following objective function (OF) to be minimized, while trial and error method could also be preferred:

$$\mathrm{OF} = \left| \frac{\mathrm{SCF_E}\,(\xi)}{\mathrm{SCF_C}} - 1 \right|_{L/a \to \infty} \tag{48}$$

where $0.0 < \xi \leq 1.0$ and $\mathrm{SCF} = \text{œ}_{\max}/\text{œ}_0 = \text{œ}_{\mathrm{A}}/\text{œ}_0$. According to Eq. (48), the numerical value of fraction coefficient is arranged such that, the stress concentration factor (SCF) obtained for Eringen's non-local theory based FE model (for $L/a = 20.0$) should be in accordance with the one reported for infinite Cosserat plates [98]. With regard to latter convergence criteria described in previous section, the optimum fraction coefficient is obtained as $\xi_{opt} = 0.3315$ with a convergence rate illustrated at Fig. 6.

Important

In post processing, nodal stresses are attained by using *direct evolution method*, in which case, stress components of each node is calculated by substituting its physical coordinates to stress field equation of corresponding element (Eq. (23) for Cosserat model, Eq. (36) for Eringen's model). For nodes that are shared by multiple elements, a single stress value is obtained by averaging contribution of surrounding elements.

After non-local material parameters of Eringen's model are fixed, the FE analysis are performed for all theories, scale ratios and discretization. The numerical values of parameters used for calculations can be found in Table 1, while corresponding SCFs are listed in Table 2.

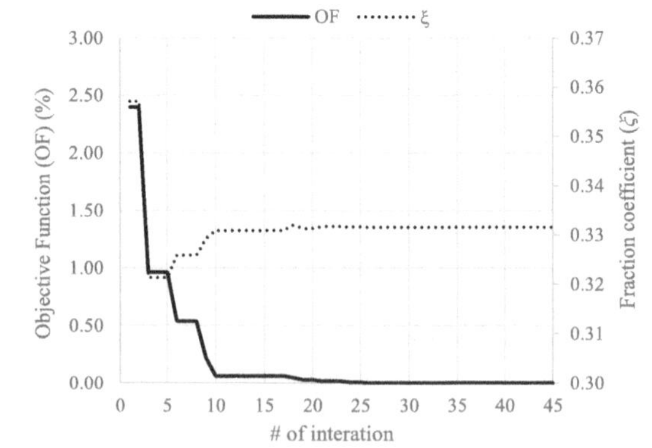

Fig. 6 Rate of convergence during optimization of fraction coefficient

Table 1 Numerical values of material properties

	G (GPa)	ν	κ/a	ξ	μ (GPa)	l_c (m)	N	χ (GPa)	γ (kN)
Local	1.0	1/3	–	–	–	–	–	–	–
Cosserat	1.0	1/3	–	–	−3.263	0.05	0.9	8.5263	10^4
Eringen	1.0	1/3	0.2	0.3315	–	–	–	–	–

Table 2 Stress concentration factors and sensitivity analysis

Models (L/a)	Theories	Analytical results	4-node element			9-node element		
			Coarse	Fine	Δ_m (%)	Coarse	Fine	Δ_m (%)
3	Local	–	4.37129	4.50118	2.89	4.38496	4.26706	2.76
	Cosserat	–	2.49640	2.48677	0.39	2.42046	2.42384	0.14
	Eringen	–	3.13088	3.16843	1.19	–	–	–
	Eringen*	–	3.13043	3.16812	1.19	–	–	–
10	Local	–	3.27776	3.35637	2.34	3.30622	3.26048	1.40
	Cosserat	–	2.28474	2.29861	0.60	2.22525	2.22562	0.02
	Eringen	–	2.20737	2.23867	1.40	–	–	–
	Eringen*	–	2.20700	2.23831	1.40	–	–	–
20	Local	3.00000	3.09164	3.30288	6.40	3.30066	3.15742	4.54
	Cosserat	2.16544	2.27173	2.26422	0.33	2.22491	2.20055	1.11
	Eringen	–	2.10095	2.16544	2.98	–	–	–
	Eringen*	–	2.10036	2.16507	2.99	–	–	–

The effect of discretization is studied by calculating the relative differences between both mesh configuration ($\Delta_m = |SCF_{coarse}/SCF_{fine} - 1|$), while superscript * in Table 2 is used to indicate the results obtained by adopting geodetical distance.

According to the results, following interferences can be made. Even though for models with high non-locality, the correction of r through determining geodetical path will become indispensable, for the presented κ/a ratio and example problem, solutions are practically not altered, as the SCFs reported for both cases suggest (i.e. Euclidean (Eringen) vs geodetical (Eringen*) path).

For all scale ratios, the presence of internal lengths leads to a decrease in stress concentration factor, which features the importance of using non-classical theories for cases with accountable size effects. This is also evident from comparative contour plots of normal stresses depicted in Fig. 7,[8] and 3D plots illustrated in Fig. 8, where the maximum stress reduces at least about %28.

[8]To achieve smooth and physically meaningful stress fields, the discontinuity of stress among linear elements are averaged by performing linear interpolation.

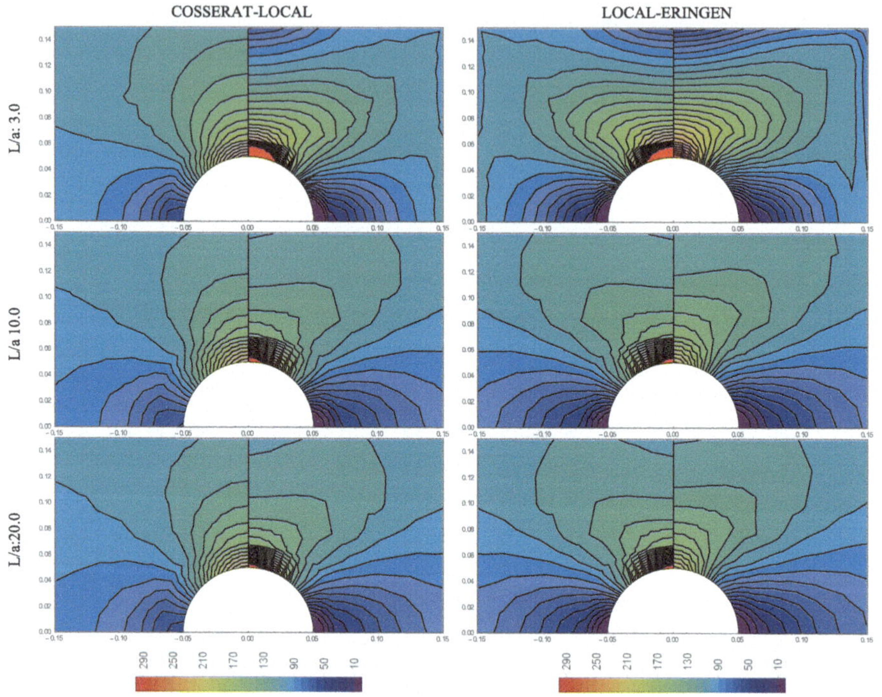

Fig. 7 Zoom-in looks of contour plots of normal stress field for 4-node, fine meshed Local, Cosserat and two-phase local/non-local Eringen's models

Similar to ones reported in [89], the dependence of SCFs on spatial discretization is weakened for models with size effects, especially for Cosserat continuum. Although, the results of second order elements are less dependent on discretization, and also closer to the analytical expressions, adopting first order elements still seems to provide a good approximation.

Aside similar diffusive character of 'explicit' two-phase local/non-local Eringen's and 'implicit' Cosserat non-local models, the variation of L/a ratio seems to have different effects on SCFs. This discrepancy can be attributed to the missing long range interactions around domain boundaries of Eringen's model (*boundary effects*), the effect of which is extended to around the hole for $L/a = 3.0$. As one can clearly see from the first rows of Figs. 8 and 9a, the reduction of stress close to upper and lower boundaries acts as an additional source of stress concentration for 'explicitly' non-local Eringen's model.

As the last point, it should be mentioned that, the results in the interior domain, sufficiently away from boundaries and hole, expected to be close to that of local elasticity as κ/L (or l_c/L) ratios are indeed small, while $\kappa/a = 0.2$ provides relatively high value of non-local parameter, evidenced by the discrepancy in stress distribution around the hole.

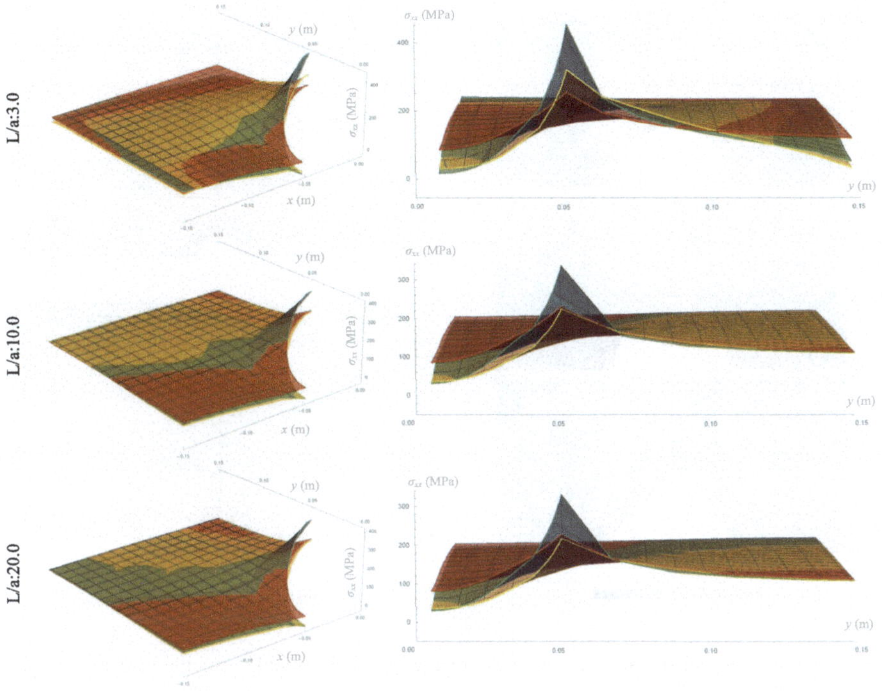

Fig. 8 Zoom-in looks of 3D plots of normal stress field for 4-node, fine meshed Local (green), Cosserat (orange) and Eringen (yellow) models

5 Conclusion

Employing non-classical continuum theories that are able to retain the memory of internal material organization with particular reference to the material's characteristic length is favourable for structures with accountable size effects. According to the works Kunin [28], Maugin [10], Eringen [12], Trovalusci [14], these theories can be distinguished as 'implicit'/'weak' and 'explicit'/'strong' non-local models depending on the interpretation of internal scale parameters. As understanding their capabilities has the utmost significance in increasing the accuracy and precision of the modelling and design studies, the present chapter focused on the comparison of two popular non-local theories; the 'implicit' Cosserat (micropolar) and 'explicit' Eringen's models through an example problem of practical importance. The calculations are performed by employing standard finite element method, while the domains are discretized using quadrilateral elements.

The material parameters of 'implicit' model are assumed to be known while the non-locality of 'explicit' model is modified to attain a stress concentration factor that is in accordance with infinite Cosserat plate. Although for larger scale ratios, the results of both non-local models are in a very good agreement, the stress field

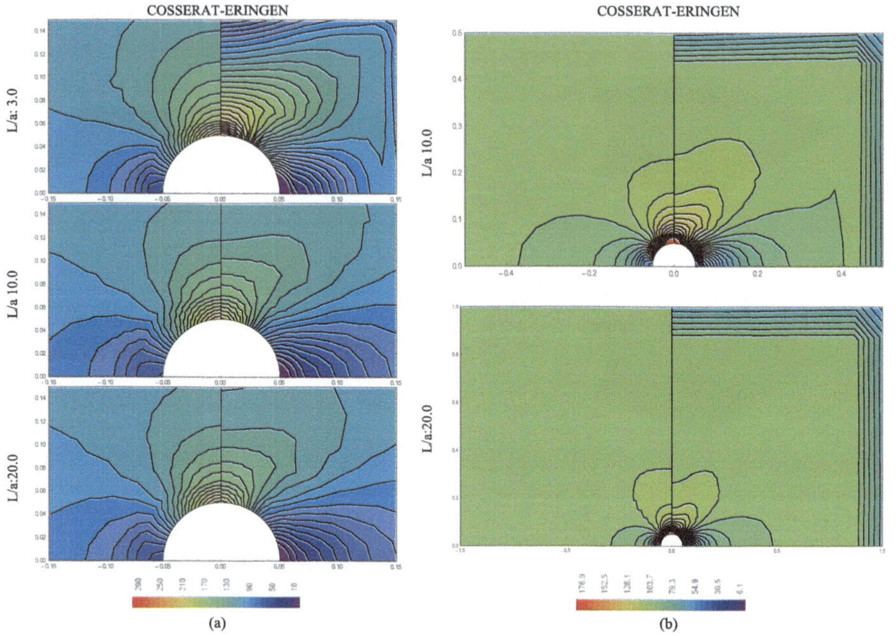

Fig. 9 **a** Zoom-in and **b** zoom-out looks of contour plots of normal stress field for 4-node, fine meshed Cosserat's and Eringen's models

at the vicinity of the hole is disturbed for $L/a = 3.0$, in which case the boundary effects of Eringen's model are evident in the entire domain. This indicates that when pores, voids or any other defects are close to boundaries, Eringen's model estimates a higher stress concentration factor than Cosserat model. Nevertheless, the critical decrease in SCFs with respect to Cauchy model features the importance of adopting a non-local theory for problems with accountable size effects. Yet, obtaining a different optimum fraction coefficient than the one reported in matrix inclusion problem [89] indicates the fact that for each sample problem the non-locality must be tailored to have some kind of equivalency between 'implicit' and 'explicit' non-local models. For considered non-local parameter, adopting Euclidean or geodetical distance during formation of the cross stiffness matrices does not change the results considerably. However, for larger R_I/a ratios, the correction of distance measure through determining the geodetical path may become very significant.

In terms of computational efficiency, the coding and implementation of Eringen's theory is relatively challenging with respect to Cauchy and Cosserat models. The necessity of accounting for the contribution of elements to each other due to Eringen's cumulative type constitutive relation intensively extends the calculations. Moreover, a special attention must be paid for numerical integration.

Although, in this study, a single parameter turns out to be suggestive enough to look for possible hints on equivalencies of both non-local models, other strategies considering global behaviour are possible, such as minimizing the differences of strain energy.

Acknowledgements This work was done when Meral Tuna was a Visiting Researcher at DISG, Sapienza University of Rome, with financial support of Italian Ministry of Education, University and Research PRIN 2015, project 2015JW9NJT (Grant No. B86J16002300001, ID: 10.13039/501100003407) "Materials with microstructure: multiscale models for the derivation non-local continua of 'implicit/weak' and 'explicit/strong"' and PRIN 2017 (Grant no. B86J16002300001) "Materials with microstructure: multiscale models for the derivation non-local continua and related numerical simulations" which are gratefully acknowledged.
Professor Patrizia Trovalusci acknowledges the support of PRIN 2017, project 2017HFPKZY (Grant No. B86J16002300001) and Sapienza Research Grants "Progetti Medi" 2017 (Grant No. B83C17001440005) and "Progetti Grandi" 2018 (Protocol No. RG1181642E3B3117).

References

1. Trovalusci P, Masiani R (1999) Material symmetries of micropolar continua equivalent to lattices. Int J Solids Struct 36:2091–2108
2. Fantuzzi N, Trovalusci P, Luciano R (2019) Multiscale analysis of anisotropic materials with hexagonal microstructure as micro-polar continua. Int J Multiscale Com 18:265–284
3. Suzuki T, Takeuchi S, Yoshinaga H (1991) Dislocation Dynamics and Plasticity. Springer-Verlag
4. Baggio C, Trovalusci P (1998) Limit analysis for no-tension and friction three dimensional discrete systems. Mech Struct Mach 26:287–304
5. Rapaport DC (2004) The art of molecular dynamics simulation. Cambridge University Press, Cambridge
6. Reccia E, Leonetti L, Trovalusci P, Cecchi A (2018) A multiscale/multidomain model for the failure analysis of masonry walls: a validation with a combined FEM/DEM approach. Int J Multiscale Com 16:325–343
7. Mindlin RD (1964) Micro-structure in linear elasticity. Arch Ration Mech An 16:51–78
8. Kunin IA (1968) The theory of elastic media with microstructure and the theory of dislocation. In Kröner E (ed) Mechanics of Generalized Continua. Springer
9. Capriz G (1989) Continua with Microstructure. Springer-Verlag
10. Maugin GA (1993) Material Inhomogeneities in Elasticity. Taylor & Francis
11. Gurtin ME (1999) Configurational Forces as Basis Concept of Continuum Physics. Springer-Verlag
12. Eringen AC (1999) Microcontinuum Field Theory. Springer
13. Eringen AC (2002) Nonlocal continuum Field Theories. Springer
14. Trovalusci P (1914) Molecular approaches for multifield continua: origins and current developments. In: Sadowski T, Trovalusci P (eds) Multiscale Modeling of Complex Materials: phenomenological, theoretical and computational aspects. Springer
15. Navier C (1827) Mémoire sur les lois de l'équilibre et du mouvement des corps solides élastiques. Mémoires de l'Academie des Sciences de l'Institut de France 7:375–393
16. Cauchy AL (1828) Sur l'équilibre et le mouvement d'un systéme de points matériels sollicités par des forces d'attraction ou de répulsion mutuelle. Exercices de Mathématiques 3:188–213
17. Poisson SD (1829) Mémoire sur l'équilibre et le mouvement des corps élastiques. Mémoires de l'Academie des Sciences de l'Institut de France 8:357–570

18. Poisson SD (1831) Mémoire sur les équations générales de l'équilibre et du mouvement des corps solides élastiques et des fluides. Journal de l'École Polytechnique 13:1–174
19. Green G (1839) On the laws of reflection and refraction of light at the common surface of two non-crystallized media. Trans Camb Philos Soc 7:1–24
20. Green G (1841) On the propagation of light in crystallized media. Trans Camb Philos Soc 7:121–140
21. Voigt W (1887) Theoretische Studien über die Elasticitätsverhältnisse der Kristalle. Abhandlungender Gesellschaft der Wissenschaften zu Göttingen— Mathematische Classe 34:1–100
22. Voigt W (1910) Lehrbuch der Kristallphysik. Teubner, B.G
23. Poincaré H (1892) *Leçons sur la Théorie de L'élasticité*. Carré
24. Trovalusci P, Capecchi D, Ruta G (2008) Genesis of the multiscale approach for materials with microstructure. Arch Appl Mech 79:981
25. Capecchi D, Ruta G, Trovalusci P (2010) From classical to Voigt's molecular models in elasticity. Arch Hist Exact Sci 64:525–559
26. Capecchi D, Ruta G, Trovalusci P (2011) Voigt and Poincaré's mechanistic-energetic approaches to linear elasticity and suggestions for multiscale modelling. Arch Appl Mech 81:1573–1584
27. Cosserat E, Cosserat F (1909) Théorie des Corps déformables. Herman et fils
28. Kunin IA (1984) On foundations of the theory of elastic media with microstructure. Int J Eng Sci 22:969–978
29. Kunin IA (2012) Elastic Media with Microstructure I: one-dimensional models. Springer Series in Solid-State Sciences. Springer
30. Toupin RA (1962) Elastic materials with couple-stresses. Arch Ration Mech An 11:385–414
31. Mindlin RD, Tiersten HF (1962) Effects of couple-stresses in linear elasticity. Arch Ration Mech An 11:415–448
32. Eringen AC, Suhubi ES (1964) Nonlinear theory of simple microelastic solidsli. Int J Eng Sci 2:189–203
33. Nowacki W (1986) Theory of Asymmetric Elasticity. Elsevier Science & Technology
34. Altenbach H, Eremeyev VA (2013) Cosserat media. In: Altenbach H, Eremeyev VA (Eds) Generalized Continua from the Theory to Engineering Applications. Springer
35. Eringen AC, Edelen DGB (1972) On nonlocal elasticity. Int J Eng Sci 10:233–248
36. Eringen AC (1966) Linear theory of micropolar elasticity. J Math Mech 15:909–923
37. Masiani R, Trovalusci P (1996) Cosserat and Cauchy materials as continuum models of brick masonry. Meccanica 31:421–432
38. Forest S, Sab K (1998) Cosserat overall modeling of heterogeneous materials. Mech Res Commun 25:449–454
39. Pau A, Trovalusci P (2012) Block masonry as equivalent micropolar continua: the role of relative rotations. Acta Mech 223:1455–1471
40. Eremeyev VA, Pietraszkiewicz W (2016) Material symmetry group and constitutive equations of micropolar anisotropic elastic solids. Math Mech Solids 21:210–221
41. Godio M, Stefanou I, Sab K, Sulem J, Sakji S (2017) A limit analysis approach based on Cosserat continuum for the evaluation of the in-plane strength of discrete media: application to masonry. Eur J Mech A-Solid 66:168–192
42. Eringen AC (1966) A unified theory of thermomechanical materials. Int J Eng Sci 4:179–202
43. Kröner E (1967) Elasticity theory of materials with long range cohesive forces. Int J Solids Struct 3:731–742
44. Krumhansl AJ (1968) Some Considerations of the Relation Between Solid State Physics and Generalized Continuum Mechanics. In Kröner E (Ed) Mechanics of Generalized Continua. Springer
45. Edelen DGB, Laws N (1971) On the thermodynamics of systems with nonlocality. Arch Ration Mech An 43:24–35
46. Eringen AC (1972) Nonlocal polar elastic continua. Int J Eng Sci 10:1–16
47. Eringen AC (1974) Theory of nonlocal thermoelasticity. Int J Eng Sci 12:1063–1077
48. Eringen AC (1977) Edge dislocation in nonlocal elasticity. Int J Eng Sci 15:177–183

49. Eringen AC (1978) Line crack subject to shear. Int J Fracture 14:367–379
50. Eringen AC (1983) On differential equations of nonlocal elasticity and solutions of screw dislocaltion and surface waves. J Appl Phys 54:4703–4710
51. Eringen AC (1987) Theory of nonlocal elasticity and some applications. Res Mech 21:313–342
52. Ghavanloo E, Rafii-Tabar H, Fazelzadeh SA (2019) Computational Continuum Mechanics of Nanoscopic Structures: Nonlocal Elasticity Approaches. Springer
53. Tuna M, Kirca M, Trovalusci P (2019) Deformation of atomic models and their equivalent continuum counterparts using Eringen's two-phase local/nonlocal model. Mech Res Commun 97:26–32
54. Altan SB (1989) Uniqueness of initial-boundary value problems in nonlocal elasticity. Int J Solids Struct 25:1271–1278
55. Polizzotto C (2001) Nonlocal elasticity and related variational principles. Int J Solids Struct 38:7359–7380
56. Pisano AA, Fuschi P (2003) Closed form solution for a nonlocal elastic bar in tension. Int J Solids Struct 40:13–23
57. Polizzotto C, Fuschi P, Pisano AA (2004) A strain-difference-based nonlocal elasticity model. Int J Solids Struct 41:2383–2401
58. Wang Q, Liew KM (2007) Application of nonlocal continuum mechanics to static analysis of micro- and nano-structures. Phys Lett A 363:236–242
59. Marotti de Sciarra F (2009) On non-local and non-homogeneous elastic continua. Int J Solids Struct 46:651–676
60. Pisano AA, Sofi A, Fuschi P (2009) Nonlocal integral elasticity: 2D finite element based solutions. Int J Solids Struct 46:3836–3849
61. Aydogdu M (2009) A general nonlocal beam theory: its application to nanobeam bending, buckling and vibration. Physica E 41:1651–1655
62. Ansari R, Sahmani S, Arash B (2010) Nonlocal plate model for free vibrations of single-layered graphene sheets. Phys Lett A 375:53–62
63. Benvenuti E, Simone A (2013) One-dimensional nonlocal and gradient elasticity: Closed-form solution and size effect. Mech Res Commun 48:46–51
64. Abdollahi R, Boroomand B (2013) Benchmarks in nonlocal elasticity defined by Eringen's integral model. Int J Solids Struct 50:2758–2771
65. Khodabakhshi P, Reddy JN (2015) A unified integro-differential nonlocal model. Int J Eng Sci 95:60–75
66. Fuschi P, Pisano AA, De Domenico D (2015) Plane stress problems in nonlocal elasticity: finite element solutions with a strain-difference-based formulation. J Math Anal Appl 431:714–736
67. Challamel N, Reddy JN, Wang CM (2016) Eringen's stress gradient model for bending of nonlocal beams. J Eng Mech 42(1):04016095
68. Tuna M, Kirca M (2016) Exact solution of Eringen's nonlocal integral model for bending of Euler-Bernoulli and Timoshenko beams. Int J Eng Sci 105:80–92
69. Tuna M, Kirca M (2016) Exact solution of Eringen's nonlocal integral model for vibration and buckling of Euler-Bernoulli beam. Int J Eng Sci 107:54–67
70. Tuna M, Kirca M (2017) Bending, buckling and free vibration analysis of euler-bernoulli nanobeams using Eringen's nonlocal integral model via nite element method. Compos Struct 179:269–284
71. Wang Y, Zhu X, Dai H-H (2016) Exact solutions for the static bending of Euler-Bernoulli beams using Eringen's two-phase local/nonlocal model. AIP Adv 6:185114
72. Zhu X, Wang Y, Dai H-H (2017) Buckling analysis of Euler-Bernoulli beams using Eringen's two-phase nonlocal model. Int J Eng Sci 116:130–140
73. Fernández-Sáez J, Zaera R (2017) Vibrations of Bernoulli-Euler beams using the two-phase nonlocal elasticity theory. Int J Eng Sci 119:232–248
74. Numanoglu HM, Akgoz B, Civalek O (2018) On dynamic analysis of nanorods. Int J Eng Sci 130:33–50
75. Meng L, Zou D, Lai H, Guo Z, He X, Xie Z, Gao C (2018) Semianalytic solution of Eringen's two-phase local/nonlocal model for Euler-Bernoulli beam with axial force. Appl Math Mech 39:1805–1824

76. Fakher M, Rahmanian S, Hosseini-Hashemi S (2019) On the carbon nanotube mass nanosensor by integral form of nonlocal elasticity. Int J Mech Sci 150:445–457
77. Abdollahi R, Boroomand B (2019) On using mesh-based and mesh-free methods in problems defined by Eringen's non-local integral model: issues and remedies. Meccanica 54:1801–1822
78. Eroglu U (2020) Perturbation approach to Eringen's local/non-local constitutive equation with applications to 1-D structures. Meccanica 55:1119–1134. https://doi.org/10.1007/s11012-020-01145-x
79. Peddieson J, Buchanan GR, McNitt RP (2003) Application of nonlocal continuum models to nanotechnology. Int J Eng Sci 41:305–312
80. Reddy JN, Pang SD (2008) Nonlocal continuum theories of beams for the analysis of carbon nanotubes. J Appl Phys 103:023511
81. Challamel N, Wang CM (2008) The small length scale effect for a nonlocal cantilever beam: a paradox solved. Nanotechnology 19:345703
82. Challamel N, Zhang Z, Wang CM, Reddy JN, Wang Q, Michelitsch T, Collet B (2014) On nonconservativeness of Eringen's nonlocal elasticity in beam mechanics: correction from a discrete-based approach. Arch Appl Mech 84:1275–1292
83. Li C, Yao L, Chen W, Li S (2015) Comments on nonlocal effects in nano-cantilever beams. Int J Eng Sci 87:47–57
84. Xu X-J, Deng Z-C, Zhang K, Xu W (2016) Observations of the softening phenomena in the nonlocal cantilever beams. Compos Struct 145:43–57
85. Fernández-Sáez J, Zaera R, Loya JA, Reddy JN (2016) Bending of Euler-Bernoulli beams using Eringen's integral formulation: a paradox resolved. Int J Eng Sci 99:107–116
86. Romano G, Barretta R (2016) Comment on the paper exact solution of Eringen's nonlocal integral model for bending of Euler-Bernoulli and Timoshenko beams by Meral Tuna and Mesut Kirca. Int J Eng Sci 109:240–242
87. Romano G, Barretta R, Diaco M, Marotti de Sciarra F (2017) Constitutive boundary conditions and paradoxes in nonlocal elastic nanobeams. Int J Mech Sci 121:151–156
88. Tuna M, Kirca M (2017) Respond to the comment letter by Romano and Barretta on the paper "Exact solution of Eringen's nonlocal integral model for bending of Euler-Bernoulli and Timoshenko beams". Int J Eng Sci 116:141–144
89. Tuna M, Leonetti L, Trovalusci P, Kirca M (2020) 'Explicit' and 'implicit' non-local continuous descriptions for a plate with circular inclusion in tension. Meccanica 55:927–944
90. Tuna M, Trovalusci P (2020) Scale dependent continuum approaches for discontinuous assemblies: 'explicit' and 'implicit' non-local models. Mech Res Commun 103:103461
91. Sokolowski M (1972) Theory of Couple-stresses in Bodies with Constrained Rotations. Springer
92. Lakes RS (1995) Experimental Methods for Study of Cosserat Elastic Solids and Other Generalized Continua. In Mü H (Ed) Continuum Models for Materials with Micro-structure. John Wiley
93. Providas E, Kattis MA (2002) Finite element method in plane Cosserat elasticity. Comput Struct 80:2059–2069
94. Polizzotto C, Fuschi P, Pisano AA (2006) A nonhomogeneous nonlocal elasticity model. Eur J Mech A-Solid 25:308–333
95. Ghosh S, Sundararaghavan V, Waas AM (2014) Construction of multidimensional isotropic kernels for nonlocal elasticity based on phonon dispersion data. Int J Solids Struct 51:392–401
96. Storn R, Price K (1997) Differential evolution-a Simple and efficient heuristic for global optimization over continuous spaces. J Global Optim 11:341–359
97. Pisano AA, Fuschi P (2018) Structural symmetry and boundary conditions for nonlocal symmetrical problems. Meccanica 53:629–638
98. Kaloni PN, Ariman T (1967) Stress concentration effects in micropolar elasticity. Z Angew Math Phys 18:136–141

Micromorphic Continuum Theory: Finite Element Analysis of 3D Elasticity with Applications in Beam- and Plate-Type Structures

Reza Ansari, Amir Norouzzadeh, and Hessam Rouhi

Abstract Due to the failure of classical elasticity to correctly model the behavior of small-scale structures as well as inhomogeneous media, a non-classical three-dimensional (3D) finite element formulation is developed on the basis of the micromorphic theory (MMT). Possessing micro-scale rotation, shear and stretch degrees of freedom (DOFs), MMT is an appropriate candidate to take the size- and microstructural-effects into consideration in mechanical problems. First, a general 3D formulation is proposed for the micromorphic solid continua which includes three stress and strain fields with 18 elastic constants. Then, the relations are written in matricized form which is advantageous for computational aims. Using the matrix-vector MMT formulation, a 3D micromorphic element with 12 DOFs (3 classical and 9 non-classical) is developed. Also, a robust scheme is used to determine the material parameters in terms of two classical constants in such a way that the positive-definiteness of the stored energy would be guaranteed. In the next step, the static deformations of micromorphic beams and plates with various kinds of edge supports are computed to reveal the efficiency of the method. The influences of length scale parameter on the bending responses of micromorphic structures with various geometrical properties are also analyzed. From comparing the results obtained from the classical and micromorphic elasticity theories, it is indicated that MMT results do not completely converge to those of the classical elasticity theory where the size does not matter. This is because of considering the micro-deformation DOFs in MMT and shows the microstructural-effects.

R. Ansari (✉) · A. Norouzzadeh · H. Rouhi
University of Guilan, Rasht, Iran
e-mail: r_ansari@guilan.ac.ir

A. Norouzzadeh
e-mail: a_norouzzadeh@webmail.guilan.ac.ir

H. Rouhi
e-mail: h_rouhi@guilan.ac.ir

© The Author(s), under exclusive license to Springer Nature Switzerland AG 2021
E. Ghavanloo et al. (eds.), *Size-Dependent Continuum Mechanics Approaches*,
Springer Tracts in Mechanical Engineering,
https://doi.org/10.1007/978-3-030-63050-8_12

1 Introduction

Based on experimental studies, the behaviours of microscopic and molecular structures are different from those of macro-scale structures, and it is not possible to properly predict their size-dependent behaviours based on the conventional continuum theories (e.g. see Refs. [1] and [2]). So, a number of non-classical theories were proposed which can capture the small-scale influences [3–7].

The micromorphic theory (MMT) is a size-dependent theory with unique features appropriate for taking the microstructure influences into account. Based on this theory, a material body is assumed to be a continuous set of deformable particles with finite size and inner structure which have 9 DOFs, in addition to the three standard DOFs of their centroid. In other words, each material particle has an extra micro-motion considered by "director" which has 9 independent components, associated with 6 DOFs for micro-deformations (shear and stretch) and 3 DOFs for micro-rotations.

There are several papers in which MMT and its micropolar counterpart (MPT) have been used in the field of solid mechanics. For example, Zhang et al. [8] proposed a finite element (FE) formulation in order to investigate the multi-body contact based on MPT. Also, Aganović et al. [9] presented the models of micropolar rods and plates. They mathematically justified their models according to the equations of linearized MPT. Pompei and Rigano [10] studied the bending response of viscoelastic plates in the context of MPT. Moreover, Jeong and Neff [11] investigated uniqueness, existence and stability of the weakest possible stress-strain equations for a linear elastic, static and isotropic micropolar model. Pietraszkiewicza and Eremeyev [12] described the strain measures in the nonlinear micropolar theory based on three approaches. Sansour et al. [13] proposed a micromorphic model for inelastic formulations at large strains in addition to for scale influences in specimen experiencing homogenous deformation. Ieşan [14] analyzed micromorphic elastic solids containing initial stresses and initial heat flux. Kumar and Kansal [15] solved the system of differential equations in the theory of micropolar thermoelastic diffusion with voids. Dos Reis and Ganghoffer [16] used homogenization techniques for discrete lattice structures and calculated the effective elastic micropolar properties. Zhang et al. [17] investigated the mechanical properties of a bimaterial strip subjected to simple shear based upon MMT. Also, using this theory, Cordero et al. [18] studied grain size influences on metal polycrystals. Altenbach and Eremeyev [19] discussed the constitutive equations of the nonlinear micropolar continuum using strain rates. Chowdhury et al. [20] proposed a state-based micropolar peridynamic theory for linear elastic solids with introducing additional micro-rotational DOFs to each material point.

Recently, developing size-dependent elements and non-classical FE approaches within the framework of MPT and MMT has attracted the attention of some researchers. For example, readers can follow the papers of Ansari and co-workers. By developing a prism element containing the micro-rotation and micro-deformation DOFs, Ansari et al. [21] presented a 3D finite element analysis (FEA) of micromorphic materials. They investigated the mechanical behavior of micromorphic plate-type structures on the basis of the Mindlin–Reissner plate theory [22]. In the framework of the geometrically nonlinear micropolar theory of Eringen, the bending characteristics of micropolar plates were studied in Ref. [23]. They also introduced a novel three-dimensional micropolar element and examined its efficiency in a wide range of applications [24]. Recently, Norouzzadeh et al. [25, 26] utilized the micromorphic continua to analyze the large deformations of small-scale shell-type structures. Also, based upon the original integral from and the differential formulation of Eringen's nonlocal elasticity combined with the micromorphic and micropolar theories, the simultaneous effects of stress-strain nonlocality and micro-scale deformations were captured in Refs. [27] and [28], respectively.

In the current work, a 3D size-dependent element is proposed based on MMT. To this end, first, MMT is generally formulated based on a new approach. Then, the obtained relations are matricized so as to develop the finite element formulations. The micromorphic beams and plates are chosen and their static bending response is studied using the proposed 3D micromorphic element. Selected numerical results are given to show the performance of element in predicting the bending response of micromorphic beams and plates under different types of boundary conditions and with various geometries. The influence of length scale parameter is illustrated, and the results calculated based on the classical elasticity theories are compared with the ones obtained based upon MMT.

2 Micromorphic Elasticity Theory

2.1 Kinematics

Based on MMT, a number of micro-elements are assumed in each macro-element whose positions are described in the current configuration using the following vector [29]

$$\boldsymbol{x}' = \boldsymbol{x} + \boldsymbol{\xi} \tag{1}$$

where $\boldsymbol{x}$ is the position vector of macro-element's centroid, and $\boldsymbol{\xi}$ shows the position of micro-element concerning macro-element's centroid. By denoting $\boldsymbol{x}$ and $\boldsymbol{\xi}$ as $\boldsymbol{X}$ and Ξ in the reference configuration, respectively, the position of micro-element in that configuration becomes

$$\boldsymbol{X}' = \boldsymbol{X} + \Xi \tag{2}$$

The deformation relation of body is thus expressed as follows using functions φ and ψ

$$\boldsymbol{x} = \varphi(\boldsymbol{X}, t), \quad \boldsymbol{\xi} = \psi(\Xi, \boldsymbol{X}, t) \tag{3}$$

Since the values of Ξ are very small, $\psi(\Xi, \boldsymbol{X}, t)$ is considered as a linear function of Ξ [29]. Hence

$$\boldsymbol{\xi} = \psi(\Xi, \boldsymbol{X}, t) = \boldsymbol{\chi}(\boldsymbol{X}, t)\,\Xi = \chi_{iI}\Xi_I \boldsymbol{e}_i \tag{4}$$

in which the second-order tensor $\boldsymbol{\chi}$ is called the micro-deformation tensor. Now, in order to obtain $d\boldsymbol{x}'$, it is required to compute $d\boldsymbol{x}$ and $d\boldsymbol{\xi}$. To this end, using Eqs. (3) and (4) one can write

$$d\boldsymbol{x} = \varphi(\boldsymbol{X} + d\boldsymbol{X}, t) - \varphi(\boldsymbol{X}, t) = \frac{\partial \varphi(\boldsymbol{X}, t)}{\partial \boldsymbol{X}} d\boldsymbol{X} = \boldsymbol{F} d\boldsymbol{X} = F_{iI} dX_I \boldsymbol{e}_i \tag{5a}$$

$$\boldsymbol{F} = \frac{\partial \varphi(\boldsymbol{X}, t)}{\partial \boldsymbol{X}} = \frac{\partial \boldsymbol{x}}{\partial \boldsymbol{X}} \tag{5b}$$

$$\begin{aligned} d\boldsymbol{\xi} &= d\boldsymbol{\chi}(\boldsymbol{X}, t)\,\Xi + \boldsymbol{\chi}(\boldsymbol{X}, t)\,d\Xi = \frac{\partial \boldsymbol{\chi}}{\partial \boldsymbol{X}} : (\Xi \otimes d\boldsymbol{X}) + \boldsymbol{\chi}(\boldsymbol{X}, t)\,d\Xi \\ &= \left(\chi_{iK,I}\Xi_K dX_I + \chi_{iI} d\Xi_I\right) \boldsymbol{e}_i \end{aligned} \tag{6}$$

where $\boldsymbol{F}$ stands for the deformation gradient tensor. Therefore, using Eq. (1) one has

$$\begin{aligned} d\boldsymbol{x}' &= d\boldsymbol{x} + d\boldsymbol{\xi} = \boldsymbol{F} d\boldsymbol{X} + \frac{\partial \boldsymbol{\chi}}{\partial \boldsymbol{X}} : (\Xi \otimes d\boldsymbol{X}) + \boldsymbol{\chi}(\boldsymbol{X}, t)\,d\Xi \\ &= \left(\left(\chi_{iK,I}\Xi_K + F_{iI}\right) dX_I + \chi_{iI} d\Xi_I\right) \boldsymbol{e}_i \end{aligned} \tag{7}$$

Accordingly,

$$\begin{aligned} dx'^2 &= d\boldsymbol{x}' \cdot d\boldsymbol{x}' = dx'_i dx'_i \\ &= \left(F_{iI}F_{iJ} + 2F_{iI}\chi_{iK,J}\Xi_K + \chi_{iK,I}\chi_{iL,J}\Xi_K\Xi_L\right) dX_I dX_J \\ &+ 2\left(F_{iI}\chi_{iJ} + \chi_{iJ}\chi_{iK,I}\Xi_K\right) dX_I d\Xi_J + \chi_{iI}\chi_{iJ} d\Xi_I d\Xi_J \\ &= \left(C_{IJ} + 2\Gamma_{IKJ}\Xi_K + \chi_{iK,I}\chi_{iL,J}\Xi_K\Xi_L\right) dX_I dX_J \\ &+ 2\left(\Psi_{IJ} + \chi_{iJ}\chi_{iK,I}\Xi_K\right) dX_I d\Xi_J + \chi_{iI}\chi_{iJ} d\Xi_I d\Xi_J \end{aligned} \tag{8}$$

where

$$C_{IJ} = F_{iI}F_{iJ}, \quad \Psi_{IJ} = F_{iI}\chi_{iJ}, \quad \Gamma_{IKJ} = F_{iI}\chi_{iK,J} \tag{9}$$

In these relations, C_{IJ} denotes the right Cauchy-Green deformation tensor; also, Ψ_{IJ} and Γ_{IKJ} are new kinematic quantities which have been made in the micromorphic theory.

Using Eq. (2), $d\boldsymbol{X}'$ is expressed as

$$d\boldsymbol{X}' = d\boldsymbol{X} + d\boldsymbol{\Xi} = (dX_I + d\Xi_I)\,\boldsymbol{E}_I \tag{10}$$

which results in

$$dX'^2 = d\boldsymbol{X}' \cdot d\boldsymbol{X}' = dX_I' dX_I' = dX_I dX_I + 2dX_I d\Xi_I + d\Xi_I d\Xi_I \tag{11}$$

Thus,

$$\begin{aligned}
&dx'^2 - dX'^2 = \left(C_{IJ} + 2\Gamma_{IKJ}\Xi_K + \chi_{iK,I}\chi_{iL,J}\Xi_K\Xi_L - \delta_{IJ}\right) dX_I dX_J \\
&+2\left(\Psi_{IJ} + \chi_{iJ}\chi_{iK,I}\Xi_K - \delta_{IJ}\right) dX_I d\Xi_J + \left(\chi_{iI}\chi_{iJ} - \delta_{IJ}\right) d\Xi_I d\Xi_J \\
&= 2\left(E_{IJ} + \Gamma_{IKJ}\Xi_K + \frac{1}{2}\chi_{iK,I}\chi_{iL,J}\Xi_K\Xi_L\right) dX_I dX_J \\
&+2\left(\mathcal{E}_{IJ} + \chi_{iJ}\chi_{iK,I}\Xi_K\right) dX_I d\Xi_J + \left(\chi_{iI}\chi_{iJ} - \delta_{IJ}\right) d\Xi_I d\Xi_J
\end{aligned} \tag{12}$$

in which E_{IJ} and $\mathcal{E}_{IJ}$ are defined as

$$\begin{aligned}
E_{IJ} &= \frac{1}{2}\left(C_{IJ} - \delta_{IJ}\right) = \frac{1}{2}\left(F_{iI}F_{iJ} - \delta_{IJ}\right), \\
\mathcal{E}_{IJ} &= \Psi_{IJ} - \delta_{IJ} = F_{iI}\chi_{iJ} - \delta_{IJ}
\end{aligned} \tag{13}$$

Ignoring the second-order terms in Eq. (12) with respect to the components of vector Ξ leads to

$$dx'^2 - dX'^2 = 2E_{IJ}dX_I dX_J + 2\mathcal{E}_{IJ}dX_I d\Xi_J + 2\Gamma_{IKJ}\Xi_K dX_I dX_J \tag{14}$$

If $\boldsymbol{U} = \boldsymbol{x} - \boldsymbol{X}$ denotes the displacement vector, $d\boldsymbol{U}$ can be formulated as

$$d\boldsymbol{U} = d\boldsymbol{x} - d\boldsymbol{X} = (\boldsymbol{F} - \boldsymbol{I})\,d\boldsymbol{X} \tag{15}$$

The deformation gradient tensor is thus written as

$$\boldsymbol{F} = \boldsymbol{I} + \frac{\partial \boldsymbol{U}}{\partial \boldsymbol{X}} \tag{16}$$

In addition, by introducing the micro–displacement tensor, $\tilde{\boldsymbol{\Phi}}$, the micro–deformation tensor can be given as [29]

$$\chi = \boldsymbol{I} + \tilde{\boldsymbol{\Phi}} \tag{17}$$

As a result, the strain tensors are written in terms of displacement vector and micro-displacement tensor as follows

$$\begin{aligned} C_{IJ} &= \delta_{IJ} + u_{J,I} + u_{I,J} + u_{K,I}u_{K,J} \\ \Rightarrow \tilde{E}_{IJ} &= \frac{1}{2}\left(u_{J,I} + u_{I,J} + u_{K,I}u_{K,J}\right) \end{aligned} \tag{18}$$

$$\begin{aligned} \Psi_{IJ} &= \delta_{IJ} + u_{J,I} + \phi_{IJ} + u_{K,I}\phi_{KJ} \\ \Rightarrow \tilde{\mathcal{E}}_{IJ} &= u_{J,I} + \phi_{IJ} + u_{K,I}\phi_{KJ} \end{aligned} \tag{19}$$

$$\tilde{\Gamma}_{IJK} = \phi_{IJ,K} + u_{L,I}\phi_{LJ,K} \tag{20}$$

in which $\tilde{E}_{IJ}$, $\tilde{\mathcal{E}}_{IJ}$ and $\tilde{\Gamma}_{IJK}$ are given by

$$\tilde{\boldsymbol{E}} = \frac{1}{2}\left(\nabla \boldsymbol{U} + \nabla^{\mathrm{T}}\boldsymbol{U} + \nabla^{\mathrm{T}}\boldsymbol{U}\nabla \boldsymbol{U}\right) \tag{21}$$

$$\tilde{\mathcal{E}} = \nabla^{\mathrm{T}}\boldsymbol{U} + \tilde{\boldsymbol{\Phi}} + \nabla^{\mathrm{T}}\boldsymbol{U}\boldsymbol{\Phi} \tag{22}$$

$$\tilde{\boldsymbol{\Gamma}} = \nabla\tilde{\boldsymbol{\Phi}} + \nabla^{\mathrm{T}}\boldsymbol{U}\nabla\tilde{\boldsymbol{\Phi}} \tag{23}$$

Within the framework of linear micromorphic theory, the above relations are rewritten as [21, 22]

$$\tilde{\boldsymbol{e}} = \frac{1}{2}\left(\nabla \boldsymbol{U} + \nabla^{\mathrm{T}}\boldsymbol{U}\right) \tag{24}$$

$$\tilde{\varepsilon} = \nabla^{\mathrm{T}}\boldsymbol{U} + \tilde{\boldsymbol{\Phi}} \tag{25}$$

$$\tilde{\gamma} = \nabla\tilde{\boldsymbol{\Phi}} \tag{26}$$

2.2 Equation of Motion

According to MMT, the balance of momentum and moment of momentum in tensor and index forms are as follows [29]

$$\begin{aligned} &\operatorname{div}\left(\tilde{\boldsymbol{\sigma}}\right) + \rho\boldsymbol{f} = \rho\ddot{\boldsymbol{u}}, \quad \tilde{\boldsymbol{\sigma}} - \tilde{\boldsymbol{s}} + \operatorname{div}\left(\tilde{\boldsymbol{m}}\right) + \rho\boldsymbol{l} = \rho\boldsymbol{j}\ddot{\phi} \\ &\sigma_{ij,j} + \rho f_i = \rho\ddot{u}_i, \quad \sigma_{ij} - s_{ij} + m_{ij,k} + \rho l_{ij} = \rho j_{ik}\ddot{\phi}_{kj} \end{aligned} \tag{27}$$

where $\tilde{\boldsymbol{\sigma}}$ and $\tilde{\boldsymbol{s}}$ are non-symmetric and symmetric stress tensors; is third-order couple stress tensor; $\boldsymbol{f}$ and $\boldsymbol{l}$ are body force vector and body couple tensor; $\boldsymbol{u}$ and ϕ are displacement vector and micro-rotation tensors and ρ is mass density. Also, $\boldsymbol{j}$ signals micro-inertia tensor that can be stated as $j_{ik} = j\delta_{ik}$ for a micro-isotropic material.

2.3 Elasticity

According to Eq. (14), the strain energy density is considered as a function of $\tilde{\boldsymbol{E}}$, $\tilde{\mathcal{E}}$ and $\tilde{\boldsymbol{\Gamma}}$. Subsequently, to write a constitutive equation between stresses and strains, the strain energy density must be stated in a quadratic form with respect to strains. Hence, it is given in the following form [25, 26]

$$\hat{W}\left(\tilde{\boldsymbol{e}}, \tilde{\varepsilon}, \tilde{\gamma}\right) = \frac{1}{2}\tilde{\gamma}\vdots\tilde{\boldsymbol{A}}\vdots\tilde{\gamma} + \frac{1}{2}\tilde{\varepsilon} : \tilde{\boldsymbol{B}} : \tilde{\varepsilon} + \frac{1}{2}\tilde{\boldsymbol{e}} : \tilde{\boldsymbol{C}} : \tilde{\boldsymbol{e}} + \tilde{\varepsilon} : \tilde{\boldsymbol{D}} : \tilde{\boldsymbol{e}} + \tilde{\boldsymbol{e}} : \tilde{\boldsymbol{F}}\vdots\tilde{\gamma} + \tilde{\varepsilon} : \tilde{\boldsymbol{G}}\vdots\tilde{\gamma} \tag{28}$$

by which the material stresses of MMT can be introduced as

$$\tilde{\boldsymbol{\sigma}} = \frac{\partial\hat{W}}{\partial\tilde{\boldsymbol{e}}} + \frac{\partial\hat{W}}{\partial\tilde{\varepsilon}} = \tilde{\boldsymbol{B}} : \tilde{\varepsilon} + \tilde{\boldsymbol{C}} : \tilde{\boldsymbol{e}} + \tilde{\varepsilon} : \tilde{\boldsymbol{D}} + \tilde{\boldsymbol{G}}\vdots\tilde{\gamma} \tag{29}$$

$$\tilde{\boldsymbol{s}} = \frac{\partial\hat{W}}{\partial\tilde{\boldsymbol{e}}} + \frac{\partial\hat{W}}{\partial\tilde{\varepsilon}} + \frac{\partial\hat{W}}{\partial\tilde{\varepsilon}^{\mathrm{T}}} = \tilde{\boldsymbol{B}} : \tilde{\varepsilon} + \tilde{\boldsymbol{e}} : \tilde{\boldsymbol{D}} + \tilde{\boldsymbol{F}}\vdots\tilde{\gamma} \tag{30}$$

$$\tilde{\boldsymbol{m}} = \frac{\partial\hat{W}}{\partial\tilde{\gamma}} = \tilde{\boldsymbol{A}}\vdots\tilde{\gamma} + \tilde{\varepsilon} : \tilde{\boldsymbol{F}} + \tilde{\boldsymbol{e}} : \tilde{\boldsymbol{G}} \tag{31}$$

In these relations, $\tilde{\boldsymbol{A}}$, $\tilde{\boldsymbol{B}}$, $\tilde{\boldsymbol{C}}$, $\tilde{\boldsymbol{D}}$, $\tilde{\boldsymbol{F}}$ and $\tilde{\boldsymbol{G}}$ are the elastic constants of material. On the basis of Eq. (28), the following symmetries are present

$$\begin{aligned} &A_{IJKLMN} = A_{LMNIJK}, \quad B_{IJKL} = B_{KLIJ}, \\ &C_{IJKL} = C_{JIKL} = C_{IJLK} = C_{KLIJ}, \quad D_{IJKL} = D_{JIKL}, \\ &G_{IJKLM} = G_{JIKLM} \end{aligned} \tag{32}$$

For an isotropic solid, $\tilde{\boldsymbol{F}}$ and $\tilde{\boldsymbol{G}}$, which are odd-order tensors, become equal to 0. Moreover, other elastic constants should be homogenous and linear functions of Kronecker delta. So one has [29]

$$\begin{aligned} A_{IJKLMN} = {} & a_1\left(\delta_{IJ}\delta_{KL}\delta_{MN} + \delta_{IN}\delta_{JK}\delta_{LM}\right) \\ & + a_2\left(\delta_{IJ}\delta_{KM}\delta_{LN} + \delta_{IK}\delta_{JN}\delta_{LM}\right) \\ & + a_3\delta_{IJ}\delta_{KN}\delta_{LM} + a_4\delta_{IL}\delta_{JK}\delta_{MN} \\ & + a_5\left(\delta_{IK}\delta_{JL}\delta_{MN} + \delta_{IM}\delta_{JK}\delta_{LN}\right) \\ & + a_6\delta_{IK}\delta_{JM}\delta_{LN} + a_7\delta_{IL}\delta_{JM}\delta_{KN} \\ & + a_8\left(\delta_{IM}\delta_{JN}\delta_{KL} + \delta_{IN}\delta_{JL}\delta_{KM}\right) \\ & + a_9\delta_{IL}\delta_{JN}\delta_{KM} + a_{10}\delta_{IM}\delta_{JL}\delta_{KN} \\ & + a_{11}\delta_{IN}\delta_{JM}\delta_{KL} \end{aligned} \tag{33a}$$

$$B_{IJKL} = (\eta - \tau)\,\delta_{IJ}\delta_{KL} + (\kappa - \sigma)\,\delta_{IK}\delta_{JL} + (\chi - \sigma)\,\delta_{IL}\delta_{JK} \tag{33b}$$

$$C_{IJKL} = \lambda\delta_{IJ}\delta_{KL} + \mu\,(\delta_{IK}\delta_{JL} + \delta_{IL}\delta_{JK}) \tag{33c}$$

$$D_{IJKL} = \tau\delta_{IJ}\delta_{KL} + \sigma\,(\delta_{IK}\delta_{JL} + \delta_{IL}\delta_{JK}) \tag{33d}$$

$$F_{IJKLM} = G_{IJKLM} = 0 \tag{33e}$$

It can be shown that [25, 26] only tensor $\tilde{\boldsymbol{A}}$ consists of length-scale parameters and there is no size-dependency in the non-classical tensors $\tilde{\boldsymbol{B}}$ and $\tilde{\boldsymbol{D}}$. So, the energy of MMT material is different from the CT counterpart even where the length-scale is negligible.

2.4 Conditions of Elastic Parameters

Clearly, the stored energy of structure must be positive. In case of the micromorphic theory with non-classical parameters, Smith [30] proposed the required conditions on elastic constants to satisfy the positive-definiteness of energy. The following constraints holds for the fourth-order elastic tensors

$$\begin{aligned}
&\mu > 0, \quad \kappa + \chi > 2\sigma, \quad \kappa - \chi > 0, \quad 3\lambda + 2\mu > 0,\\
&(\kappa + \chi - 2\sigma)\,\mu > 2\sigma^2, \quad \kappa + \chi + 3\eta > 3\tau + 2\sigma,\\
&(\kappa + \chi + 3\eta - 3\tau - 2\sigma)\,(3\lambda + 2\mu) > (3\tau + 2\sigma)^2
\end{aligned} \tag{34}$$

Also, one should consider the following conditions for the sixth-order tensor

$$\begin{aligned}
&a_4 + 2a_{11} > |a_5 + a_6 + a_7|\,,\\
&a_4 - a_{11} > \sqrt{(a_5 - a_6)^2 + (a_6 - a_7)^2 + (a_7 - a_5)^2/\sqrt{2}}
\end{aligned} \tag{35}$$

where

$$\begin{aligned}
&\boldsymbol{a} = \left[a_{ij}\right]_{3\times 3},\\
&a_{11} = 3a_1 + a_4 + a_6 + a_8 + a_9,\\
&a_{12} = a_3 + a_7 + 3a_8 + a_{10} + a_{11}, \quad a_{13} = a_2 + a_5 + 3a_9 + a_{10} + a_{11},\\
&a_{21} = a_1 + a_7 + 3a_8 + a_9 + a_{11}, \quad a_{22} = 3a_3 + a_4 + a_5 + a_8 + a_{10},\\
&a_{23} = a_2 + a_6 + a_9 + 3a_{10} + a_{11}, \quad a_{31} = a_1 + a_5 + a_8 + 3a_9 + a_{11},\\
&a_{32} = a_3 + a_6 + a_8 + 3a_{10} + a_{11}, \quad a_{33} = 3a_2 + a_4 + a_7 + a_9 + a_{10}
\end{aligned} \tag{36}$$

and $\boldsymbol{C}_a$ is the cofactor of $\boldsymbol{a}$.

In this stage, by selecting several coefficients, the inequality conditions are converted to a set of equality constraints. For determined values of λ, μ and κ, one arrives at

$$\begin{aligned}
\chi &= \bar{a}\kappa, \\
\sigma &= \left(\frac{\bar{a}+1}{2}\right)\bar{b}\kappa, \\
\eta &= \frac{\bar{c}}{3}\left(\frac{1-\bar{d}}{\bar{d}^2}(3\lambda+2\mu) - (\bar{a}+1)\kappa\right), \\
\tau &= \frac{1}{3}\left((\bar{a}+1)\left(\bar{d}(1-\bar{c})-\bar{b}\right)\kappa + \bar{c}\bar{d}\frac{1-\bar{d}}{\bar{d}^2}(3\lambda+2\mu)\right)
\end{aligned} \tag{37}$$

in which $\bar{a}$, $\bar{b}$, $\bar{c}$ and $\bar{d}$ are coefficients in the interval [0, 1] for positive values of all parameters. In addition, if $\kappa = \alpha\mu$, then $\alpha < 2\left(1-\bar{b}\right)/(\bar{a}+1)\bar{b}^2$.

3 Matrix-Vector Representation

In this section, the strain energy density is restated via matrix relations for the computational purposes. It can be written as

$$\hat{W} = \frac{1}{2}\left(\boldsymbol{\gamma}^{\mathrm{T}}\boldsymbol{A}\boldsymbol{\gamma} + \boldsymbol{\varepsilon}^{\mathrm{T}}\boldsymbol{B}\boldsymbol{\varepsilon} + \boldsymbol{e}^{\mathrm{T}}\boldsymbol{C}\boldsymbol{e} + \boldsymbol{e}^{\mathrm{T}}\boldsymbol{D}\boldsymbol{\varepsilon} + \boldsymbol{\varepsilon}^{\mathrm{T}}\boldsymbol{D}^{\mathrm{T}}\boldsymbol{e}\right) \tag{38}$$

where

$$\boldsymbol{e} = \left[e_{11}\ e_{22}\ e_{33}\ 2e_{23}\ 2e_{31}\ 2e_{12}\right]^{\mathrm{T}} \tag{39}$$

$$\boldsymbol{\varepsilon} = \left[\varepsilon_{11}\ \varepsilon_{22}\ \varepsilon_{33}\ \varepsilon_{23}\ \varepsilon_{32}\ \varepsilon_{31}\ \varepsilon_{13}\ \varepsilon_{12}\ \varepsilon_{21}\right]^{\mathrm{T}} \tag{40}$$

$$\boldsymbol{\gamma} = \begin{bmatrix}\boldsymbol{\gamma}_1\\ \boldsymbol{\gamma}_2\\ \boldsymbol{\gamma}_3\\ \boldsymbol{\gamma}_4\end{bmatrix}_{27\times 1}$$

$$\begin{aligned}
\text{where}\quad \boldsymbol{\gamma}_1 &= \left[\gamma_{111}\ \gamma_{221}\ \gamma_{122}\ \gamma_{212}\ \gamma_{331}\ \gamma_{133}\ \gamma_{313}\right]^{\mathrm{T}}, \\
\boldsymbol{\gamma}_2 &= \left[\gamma_{222}\ \gamma_{112}\ \gamma_{211}\ \gamma_{121}\ \gamma_{332}\ \gamma_{233}\ \gamma_{323}\right]^{\mathrm{T}}, \\
\boldsymbol{\gamma}_3 &= \left[\gamma_{333}\ \gamma_{113}\ \gamma_{311}\ \gamma_{131}\ \gamma_{223}\ \gamma_{322}\ \gamma_{232}\right]^{\mathrm{T}}, \\
\boldsymbol{\gamma}_4 &= \left[\gamma_{231}\ \gamma_{321}\ \gamma_{312}\ \gamma_{132}\ \gamma_{123}\ \gamma_{213}\right]^{\mathrm{T}}
\end{aligned} \tag{41}$$

The matrices $\boldsymbol{A}$, $\boldsymbol{B}$, $\boldsymbol{C}$ and $\boldsymbol{D}$ are expressed in Appendix A. Using the following displacement vector

$$\boldsymbol{U} = \left[u_1\ u_2\ u_3\right]^{\mathrm{T}} \tag{42}$$

and the following micro-displacement vector

$$\boldsymbol{\Phi} = \left[\phi_{11}\ \phi_{22}\ \phi_{33}\ \phi_{23}\ \phi_{32}\ \phi_{31}\ \phi_{13}\ \phi_{12}\ \phi_{21}\right]^{\mathrm{T}} \tag{43}$$

the vectors $\boldsymbol{e}$, ε and γ are expressed as

$$\boldsymbol{e} = \boldsymbol{E}_1\boldsymbol{U} \tag{44}$$

$$\varepsilon = \boldsymbol{\Phi} + \boldsymbol{E}_2\boldsymbol{U} \tag{45}$$

$$\gamma = \boldsymbol{E}_3\boldsymbol{\Phi} \tag{46}$$

The differential operators of the preceding relations are presented in Appendix B. The variations of Eqs. (44)–(46) are written as

$$\delta\boldsymbol{e} = \boldsymbol{E}_1\delta\boldsymbol{U} \tag{47}$$

$$\delta\varepsilon = \delta\boldsymbol{\Phi} + \boldsymbol{E}_2\delta\boldsymbol{U} \tag{48}$$

$$\delta\gamma = \boldsymbol{E}_3\delta\boldsymbol{\Phi} \tag{49}$$

The strain energy density given in Eq. (38) becomes

$$\hat{W} = \frac{1}{2}\left(\gamma^{\mathrm{T}}\boldsymbol{A}\gamma + \varepsilon^{\mathrm{T}}\boldsymbol{B}\varepsilon + \boldsymbol{e}^{\mathrm{T}}\boldsymbol{C}\boldsymbol{e} + \boldsymbol{e}^{\mathrm{T}}\boldsymbol{D}\varepsilon + \varepsilon^{\mathrm{T}}\boldsymbol{D}^{\mathrm{T}}\boldsymbol{e}\right) \tag{50}$$

whose variation is

$$\delta\hat{W} = \delta\gamma^{\mathrm{T}}\boldsymbol{A}\gamma + \delta\varepsilon^{\mathrm{T}}\boldsymbol{B}\varepsilon + \delta\boldsymbol{e}^{\mathrm{T}}\boldsymbol{C}\boldsymbol{e} + \delta\boldsymbol{e}^{\mathrm{T}}\boldsymbol{D}\varepsilon + \delta\varepsilon^{\mathrm{T}}\boldsymbol{D}^{\mathrm{T}}\boldsymbol{e} \tag{51}$$

and can be stated as

$$\begin{aligned}\delta\hat{W} &= \delta\boldsymbol{\Phi}^{\mathrm{T}}\boldsymbol{E}_3^{\mathrm{T}}\boldsymbol{A}\boldsymbol{E}_3\boldsymbol{\Phi} + \left(\delta\boldsymbol{\Phi}^{\mathrm{T}} + \delta\boldsymbol{U}\boldsymbol{E}_2^{\mathrm{T}}\right)\boldsymbol{B}\left(\boldsymbol{\Phi} + \boldsymbol{E}_2\boldsymbol{U}\right)\\ &+ \delta\boldsymbol{U}^{\mathrm{T}}\boldsymbol{E}_1^{\mathrm{T}}\boldsymbol{C}\boldsymbol{E}_1\boldsymbol{U} + \delta\boldsymbol{U}^{\mathrm{T}}\boldsymbol{E}_1\boldsymbol{D}\left(\boldsymbol{\Phi} + \boldsymbol{E}_2\boldsymbol{U}\right)\\ &+ \left(\delta\boldsymbol{\Phi}^{\mathrm{T}} + \delta\boldsymbol{U}\boldsymbol{E}_2^{\mathrm{T}}\right)\boldsymbol{D}^{\mathrm{T}}\boldsymbol{E}_1\boldsymbol{U}\end{aligned} \tag{52}$$

The variation of strain energy is finally written as

$$\delta W = \int_V \delta\hat{W} dV \tag{53}$$

4 Finite Element Formulation

In this section, the relations of the 3D micromorphic element shown in Fig. 1 are derived. With considering the shape functions of $\boldsymbol{N}$ and $\bar{\boldsymbol{N}}$, the vectors $\boldsymbol{U}$ and $\boldsymbol{\Phi}$ can be interpolated as

$$\boldsymbol{U} = \boldsymbol{N}\boldsymbol{d}_1, \quad \boldsymbol{\Phi} = \bar{\boldsymbol{N}}\boldsymbol{d}_2 \tag{54}$$

By defining

$$\boldsymbol{B}_i = \boldsymbol{E}_i\boldsymbol{N}, \quad \bar{\boldsymbol{B}}_i = \boldsymbol{E}_i\bar{\boldsymbol{N}} \tag{55}$$

$\delta\hat{W}_e$ in each element is

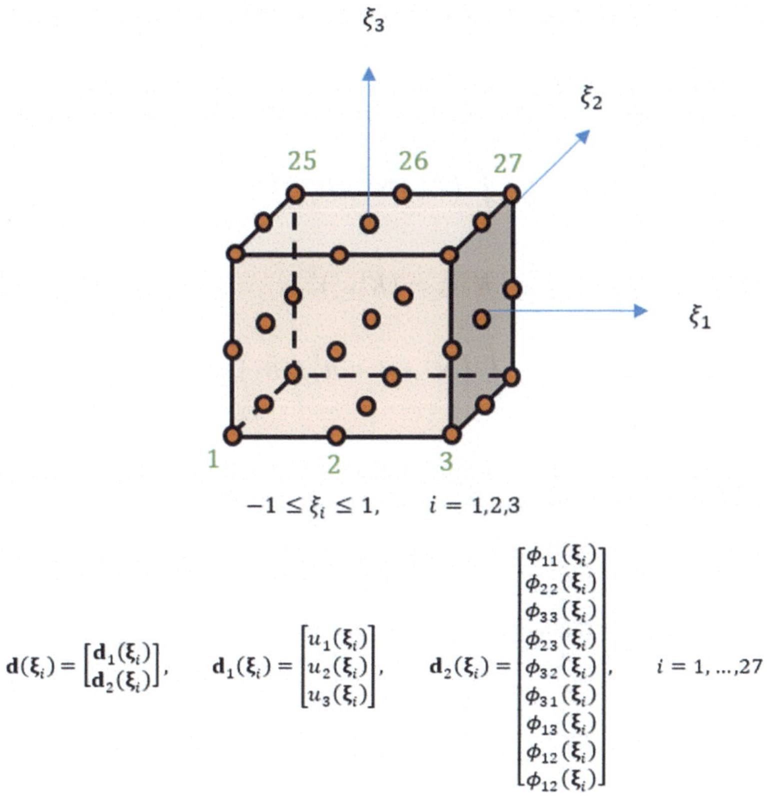

Fig. 1 Schematic view of the 3D micromorphic element

$$\delta\hat{W}_e = \delta\boldsymbol{d}_2^{\mathrm{T}}\bar{\boldsymbol{B}}_3^{\mathrm{T}}\boldsymbol{A}\boldsymbol{B}_3\boldsymbol{d}_2 + \left(\delta\boldsymbol{d}_2^{\mathrm{T}}\bar{\boldsymbol{N}} + \delta\boldsymbol{d}_1\boldsymbol{B}_2^{\mathrm{T}}\right)\boldsymbol{B}\left(\bar{\boldsymbol{N}}\boldsymbol{d}_2 + \boldsymbol{B}_2\boldsymbol{d}_1\right) + \delta\boldsymbol{d}_1^{\mathrm{T}}\boldsymbol{B}_1^{\mathrm{T}}\boldsymbol{C}\boldsymbol{B}_1\boldsymbol{d}_1 + \delta\boldsymbol{d}_1^{\mathrm{T}}\boldsymbol{B}_1\boldsymbol{D}\left(\bar{\boldsymbol{N}}\boldsymbol{d}_2 + \boldsymbol{B}_2\boldsymbol{d}_1\right) + \left(\delta\boldsymbol{d}_2^{\mathrm{T}}\bar{\boldsymbol{N}} + \delta\boldsymbol{d}_1\boldsymbol{B}_2^{\mathrm{T}}\right)\boldsymbol{D}^{\mathrm{T}}\boldsymbol{B}_1\boldsymbol{d}_1 \tag{56}$$

Also, for the strain energy one can write

$$\delta W_e = \int\limits_{V_e} \delta\hat{W}_e J\, dV = \delta\boldsymbol{d}_1^{\mathrm{T}}\left(\boldsymbol{K}_{11}\boldsymbol{d}_1 + \boldsymbol{K}_{12}\boldsymbol{d}_2\right) + \delta\boldsymbol{d}_2^{\mathrm{T}}\left(\boldsymbol{K}_{21}\boldsymbol{d}_1 + \boldsymbol{K}_{22}\boldsymbol{d}_2\right) = \delta\boldsymbol{d}^{\mathrm{T}}\boldsymbol{K}\boldsymbol{d} \tag{57}$$

where

$$\boldsymbol{d} = \begin{bmatrix}\boldsymbol{d}_1\\ \boldsymbol{d}_2\end{bmatrix}, \quad \boldsymbol{K} = \begin{bmatrix}\boldsymbol{K}_{11} & \boldsymbol{K}_{12}\\ \boldsymbol{K}_{21} & \boldsymbol{K}_{22}\end{bmatrix} \tag{58}$$

and the components of the stiffness matrix is given by

$$\boldsymbol{K}_{11} = \int\limits_{V_e}\left(\boldsymbol{B}_1^{\mathrm{T}}\boldsymbol{C}\boldsymbol{B}_1 + \boldsymbol{B}_2^{\mathrm{T}}\boldsymbol{B}\boldsymbol{B}_2 + \boldsymbol{B}_1^{\mathrm{T}}\boldsymbol{D}\boldsymbol{B}_2 + \boldsymbol{B}_2^{\mathrm{T}}\boldsymbol{D}^{\mathrm{T}}\boldsymbol{B}_1\right)dV \tag{59}$$

$$\boldsymbol{K}_{12;s} = \int\limits_{V_e}\left(\boldsymbol{B}_2^{\mathrm{T}}\boldsymbol{A}\bar{\boldsymbol{N}} + \boldsymbol{B}_1^{\mathrm{T}}\boldsymbol{D}\bar{\boldsymbol{N}}\right)dV \tag{60}$$

$$\boldsymbol{K}_{21;s} = \left(\boldsymbol{K}_{12;s}^{l}\right)^{\mathrm{T}} \tag{61}$$

$$\boldsymbol{K}_{22;s} = \int\limits_{V_e}\left(\bar{\boldsymbol{N}}^{\mathrm{T}}\boldsymbol{A}\bar{\boldsymbol{N}} + \bar{\boldsymbol{B}}_3^{\mathrm{T}}\boldsymbol{H}\bar{\boldsymbol{B}}_3\right)dV \tag{62}$$

If one denotes the force vector of element by $\boldsymbol{F}$, the final algebraic equations of the finite element micromorphic formulation is written as follows

$$\boldsymbol{K}\boldsymbol{d} = \boldsymbol{F} \tag{63}$$

Performing the assemblage process over elements (e)

$$\mathbb{K} = \sum^{e}\boldsymbol{K}, \quad \mathbb{F} = \sum^{e}\boldsymbol{F} \tag{64}$$

one has

$$\mathbb{K}\boldsymbol{d} = \mathbb{F} \tag{65}$$

and the vector of nodal variables $\boldsymbol{d}$ is obtained.

5 Results and Discussion

Here, the proposed 3D element is utilized to investigate the bending behaviour of micromorphic beams and plates. The schematic views of discretized beam and plate are shown in Fig. 2. The CC, CF, CS and SS boundary conditions for beams; and CCCC, CCSS, CSCS and SSSS boundary conditions for plates are considered. Note that C, S and F denote clamped, simply-supported and free, respectively. The set of essential boundary conditions for beams and plates are respectively defined as follows [22]

$$\begin{aligned} \mathrm{C}:&\ \ x = 0, a \ \ u_1, u_2, u_3, \phi_{ij} = 0 \\ \mathrm{S}:&\ \ x = 0, a \ \ u_2, u_3, \phi_{ij} = 0 \end{aligned} \tag{66}$$

$$\begin{aligned} \mathrm{C}:&\ \ \begin{cases} x = 0, a \ \ u_1, u_2, u_3, \phi_{ij} = 0 \\ y = 0, b \ \ u_1, u_2, u_3, \phi_{ij} = 0 \end{cases} \\ \mathrm{S}:&\ \ \begin{cases} x = 0, a \ \ u_2, u_3, \phi_{ij} = 0 \\ y = 0, b \ \ u_1, u_3, \phi_{ij} = 0 \end{cases} \end{aligned} \tag{67}$$

It should be mentioned that no constraint should be considered for the free ends or edges.

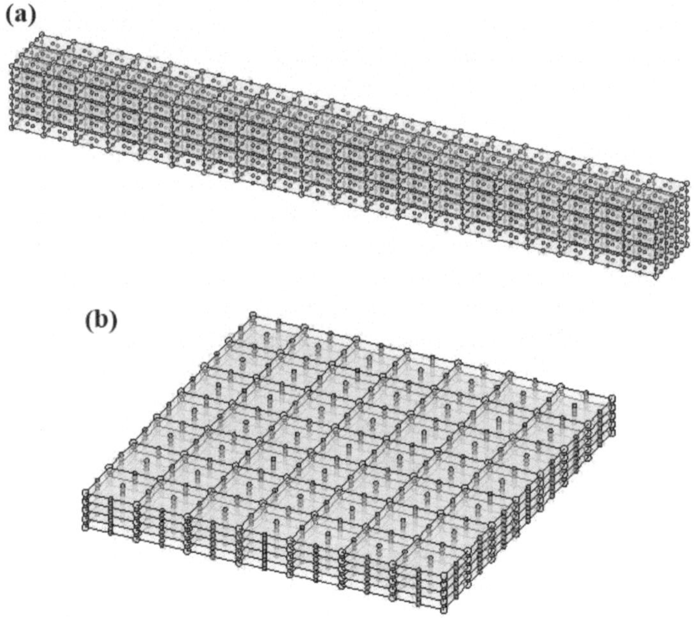

Fig. 2 Schematic view of discretized 3D micromorphic **a** beam and **b** plate

From Sect. 2, it was found that by choosing appropriate values for the coefficients $\bar{a}, \bar{b}, \bar{c}$ and $\bar{d}$, elastic constants of MMT can be specified with guaranteeing compliance with the conditions. By considering [31]

$$G = \frac{2\mu + \kappa}{2}, \quad \nu = \frac{\lambda}{2\lambda + 2\mu + \kappa}, \quad N^2 = \frac{\kappa}{2(\mu + \kappa)} \tag{68}$$

the following relations are given

$$\lambda = \frac{2G\nu}{1 - 2\nu}, \quad \mu = \frac{G(1 - 2N^2)}{1 - N^2}, \quad \kappa = \frac{2GN^2}{1 - N^2} \tag{69}$$

In the following, the set of coefficients are selected as $\bar{a} = 0.1$, $\bar{b} = 0.5$, $\bar{c} = 0.1$ and $\bar{d} = 0.1$. The polyurethane foam is assumed in this study with [31] $G = 104$ MPa, $\nu = 0.44$, $N^2 = 0.04$.

Also, in case of the six-order elastic tensor, it is considered that $a_i = 0, i = 1 : 11$, $i \neq 7$ and $a_4 = \gamma$, where $\gamma = 4Gl^2$ and l is the characteristic bending length equal to 0.327 mm.

The side length dimensions for plates are assumed to be $a = b = L$, and $a = L$, $b = h$ for beams. Besides, the transverse uniform distributed load is considered as $P = \frac{1}{L} N/m^3$.

The convergence study of the proposed 3D finite element analysis of beam- and plate-type micromorphic structures are respectively given in Tables 1 and 2. The maximum deflections are reported for different types of boundary conditions as well as different number of nodes in three dimensions. As a result, an appropriate convergence behaviour can be seen in all MMT beams and plates.

In Figs. 3 and 4, the non-dimensional maximum deflection of micromorphic beams and plates is plotted versus h/l (non-dimensional length scale parameter) based on both micromorphic and classical elasticity theories. The results of these figures are given for three values of a/h. The considerable effect of size on the bending of beams and plates is clearly seen especially for small values of h/l. An important

Table 1 Non-dimensional maximum deflection ($(\lambda + 2\mu + \kappa)\, w/PL^4 \times 10^2$) of beams

Boundary conditions	Number of nodes in each direction			
	[17, 3, 5]	[21, 5, 7]	[25, 5, 7]	[27, 7, 7]
CC	0.6167	0.7380	0.7614	0.7638
CF	21.3853	26.9330	29.2844	30.5710
CS	1.2349	1.3839	1.4394	1.4479
SS	3.1704	3.2466	3.2772	3.2885

Table 2 Non-dimensional maximum deflection ($(\lambda + 2\mu + \kappa) w/PL^4 \times 10^3$) of plates

Boundary conditions	Number of nodes in each direction			
	[13, 13, 5]	[15, 15, 7]	[15, 15, 9]	[17, 17, 9]
CCCC	3.0028	3.2871	3.3617	3.3702
CCSS	4.5750	4.8112	4.8873	4.8916
CSCS	5.1448	5.3209	5.4014	5.4189
SSSS	9.0337	9.1794	9.2236	9.2340

observation here is that the predictions of MMT do not completely converge to the results of CT at large values of h/l, although the discrepancy between two sets of results diminishes significantly. It can be explained by the fact that in MMT, the director is considered to be deformable, i.e. it includes micro-stretches and micro-shears in addition to micro-rotations. Accordingly, the energy of material particles obtained from MMT differs from that calculated based on CT. It should be noted that the results of MPT should converge to those of CT at large scales since the director is considered to be rigid in that theory.

Figures 5 and 6 show the effect of dimensionless length scale parameter on the deflection of micromorphic beams and plates. In these figures, the dimensionless deflection is plotted along the length of beams/plates for different values of h/l. The classical results are also presented for the comparison purpose. It is observed that there is a large difference between the prediction of MMT and its classical counterpart when the thickness of structure becomes equal to the characteristic bending length. As the thickness increases, the difference tends to decrease. Again, it is seen that the difference between two theories does not completely disappear at large sizes. According to Figs. 5 and 6, the dimensionless deflection of structure increases with increasing h/l.

As the last set of numerical results, graphically presented in Figs. 7 and 8 is the variation of dimensionless maximum deflection of micromorphic beams and plates with the length-to-thickness ratio corresponding to three values of h/l. Again, for comparison purpose, the results of the classical continuum theory are given in addition to the MMT predicted values. Corresponding to the four types of boundary conditions of the beam- and plate-type structures, one can find that as the length-to-thickness ratio increases, the dimensionless maximum deflection decreases. This reduction is more prominent as a smaller value for h/l is selected.

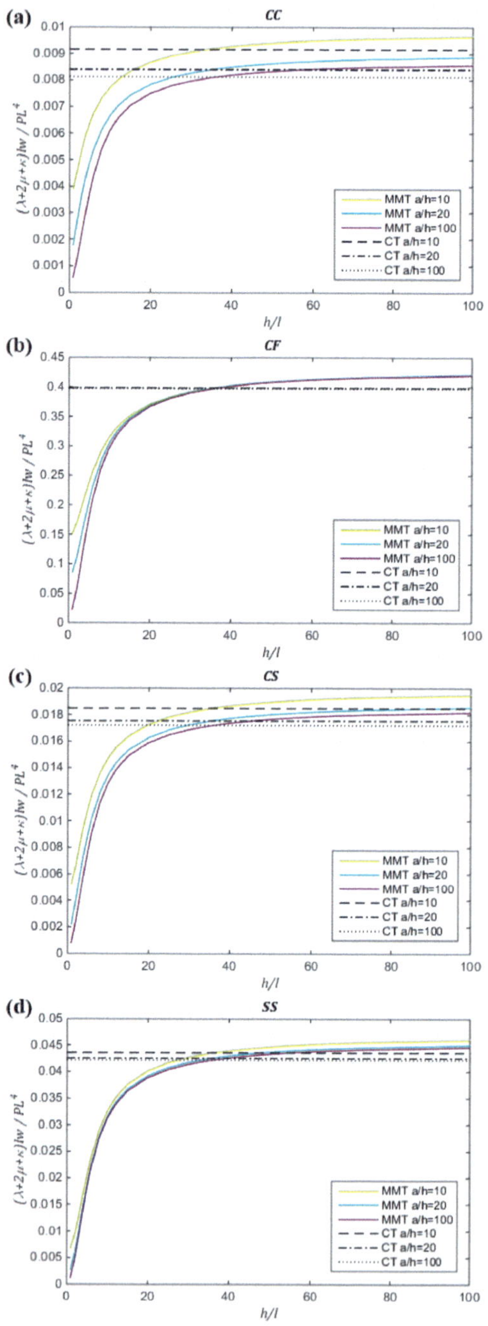

Fig. 3 Effect of h/l on the non-dimensional maximum deflection of beams for various values of a/h

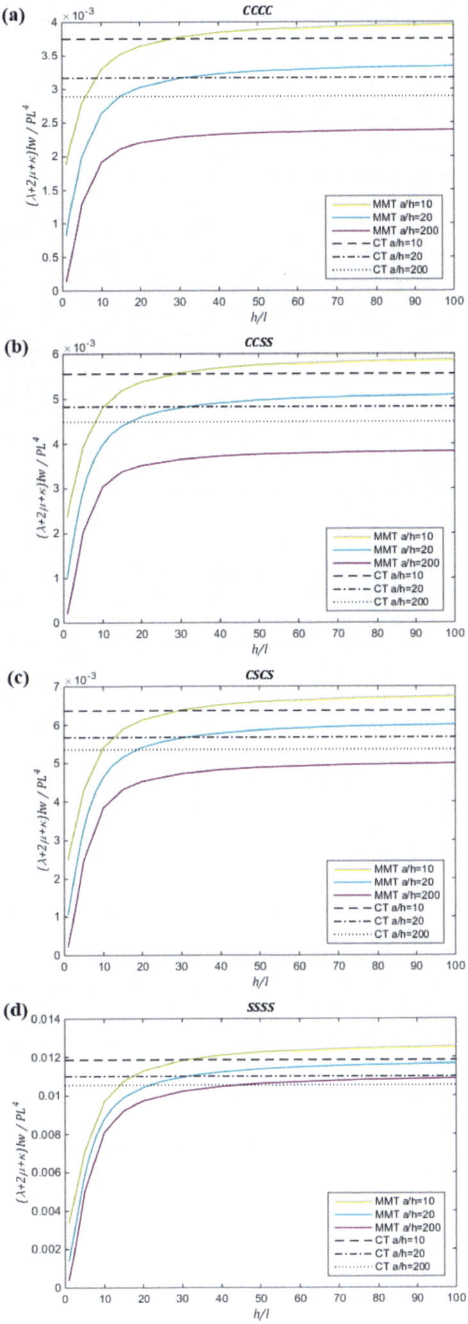

Fig. 4 Effect of h/l on the non-dimensional maximum deflection of plates for various values of a/h

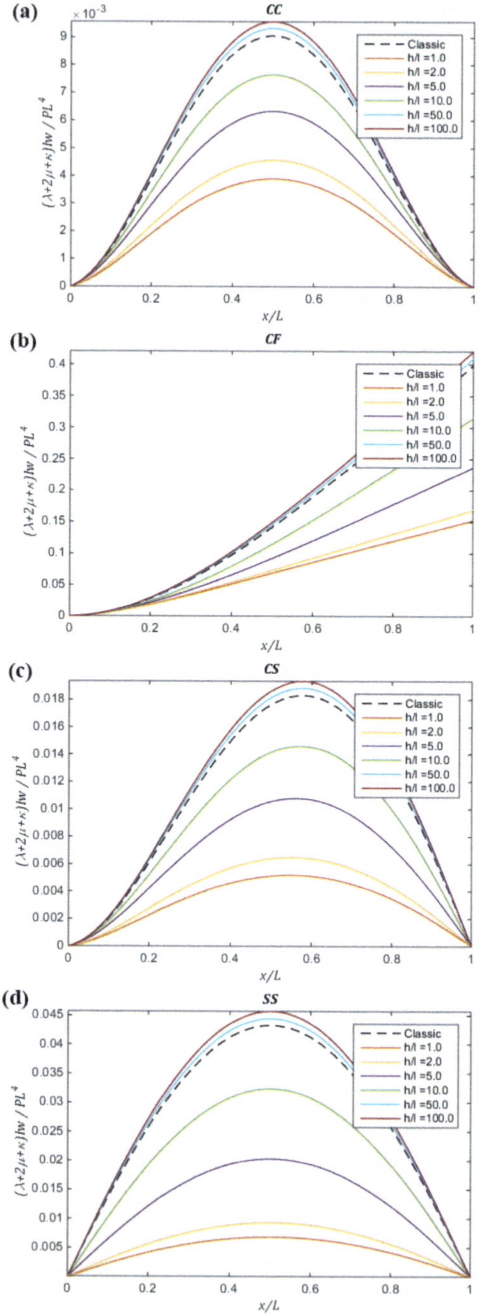

Fig. 5 Variation of non-dimensional deflection of beams with non-dimensional length for different values of h/l $(a/h = 10)$

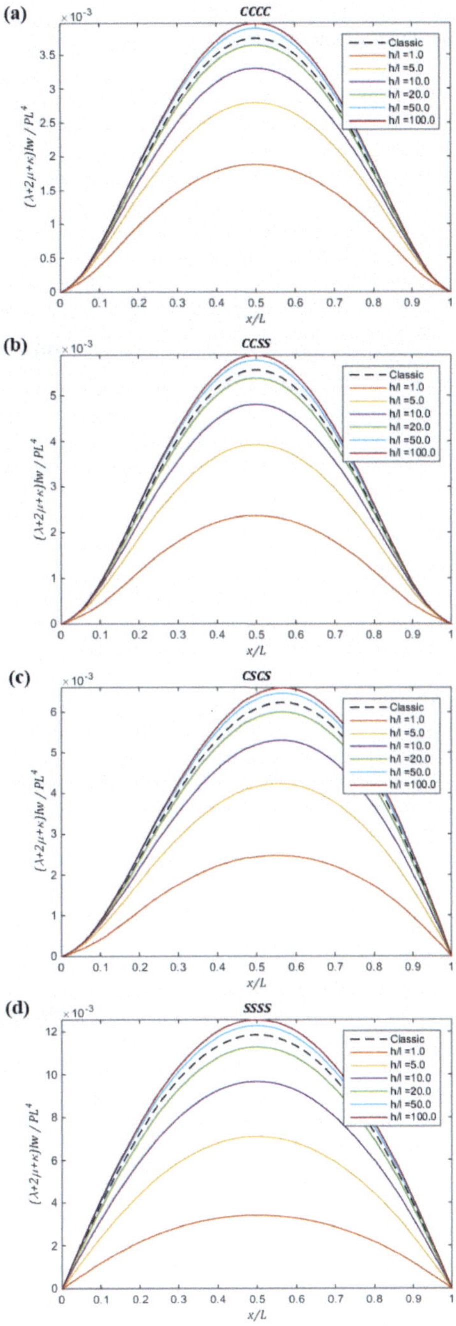

Fig. 6 Variation of non-dimensional deflection of plates with non-dimensional length for different values h/l ($a/h = 10$)

Fig. 7 Effect of a/h on the non-dimensional maximum deflection of beams for various values of h/l

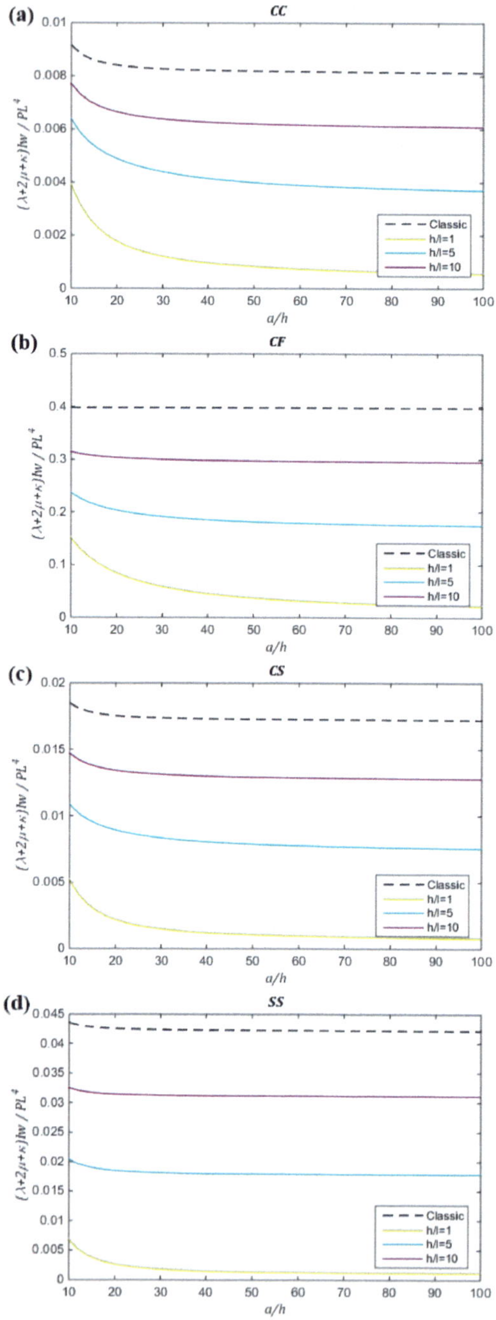

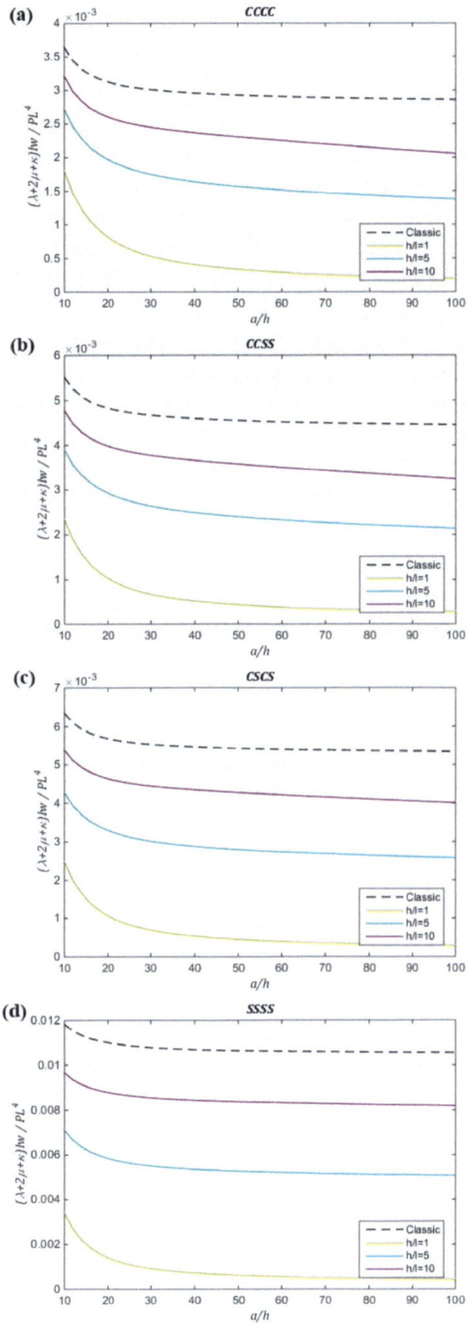

Fig. 8 Effect of a/h on the non-dimensional maximum deflection of plates for various values of h/l

6 Conclusions

With the aim of investigating the mechanical behavior of small-scale engineering structures, a new 12-DOF three-dimensional size-dependent micromorphic element was introduced in this research. Based on MMT, the element takes micro-rotation and micro-deformation DOFs of material particles into account. To derive the governing equations, a new and general formulation was developed for MMT which can be easily used in the finite element analyses. In order to show the reliability of the approach, it was used in the bending problems of micromorphic beams and plates. Numerical examples were provided for various boundary conditions, h/l and a/h. It was found that there is a considerable difference between the results of MMT and the ones obtained based on CT at small scales. Moreover, it was observed that as the dimensionless length scale parameter increases, the difference decreases, but does not completely vanish. The reason is that in MMT, the director is deformable which contains micro-stretches as well as micro-shears, and hence the associated energy in material particles differs from that of CT. It was also seen that the structure becomes stiffer when the dimensionless length scale parameter decreases.

Appendix A: Elastic Matrices of MMT

$$
\begin{aligned}
\boldsymbol{A} &= \begin{bmatrix} \boldsymbol{A}_{11} & 0 \\ 0 & \boldsymbol{A}_{22} \end{bmatrix}_{27\times 27} \\
\boldsymbol{A}_{11} &= \boldsymbol{I}_3 \otimes \begin{bmatrix} a & \boldsymbol{a} & \boldsymbol{a} \\ & \bar{\boldsymbol{A}}_{11} & \frown{\boldsymbol{A}}_{11} \\ \text{sym.} & & \bar{\boldsymbol{A}}_{11} \end{bmatrix} \\
a &= 2\left(a_1 + a_2 + a_5 + a_8\right) + a_3 + a_4 + a_6 + a_7 + a_9 + a_{10} + a_{11} \\
\boldsymbol{a} &= \begin{bmatrix} a_1 + a_2 + a_3 & a_1 + a_4 + a_5 & a_2 + a_5 + a_6 \end{bmatrix} \\
\bar{\boldsymbol{A}}_{11} &= \begin{bmatrix} a_3 + a_7 + a_{10} & a_1 + a_8 + a_{11} & a_2 + a_8 + a_9 \\ & a_4 + a_7 + a_9 & a_5 + a_8 + a_{10} \\ \text{sym.} & & a_6 + a_7 + a_{11} \end{bmatrix} \\
\frown{\boldsymbol{A}}_{11} &= \begin{bmatrix} a_3 & a_1 & a_2 \\ & a_4 & a_5 \\ \text{sym.} & & a_6 \end{bmatrix} \\
\boldsymbol{A}_{22} &= \begin{bmatrix} a_7 & a_{10} & a_8 & a_{11} & a_8 & a_9 \\ & a_7 & a_9 & a_8 & a_{11} & a_8 \\ & & a_7 & a_{10} & a_8 & a_{11} \\ & & & a_7 & a_9 & a_8 \\ & & & & a_7 & a_{10} \\ \text{sym.} & & & & & a_7 \end{bmatrix}
\end{aligned}
\tag{70}
$$

$$\boldsymbol{B}=\begin{bmatrix}\boldsymbol{B}_{11} & 0\\ 0 & \boldsymbol{B}_{22}\end{bmatrix}$$

$$\boldsymbol{B}_{11}=\begin{bmatrix}b_1+b_2+b_3 & b_1 & b_1\\ & b_1+b_2+b_3 & b_1\\ \text{sym.} & & b_1+b_2+b_3\end{bmatrix}$$

$$\boldsymbol{B}_{22}=\boldsymbol{I}_3\otimes\begin{bmatrix}b_2 & b_3\\ b_3 & b_2\end{bmatrix} \tag{71}$$

$$\boldsymbol{C}=\begin{bmatrix}\boldsymbol{C}_{11} & 0\\ 0 & \boldsymbol{C}_{22}\end{bmatrix},\quad \boldsymbol{C}_{11}=\begin{bmatrix}c_1+2c_2 & c_1 & c_1\\ & c_1+2c_2 & c_1\\ \text{sym.} & & c_1+2c_2\end{bmatrix},\quad \boldsymbol{C}_{22}=c_2\boldsymbol{I}_3 \tag{72}$$

$$\boldsymbol{D}=\begin{bmatrix}\boldsymbol{D}_{11} & 0\\ 0 & \boldsymbol{D}_{22}\end{bmatrix},\quad \boldsymbol{D}_{11}=\begin{bmatrix}d_1+2d_2 & d_1 & d_1\\ & d_1+2d_2 & d_1\\ \text{sym.} & & d_1+2d_2\end{bmatrix},\quad \boldsymbol{D}_{22}=d_2\boldsymbol{I}_3\otimes\begin{bmatrix}1 & 1\end{bmatrix} \tag{73}$$

where 0 denotes zero matrix, $\boldsymbol{I}_n$ is the n-th order identity matrix, $\otimes$ is the symbol of the Kronecker product and

$$c_1=\lambda,\quad c_2=\mu,\quad b_1=\eta-\tau,\quad b_2=\kappa-\sigma,\quad b_3=\chi-\sigma,\quad d_1=\tau,\quad d_2=\sigma \tag{74}$$

Appendix B: Differential Operators

By introducing

$$\partial_i=\frac{\partial}{\partial x_i}$$

$$\boldsymbol{e}_i=\begin{bmatrix}\delta_{i1} & \delta_{i2} & \delta_{i3}\end{bmatrix},\quad \delta_{ij}=\begin{cases}1 & i=j\\ 0 & i\neq j\end{cases} \tag{75}$$

one has

$$\boldsymbol{E}_1=\begin{bmatrix}\partial_1 & 0 & 0\\ 0 & \partial_2 & 0\\ 0 & 0 & \partial_3\\ 0 & \partial_3 & \partial_2\\ \partial_3 & 0 & \partial_1\\ \partial_2 & \partial_1 & 0\end{bmatrix},\quad \boldsymbol{E}_2=\begin{bmatrix}\partial_1 & 0 & 0\\ 0 & \partial_2 & 0\\ 0 & 0 & \partial_3\\ 0 & 0 & \partial_2\\ 0 & \partial_3 & 0\\ \partial_3 & 0 & 0\\ 0 & 0 & \partial_1\\ 0 & \partial_1 & 0\\ \partial_2 & 0 & 0\end{bmatrix},\quad \boldsymbol{E}_3=\begin{bmatrix}\boldsymbol{E}_{1;3}\\ \boldsymbol{E}_{2;3}\\ \boldsymbol{E}_{3;3}\\ \boldsymbol{E}_{4;3}\end{bmatrix} \tag{76}$$

$$\boldsymbol{E}_{1;3} = \begin{bmatrix} \partial_1 & 0 & 0 & 0 & 0 & 0 & 0 & 0 & 0 \\ 0 & \partial_1 & 0 & 0 & 0 & 0 & 0 & 0 & 0 \\ 0 & 0 & 0 & 0 & 0 & 0 & 0 & \partial_2 & 0 \\ 0 & 0 & 0 & 0 & 0 & 0 & 0 & 0 & \partial_2 \\ 0 & 0 & \partial_1 & 0 & 0 & 0 & 0 & 0 & 0 \\ 0 & 0 & 0 & 0 & 0 & 0 & \partial_3 & 0 & 0 \\ 0 & 0 & 0 & 0 & 0 & \partial_3 & 0 & 0 & 0 \end{bmatrix} \tag{77}$$

$$\boldsymbol{E}_{2;3} = \begin{bmatrix} 0 & \partial_2 & 0 & 0 & 0 & 0 & 0 & 0 & 0 \\ \partial_2 & 0 & 0 & 0 & 0 & 0 & 0 & 0 & 0 \\ 0 & 0 & 0 & 0 & 0 & 0 & 0 & 0 & \partial_1 \\ 0 & 0 & 0 & 0 & 0 & 0 & 0 & \partial_1 & 0 \\ 0 & 0 & \partial_2 & 0 & 0 & 0 & 0 & 0 & 0 \\ 0 & 0 & 0 & \partial_3 & 0 & 0 & 0 & 0 & 0 \\ 0 & 0 & 0 & 0 & \partial_3 & 0 & 0 & 0 & 0 \end{bmatrix} \tag{78}$$

$$\boldsymbol{E}_{3;3} = \begin{bmatrix} 0 & 0 & \partial_3 & 0 & 0 & 0 & 0 & 0 & 0 \\ \partial_3 & 0 & 0 & 0 & 0 & 0 & 0 & 0 & 0 \\ 0 & 0 & 0 & 0 & 0 & \partial_1 & 0 & 0 & 0 \\ 0 & 0 & 0 & 0 & 0 & 0 & \partial_1 & 0 & 0 \\ 0 & \partial_3 & 0 & 0 & 0 & 0 & 0 & 0 & 0 \\ 0 & 0 & 0 & 0 & \partial_2 & 0 & 0 & 0 & 0 \\ 0 & 0 & 0 & \partial_2 & 0 & 0 & 0 & 0 & 0 \end{bmatrix} \tag{79}$$

$$\boldsymbol{E}_{4;3} = \begin{bmatrix} 0 & 0 & 0 & \partial_1 & 0 & 0 & 0 & 0 & 0 \\ 0 & 0 & 0 & 0 & \partial_1 & 0 & 0 & 0 & 0 \\ 0 & 0 & 0 & 0 & 0 & \partial_2 & 0 & 0 & 0 \\ 0 & 0 & 0 & 0 & 0 & 0 & \partial_2 & 0 & 0 \\ 0 & 0 & 0 & 0 & 0 & 0 & 0 & \partial_3 & 0 \\ 0 & 0 & 0 & 0 & 0 & 0 & 0 & 0 & \partial_3 \end{bmatrix} \tag{80}$$

References

1. Nasr Esfahani M, Erdem Alaca B (2019) A review on size-dependent mechanical properties of nanowires. Adv Eng Mater 12:1900192
2. Taloni A, Vodret M, Costantini G, Zapperi S (2018) Size effects on the fracture of microscale and nanoscale materials. Nature Rev Mater 3:211–224
3. Farajpour A, Ghayesh MH, Farokhi H (2018) A review on the mechanics of nanostructures. Int J Eng Sci 133:231–263
4. Thai HT, Vo TP, Nguyen TK, Kim SE (2017) A review of continuum mechanics models for size-dependent analysis of beams and plates. Compos Struct 177:196–219
5. Hosseini M, Hadi A, Malekshahi A, Shishesaz M (2018) A review of size-dependent elasticity for nanostructures. J Comput Appl Mech 49:197–211
6. Wang KF, Wang BL, Kitamura T (2016) A review on the application of modified continuum models in modeling and simulation of nanostructures. Acta Mech Sin 32:83–100

7. Wang J, Huang Z, Duan H, Yu S, Feng X, Wang G, Zhang W, Wang T (2011) Surface stress effect in mechanics of nanostructured materials. Acta Mech Solida Sin 24:52–82
8. Zhang H, Wang H, Wriggers P, Schrefler B (2005) A finite element model for contact analysis of multiple Cosserat bodies. Comput Mech 36:444–458
9. Aganović I, Tambača J, Tutek Z (2006) Derivation and justification of the models of rods and plates from linearized three-dimensional micropolar elasticity. J Elast 84:131–152
10. Pompei A, Rigano MA (2006) On the bending of micropolar viscoelastic plates. Int J Eng Sci 44:1324–1333
11. Jeong J, Neff P (2008) Existence, uniqueness and stability in linear Cosserat elasticity for weakest curvature conditions. Math Mech Solids 15:78–95
12. Pietraszkiewicz W, Eremeyev V (2009) On natural strain measures of the non-linear micropolar continuum. Int J Solids Struct 46:774–787
13. Sansour C, Skatulla S, Zbib H (2010) A formulation for the micromorphic continuum at finite inelastic strains. Int J Solids Struct 47:1546–1554
14. Ieşan D (2011) Micromorphic elastic solids with initial stresses and initial heat flux. Int J Eng Sci 49:1350–1356
15. Kumar R, Kansal T (2012) Fundamental solution in the theory of micropolar thermoelastic diffusion with voids. Comput Appl Math 31:169–189
16. Dos Reis F, Ganghoffer JF (2012) Construction of micropolar continua from the asymptotic homogenization of beam lattices. Comput Struct 112–113:354–363
17. Zhang ZH, Nie JF, Liu ZL, Gao Y, Zhuang Z (2012) Analytical and numerical studies on simple shear of a bimaterial strip by using elastic micromorphic theory. Mech Res Commun 39:44–50
18. Cordero NM, Forest S, Busso EP (2013) Micromorphic modelling of grain size effects in metal polycrystals. GAMM-Mitteilungen 36:186–202
19. Altenbach H, Eremeyev VA (2014) Strain rate tensors and constitutive equations of inelastic micropolar materials. Int J Plast 63:3–17
20. Chowdhury SR, Rahaman MM, Roy D, Sundaram N (2015) A micropolar peridynamic theory in linear elasticity. Int J Solids Struct 59:171–182
21. Ansari R, Bazdid-Vahdati M, Shakouri A, Norouzzadeh A, Rouhi H (2017) Micromorphic prism element. Math Mech Solids 22:1438–1461
22. Ansari R, Bazdid-Vahdati M, Shakouri A, Norouzzadeh A, Rouhi H (2016) Micromorphic first-order shear deformable plate element. Meccanica 51:1797–1809
23. Ansari R, Shakouri A, Bazdid-Vahdati M, Norouzzadeh A, Rouhi H (2017) A nonclassical finite element approach for the nonlinear analysis of micropolar plates. J Comput Nonlinear Dynam 12:011019
24. Ansari R, Norouzzadeh A, Shakouri A, Bazdid-Vahdati M, Rouhi H (2018) Finite element analysis of vibrating micro-beams and-plates using a three-dimensional micropolar element. Thin-Walled Struct 124:489–500
25. Norouzzadeh A, Ansari R, Darvizeh M (2019) Large elastic deformation of micromorphic shells. Part I: Variational formulation. Math Mech Solids 24:3920–3956
26. Norouzzadeh A, Ansari R, Darvizeh M (2019) Large elastic deformation of micromorphic shells. Part II. Isogeometric analysis. Math Mech Solids 24:3753–3778
27. Norouzzadeh A, Faraji-Oskouie M, Ansari R, Rouhi H (2019) Integral and differential nonlocal micromorphic theory. Eng Comput 37:566–590
28. Faraji-Oskouie M, Norouzzadeh A, Ansari R, Rouhi H (2019) Bending of small-scale Timoshenko beams based on the integral/differential nonlocal-micropolar elasticity theory: a finite element approach. Appl Math Mech 40:767–782
29. Eringen AC (2012) Microcontinuum field theories: I. Foundations and solids. Springer
30. Smith AC (1968) Inequalities between the constants of a linear micro-elastic solid. Int J Eng Sci 6:65–74
31. Lakes R (1986) Experimental microelasticity of two porous solids. Int J Solids Struct 22:55–63

Peridynamic Modeling of Laminated Composites

Erdogan Madenci and Mehmet Dorduncu

Abstract Damage growth in composites involves complex and progressive failure modes. Current computational tools are incapable of predicting failure in composite materials in a unified manner mainly due to their mathematical structure. However, the peridynamic (PD) theory removes these obstacles by taking into account nonlocal interactions between material points. The PD theory simply replaces the displacement derivatives in the equilibrium equations. The PD enables the modeling of progressive failure during deformation through the removal of nonlocal PD interactions (bonds). These bonds enable the interaction of material points within each ply as well as their interaction with other material points in the adjacent plies. The solution continues by using standard explicit time integration techniques until final failure. This chapter presents PD modeling approaches namely bond-based, ordinary state-based, and PD differential operator for predicting progressive damage in fiber-reinforced composite materials under general loading conditions. The predictions from the PD models agree with the experimental observations published in the literature.

1 Introduction

Composites exhibit distinct strength properties along different orientations. Therefore, they are extensively used in many engineering fields, especially in the commercial and military aerospace structures due to their load-carrying capability and light weights. However, their numerical modeling for failure prediction is still a challenging task due to their complex damage patterns and heterogeneous nature. Depending on the fiber orientation, loading, and boundary conditions, the damage may occur

E. Madenci (✉)
Department of Aerospace and Mechanical Engineering, University of Arizona, Tucson, AZ 85715, USA
e-mail: madenci@email.arizona.edu

M. Dorduncu
Department of Mechanical Engineering, Erciyes University, 38039 Kayseri, Turkey
e-mail: dorduncu@erciyes.edu.tr

© The Author(s), under exclusive license to Springer Nature Switzerland AG 2021
E. Ghavanloo et al. (eds.), *Size-Dependent Continuum Mechanics Approaches*,
Springer Tracts in Mechanical Engineering,
https://doi.org/10.1007/978-3-030-63050-8_13

in the form of matrix cracking, fiber breakage, fiber kinking, and delamination. Therefore, accurate modeling and evaluation of damage initiation and progression in composite structures under complex loading conditions is a primary design concern.

Although the finite element method (FEM) is commonly employed for predicting progressive failure in composite, it faces difficulties when predicting all possible failure modes in a unified manner. This difficulty arises due the presence of undefined derivatives of the displacement fields along crack surfaces or at crack tips in the governing equation of the FEM. The virtual crack closure technique (VCCT) [1–3] and cohesive zone model (CZM) [4–6] were developed to improve the shortcomings of the FEM for predicting failures in composites. Rybicki and Kanninen [7] proposed the virtual crack closure technique (VCCT) for predicting delamination growth. However, the VCCT requires a pre-defined location of a crack and is dependent on the mesh near the crack tip. Therefore, it becomes challenging to predict the crack initiation and its exact location and size a priori [2, 3, 11–13].

The CZM introduced by Dugdale [8] and Barenblatt [9] is widely used for failure predictions in composites. In the CZM, material interfaces are modeled through a cohesive traction-displacement relation. The tractions become zero when the opening displacement (separation) reaches a critical value. The cohesive zone elements are placed along the element boundaries; hence, crack growth occurs only between the elements. Therefore, the material response exhibits characteristics of both traditional and cohesive zone elements; the cohesive elements are only introduced to produce fracture behavior. The concept of eXtended Finite Element Method (XFEM) introduced for modeling cracks and crack growth eliminates remeshing process as in the traditional FE analysis [10, 11]. Unlike the CZM, it permits the cracks to propagate on any surface within an element. Although the XFEM is one of the robust techniques for fracture problems, it depends on external criteria for injection of discontinuous displacement enrichment functions.

The peridynamic (PD) theory [12–14] eliminates the shortcomings of the existing methods, in particular the difficulties associated with modeling initiation and growth of multiple discontinuities (cracks) in solids. The equilibrium equation of PD theory consists of an integral expression rather than its differential form. Thus, these equations exist in the domain even in the presence of cracks. The damage in the PD theory is introduced through the removal of interactions between material points, so that their corresponding contributions in the integral representation are simply removed. The PD theory is suitable for modeling fracture mechanism for composite structures under general dynamic and static loading conditions because it is capable of capturing all failure modes without simplifying assumptions. The PD equations are classified as "Bond-based (BB)", "Ordinary State-Based (OSB)" and "Nonordinary State-Based (NOSB)" PD. The BB PD does not distinguish between dilatation and distortional parts of deformation with a constraint on the Poisson's ratio (1/3 for 2D and 1/4 for 3D) [15, 16]. The pairwise interaction of points in BB PD is located in the same domain of interaction. In the OSB PD, the two interacting points have their own domains of interaction without this constraint. The NOSB PD considers a more general representation of the continuum mechanics by employing the stress-strain relations. Therefore, the material constitutive matrix can be used rather than determining the material parameters for the bonds between the material points as in BB and OSB PD.

The PD theory was successfully applied to simulate crack propagation in composites. Askari et al. [17] and Xu et al. [18, 19] predicted damages in notched laminated composites subjected to low-velocity impact and quasi-static in-plane loads. Xu et al. [18] distinguished the bonds as fiber and matrix bonds and aligned the grid of a lamina with the fiber orientation. Kilic and Madenci [20] proposed a PD model for composites by identifying the material points as fiber and matrix. They predicted the fiber, matrix, and delamination failures in various laminates with a pre-existing central crack under tension. Oterkus and Madenci [14, 21] presented a study on the damage response of fiber-reinforced composite materials under mechanical and thermal loading conditions by using the BB PD theory. They specified four different types of bonds between the material points: fiber, matrix, inter-lamina normal and shear bonds and considered both lamina and laminate with/without pre-existing crack under uniaxial tension and uniform temperature change. Hu et al. [22] developed a homogenization-based PD model for modeling damage in fiber-reinforced composites. They investigated the response of a lamina with a pre-existing crack under tension due to dynamic loading. Oterkus et al. [23] coupled PD with FEM to predict the failure loads in a curved, stiffened composite panel with a central slit subjected to uniaxial loading and an internal pressure. Ghajari et al. [24] developed an orthotropic elastic material model in the framework of BB PD. Hu et al. [25] proposed a BB PD model for progressive failure damages of fiber-reinforced composite laminate by introducing fiber bond and matrix bonds as well as longitudinal and transverse interlayer bonds. They assigned distinct micro-elastic moduli to these bonds to describe the anisotropy, and determined their values based on the concepts from mechanics of laminated composites. Hu and Madenci [26] presented a different BB PD approach for modeling composite laminates with arbitrary fiber orientation and stacking sequence. Their approach enabled the evaluation of stress and strain fields in each ply of the laminate. Therefore, it permits the use of existing stress- or strain-based failure criteria for damage prediction. They intrdouced in-plane normal, in-plane shear, transverse normal and transverse shear PD bonds to achieve the orthotropic material behavior. Diyaroglu et al. [27] used BB PD for the failure analysis of composite laminates under explosive loading. Sadowski and Pankowski [28] proposed a PD model to investigate nano indentation of ceramic composites. Ren et al. [29] presented a discontinuous Galerkin weak form for BB PD models to predict the damage of fiber-reinforced composite laminates. Diana and Casolo [30] proposed a 2D orthotropic BB PD model for linearly elastic solids. They used four independent elastic moduli for modeling non-isotropic and orthotropic materials. Also, they obtained the stiffness of each bond by using continuous functions of bond orientation in the principal material axes.

Madenci and Oterkus [14] introduced the PD laminate theory within the realm of the OSB PD for a fiber-reinforced laminates with explicit expressions for PD material parameters. Colavito et al. [31] investigated the applicability of OSB PD in predicting the residual strength of notched composite laminates under different loadings. They compared the PD predictions for peak failure loads and damage patterns with those from experimental observations. Jiang and Wang [32, 33] extended the OSB PD model by Madenci and Oterkus [14] to predict the tensile strength of fiber-reinforced composite laminates with an open-hole. Zhang and Qiao [34] developed the OSB PD model for elastic and fracture analysis of 2D orthotropic materials. This approach was verified for an orthotropic lamina under stress load and the compact tension and single edge-notched tension (SENT) tests. Gao and Oterkus [35] extended the OSB PD approach introduced by Madenci and Oterkus [14] for fully coupled thermo-mechanical analysis of composite laminates. They examined the failure behavior of single layer and multi-layer models with pre-existing cracks. Radel et al. [36] used the OSB PD for failure analysis of cementitious composites and ceramics. They also used a novel energy-based failure criterion for linear peridynamic solid (LPS) materials.

Hattori et al. [37] introduced another NOSB PD formulation for modeling general anisotropic materials. They used the constitutive matrix for the material properties of each material point. Yaghoobi and Chorzepa [38] applied the NOSB PD to model damage in fibre-reinforced concrete. Shang et al. [39] employed the NOSB PD approach to determine damage due to machining of carbon-fiber-reinforced polymer (CFRP) composites. They used both the maximum stress failure criteria and Hashin failure criteria to describe failure characteristic of composites. Madenci et al. [40] combined the PeriDynamic Differential Operator (PDDO) and Classical Laminate Theory (CLT) to predict damage in laminated composite structures.

This chapter provides a general perspective the PD theory and its applications to composite laminates. It presents the basic concept of PD theory and the derivation of governing equations for BB, OSB and NOSB PD. Subsequently, it describes the derivation of the force density vectors for composite laminates. The numerical results concern the applications of BB, OSB and NOSB PD for predicting damage in composite structures.

2 Fundamentals of Peridynamics

As introduced by Silling [12, 13], the PD theory is a reformulation of the classical continuum equations of motion. It replaces the partial differential equations of motion with integro-differential equations. The equations of motion from classical continuum theory can be expressed as

$$\rho\left(\mathbf{x}\right)\ddot{\mathbf{u}}\left(\mathbf{x},t\right)=\mathbf{L}(\mathbf{x},t)+\mathbf{b}\left(\mathbf{x},t\right) \tag{1}$$

in which ρ is the mass density, $\mathbf{u}(\mathbf{x},t)$ is the displacement vector, $\mathbf{L}(\mathbf{x},t)$ is the internal force vector, and $\mathbf{b}(\mathbf{x},t)$ is the body force density vector. The internal force vector, $\mathbf{L}(\mathbf{x},t)$ is defined as

$$\mathbf{L}(\mathbf{x},t)=\nabla\cdot\boldsymbol{\sigma}(\mathbf{x},t) \tag{2}$$

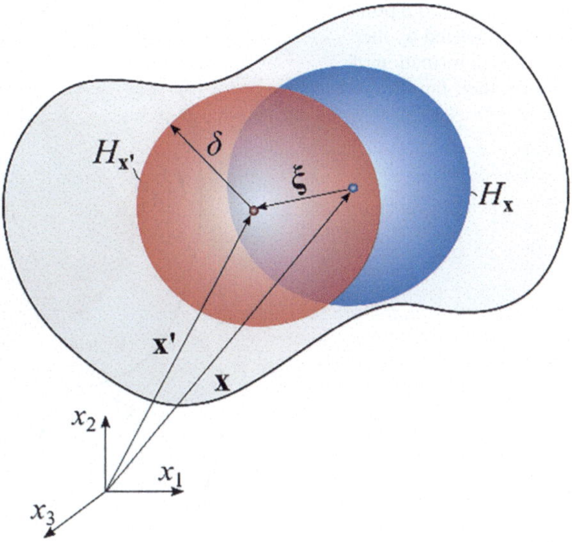

Fig. 1 Interaction between material points $\mathbf{x}$ and $\mathbf{x}'$ in the undeformed configuration

where $\cdot$ is dot product and $\boldsymbol{\sigma}(\mathbf{x}, t)$ is Cauchy's stress tensor at point $\mathbf{x}$. The position vector $\mathbf{x}$ defines the location of each point with respect to the reference configuration at time t.

In the PD theory, the divergence of the stress tensor in Eq. (2) is replaced with an integral that accounts for all the nonlocal force interactions that exists between a material point and all the material points within its horizon. The PD equations of motion can be expressed in terms of the force density vector, $\mathbf{t}$, that exists between material points, as

$$\rho(\mathbf{x})\,\ddot{\mathbf{u}}(\mathbf{x}, t) = \int\limits_{H_{\mathbf{x}}} \left(\mathbf{t}\left(\mathbf{u}' - \mathbf{u}, \mathbf{x}' - \mathbf{x}, t\right) - \mathbf{t}'\left(\mathbf{u} - \mathbf{u}', \mathbf{x} - \mathbf{x}', t\right)\right) dH_{\mathbf{x}'} + \mathbf{b}(\mathbf{x}, t) \tag{3}$$

where the material point $\mathbf{x}$ interacts with material point $\mathbf{x}'$ within its domain of interaction (family) of material points, $H_{\mathbf{x}}$ as shown in Fig. 1. The interaction domain $H_{\mathbf{x}}$ of material point $\mathbf{x}$ is defined by its *horizon*, δ. Material points $\mathbf{x}'$, located within the interaction domain $H_{\mathbf{x}}$ are called the *family members* of $\mathbf{x}$. The initial distance between the material points $\mathbf{x}$ and $\mathbf{x}'$ is $\xi = \mathbf{x}' - \mathbf{x}$.

The interaction force or peridynamic force between material points $\mathbf{x}$ and $\mathbf{x}'$ can be expressed as $\mathbf{t}(\mathbf{x}' - \mathbf{x}, \mathbf{u}' - \mathbf{u})$, and it is a function of the relative position vector, $\mathbf{x}' - \mathbf{x}$, and relative displacement vector, $\mathbf{u}' - \mathbf{u}$. They must satisfy the balance of angular momentum expressed as

Fig. 2 PD material points $\mathbf{x}$ and $\mathbf{x}'$ influenced by the collective deformation of others in their families

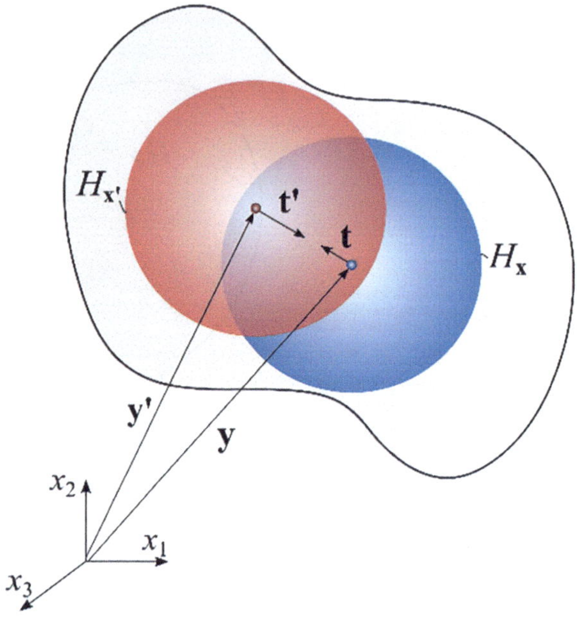

$$\int_{H_{\mathbf{x}}} (\mathbf{y}' - \mathbf{y}) \times \mathbf{t}\, dH_{\mathbf{x}'} = 0 \tag{4}$$

As shown in Fig. 2, the force density vectors, $\mathbf{t}$ and $\mathbf{t}'$ can be parallel to the relative position vector in the deformed state with unequal magnitudes, and automatically satisfy the balance of linear and angular momentum. This particular form of the force density vectors is referred to as Ordinary State-Based (OSB) PD.

The force density vectors $\mathbf{t}(\mathbf{x}' - \mathbf{x}, \mathbf{u}' - \mathbf{u}, t)$ and $\mathbf{t}'(\mathbf{x}' - \mathbf{x}, \mathbf{u}' - \mathbf{u}, t)$ can be related to the strain energy density function, W. Considering the requirement on their direction in order to satisfy the balance of angular momentum, Eq. (4), they can be expressed as

$$\mathbf{t}\left(\mathbf{u}' - \mathbf{u}, \mathbf{x}' - \mathbf{x}, t\right) = \frac{\partial W(\mathbf{x})}{\partial\left(|\mathbf{y}' - \mathbf{y}|\right)} \frac{\mathbf{y}' - \mathbf{y}}{|\mathbf{y}' - \mathbf{y}|} \tag{5a}$$

$$\mathbf{t}'\left(\mathbf{u} - \mathbf{u}', \mathbf{x} - \mathbf{x}', t\right) = \frac{\partial W(\mathbf{x}')}{\partial\left(|\mathbf{y} - \mathbf{y}'|\right)} \frac{\mathbf{y}' - \mathbf{y}}{|\mathbf{y}' - \mathbf{y}|} \tag{5b}$$

where $\mathbf{y}$ and $\mathbf{y}'$ are the position vector of points $\mathbf{x}$ and $\mathbf{x}'$ in the deformed configuration, respectively. However, the explicit determination of the force density vectors requires the PD representation of the strain energy density function.

As a special case, shown in Fig. 3, the force density vectors in Eq. (3) can be equal and in opposite directions, and the PD equilibrium equation reduces to

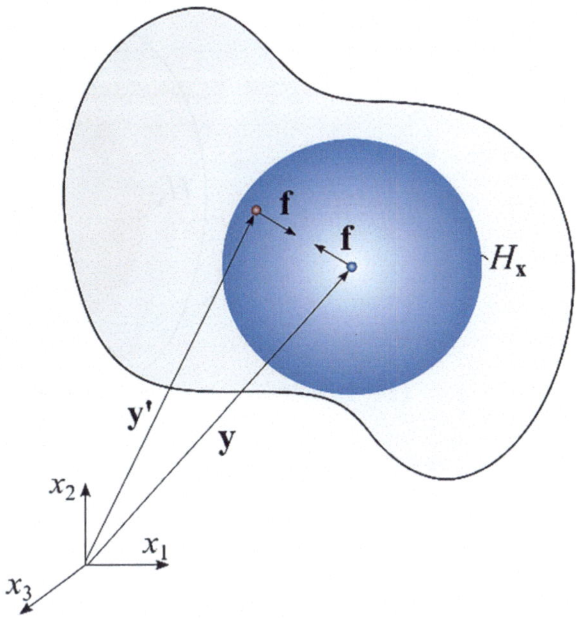

Fig. 3 Deformation of PD material points **x** and **x**′, and developing equal and opposite pairwise force densities in the deformed state

$$\rho(\mathbf{x})\,\ddot{\mathbf{u}}(\mathbf{x},t) = \int_{H_x} \mathbf{f}\left(\mathbf{x}' - \mathbf{x}, \mathbf{u}' - \mathbf{u}\right) dH_{\mathbf{x}} + \mathbf{b}(\mathbf{x},t) \tag{6}$$

This form of the PD equilibrium equation is referred to as the Bond-Based (BB) PD model. Although it is rather simple, it suffers from the reduction of independent engineering material constants.

As a general case, shown in Fig. 4, the force density vectors, $\mathbf{t}(\mathbf{x}, t)$ and $\mathbf{t}(\mathbf{x}', t)$, acting on the material points can be in arbitrary directions with unequal magnitudes. This form of the force density vectors is referred to as Non-Ordinary State-Based (NOSB) PD. It is a more generalized approach and the force density vector at a point x can be associated with the first Piola-Kirchhoff stress, $\mathbf{P}(\mathbf{x})$, and shape, $\mathbf{K}(\mathbf{x})$, tensors. Therefore, it invokes the material constitutive matrix from classical continuum mechanics. As derived by Silling et al. [13], the force density vector can be expressed as

$$\mathbf{t}(\mathbf{x}, t) = w\left(|\mathbf{x}' - \mathbf{x}|\right) \mathbf{P}(\mathbf{x})\mathbf{K}^{-1}(\mathbf{x}) \tag{7a}$$

$$\mathbf{P} = \det(\mathbf{F})\,\boldsymbol{\sigma}\mathbf{F}^{-\mathrm{T}} = \mathbf{F}\mathbf{S} \tag{7b}$$

$$\mathbf{K}(\mathbf{x}) = \int_{H_x} w\left(|\mathbf{x}' - \mathbf{x}|\right)\left((\mathbf{x}' - \mathbf{x}) \otimes (\mathbf{x}' - \mathbf{x})\right) dV_{\mathbf{x}'} \tag{7c}$$

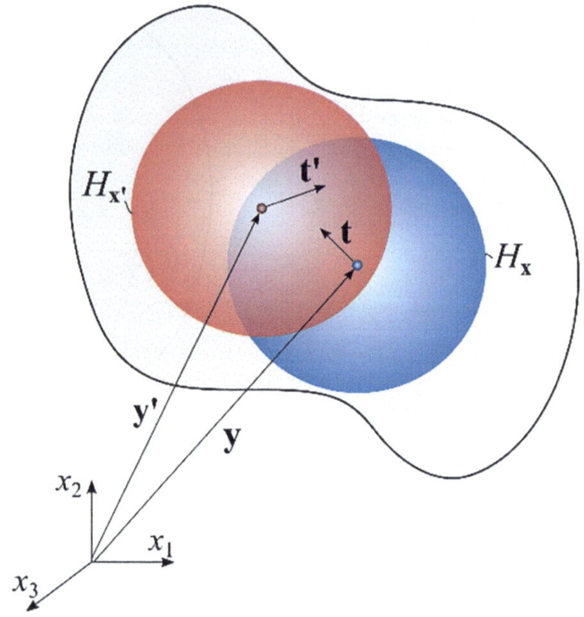

Fig. 4 Deformation of PD material points $\mathbf{x}$ and $\mathbf{x}'$, and developing force densities in arbitrary directions

where $w\left(|\mathbf{x}' - \mathbf{x}|\right)$ is the weight function specifying the degree of interaction between the material points, $\mathbf{S}$ is the second Piola-Kirchhoff stress tensor and σ is the Cauchy stress tensor, and the symbol $\otimes$ denotes the dyadic product of two vectors. The PD form of the deformation gradient tensor, $\mathbf{F}$ can be derived as [13]

$$\mathbf{F}(\mathbf{x}) = \left(\int\limits_{H_{\mathbf{x}}} w\left(|\mathbf{x}' - \mathbf{x}|\right) \left(\mathbf{y}' - \mathbf{y}\right) \otimes \left(\mathbf{x}' - \mathbf{x}\right) dV_{\mathbf{x}'} \right) \mathbf{K}^{-1}(\mathbf{x}) \tag{8}$$

Material damage in PD is introduced through elimination of interactions (micropotentials) among the material points. Damage is reflected in the equations of motion by removing the force density vectors between the material points in an irreversible manner. As a result, the load is redistributed among the material points in the body, leading to progressive damage growth in an autonomous fashion.

In order to include damage initiation in the material response, the material constants are modified through the failure (status) parameter μ as [41]

$$\mu\left(\mathbf{x}' - \mathbf{x}, t\right) = \begin{cases} 1 \text{ if } s\left(\mathbf{x}' - \mathbf{x}, t'\right) < s_c \text{ for all } 0 < t' < t \\ 0 \text{ otherwise} \end{cases} \tag{9}$$

where s_c is the critical stretch value related to the energy release rate, G. The total stretch, s between two material points, $\mathbf{x}$ and $\mathbf{x}'$ can be defined

$$s = \frac{|\mathbf{y}' - \mathbf{y}| - |\mathbf{x}' - \mathbf{x}|}{|\mathbf{x}' - \mathbf{x}|} \tag{10}$$

During the solution process, the displacements of each material point, as well as the stretch, s, between pairs of material points, $\mathbf{x}$ and $\mathbf{x}'$, are computed and monitored. When the stretch between these material points exceeds its critical stretch, failure occurs; thus, the history-dependent scalar-valued function μ is zero, rendering the associated part of the force density vector to be zero.

As suggested by Silling and Askari [41], and later extended by Madenci and Oterkus [14], the critical stretch can be determined by equating the strain energy required to eliminate all interactions across a new crack surface to the mode-I critical energy release rate G_{Ic}. Local damage at a point is defined as the weighted ratio of the number of eliminated interactions to the total number of initial interactions of a material point with its family members. The local damage at a point can be quantified as [41]

$$\varphi(\mathbf{x}, t) = 1 - \frac{\int_{H_{\mathbf{x}}} \mu(\mathbf{x}' - \mathbf{x}, t) dH_{\mathbf{x}'}}{\int_{H_{\mathbf{x}}} dH_{\mathbf{x}'}} \tag{11}$$

The local damage ranges from zero to one. When the local damage is one, all the interactions initially associated with the point have been eliminated, while a local damage of zero means that all interactions are intact. The measure of local damage is an indicator of possible crack formation within a body.

3 Peridynamic Modeling of a Laminate

Each fiber-reinforced composite lamina of a laminate shown in Fig. 5 is idealized as a two-dimensional structure with the directional dependency of the interactions between the PD material points. As shown in Fig. 6, the material point (q) represents material points that interact with material point (k) only along the fiber direction with an orientation angle of θ in reference to the x-axis. Similarly, material point (r) represents material points that interact with material point (k) only along the transverse direction. However, the material point (p) represents material points that interact with material point (k) in any direction, including the fiber and transverse directions. The orientation of a PD interaction between the material point (k) and the material point (p) is defined by the angle ϕ with respect to the x-axis. The domain of integration, $H_{(k)}$ shown in Fig. 6 is a disk with radius δ and thickness h. The material points in a particular lamina interact with the other material points of immediate neighboring laminae above and below it.

As shown in Fig. 5, the reference coordinate system (x, y, z) is located on the mid-plane of the laminate. The laminate thickness, h is given by

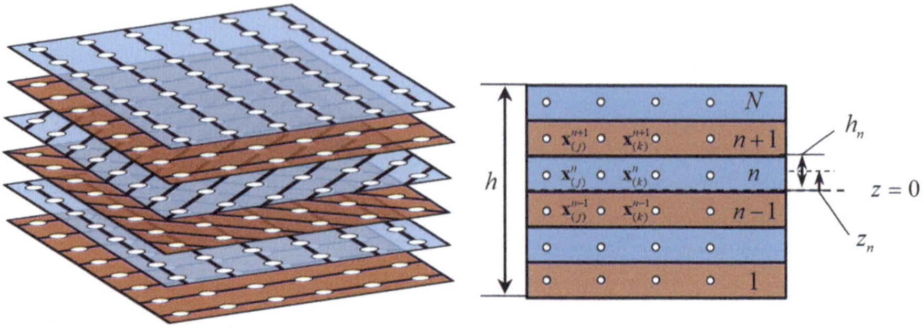

Fig. 5 Elevation of each lamina in laminate and PD material points

Fig. 6 PD horizon for a family of material points and their interactions in a lamina

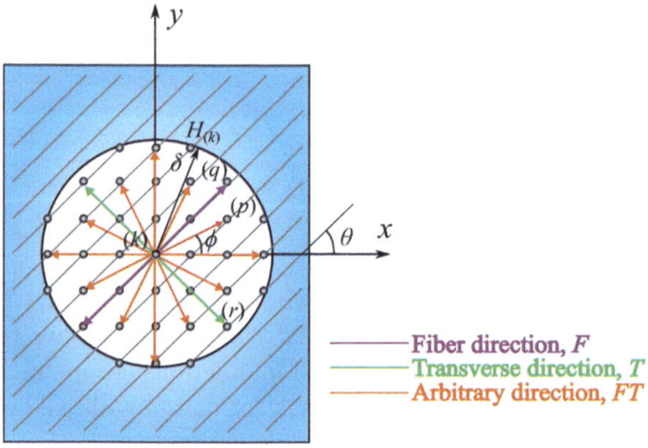

$$h = \sum_{n=1}^{N} h_n \tag{12}$$

where N is the total number of lamina in the stacking sequence, and h_n is the thickness of nth lamina. With respect to the mid-plane, the position of each lamina, z_n is defined as

$$z_n = -\frac{h}{2} + \sum_{m=1}^{n-1} h_m + \frac{1}{2} h_n \tag{13}$$

As derived by Madenci and Oterkus [14], the equation of motion for material point $\mathbf{x}^n_{(k)}$ located on the nth layer of a laminate with N layers can be expressed as

$$
\begin{aligned}
\rho^n_{(k)}\ddot{\mathbf{u}}^n_{(k)} = & \sum_{j=1} \left[\mathbf{t}^n_{(k)(j)}\left(\mathbf{u}^n_{(j)} - \mathbf{u}^n_{(k)}, \mathbf{x}^n_{(j)} - \mathbf{x}^n_{(k)}, t\right)\right] V^n_{(j)} \\
& - \sum_{j=1} \left[\mathbf{t}^n_{(j)(k)}\left(\mathbf{u}^n_{(k)} - \mathbf{u}^n_{(j)}, \mathbf{x}^n_{(k)} - \mathbf{x}^n_{(j)}, t\right)\right] V^n_{(j)} \\
& + \sum_{m=n+1,n-1} \mathbf{p}^{(n)(m)}_{(k)} V^m_{(k)} + \sum_{m=n+1,n-1} \sum_{j=1} \mathbf{q}^{(n)(m)}_{(k)(j)} V^m_{(j)} + \mathbf{b}^n_{(k)}
\end{aligned}
\tag{14}
$$

where the material point, $\mathbf{x}^n_{(k)}$ on the nth layer is associated with an incremental volume, $V^n_{(k)}$ and a mass density of $\rho^n_{(k)}$. With respect to a Cartesian coordinate system, the material point $\mathbf{x}^n_{(k)}$ experiences displacement, $\mathbf{u}^n_{(k)}$, and its location is described by the position vector, $\mathbf{y}_{(k)}$ in the deformed state. The displacement and body load vectors at material point, $\mathbf{x}^n_{(k)}$, are represented by $\mathbf{u}^n_{(k)}$ and $\mathbf{b}^n_{(k)}$, respectively. The motion of a material point conforms to the Lagrangian description of motion.

Arising from in-plane deformation, $\mathbf{t}^n_{(k)(j)}$ represents the force density that material point, $\mathbf{x}^n_{(j)}$ exerts up on material point, $\mathbf{x}^n_{(k)}$. The force density vectors, $\mathbf{p}^{(n)(m)}_{(k)}$ and $\mathbf{q}^{(n)(m)}_{(k)(j)}$ with $m = (n+1), (n-1)$ develop due to the transverse normal and transverse shear deformations, respectively, between the material points $\mathbf{x}^n_{(k)}$ and $\mathbf{x}^m_{(k)}$. The explicit form of the force density vectors, $\mathbf{t}^n_{(k)(j)}$, $\mathbf{p}^{(n)(m)}_{(k)}$ and $\mathbf{q}^{(n)(m)}_{(k)(j)}$ associated with in-plane and transverse deformations, respectively, are explicitly derived by Madenci and Oterkus [14] in the form

$$
\begin{aligned}
\mathbf{t}^{(n)}_{(k)(j)} = & \, 2\delta a d \frac{\Lambda^n_{(k)(j)}}{\left|\mathbf{x}^n_{(j)} - \mathbf{x}^n_{(k)}\right|} \theta^n_{(k)} \frac{\mathbf{y}^n_{(j)} - \mathbf{y}^n_{(k)}}{\left|\mathbf{y}^n_{(j)} - \mathbf{y}^n_{(k)}\right|} \\
& + 2\delta\left(\mu_F b_F + b_{FT} + \mu_T b_T\right) s^{(n)}_{(k)(j)} \frac{\mathbf{y}^n_{(j)} - \mathbf{y}^n_{(k)}}{\left|\mathbf{y}^n_{(j)} - \mathbf{y}^n_{(k)}\right|}
\end{aligned}
\tag{15}
$$

$$
\mathbf{p}^{(n)(m)}_{(k)} = 2b_N\left(\left|\mathbf{y}^m_{(k)} - \mathbf{y}^n_{(k)}\right| - \left|\mathbf{x}^m_{(k)} - \mathbf{x}^n_{(k)}\right|\right) \frac{\mathbf{y}^m_{(k)} - \mathbf{y}^n_{(k)}}{\left|\mathbf{y}^m_{(k)} - \mathbf{y}^n_{(k)}\right|}
\tag{16}
$$

$$
\begin{aligned}
\mathbf{q}^{(n)(m)}_{(k)(j)} = & \, 2b_S\left(\left|\mathbf{y}^m_{(j)} - \mathbf{y}^n_{(k)}\right| - \left|\mathbf{x}^m_{(j)} - \mathbf{x}^n_{(k)}\right|\right) \frac{\mathbf{y}^m_{(j)} - \mathbf{y}^n_{(k)}}{\left|\mathbf{y}^m_{(j)} - \mathbf{y}^n_{(k)}\right|} \\
& - 2b_S\left(\left|\mathbf{y}^m_{(k)} - \mathbf{y}^n_{(j)}\right| - \left|\mathbf{x}^m_{(k)} - \mathbf{x}^n_{(j)}\right|\right) \frac{\mathbf{y}^m_{(j)} - \mathbf{y}^n_{(k)}}{\left|\mathbf{y}^m_{(j)} - \mathbf{y}^n_{(k)}\right|}
\end{aligned}
\tag{17}
$$

$$s_{(k)(j)}^{(n)} = \frac{\left|\mathbf{y}_{(j)}^{n} - \mathbf{y}_{(k)}^{n}\right| - \left|\mathbf{x}_{(j)}^{n} - \mathbf{x}_{(k)}^{n}\right|}{\left|\mathbf{x}_{(j)}^{n} - \mathbf{x}_{(k)}^{n}\right|} \tag{18}$$

in which a, d, b_F, b_T and b_{FT} are the PD material parameters, and μ_F and μ_T are defined as

$$\mu_F = \begin{cases} 1 & (\mathbf{x}_{(j)} - \mathbf{x}_{(k)}) \parallel \text{fiber direction} \\ 0 & \text{otherwise} \end{cases} \tag{19}$$

and

$$\mu_T = \begin{cases} 1 & (\mathbf{x}_{(j)} - \mathbf{x}_{(k)}) \perp \text{fiber direction} \\ 0 & \text{otherwise} \end{cases} \tag{20}$$

The parameter $\theta_{(k)}$ at material point $\mathbf{x}_{(k)}^{n}$ is defined as

$$\theta_{(k)}^{n} = d \sum_{j=1}^{\infty} \frac{\delta}{\left|\mathbf{x}_{(j)}^{n} - \mathbf{x}_{(k)}^{n}\right|} \left(\left|\mathbf{y}_{(j)}^{n} - \mathbf{y}_{(k)}^{n}\right| - \left|\mathbf{x}_{(j)}^{n} - \mathbf{x}_{(k)}^{n}\right|\right) \Lambda_{(k)(j)}^{n} V_{(j)}^{n} \tag{21}$$

The parameter $\Lambda_{(k)(j)}^{n}$ is defined as

$$\Lambda_{(k)(j)}^{n} = \frac{\mathbf{y}_{(j)} - \mathbf{y}_{(k)}}{\left|\mathbf{y}_{(j)} - \mathbf{y}_{(k)}\right|} \cdot \frac{\mathbf{x}_{(j)} - \mathbf{x}_{(k)}}{\left|\mathbf{x}_{(j)} - \mathbf{x}_{(k)}\right|} \tag{22}$$

It is $\Lambda_{(k)(j)}^{n} \approx 1$ for small deformation. The PD material parameters can then be obtained in terms of engineering material constants by considering simple loading conditions and equating the PD strain energy density to the strain energy density from the classical continuum mechanics. The PD material parameters, a, d, b_F, b_T and b_{FT}, related to in-plane deformations can be obtained by considering four different loading conditions: simple shear, uniaxial stretch in fiber direction, uniaxial stretch in transverse direction, and biaxial stretch. As derived by Madenci and Oterkus [14], the parameters can be related to the four independent material constants of elastic modulus in the fiber direction, E_{11}, elastic modulus in the transverse direction, E_{22}, in-plane shear modulus, G_{12}, and in-plane Poisson's ratio, ν_{12}, as

$$a = \frac{1}{2}\left(Q_{12} - Q_{66}\right) \tag{23a}$$

$$d = \frac{2}{\pi h \delta^3} \tag{23b}$$

$$b_F = \frac{(Q_{11} - Q_{12} - 2Q_{66})}{2\delta\left(\sum_{j=1}^{J}\left|\mathbf{x}_{(j)}^n - \mathbf{x}_{(k)}^n\right| V_{(j)}\right)} \tag{23c}$$

$$b_T = \frac{(Q_{22} - Q_{12} - 2Q_{66})}{2\delta\left(\sum_{j=1}^{N}\left|\mathbf{x}_{(j)}^n - \mathbf{x}_{(k)}^n\right| V_{(j)}\right)} \tag{23d}$$

$$b_{FT} = \frac{6Q_{66}}{\pi h \delta^4} \tag{23e}$$

where

$$Q_{11} = \frac{E_{11}}{1 - \nu_{12}\nu_{21}}, \; Q_{12} = \frac{\nu_{12}E_{22}}{1 - \nu_{12}\nu_{21}}, \; Q_{22} = \frac{E_{22}}{1 - \nu_{12}\nu_{21}}, \; Q_{66} = G_{12} \tag{24}$$

The fiber and transverse interactions exist only in the directions parallel and transverse to the fiber, and therefore, are limited to $j = 1, J$ with J indicating the number of interactions. The arbitrary interactions exist in all directions with N number of interactions.

The PD material parameters, b_N and b_S, associated with transverse deformations can be obtained by considering two different loading conditions: transverse normal stretch and simple transverse shear. As derived by Madenci and Oterkus [14], these parameters can be determined as

$$b_N = \frac{E_m}{\hat{\delta}\left[(h_{n+1} + h_n) V_{(k)}^{n+1} + (h_{n-1} + h_n) V_{(k)}^{n-1}\right]} \tag{25a}$$

$$b_S = \frac{G_m}{8\pi\tilde{\delta}(\Delta_1 + \Delta_2)} \tag{25b}$$

$$\Delta_1 = \left(\frac{h_{n+1} + h_n}{2}\right)^3 \left(\frac{\delta^2 + 2\left(\frac{h_{n+1}+h_n}{2}\right)^2}{\sqrt{\delta^2 + \left(\frac{h_{n+1}+h_n}{2}\right)^2}} - (h_{n+1} + h_n)\right) \tag{25c}$$

$$\Delta_2 = \left(\frac{h_{n-1} + h_n}{2}\right)^3 \left(\frac{\delta^2 + 2\left(\frac{h_{n-1}+h_n}{2}\right)^2}{\sqrt{\delta^2 + \left(\frac{h_{n-1}+h_n}{2}\right)^2}} - (h_{n-1} + h_n)\right) \tag{25d}$$

where E_m and G_m represent the Young's modulus and shear modulus of the matrix material. The parameter, $\hat{\delta}$ represents the horizon in the thickness direction, and $\tilde{\delta}$ is defined as $\tilde{\delta} = \sqrt{\delta^2 + \hat{\delta}^2}$.

As the force density vectors must be equal in magnitude for BB PD, the parameter a and the parameter b_T should both vanish. This requirement leads to the constraint equations, previously derived by Oterkus and Madenci [21], as $Q_{12} = Q_{66}$ and $Q_{22} = 3Q_{12}$.

The non-vanishing PD parameters, b_F and b_{FT} in the fiber and remaining directions, respectively, also recover the expressions derived by Oterkus and Madenci [21] as

$$b_F = \frac{(Q_{11} - Q_{22})}{2\delta\left(\sum_{j=1}^{N}\left|\mathbf{x}_{(j)}^n - \mathbf{x}_{(k)}^n\right| V_{(j)}\right)} \tag{26a}$$

$$b_{FT} = \frac{6Q_{66}}{\pi h \delta^4} \tag{26b}$$

Therefore, the force density vector, $\mathbf{f}_{(k)(j)}^n$between material points, $\mathbf{x}_{(j)}^n$ and $\mathbf{x}_{(k)}^n$ can be defined as

$$\mathbf{f}_{(k)(j)}^{(n)} = \delta b s_{(k)(j)}^{(n)} \frac{\mathbf{y}_{(j)}^n - \mathbf{y}_{(k)}^n}{\left|\mathbf{y}_{(j)}^n - \mathbf{y}_{(k)}^n\right|} \tag{27}$$

where b represents a PD material parameter and can be defined as

$$b = \begin{cases} b_F + b_{FT} & \phi = \theta \\ b_{FT} & \phi \neq \theta \end{cases} \tag{28}$$

in which the material parameter b_F concerns the interaction of material points only in the fiber direction. The interaction of material points in all other directions within a lamina is governed by the material parameter, b_{FT}. As shown in Fig. 7, the material point (k) represents material points that interact with material point (q) only along the fiber direction. However, the material point (p) represents material points that interact with material point (k) in any direction, including the fiber direction. The orientation of a PD bond between the material point (k) and the material point (p) is defined by the angle ϕ with respect to the x-axis.

In the case of NOSB PD, the force density vector under small strain assumptions can be expressed as

$$\mathbf{t}(\mathbf{x}, t) = w\left(|\mathbf{x}' - \mathbf{x}|\right) \mathbf{C}\varepsilon(\mathbf{x})\mathbf{K}^{-1}(\mathbf{x}) \tag{29a}$$

$$\boldsymbol{\varepsilon}(\mathbf{x}) = \frac{1}{2}\left(\mathbf{F}(\mathbf{x}) + \mathbf{F}^T(\mathbf{x})\right) - \mathbf{I} \tag{29b}$$

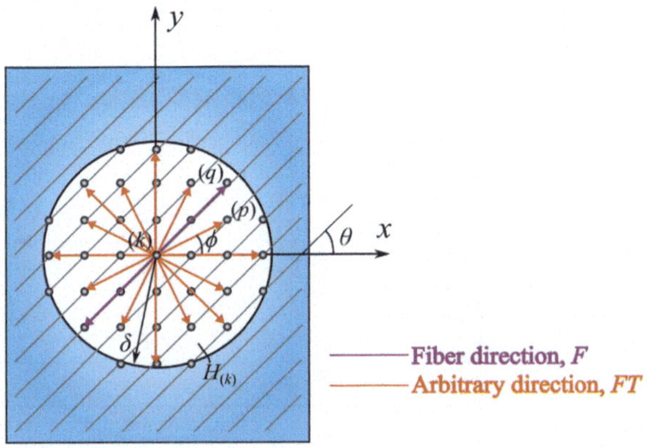

Fig. 7 PD horizon for a lamina with a fiber orientation of θ and PD bonds between material point (k) and other material points within its horizon

where **C** is the orthotropic material property matrix, and **I** represents the identity tensor [37, 39].

The damage is introduced through the removal of a PD bond between material points. Failure prediction in BB and OSB requires the critical stretch values for fiber breakage and matrix cracking associated with in-plane deformation and delamination associated with transverse deformations. Their critical values can be determined based on the experimental measurements. Determination of these critical stretch parameters are explained in Oterkus and Madenci [21] and Colavito [42]. In the case NOSB PD, the stress state in a bond can be defined as the average stress or strain between the material points $\mathbf{x}$ and $\mathbf{x}'$. When the average stress exceeds its threshold, the bond between material points $\mathbf{x}$ and $\mathbf{x}'$ is removed [37, 39].

4 Numerical Results

This section presents numerical results concerning composite structures under different boundary and loading conditions for BB, OSB, and NOSB PD models.

4.1 *Bond-Based PD*

4.1.1 Failure in a Laminate with a Crack Under Tension

This model previously considered by Kilic et al. [20] is shown in Fig. 8. The laminate has a length of $l = 10.16$ mm and a width of $w = 5.08$ mm. The pre-existing crack is located at the center along y-axis with a length of $a_0 = 1.27$ mm. The crack is introduced by breaking all the interactions passing through crack surfaces at the start of the simulation. The laminate has a stacking sequence of [0°/45°/0°]. The thickness

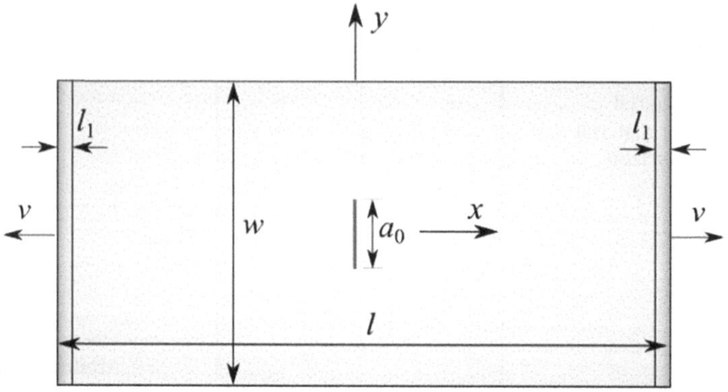

Fig. 8 Geometry and velocity boundary conditions in a laminate with a center crack

of each lamina is assumed to be 0.1651 mm. Each lamina is assumed to possess the elastic properties of T800/3900-2 pre-preg tape reported by Satyanarayana et al. [43]. The critical stretches for fiber and matrix are taken to be 0.004 and 0.005, respectively, Colavito et al. [44, 45].

Uniform tension is applied gradually by using velocity boundary conditions, v, along both vertical ends over the volumes, with length $l_1 = 0.097$ mm as shown in Fig. 7. The boundaries parallel to the x-axis at $y = -l/2$ and $y = l/2$ are free of any constraints.

Figure 9 shows the damage on the outer surface of the first ply, the damage along the interfaces between the first and second plies and between the second and third plies, in addition to the damage pattern on the outer surface of the third ply. The damage pattern displays some delamination at the interfaces due to different ply orientations, as presented in Fig. 9b, c. Similar to the experimental observations [20], the delamination pattern is asymmetric because of the presence of a [45°] ply. Mathematically expected but physically unrealistic symmetry of the damage propagation in the top and bottom plies does not emerge in the simulations due to the random distribution of fibers, as shown in Fig. 9a, d. Furthermore, the multiple splitting around the crack tip is consistent with the experimental observation.

4.1.2 Failure in a Lamina with a Hole Under Tension

This model previously considered by Hu and Madenci [26] is shown in Fig. 10. The length and width of the lamina are defined by $L = 138.43$ mm and $W = 38.1$ mm, respectively, with thickness $h = 1.0$ mm. The hole diameter is $d = 6.35$ mm. Four different fiber orientations of 0°, 30°, 60° and 90° are considered in the simulation. The lamina is made of IM-7/977-3 carbon fiber composite with Young's moduli of $E_1 = 164.3$ GPa, $E_2 = E_3 = 8.977$ GPa, and shear moduli of $G_{12} = G_{13} = 5.02$ GPa and $G_{23} = 3.126$ GPa, and Poisson's ratio of $\nu_{12} = \nu_{13} = 0.32$ and $\upsilon_{23} = 0.436$.

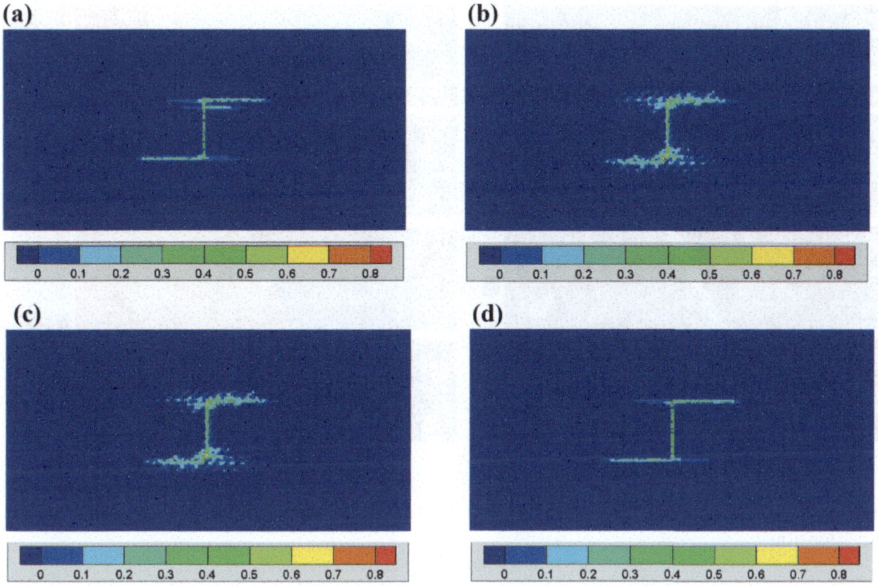

Fig. 9 Local damage in a [0°/45°/0°] laminate: **a** outer surface of first lamina; **b** interface between first and second laminae; **c** interface between second and third laminae; **d** outer surface of third lamina

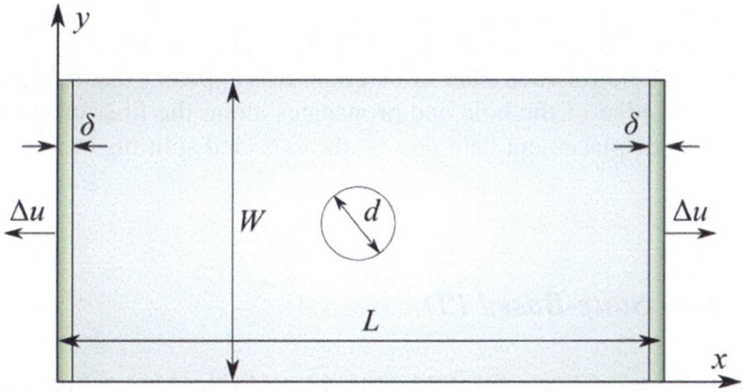

Fig. 10 Computational model of a lamina with a hole under tension

Its density is $\rho = 1603\,\text{kg/m}^3$. The strength properties are specified as 2.9 GPa 1.68 GPa, 100 MPa, 247 MPa, and 80 MPa for X^T, X^C, Y^T, Y^C, and S, respectively. The critical energy release rates are specified as $G_{Ic} = 0.256\,\text{N/mm}$ and $G_{IIc} = 0.6499\,\text{N/mm}$. Both the stress- and energy-based failure criteria are employed to evaluate fiber, matrix and delamination damage in composite laminates.

The PD simulations are achieved by using a combined implicit and explicit solution schemes. Implicit solver is employed until the first bond breakage. During

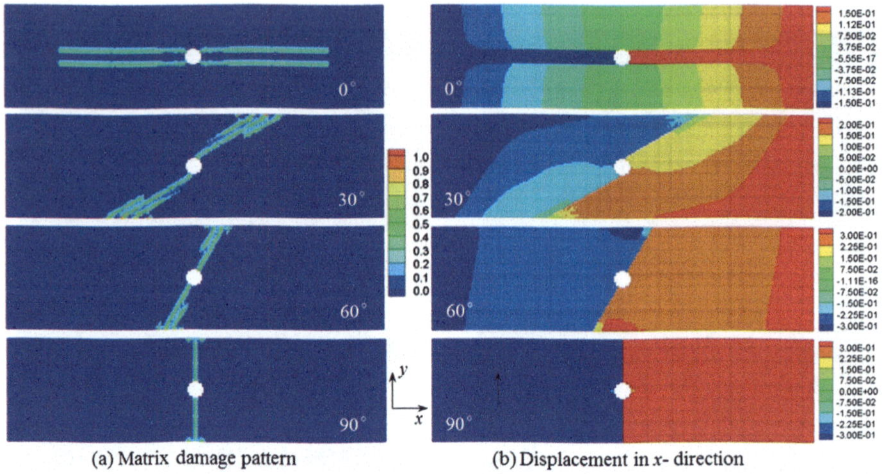

Fig. 11 PD predictions of matrix cracking and displacement discontinuities in laminae with a hole under tension

the implicit simulation, incremental step size is specified as $\Delta u = 0.001\,\text{mm}$. The analysis continues with the explicit solver until complete failure. A constant speed $\dot{u} = 500\,\text{mm/s}$ and time step size $\Delta t = 1.0 \times 10^{-7} s$ are specified in the explicit simulation.

Figure 11 shows PD predictions of matrix damage pattern and the corresponding displacement field for each fiber orientation. It is apparent that matrix cracking initiates from the edge of the hole and propagates along the fiber orientation. The discontinuities in displacement field present the expected splitting of material along the fibers.

4.2 *Ordinary State-Based PD*

4.2.1 Notched Quasi-Isotropic AS4/3501-6 Laminates Under Tension

This model previously considered by Colavito et al. [31] is shown in Fig. 12. The layup for all of the composite panels is $[45°/0°/-45°/90°]_{2S}$. Each laminate specimen is $L = 118.5\,\text{mm}$ by $W = 38.1\,\text{mm}$ and $h = 2.08\,\text{m}$ with a central through thickness hole. Four different diameters $d = 2.00$, 3.81,6.35 and 9.55 mm of central holes are considered. The material properties of AS4/3501-6 carbon/epoxy lamina are $E_{11} = 142$ GPa, $E_{22} = 10.3$ GPa, $G_{12} = 4.5$ GPa, and $\nu_{12} = 0.27$. The strength properties of the lamina are $\sigma_{1t} = 2280$ MPa, $\sigma_{1c} = -1440$ MPa, $\sigma_{2t} = 57$ MPa, $\sigma_{2c} = -228$ MPa, and $\tau_{12} = 100$ MPa.

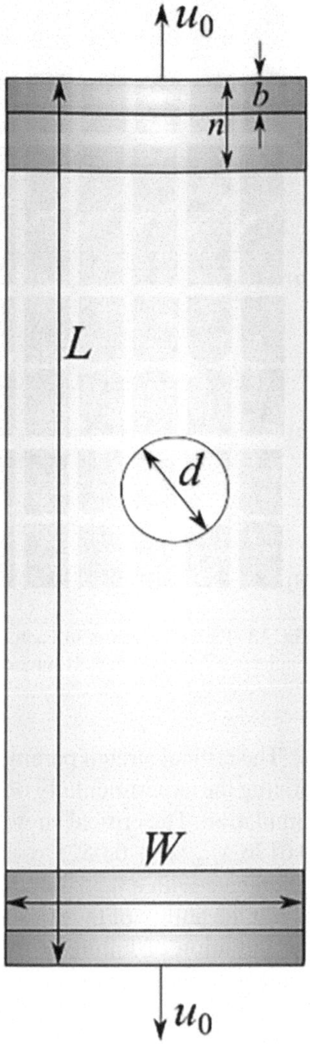

Fig. 12 Dimension and loading conditions for notched quasi-isotropic AS4/3501-6 laminates

Each laminate is discretized into a grid with 156 material points in the length direction, 50 material points in the width direction, and 16 material points in the thickness direction. The in-plane material point spacing is $\Delta x = L/156 = 0.76\,\text{mm}$. The nominal ply thickness is $t_k = 0.13\,\text{mm}$ resulting in a total laminate thickness of $h = 2.08\,\text{m}$. The horizon radius is specified as $\delta = 3.015\Delta x$. The loading is applied by specifying an end displacement u_0 to a volumetric region of length $b = 10\,\text{mm}$ at both the top and bottom edge of the model in incremental steps of 0.5 mm. A no-fail region is enforced in the volumetric region of length $n = 24\,\text{mm}$. Convergence for each step is ensured to enforce the quasi-static loading conditions.

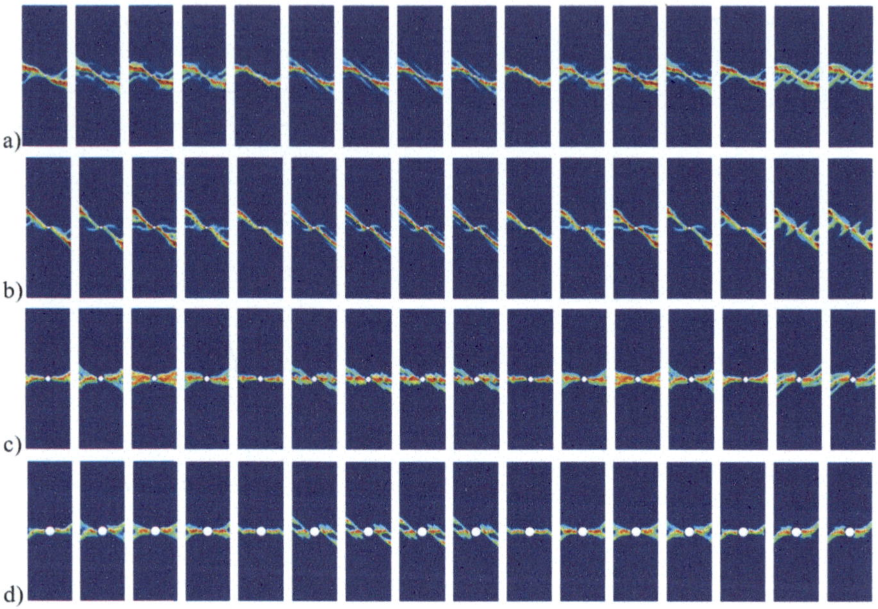

Fig. 13 Failure patterns in each ply of a $[45°/0°/-45°/90°]_{2S}$ laminate with a central hole of diameter **a** 2.00 mm, **b** 3.81 mm, **c** 6.35 mm, and **d** 9.55 mm subjected to tensile loading

The critical stretch parameters are obtained through the inverse approach by calibrating the experimentally observed peak failure loads to those obtained from the PD simulation. The critical stretch values obtained using the inverse approach are $s_{cmt} = 0.0176$, $s_{cmc} = -0.0528$, $s_{cft} = 0.01882$, and $s_{cfc} = -0.01189$. A PD simulation of a laminate with a 6.35 mm diameter central hole subjected to tensile loading is used to test the ability of these values to capture the failure load. The simulation predicted the initiation of failure at the central hole followed closely by an accumulation of damage throughout the entire laminate. The predicted failure load is 13651 N, which is significantly lower than that of the experimentally observed value.

Figure 13 shows the final damage patterns for each ply in the laminate composite subjected to a tensile load for the four different diameter central hole sizes. The damage pattern transitions from one that is aligned with the 45° angle with respect to the principle coordinate system for the small diameter hole, to one aligned with the 90° angle for the larger hole sizes. The specimen fracture pattern can be obtained by plotting the minimum damage at each material point in the thickness direction. The fracture for each of the specimens is plotted in Fig. 14. The resulting fracture patterns transition from a slant to a flat failure mode as the hole size increases.

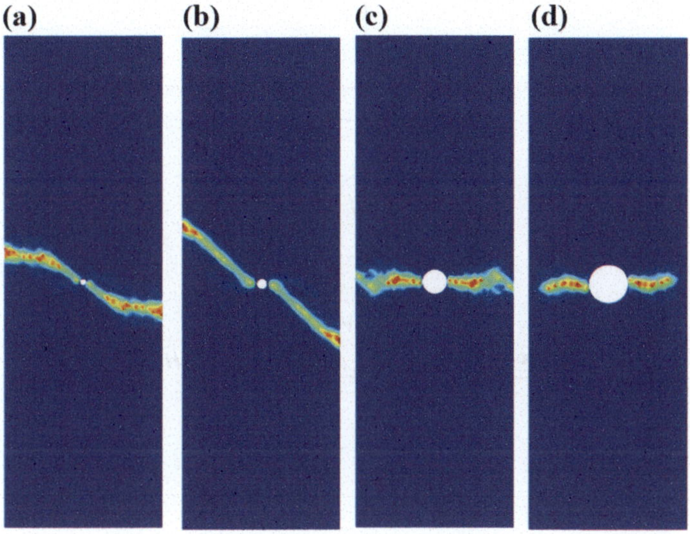

Fig. 14 Failure patterns for a $[45°/0°/-45°/90°]_{2S}$ laminate with a central hole of diameter **a** 2.00 mm, **b** 3.81 mm, **c** 6.35 mm, and **d** 9.55 mm subjected to tensile loading

4.3 Non-ordinary State-Based PD

4.3.1 Edge Crack in an Anisotropic Plate with Inclusion and Hole

This model previously considered by Hattori et al. [37] is a rectangular anisotropic plate with dimensions $w = h = 20$ mm and an edge crack of length $a = 4$ mm as shown in Fig. 15. The plate has a circular inclusion and an open hole with radius $r = 4.5$ mm. They are located at $b = 8$ mm from above and below the center of the plate. The plate is subjected to two different initial velocities defined across the plate, and given by $v = 25y/(2h)$ m/s and $v = 50y/(2h)$ m/s. The density of the plate is $\rho_p = 1600 \text{kg/m}^3$ and the density of the inclusion is $\rho_i = 5670 \text{kg/m}^3$. The fibers in the plate are rotated by an angle θ_1 with respect to the horizontal axis, while the inclusion represents an orthotropic material ($\theta_2 = 0°$). The tensile strength in the fiber, matrix and shear direction of the inclusion are specified as $\sigma_{Lu} = 2100$ MPa, $\sigma_{Tu} = 120$ MPa and $\tau_{LTu} = 135$ MPa, respectively. The interface strength parameters between the plate and the inclusion are the same as that of the tensile strength parameters of the plate.

Figures 16, 17, 18 show the crack propagation for $\theta_1 = 0°, 45°, 90°$, respectively. The different orientation of the material properties provides different crack propagation paths. Also, the loading conditions affect the crack pattern. For the case of $\theta_1 = 0°$, Fig. 16 illustrates the damage in the inclusion, and a pair of parallel cracks originate from the hole. For the case of $\theta_1 = 45°$, limited damage develops at the interface of the plate and inclusion. Crack propagations from the edge crack and the

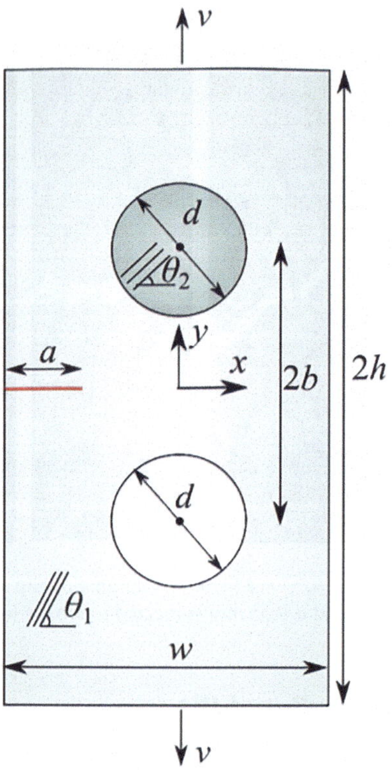

Fig. 15 Edge crack in an anisotropic plate with inclusion and hole

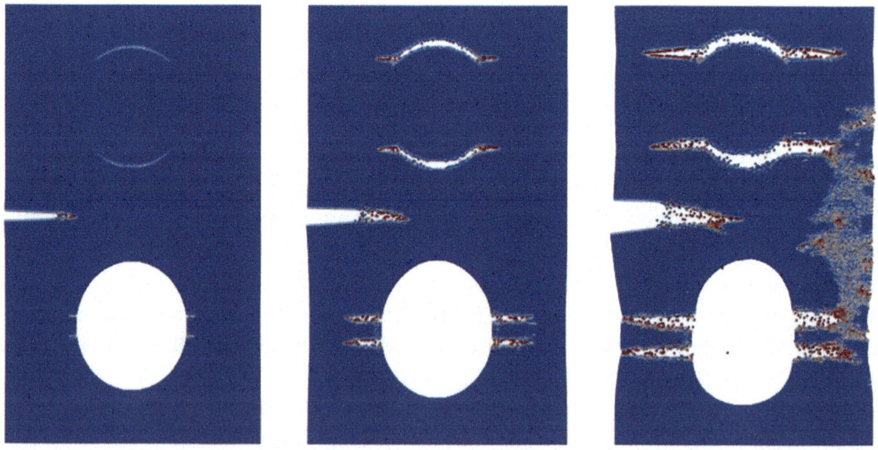

(a) $t = 3.57\mu s$ (b) $t = 4.94\mu s$ (c) $t = 6.98\mu s$

Fig. 16 Crack propagation for $\theta_1 = 0°$ with $v = 50\,\text{m/s}$ [37]

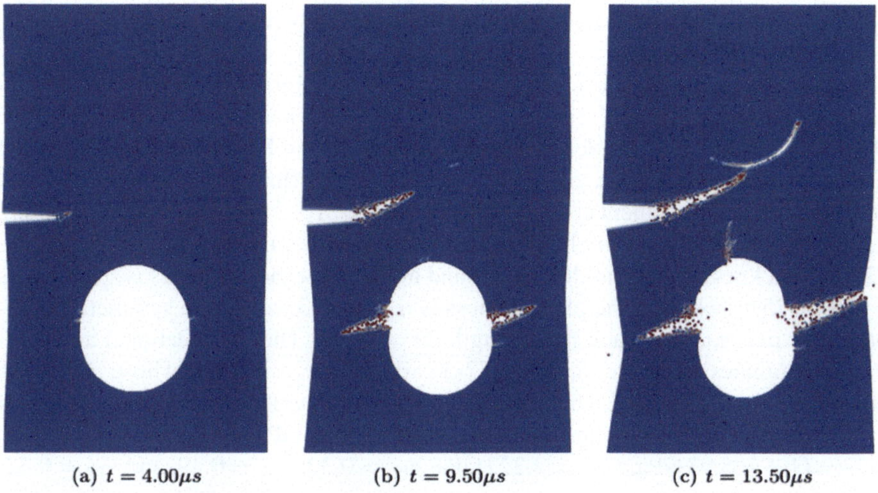

Fig. 17 Crack propagation for $\theta_1 = 45°$ with $v = 25\,\text{m/s}$ [37]

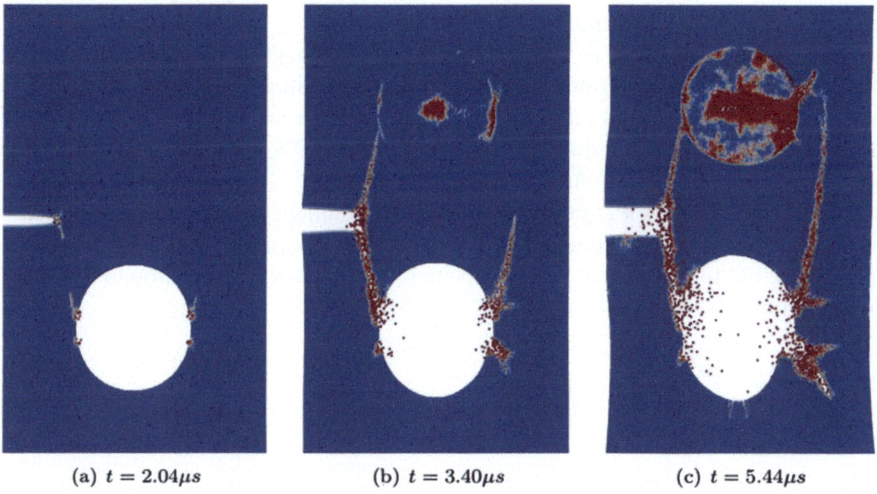

Fig. 18 Crack propagation for $\theta_1 = 90°$ with $v = 50\,\text{m/s}$ [37]

hole are along the fiber orientation as shown in Fig. 17. For the case of $\theta_1 = 90°$, Fig. 18 shows almost vertical crack propagation, reaching both the inclusion and the hole. Also, significant damage develops in the inclusion due to a sufficiently high applied velocity.

4.3.2 Laminates with a Pre-existing Crack Under Uniform Stretch

This model previously considered by Madenci et al. [40] is based on the Classical Laminate Theory (CLT) in conjunction with Peridynamic Differential Operator (PDDO) introduced by Madenci et al. [46]. As shown in Fig. 19, the unidirectional laminates and a symmetric cross-ply laminate with a through-the-thickness crack subjected to a displacement constraint of $u_0 = 1.0 \times 10^{-3}$ m in the x-direction along its left and right edges. The length and width of the laminates are $L = W = 1$ m. The ply thickness for the unidirectional and non-symmetric laminates is specified as $t_k = 0.0025$ m, whereas the ply thickness is $t_k = 0.0033$ m for the symmetric laminate. The pre-existing crack has a length $2a = 0.2$ m. The unidirectional and symmetric laminates are made of IM-7/977-3 carbon fiber composite whose properties are specified in Sect. 4.2.1.The non-symmetric laminate is made of Gr/Epoxy with orthotropic material properties specified as $E_1 = 157.9$ GPa, $E_2 = E_3 = 9.584$ GPa, $G_{12} = G_{13} = 5.93$ GPa, $G_{23} = 3.227$ GPa, $\nu_{12} = 0.32$, $\nu_{23} = 0.49$, and $\nu_{13} = 0.32$.

The laminate is discretized by using a uniform grid spacing of $\Delta x = \Delta y = \Delta = L/100$ m resulting in 30,000 degrees of freedom. The family member of each material point is constructed by the horizon size of $\delta = 4\Delta$. The corresponding crack growth patterns from the pre-existing crack are shown in Fig. 20. The lamina with a $\alpha = 0°$ fiber orientation fails due to splitting arising from shear stress in the matrix. As expected, the crack propagates co-linearly with the initial crack for a lamina with $\alpha = 90°$ fiber orientation. In the case of a lamina with $\alpha = 45°$ fiber orientation, the crack grows along the fiber direction, creating a $\alpha = 45°$ kink angle. In all cases, the crack propagates parallel to the fiber direction.

Figure 21 shows the PD predictions of matrix damage pattern in the symmetric laminate of [0°/90°/0°]. The matrix cracking initiates from the crack tips and propagates along the fiber orientation. As expected, the splitting occurs around the crack tip in the top and bottom layers.

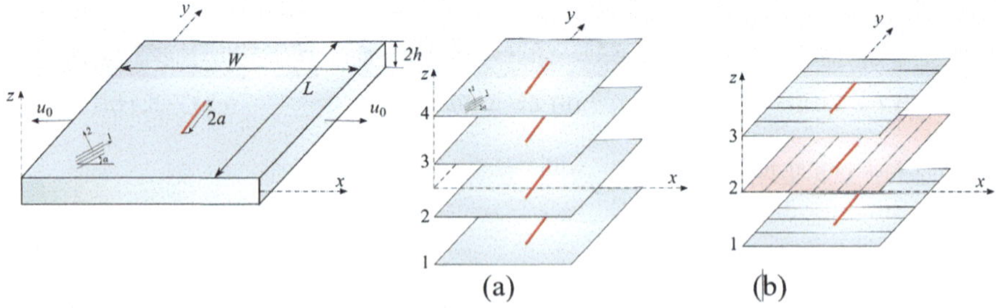

Fig. 19 Laminates under uniform stretch: **a** unidirectional, and **b** cross-ply

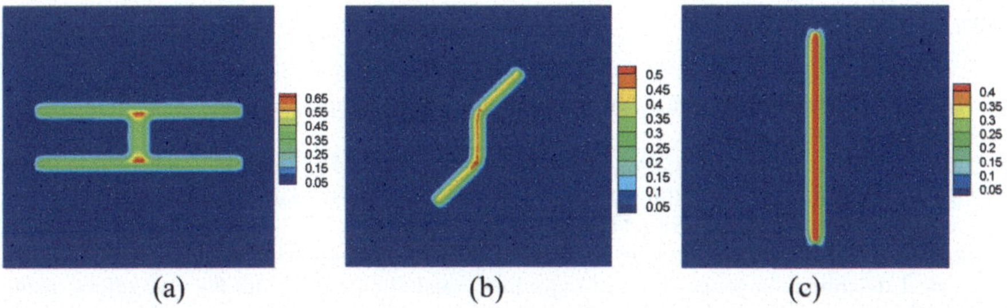

Fig. 20 Damage propagation in unidirectional laminates **a** $\alpha = 0°$, **b** $\alpha = 45°$, and **c** $\alpha = 90°$

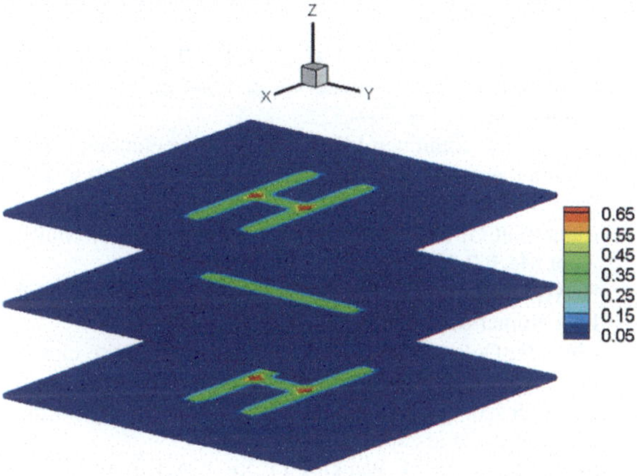

Fig. 21 Damage propagation in a [0°/90°/0°] laminate with a pre-existing crack

5 Conclusions

This chapter provides an overview of PD modeling of composite laminates with specific applications highlighting the differences in BB, OSB, and NOSB PD. The PD models successfully predicts the damage growth patterns in fiber-reinforced laminates while considering the distinct properties of the fiber and matrix, as well as of the interlayer material between the plies. The predictions capture the correct failure mechanisms of matrix cracking, fiber breakage, and delamination without resorting to any special treatments, and agree with the experimental observations published in the literature. The PD models offer a reliable model for progressive failure and strength prediction of composite laminates with arbitrary layup under complex loading conditions.

References

1. Krueger R (2004) Virtual crack closure technique: history, approach, and applications. Appl Mech Rev 57:109–143
2. Liu PF, Hou SJ, Chu JK, Hu XY, Zhou CL, Liu YL, Zheng JY, Zhao A, Yan L (2011) Finite element analysis of postbuckling and delamination of composite laminates using virtual crack closure technique. Compos Struct 93:1549–1560
3. Shokrieh MM, Rajabpour-Shirazi H, Heidari-Rarani M, Haghpanahi M (2012) Simulation of mode I delamination propagation in multidirectional composites with R-curve effects using VCCT method. Comput Mater Sci 65:66–73
4. Borg R, Nilsson L, Simonsson K (2001) Simulation of delamination in fiber composites with a discrete cohesive failure model. Compos Sci Tech 61:667–677
5. Yang Q, Cox B (2005) Cohesive models for damage evolution in laminated composites. Int J Fract 133:107–137
6. Zhao L, Gong Y, Zhang J, Chen Y, Fei B (2014) Simulation of delamination growth in multidirectional laminates under mode I and mixed mode I/II loadings using cohesive elements. Compos Struct 116:509–522
7. Rybicki EF, Kanninen MF (1977) A finite element calculation of stress intensity factors by a modified crack closure integral. Eng Fract Mech 9:931–938
8. Dugdale DS (1960) Yielding of steel sheets containing slits. J Mech Phys Solids 8:100–104
9. Barenblatt GI (1962) The mathematical theory of equilibrium cracks in brittle fracture. Adv Appl Mech 7:55–129
10. Belytschko T, Black T (1999) Elastic crack growth in finite elements with minimal remeshing. Int J Numer Method Eng 45:601–620
11. Moes N, Dolbow J, Belytschko T (1999) A finite element method for crack growth without remeshing. Int J Numer Method Eng 46:131–150
12. Silling S (2000) Reformulation of elasticity theory for discontinuities and long-range forces. J Mech Phys Solids 48:175–209
13. Silling S, Epton M, Weckner O, Xu J, Askari E (2007) Peridynamic states and constitutive modeling. J Elast 88:151–184
14. Madenci E, Oterkus E (2014) Peridynamic theory and its applications. Springer
15. Madenci E, Dorduncu M, Phan N, Gu X (2019) Weak form of bond-associated non-ordinary state-based peridynamics free of zero energy modes with uniform or non-uniform discretization. Eng Fract Mech 218:106613
16. Madenci E, Dorduncu M, Barut A, Phan N (2018) A state-based peridynamic analysis in a finite element framework. Eng Fract Mech 195:104–128
17. Askari E, Xu J, Silling S (2006) Peridynamic analysis of damage and failure in composites. In: 44th AIAA/ASME/ASCE/AHS/ASC aerospace sciences meeting and exhibit, Reno, NV. AIAA 2006-88
18. Xu J, Askari A, Weckner O, Razi H, Silling S (2007) Damage and failure analysis of composite laminates under biaxial loads. 48th AIAA/ASME/ASCE/AHS/ASC structures, structural dynamics, and materials conference, Honolulu, HI. AIAA 2007–2315
19. Xu J, Askari A, Weckner O, Silling S (2008) Peridynamic analysis of impact damage in composite laminates. J Aerospace Eng 21:187–194
20. Kilic B, Agwai A, Madenci E (2009) Peridynamic theory for progressive damage prediction in center-cracked composite laminates. Compos Struct 90:141–151
21. Oterkus E, Madenci E (2012) Peridynamic analysis of fiber-reinforced composite materials. J Mech Mater Struct 7:45–84
22. Hu W, Ha YD, Bobaru F (2011) Modeling dynamic fracture and damage in a fiber-reinforced composite lamina with peridynamics. Int J Multiscale Comput Eng 9:707–726
23. Oterkus E, Madenci E, Weckner O, Silling S, Bogert P, Tessler A (2012) Combined finite element and peridynamic analyses for predicting failure in a stiffened composite curved panel with a central slot. Compos Struct 94:839–850

24. Ghajari M, Iannucci L, Curtis P (2014) A peridynamic material model for the analysis of dynamic crack propagation in orthotropic media. Comput Method Appl Mech Eng 276:431–452
25. Hu YL, Yu Y, Wang H (2014) Peridynamic analytical method for progressive damage in notched composite laminates. Compos Struct 108:801–810
26. Hu YL, Madenci E (2016) Bond-based peridynamic modeling of composite laminates with arbitrary fiber orientation and stacking sequence. Compos Struct 153:139–175
27. Diyaroglu C, Oterkus E, Madenci E, Rabczuk T, Siddiq A (2016) Peridynamic modeling of composite laminates under explosive loading. Compos Struct 144:14–23
28. Sadowski T, Pankowski B (2016) Peridynamical modelling of nanoindentation in ceramic composites. Solid State Phenom 254:55–59
29. Ren B, Wu CT, Seleson P, Zeng D, Lyu D (2018) A peridynamic failure analysis of fiber-reinforced composite laminates using finite element discontinuous Galerkin approximations. Int J Fract 214:49–68
30. Diana V, Casolo S (2019) A full orthotropic micropolar peridynamic formulation for linearly elastic solids. Int J Mech Sci 160:140–155
31. Colavito KW, Barut A, Madenci E, Phan N (2013) Residual strength of composite laminates with a hole by using peridynamic theory. 54th AIAA/ASME/ASCE/AHS/ASC structures, structural dynamics and materials conference, Boston, MA, AIAA-2013-1761
32. Jiang XW, Wang H (2018) Ordinary state-based peridynamics for open-hole tensile Strength prediction of fiber-reinforced composite laminates. J Mech Mater Struct 13:53–82
33. Jiang XW, Wang H, Shijun G (2019) Peridynamic open-hole tensile strength prediction of fiber-reinforced composite laminate using energy-based failure criteria. Adv Mater Sci Eng 2019:7694081
34. Zhang H, Qiao P (2019) A state-based peridynamic model for quantitative elastic and fracture analysis of orthotropic materials. Eng Fract Mech 206:147–171
35. Gao Y, Oterkus S (2019) Fully coupled thermomechanical analysis of laminated composites by using ordinary state based peridynamic theory. Compos Struct 207:397–424
36. Rädel M, Willberg C, Krause D (2019) Peridynamic analysis of fibre-matrix debond and matrix failure mechanisms in composites under transverse tensile load by an energy-based damage criterion. Compos. Part B 158:18–27
37. Hattori G, Trevelyan J, Coombs WM (2018) A non-ordinary state-based peridynamics framework for anisotropic materials. Comput Method Appl Mech Eng 339:416–442
38. Yaghoobi A, Chorzepa MG (2015) Meshless modeling framework for fiber reinforced concrete structures. Comput Struct 161:43–54
39. Shang S, Qin X, Li H, Cao X (2019) An application of non-ordinary state-based peridynamics theory in cutting process modelling of unidirectional carbon fiber reinforced polymer material. Compos Struct 226:111194
40. Madenci E, Dorduncu M, Phan N (2018) Peridynamics for progressive failure analysis of composites. 33rd technical conference of the American society for composites, 599–613
41. Silling S, Askari E (2005) A meshfree method based on the peridynamic model of solid mechanics. Comput Struct 83:1526–1535
42. Colavito KW (2013) Peridynamics for failure and residual strength prediction of fiber-reinforced composites. PhD Thesis, University of Arizona
43. Satyanarayana A, Bogert PA, Chunchu PB (2007) The effect of delamination on damage path and failure load prediction for notched composite laminates. 48th AIAA/ASME/ASCE/AHS/ASC structures, structural dynamics, and materials conference, Honolulu, HI. AIAA 2007-1993
44. Colavito KW, Kilic B, Celik E, Madenci E, Askari E, Silling S (2007) Effects of nanoparticles on stiffness and impact strength of composites. 48th AIAA/ASME/ASCE/AHS/ASC structures, structural dynamics, and materials conference, Honolulu, HI. AIAA 2007-2021
45. Colavito KW, Kilic B, Celik E, Madenci E, Askari E, Silling S (2007) Effect of void content on stiffness and strength of composites by a peridynamic analysis and static indentation test. 48th AIAA/ASME/ASCE/AHS/ASC structures, structural dynamics, and materials conference, Honolulu, HI. AIAA 2007-2257

46. Madenci E, Barut A, Futch M (2016) Peridynamic differential operator and its applications. Comput Method Appl Mech Eng 304:408–451

Nonlocal Approaches to the Dynamics of Metamaterials

Giuseppe Failla and Esmaeal Ghavanloo

Abstract A metamaterial is a broad concept of engineered material with outstanding properties not existing in conventional materials. Research on metamaterials is strongly interdisciplinary and includes materials science, physics, vibration and acoustic engineering, optics. In the last decades, a considerable number of studies focused on design, experimental characterization and computational modelling of metamaterials and, among others, nonlocal theories soon revealed a great potential of capturing essential behaviour at relatively-low computational costs. This Chapter aims to provide a concise review of nonlocal theories as applied to metamaterials, with special consideration given to vibrations and dynamics.

1 Introduction

The concept of metamaterial represents an emerging and challenging frontier in materials science, with a remarkable number of potential applications in vibration mitigation and isolation, impact absorption and wave guides. The term "metamaterials" is attributed to artificially-structured materials, typically inhomogeneous elastic media composed of periodic arrays of inclusions embedded in a matrix; different matrices and inclusions are studied in solid–solid composite systems but also in mixed solid–fluid and fluid–fluid ones [1].

The microstructure of metamaterials can be tailored to provide unique properties as band-stop filtering, redirection, channelling and multiplexing, which cannot be usually obtained by conventional materials [2, 3]. Acoustic metamaterials are a special class of composites that exhibit remarkable acoustic properties ranging

G. Failla (✉)
Department of Civil, Energy, Environmental and Materials Engineering,
University of Reggio Calabria, 89124 Reggio Calabria, Italy
e-mail: giuseppe.failla@unirc.it

E. Ghavanloo
School of Mechanical Engineering, Shiraz University, 71963-16548 Shiraz, Iran
e-mail: ghavanloo@shirazu.ac.ir

© The Author(s), under exclusive license to Springer Nature Switzerland AG 2021
E. Ghavanloo et al. (eds.), *Size-Dependent Continuum Mechanics Approaches*,
Springer Tracts in Mechanical Engineering,
https://doi.org/10.1007/978-3-030-63050-8_14

from near zero transmissibility [4], enhanced absorption [5], negative dynamic mass density/bulk modulus [6, 7], negative refractive index [8, 9], super anisotropy, zero rigidity [10] etc. These exotic phenomena have numerous potential applications such as low frequency noise attenuation [5], isolation of civil structures from seismic waves [11], super lenses with a resolution beyond the Rayleigh limit [9, 12], wave guides that can be used to channel acoustic waves, etc. Optical metamaterials are a novel class of materials, typically featuring a spatially distributed permittivity that is periodic in space, which prove capable of controlling the propagation of light in a way inaccessible by natural materials [13–16].

Wave propagation in metamaterials exhibits remarkable properties that are typical of periodic media such as heterogeneous layered structures or crystals. Essentially, these properties arise from dispersion effects attributable to wave reflection and refraction from micro-constituent interfaces, which become significant when the microstructural size is comparable to the length of the travelling wave. Successive reflections and refractions of the waves at the component interfaces result in the formation of a complicated sequence of pass and stop frequency bands [17]. In this context, the most outstanding properties of metamaterials are related essentially to Bragg scattering and local resonance phenomena [18], with the first depending on the perfect medium periodicity and the latter on the dynamics of the embedded inclusions [3]. Bragg scattering is dominant for wavelengths of the propagating wave of the same order as the size of microstructural phases and local resonance prevails at larger wavelengths [3]; that is, local resonance occurs at lower frequency regime. For instance, a typical unit cell microstructure of a locally-resonant acoustic metamaterial may consist of a matrix with an embedded inclusion or a substructure [18]. The inclusions/substructures consist of two parts, a central region with high mass density supported by a surrounding highly compliant region (e.g. rubber coated lead inclusions) [4]. This enables the unit cell to exhibit low frequency localized vibration modes that strongly couple to the long wavelength propagating wave in the matrix at the resonance frequency. The strong coupling around this frequency is what is responsible for the local resonance phenomena [4, 6].

Bragg scattering and local resonance produce the band-stop filtering typical of metamaterials; this phenomenon is nonlocal in nature, as it stems from strong interactions between waves and microstructure [19, 20]. Here lies a typical difference between natural materials and metamaterials: while the critical dimensions of natural materials may be of a fraction of one nanometer and the disparate length scales between critical feature size and operational wavelength for natural materials substantiates their treatment by local constitutive relations, the critical length scale of metamaterials may be comparable with the wavelength of the travelling signal and, for this reason, the assumption of local medium is not applicable and nonlocal effects can no longer be neglected.

As for wave propagation in metamaterials, the term "nonlocal" means coupling between the responses at non-neighbouring points induced by the travelling waves. This coupling is translated into a corresponding analytical/computational model in different ways. Among others, the most established approaches involve phenomenological models, high-order homogenization methods or averaging techniques. Nonlocal coupling may be introduced explicitly, for instance in terms of constitutive relations, or may result from mathematical models adopted for response variables at the microstructural level, typically the displacement field. The nonlocal formulations proposed in the literature exhibit different frequency ranges of validity and different capabilities in capturing wave dispersion phenomena of metamaterials. Although several review articles [21–23] on the modelling of metamaterials may be found in the literature, reviews that extensively deal with the nonlocal models of metamaterials have not been presented so far. In this Chapter, we will survey the various nonlocal approaches concerning the mechanical behaviour of the metamaterials. As the literature on the metamaterials is rich and continuously growing, here we review only some instances with focus on vibrations and dynamics.

2 Phenomenological Models

Classical continuum mechanics is one of the most important and widely used tools to describe the mechanical behaviour of materials. However, it is now well understood that it cannot always predict well experimentally-observed phenomena in natural as well as engineered materials. In order to overcome the limitations of classical continuum mechanics, various types of enriched continuum approaches have been proposed, which can be divided into three main categories. In the first category, the medium is treated as a collection of deformable or rigid particles that are endowed with additional degrees of freedom. Microcontinuum theories like micropolar, microstretch and micromorphic elasticity are the most well-known approaches of the first category. In the second category, the classical continuum mechanics is enriched with higher-order gradients of the displacement, strain and velocity. Two well-known approaches in this context are strain gradient and couple stress theories. The third category includes those theories based on long-range interactions (e.g., Peridynamics nonlocal theories). The three categories offer a phenomenological description of the microstructure effects on wave propagation in metamaterials. That is, the models are usually formulated at the macroscale, while the parameters describing wave propagation have a phenomenological character and shall be calibrated experimentally.

The micromorphic theory was introduced by Eringen [24, 25] and is considered one of the pioneering top-down microscale model [26]. This theory can be also simplified to the micropolar theory and microstretch theory. The classical micromorphic theory can be seen as a penalty formulation of gradient elasticity, see e.g. Refs. [27, 28]. Its constitutive equations may involve more than 1000 material coefficients in the general anisotropic case and, even for isotropic materials, the constitutive equations contain 18 material coefficients [29]. This, indeed, limited the applicability of the original micromorphic theory. In recent years, various simplified versions of the classical micromorphic theory were developed. Neff et al. [29] developed a relaxed version of the classical micromorphic model, named relaxed micromorphic model, which involves a drastically reduced numbers of constitutive coefficients with respect to the classical micromorphic theory. The relaxed micromorphic model may be thought as a subset of the classical model where the curvature depends only on the micro-dislocation tensor, the local response depends on the symmetric part of the elastic distortion, while full kinematical degrees of freedom are retained, namely the displacement field of the macroscopic material points and micro-distortion (plastic distortion) accounting for deformations associated with the microstructure of the continuum itself. On this basis, a proper decomposition of the strain energy density was introduced, where only the curl of the microstrain tensor is considered instead of the whole gradient, as is typically done in classical micromorphic models. This assumption was explicitly made in order to account for microstructures inside the considered continuum which form weaker connections at the microscopic level, thus allowing for the description of band gaps. Nonlocal effects were introduced via including higher-order derivatives of the micro-distortion tensor in the strain energy density. Specifically, a six-parameter model was derived, which was capable of capturing the presence of band gaps in microstructured materials [30]. Experimental-based estimation of the model parameters was pursued in a subsequent publication [31]. Madeo et al. [32] also proposed a comprehensive and rigorous treatment of the jump conditions to be enforced at surfaces of discontinuity in relaxed micromorphic continua, which could serve as a basis to set jump conditions at internal surfaces embedded in these media. In-depth investigations on the relaxed micromorphic model were carried out until very recently, on both formulation and identification of the model parameters [33–35]. Furthermore, Shaat [36] developed the reduced micromorphic model which is a dynamic formulation of Forest's microstrain continuum [28]. The reduced micromorphic model can capture all microstructural deformation patterns but with lower number of microstructural degrees of freedom. Now, several studies have shown that the microcontinuum formulations based on micropolar [37, 38], relaxed micromorphic [39–41], reduced micromorphic [42–44] and micro-dilatation theories [45, 46] can effectively model the mechanics of metamaterials. In this context, it is noteworthy that Bigoni and Drugan [47] provided a general methodology to determine the moduli of homogeneous micropolar elastic materials that best approximate heterogeneous Cauchy-elastic materials; the methodology was based on the Taylor-series expansion of the applied load and applications were proposed to investigate the three-dimensional deformations of a dilute suspension of Cauchy, linear and isotropic elastic spherical inclusions, as well as the two-dimensional deformations of circular cylindrical inclusions in a Cauchy, linear and isotropic elastic matrix. A final note regarding micromorphic theories concerns a recent study investigating the topological characteristics of a micromorphic model of metamaterials, providing evidence for the formation of band gaps [48].

Certainly, an early and seminal contribution in the field was given by Mindlin in 1964 [49], who established a linear elastic theory with microstructure. The work revealed the key role played by the microstructure in dispersive behaviour of elastic waves in heterogeneous materials and set the bases for the formulation of several generalized continuum models describing wave dispersion in such materials [50, 51]. Specifically, the Mindlin microstructure model is equivalent to Eringen's micromorphic model in the linear case. Further, Mindlin [52] developed a second-order strain gradient theory in which the strain energy is a function of the strain and its first and second gradients. Then, Mindlin and Eshel [53] reduced this theory to a first-order strain gradient theory in which the strain energy is a function of the strain and its first gradient only. Later on, a simplified version of the first strain gradient theory was developed [54]. Another version–called the modified strain gradient theory–was developed by Lam et al. [55]; it is also known as modified couple stress theory and, in this respect, see Ref. [56]. The fourth-order gradient elasticity model (with fourth-order spatial, fourth-order temporal and mixed spatial–temporal derivative terms) was investigated by Askes et al. [57], Metrikine [58], Pichugin et al. [59]. In the following, we will briefly review some of the most representative works on applications of the gradient theory to metamaterials.

The use of higher-order gradient continuum models to simulate wave dispersion in periodic lattice materials was investigated by Lombardo and Askes [60]. Dontsov et al. [61] provided an insight into the capability of a gradient elastic approach to predict the dispersion relation of one-dimensional elastic layered composites up to the second pass band, provided that the length-scale parameters associated with the higher-order gradients are appropriately calibrated. Anisotropic and dispersive wave propagation in hexagonal and hexachiral lattices was investigated by Rosi and Auffray [62] in the framework of linear strain gradient elasticity. Rosi et al. [63] revealed that the validity domain of the strain gradient elasticity theory is sufficiently large and that this theory is useful in practical applications. Khakalo and coworkers [64, 65] indicated that the strain gradient theories are applicable for the homogenization of the microstructure of various lattice metamaterials. In another study, a novel approach for the identification of equivalent second-gradient Mindlin solids was proposed for higher-order elastic material equivalent to a heterogeneous (periodic) Cauchy elastic composite [66]. Indeed, the identification of the length-scale parameters in multidimensional problems poses several challenges and various approaches were proposed to this aim [67].

In addition to the microcontinuum and strain gradient theories, other enriched continuum models were proposed and used to predict wave propagation in the metamaterials. Huang and coworkers [68, 69] developed and utilized a multi-displacement continuum model and showed that the model is capable of describing the dispersive behaviour of metamaterials. In another study, Liu et al. [70] developed a new multi-displacement microstructure continuum by introducing both multi-displacement variables and micro-deformation variables. They used the proposed model to investigate wave propagation in two-dimensional anisotropic elastic metamaterials. Zhou and coworkers [71–73] derived and compared three kinds of enriched continuum models, namely the multi-displacement continuum model, the gradient continuum model and the nonlocal gradient continuum model for wave propagation in one- and two-dimensional metamaterials. Recently, the bi-Helmholtz nonlocal continuum model with effective mass density was proposed to predict the wave dispersion relations of one-dimensional infinite acoustic metamaterials with long-range interactions [74].

As for optical metamaterials, effective material parameters were obtained to describe the propagation of light through optical metamaterials by nonlocal constitutive models. They describe the material at an effective level, avoiding the almost prohibitive computational costs required by a fine description of the material at the unit cell level. Mnasri et al. [19, 75] proposed a nonlocal constitutive approach to capture light propagation (reflection and transmission) through optical metamaterials, which involves a frequency-domain, fourth-order differential model relating electric displacement field D and electric field E. Frequency-dependent nonlocal parameters associated with the fourth-order terms were obtained by a fitting procedure targeting reflection and transmission coefficients. Focusing on two examples of metamaterial slabs, made by dielectric spheres on a cubic lattice surrounded by vacuum or by a fishnet rectangular structure, the nonlocal model exhibited good agreement with full-wave simulation data obtained by Fourier modal method, showing strong spatial dispersion. For the approach to be meaningful, the authors pointed out that the period of the metamaterial shall be smaller than half the operational wavelength, i.e. the metamaterial shall be subwavelength.

3 High-Order Homogenization Methods

A well-established approach to capture wave propagation dispersive phenomena in metamaterials is homogenization, whose aim is to replace an actual heterogeneous medium by a continuous equivalent one giving the average behaviour of the medium. Research in this field moved from the observation that first-order homogenization methods are typically limited to low frequencies or long wavelengths but are not suitable to capture dispersion phenomena that generally occur for high-frequency waves. Indeed, the fundamental characteristics of wave propagation in metamaterials arise from dispersion effects attributable to wave reflection and refraction from micro-constituent interfaces, which become significant when the microstructural size is comparable to the length of the travelling wave. These phenomena generally occur for high-frequency waves which go beyond the long wavelength tackled by first-order homogenization methods. It is well understood that first-order homogenization methods and resulting effective media cannot predict reflection and refraction of stress waves causing wave dispersion and attenuation with a material microstructure [76, 77].

Typical high-order homogenization approaches of metamaterials are multi-scale approaches where scales are identified by asymptotic expansion or additive decomposition. Alternatively, there exist high-order homogenization methods producing macroscopic equivalent continua endowed with nonlocal terms, such as spatial high-order differential operators or nonlocal time-dependent terms modelling memory effects. As compared with phenomenological models, the key feature of nonlocal models derived by high-order homogenization methods is that the nonlocal parameters are identified, in general, in the homogenization process itself. The most relevant high-order homogenization methods are summarized below.

3.1 Asymptotic Expansions

A most straightforward high-order homogenization approach relies on asymptotic expansions. Making the key hypothesis of scale separation or long wavelengths and adopting a number of additional assumptions, asymptotic elastodynamic homogenization approaches consist in first expanding and computing the relevant local (or microscopic) fields term by term up to, theoretically speaking, getting an arbitrarily high accuracy and, next, constructing the macroscopic fields and the effective constitutive properties by carrying out appropriate volume averages over a unit cell [78]. Building on the first studies proving the asymptotic approach as a valuable tool to capture local elastic resonance mechanics [79–81], various formulations were developed for the asymptotic elastodynamic homogenization of metamaterials and, more generally, periodically inhomogeneous media.

Asymptotic expansions rely on the assumption of separation of scales that, generally speaking, holds true in the long wavelength regime. The frequency ranges of validity of the asymptotic approaches proposed in the literature, however, vary depending on the specific formulation adopted, as briefly detailed below.

Boutin [82] and Triantafyllidis and Bardenhagen [83] developed a formal asymptotic approach to derive higher-order strain gradient stress-strain relations for homogenized media. An infinite-order stress-strain relation was derived in terms of the small parameter ε giving the ratio of the scale of the microstructure (e.g. the size of the periodicity cell) to the outer length scale (e.g. the size of the deformed domain) and involved averaged terms over the periodicity cell; specifically, the stress-strain relation mirrors the typical displacement solution built by a two-scale asymptotic approach in terms of a rapid (oscillating) variable and a slow (modulating) variable. The work by Boutin [82] and Triantafyllidis and Bardenhagen [83] inspired further developments by Smyshlyaev and Cherednichenko [84], who rigorously derived higher-order, strain gradient constitutive equations for homogenized media via a combination of variational and asymptotic techniques for an infinitely extended periodic elastic medium, with periodicity cell of a small size and in the presence of a fixed body force. The coefficients of these equations are explicitly related to solutions of higher-order unit cell problems. The structure of the variationally derived higher-order homogenized constitutive relations was found to be in agreement with those proposed by phenomenological strain gradient theories (see, e.g., Refs. [52, 85, 86]). With respect to the purely asymptotic approach by Boutin [82] and Triantafyllidis and Bardenhagen [83], the combined variational-asymptotic approach ensured better approximation to the actual solution when the small parameter characterising the separation of scales is "small but not too small", which is exactly the case when the scale effects are observed for the "larger" and "smaller" scales becoming comparable. Also, it guarantees that the associated "truncated" higher-order constitutive relations are elliptic, i.e. that the related higher-order effective tensor derived from this asymptotic analysis is positive definite, with considerable advantages when implementing a direct numerical solution of the higher-order equations, for example using finite elements. A combined asymptotic-variational approach was pursued also by Bacigalupo and Gambarotta [87] to build a second-order strain homogenized model. The key idea was to describe the multi-scale kinematics by suitable microscale fluctuation functions of the displacement field depending on the material microstructure, to be obtained from the solution of a recurrent sequence of cell boundary value problems derived through the asymptotic procedure. The microscale fluctuation functions were obtained as the superposition of static and dynamic contributions, the first depending on the elastic properties of the phases and the latter on the elastic and mass properties of the phases. Specifically, the dynamic contribution was found to be proportional to even powers of the phase velocity, which made the microscale fluctuation functions to depend on the direction of propagation, namely angular frequency and wave vector. The equation of motion at the macroscale were obtained by Hamilton's principle, with higher-order elastic moduli and inertial terms depending on dynamic correctors. The model was implemented for bi-material periodically-layered composites with orthotropic phases, featuring an orthotropy axis parallel to the layering direction, providing provided accurate approximations of the lowest acoustic branch of the dispersion curve.

Andrianov et al. [17] developed a classical high-order asymptotic approach leading to an effective high-order strain gradient medium, which can model well dispersive behaviours and size effects near the acoustic branches, independently of the order of the asymptotic approximations used. The approach relied on the assumption that stiffness and mass parameters of the components are of the same asymptotic order.

Some asymptotic approaches were proposed for high-contrast elastic composites. For instance, Aurialt and co-workers [88, 89] proposed an asymptotic approach to build the equivalent macroscopic models of high-contrast two-and three-constituent elastic composites, the former made of high-rigidity connected solid and soft inclusion, the latter of hard inclusions coated with a soft material and embedded in a connected stiff material. They showed that, if the stiffness contrast between soft and connected stiff materials is of the order of the squared scale-separation parameter, the propagation of long waves induces local resonance within the soft medium (in the two-constituent case) and within the hard inclusions/soft material system (in the three-constituent case). Specifically, the high-rigidity solid acts as carrying structure for the long wavelength, provided that is connected and the wavelength is large with respect to the period size; within this frequency range, the soft inclusion (possibly containing hard unconnected inclusions) acts as a resonant system. Therefore, the macroscopic model exhibits a series of inner-resonance cut-off frequencies and each band gap is associated with an eigenvalue problem of the resonating system with homogeneous Dirichlet condition along the boundary with the carrying medium. The effective mass density was found to depend nonlinearly on the frequency and, as a result, the governing equations at the macroscopic scale were nonlocal in time. Numerical applications were presented for a high-contrast stratified composite made of isotropic elastic plates proving the existence of band gaps. In this respect, results agreed with previous findings on high-contrast composites existing in the literature, made of epoxy matrix-duralumin cylinders [90] or epoxy matrix embedding lead spheres coated by silicon rubber [4, 91, 92]. An asymptotic approach for high-contrast periodic elastic composites was proposed also by Smyshlyaev [93], focusing on highly-contrasting and possibly highly anisotropic stiffness and moderately contrasting density of the phases. In particular, Smyshlyaev [93] showed that anisotropy allows to vary the number of propagating modes with the direction and achieve, in this manner, a directional localization if the inclusion phase is inter-connected. Two-scale effective elastodynamic equations were obtained with coupled micro and macro components, which were nonlocal in both time and space. Approaches developed in these studies [88, 89, 93] capture frequency band gaps resulting from local resonance phenomena in the limit of the long wavelength assumption, i.e. for long wavelengths at which the scale separation holds true. By these approaches, therefore, high-frequency diffraction phenomena typical of periodic elastic composites cannot be handled as, e.g., Rayleigh scattering where wavelengths are not very large compared to the characteristic period of the composite, as well as Bragg scattering, which appears in high-frequency phononic crystals when wavelengths are comparable with the period size. Still under the assumption of scale separation, a rigorous demonstration of nonlocality inherent to the homogenization of high-contrast composites was provided by Cherednichenko et al. [94], who focused on a composite consisting of a matrix of a certain main material, which is assumed to be connected, and highly anisotropic fibres included periodically into the matrix, with large contrast between the conductivity along the fibres and the conductivities in the transverse directions. They provided mathematical evidence that the homogenized equation is an integro-differential one, displaying nonlocality along the fibres. Specifically, the kernel of the emerging integral operator was expressed explicitly in terms of the Green function on the fibre, while the local part was determined as in classical homogenization theory. Further evidence on spatial and time nonlocal effects may be found in various works [95–100].

Fish and coworkers proposed a mathematical homogenization theory with multiple spatial and temporal scales [101–103] for media with periodic microstructure. An asymptotic expansion was used for the displacement field while fast spatial scale and a slow temporal scale were introduced to account for the rapid spatial fluctuations as well as to capture the long-term behaviour of the homogenized solution. By this approach, the authors successfully addressed the issue of conventional homogenization approaches where, as the observation time window increases, higher-order terms may become close to or even larger than the leading-order term, causing the asymptotic expansion to be no longer valid. Using the averaging process over the period of the medium to eliminate the multiple temporal scales, the formulation led to spatial nonlocal equations [101, 102], for which a C0-continuous fnite element formulation was proposed with stabilization for all frequency excitations, independently of either the unit cell size or the fnite element discretization [103]; specifically, the finite element formulation relied on a three-field Hamilton variational principle involving displacements as well as nonlocal stress and strain. In this context, the mass matrix consists of the classical mass matrix (consistent or lumped), as well as a dispersion-induced mass, which depends on the microstructural properties and the relative size of the microstructure compared to the component size. While for low strain rates the second term is negligible, the added mass term gives rise to lower deformation energy absorption at high strain rates, in agreement with experimental observation in crash tests on composite structures [103]. Issues of stability and mathematical consistency of the proposed approach were addressed by Fish and coworkers in various publications [104], e.g. starting from weak forms of homogenized macroscopic equations at different scales, the authors identified boundary conditions and secularity constrains ensuring the uniform validity of asymptotic expansions. On this basis, they used a finite-element semi-discretization in space along with an analytical solution for slow time scales and Padé approximation for the fast time scale to investigate wave propagation problems in semi-infinite and finite domains [104].

Hu and Oskay [105] proposed a spatial-temporal nonlocal homogenization model based on asymptotic expansion of the displacement field up to the eight order to address transient anti-plane shear wave propagation in viscoelastic composites. The displacement field at each order was expressed in terms of a macroscopically constant displacement field and a series of locally varying fields with zero average over the unit cell, which depend on lower-order macroscopic strain [106]. Operating in the Laplace domain, macroscale momentum balance equations were obtained in the macroscopic displacement at various orders by averaging over the unit cell. Time-nonlocality was introduced by further manipulations on the homogenized constitutive tensors, involving a Laplace variable dependent scalar term to be identified under stability constraints on the solution. Linear superposition of the macroscale momentum balance equations provided a single momentum balance equation; on neglecting terms higher than four for practical purposes a fourth-order partial differential equation was derived formally equivalent to gradient elasticity ones, spatial nonlocal, temporal nonlocal and mixed spatial-temporal nonlocal terms [57–59, 61]. The unique feature of the proposed approach is that all model parameters were directly computed from the microstructure equilibrium. In view of the difficulties in setting high-order boundary conditions, a reduced-order model in the form of a second-order partial differential equation is then proposed for efficient transient wave propagation analysis, retaining the dispersive character of the original nonlocal model through the effective stiffness tensor. Transient shear wave propagation in two-dimensional domain with periodic elastic and viscoelastic microstructure was investigated and the proposed models were verified against direct numerical simulations using finite element modelling of the cells. Numerical results were presented for an elastic bi-material layered microstructure composed of aluminium and steel and viscoelastic composite with the microstructure that has elastic circular inclusion embedded in viscoelastic matrix. The spatial-temporal nonlocal homogenization model proved to accurately capture shear wave dispersion in the first pass band and attenuation within the first stop band. Higher-order asymptotic expansions and temporal nonlocal term are found to be critical in extending the applicability of the asymptotic homogenization to shorter wavelength regime; specifically, the temporal nonlocal term is demonstrated to be important in accurately capturing the initiation of the first stop band. Applications were proposed by the same authors for one-dimensional propagation in elastic and viscoelastic periodic layered media [107]. The model proposed by Hu and Oskay [105, 107] improved the results previously obtained using lower-order expansions of the displacement field and no temporal nonlocal term [108, 109].

Finally, it is noteworthy that a two-scale asymptotic procedure was developed by Craster and co-workers [110] for wave propagation in a doubly periodic inhomogeneous medium, with focus on standing waves, periodic with the period or double period of the cell. As the frequencies of these standing waves do not belong to the low-frequency range of validity covered by the classical homogenization theory, the procedure was termed as "high-frequency homogenization" . On adopting an asymptotic representation of the displacement field where variations at the microscale and macroscale are accounted for also in the leading term, a hierarchy of separate second-order differential equations was obtained at various orders and associated boundary conditions were derived from the cell periodicity. While the eigenvalue problem at the leading order provides the standing wave solutions, the next-order differential equations involve coefficients given as integrals of the standing wave solutions. The procedure proved successful in capturing dispersion phenomena at frequencies in the vicinity of the standing waves and the final form of the effective model exhibited similarities to high-frequency long-wave equations in asymptotic theories, for elastic and acoustic waveguides, which are valid in the vicinity of thickness resonances in homogeneous plates or shells. Various applications of high-frequency asymptotic approaches were proposed by Craster and co-workers [111, 112]. In this context, the work by Boutin et al. [113] is also relevant, which focuses on the modelling of large scale modulations of high frequency mechanical waves propagating in periodic elastic composite media.

3.2 *Other Higher-Order Homogenization Techniques*

As any asymptotic expansion approach relies on the principle of scale separation, which holds true for wavelengths longer than the typical microstructural size, alternative higher-order homogenization techniques were developed for those applications where the separation of scales is no longer applicable, as is typically the case for wave propagation with wavelength of the order of inclusions dimensions. Some of the relevant techniques in this context are described below.

For heterogeneous materials with periodic microstructure, Yvonnet and Bonnet [114] introduced a consistent nonlocal homogenization, which involves the concept of mesoscopic strain and stress fields defined by integral operators acting as low-pass filters on the fine scale fluctuations and calculated over the whole medium. Here, a mesoscopic model is intended as halfway between a fully microscopic representation and a fully macroscopic homogenized one using effective tensors and is expressively meant to address behaviours where scales cannot be separated. On assuming that the (total) microscopic strain at the local scale is the superposition of mesoscopic strain and fine scale fluctuations, the formulation involves a nonlocal constitutive relation at the mesoscopic scale, which is obtained from a localization problem over a unit cell where the strain at the local scale is expressed in terms of the mesoscopic strain by an integral nonlocal operator. In other terms, the proposed approach extends the classical linear homogenization by replacing averaging operators with integral operators (the low-pass filters) and localization tensors by nonlocal operators. A Gaussian function and least-square polynomials were used as low-pass filters defining the mesoscopic fields; specifically, the least-square polynomials were used to address boundary effects induced by the Gaussian convolution when applied to finite domains for non-periodic mesoscopic strain fields defined over the unit cell [115]. For a known mesoscopic strain field, the theory was implemented approximating the mesoscopic strain as a linear combination of basis functions, so that the response at the local scale is evaluated by means of a finite number of transformation tensors computed over the unit cell [114, 115]. Later on, the authors adapted the theory to a classical displacement-based finite element formulations of elasticity, to overcome the disadvantages of the original formulation directly based on nodal strains. In this context, the nonlocal constitutive law at the mesoscopic scale is expressed by a discrete displacement-based scheme, while the microscopic local fields at all points are reconstructed from the nodal response at the mesoscale coarse grid using appropriate localization operators [116]. If compared with phenomenological nonlocal continuum formulations, the theory represents a systematic approach to construct nonlocal relations for the homogenized medium that is consistent with the actual microstructural constituents.

Filonova et al. [117] proposed a dispersive computational continuum inspired by the C2 formulation [118], which employs an additive decomposition of the displacement field into a coarse-scale displacement and a fine-scale perturbation, with the former being approximated by a quadratic function over the unit cell domain. The coarse-scale problem is formulated in a weak form over the material domain a nonlocal quadrature approach is used to approximate the integrals of the weak form over coarse-scale elements defined as disjoint union of computational unit cells. Strong and weak forms were derived also for the fine-scale problem. The dispersion analysis was conducted analytically and numerically on an elastic periodic medium. The analytical solution was obtained based on the Floquet-Bloch wave solution, solving the fine scale problem in closed form and assuming the coarse-scale solution in Floquet-Bloch form. The numerical solution, instead, was obtained by modal analysis of the discretized coupled fine-coarse scale problems. Comparison with the reference solution obtained by Floquet-Bloch theory revealed remarkable accuracy of the proposed approach, especially in the case of unit cell sizes being either half or equal to the size of the coarse-scale element. The advantage of the approach is that is free of scale separation and accounts for finite-size microstructure by the nonlocal quadrature.

Extensive work to overcome inherent limitations in the scale-separation based approaches was carried out by Geers and co-workers [119–121]. Sridhar et al. [119] developed a multi-scale computational homogenization approach. The method may be cast within the general first-order computational homogenization framework establishing the coupling between microscale and macroscale problems [120, 121]. The key feature is a Craig-Bampton mode decomposition [122], according to which the dynamic response of a representative volume element is decomposed into two parts, the quasi-static response giving the instantaneous response and internal dynamics representing the inertial response due to local resonance of the heterogeneities; specifically, the internal dynamics is obtained by an optimum eigen-mode basis. The solution decomposition relies on the relaxed scale separation principle, i.e. that the shortest characteristic wavelengths pertinent to core matrix and heterogeneities are, respectively, much larger and just larger than the size of the microstructural phases constituting the core matrix and heterogeneities; since the microstructural lengths can scale with the wavelengths associated with heterogeneities, possible micro-inertia effects can be incorporated in the model. Upon using the Craig-Bampton approach to derive the governing equations of the representative volume element, typical scale-transition relations were applied to transform the reduced balance of momentum into homogenized macroscopic continuum equations, enriched by additional kinematic degrees of freedom representing the internal dynamics of the microstructure; indeed, these kinematic degrees of freedom are the generalized amplitudes associated to the local resonance modes. The eigen-mode problem is solved off-line, eliminating the expensive solution of the microscale problems at each time step, which is typically used in a computational homogenization approach. The macroscale on-line solution procedure therefore reduces to a single-scale enriched problem accounting for the effect of microscale dynamics. The proposed approach was combined with infinite-element-method techniques for multiscale problems, thus allowing complex microstructure topologies to be incorporated within infinite macrostructure geometries, considering arbitrary transient excitation and sophisticated boundary conditions. The approach was validated against direct numerical simulation with infinite element method; considering a one-dimensional compressional wave propagation test on epoxy matrix with an embedded lead inclusion coated with soft rubber, the authors demonstrated the strong influence of local resonance on the macroscopic dynamics beyond the low frequency (long wavelength) quasi-static regime. As pointed out by the authors, the formulation is appealing as lends itself to incorporate moderate nonlinearities, damping as well as material nonlocality of second-order gradient type. Further insight into the formulation was provided in a successive work by Sridhar et al. [123]; using a plane wave transform to obtain the dispersion eigenvalue problem of the enriched continuum, the authors investigated the influence of the inclusion symmetry on the dispersion characteristics and, based on the obtained dispersion spectrum, proposed a procedure to assess the homogenizability of the problem in the frequency range of analysis. Further, the suitability of the formulation to describe negative stiffness and double negative effects (i.e. simultaneous negative effective mass density and stiffness) was also demonstrated [3].

Recently, a great deal of attention was devoted to periodic material microstructures featuring a rigid phase with dominant volumetric fraction and a soft phase with a vanishing volume fraction, as is typically the case of granular materials, masonry-like biological and nacreous bioinspired heterogeneous composites; the interest is motivated by studies showing the interesting phononic properties of these materials, in particular the existence of band gaps [124–126] and the possibility to obtain devices for vibration reduction and isolation [127]. Although a rigorous treatment of wave propagation through these materials may be made using the infinite element Bloch approach, approximate models are appealing for control and optimization of the acoustic performances. On modelling the material as an ensemble of rigid blocks (rigid phase) each equipped with in-plane translation and rotation, interconnected by elastic interfaces of vanishing thickness (soft phase), a Lagrangian approach provides the motion equations. Further, if the characteristic size of the blocks is negligible with respect to the size of the whole body, the discrete system may be approximated by a homogeneous equivalent nonlocal continuum; specifically, nonlocal constitutive models including geometric and material length scales were proposed to account for the influence of the block size on wave propagation and, in this context, the presence of rotational degrees of freedom of the blocks resulted in Cosserat micropolar equivalent continua. Dispersive propagation of harmonic waves in granular materials was studied by Suiker et al. [128], Suiker and de Borst [129], who derived a second-order micropolar continuum and compared the results with those from the Lagrangian model. Bacigalupo and Gambarotta [130] developed the micropolar equivalent continuum of a two-dimensional blocky material with periodic microstructure, consisting of equal rigid blocks of polygonal centro-symmetric shape with mass and gyroscopic inertia, connected to each other by homogeneous linearly-elastic interfaces. The micropolar model was obtained by two different approaches, namely by a continualization of the Lagrangian discrete motion equations in which the generalized displacements of the blocks are represented by a second-order Taylor expansion of the generalized macro-displacement field and, alternatively, by an alternative homogenization approach based on an extended Hamiltonian principle, the latter being based on a proper representation of the elastic potential energy that is related to the Hill-Mandel macro homogeneity condition and involves a second-order expansion of the generalized displacement field, as suggested by Bazant and Christensen [131] for rectangular frames. The two homogenization approaches provided the same Cosserat model where, as a result of the assumption of centro-symmetric blocks, the constitutive equations of the equivalent micropolar continuum were derived in a general form with uncoupled strain and curvature. While the fourth-order elasticity tensor associated with the micropolar strains is positive defined, the positive definiteness of the second-order symmetric tensor associated with the curvature vector was not guaranteed but was found to depend both on the ratio of the tangent to the normal local stiffness and on the block shape. From the results obtained on blocky materials with rhombic and hexagonal tilings, the micropolar model turned out to be particularly accurate to describe dispersive functions for wavelengths greater than 3-4 times the characteristic dimension of the block and proved capable of capturing rather complex dispersive functions, due to the interaction between optical and acoustic branches. Despite the major issue that the positive definiteness of the second-order tensor cannot be guaranteed, existence and uniqueness of the solution for the dynamic problems under study was verified through the ellipticity condition of Legendre-Hadamard, i.e. assessing that the wave phase velocity is real in the elastic micropolar continuum. Further examples of continualization approaches to derive micropolar equivalent continua for lattice metamaterials may be found in other works by Bacigalupo and Gambarotta [132, 133].

Among the high-order homogenization approaches, Torrent et al. [134] and Ponge et al. [135] developed a homogenization theory based on the Plane Wave Expansion (PWE) method, which defines a set of nonlocal generalized effective parameters for periodic acoustic composites. The theory extracts the average fields from the generalized eigenvalue equation, and defines a set of generalized constitutive parameters. Specifically, the authors demonstrated frequency dependence and nonlocality of both the effective mass density and bulk modulus, with the additional complexity that the mass density was also found to be anisotropic. A nonlocal, PWE-based homogenization approach to the dynamics of metamaterials was formulated also by Flores Méndez et al. [136, 137].

4 Averaging Techniques

Nonlocal effective constitutive relations for metamaterials were derived by Willis [138, 139]. The concept of effective constitutive relations stems from early seminal works on the dynamics of media with microstructure below the scale of measurement [140–142], where a volume ensemble averaging approach led to an *effective mass density* and an *effective modulus* given by nonlocal operators in space and time. Willis pointed out that, in general, the ensemble averaged stress is related to ensemble averaged strain and ensemble averaged velocity, as the ensemble averaged momentum density is related not only to ensemble averaged velocity but also to ensemble averaged strain. Specifically, the *effective mass density* was obtained as a second-order tensor operator reducing to mean scalar mass density only in the quasi-static limit. For electrodynamics and elastodynamics of metamaterials, nonlocal effective constitutive relations were obtained in two forms [138, 139] and, in both, ensemble averaged *effective* variables were introduced as weighted averages of the local response fields. This is a particularly desirable feature for metamaterials where, e.g., *effective* strain and velocity may be related to weighted mean of the displacement u through a weighing function of position reflecting some appropriate aspect of the microstructure: for example, it could be uniform over the matrix phase of a matrix-inclusion composite, and zero over the regions occupied by inclusions. Willis compared the two effective constitutive relations with focus on one-dimensional wave propagation in a periodic medium, whose effective properties were derived using Green's function of the motion equation [143]. He discussed the issue of obtaining a unique response when the effective constitutive relations are formulated in terms of weighted averages and showed that uniqueness can be obtained introducing an additional term in the constitutive equation, which takes the meaning of an inelastic strain [139]. For periodic media he showed that the proposed ensemble averaging can be implemented exploiting the Floquet-Bloch structure of the wave fields, meaning that the fields have only to be constructed from a single realization of the medium; this is particularly suitable for metamaterials, as they are generally designed to feature a periodic microstructure [139]. Another considerable aspect is that the effective medium formulation is self-adjoint if the original problem is self-adjoint [139]; in the case of self-adjointness, variational characterizations of the effective response are feasible [144]. Willis' theory was suitable for elastic and viscoelastic media, periodically or randomly inhomogeneous media.

Since its publication, Willis' theory inspired a great deal of work on elastodynamics homogenization [145–150]. Nassar et al. [151] revisited Willis' theory suggesting a few necessary conditions for physically meaningful applications and, moreover, to make the effective constitutive parameters computable by the finite-element method; specifically, this was made by expressing the effective constitutive parameters in terms of localization tensors expressed in terms of the relevant Green's functions, for a given pair of frequency/wave number. The work was complemented by the analytical study of a simple one-dimensional, periodic discrete system which provided physical insight into Willis' theory. In a subsequent work, the same authors unveiled the strong connections between Willis' theory and a number of asymptotic elastodynamic homogenization approaches [78]. A relation between Willis' theory and asymptotic homogenization was discussed by Meng and Guzina [152]. For this purpose, the authors built an asymptotic homogenization of Willis' model itself and showed that, under the assumption of long wavelength-low frequency waves, the second-order impedance functions obtained by the so-approximated Willis' model and the standard asymptotic approach differ by a modulation factor expressible as a polynomial in the wavenumber-frequency domain. The inconsistency was attributed to the fact that the standard asymptotic approach is generally restricted to the free-wave solution and, as such, does not account for a body-source term that, instead, is generally involved in Willis' homogenization approach.

A generalization of Willis' theory was later proposed by Nassar et al. [153], with the purpose of improving the quality of and reducing the error committed during the upscaling process, especially at high frequencies. New kinematical degrees of freedom were taken into account so as to describe some short-wavelength components of the microscopic displacement field which become dominant at high frequencies. The new degrees of freedom were introduced in conjunction with rapidly oscillating body forces as microscopic and macroscopic loadings. Consistency in terms of energy was established by an energy equivalency principle, which is a balance between the microscopic and macroscopic virtual works and proves to yield a generalized version of the well-known Hill–Mandel lemma. For a one-dimensional two-phase string, an analytical long wavelength-low frequency asymptotic approximation to the effective motion equation proved to capture simultaneously the acoustic and the first optical branch of the microscopic dispersion curve.

Still within the general context of averaging techniques, a general multiscale framework was developed by Srdihar et al. [154]. The key concept here is to use the Floquet-Bloch transform [155, 156] to perform a spectral decomposition in micro (-fast) and macro (-slow) scales of any function defined on a coordinate space with an underlying lattice. The Floquet-Bloch transform decomposition is fully general and does not require any assumption of reasonably large scale separation that, instead, is typical of a standard homogenization approach based on additive scale decomposition. The multiscale framework involves homogenization and localization. Homogenization is performed via a Floquet-Bloch average that combines Floquet-Bloch transform, volume average over the unit cell domain and inverse Floquet-Bloch transform. The volume average is defined with respect to a family of particular projection functions that, for a linearly-elastic periodic microstructure, are taken as the

Floquet-Bloch eigenvectors computed via dispersion analysis of the unit cell [157]; specifically, the optimal choice is to include the Floquet-Bloch eigenvectors at high symmetry points of the Brillouin zone in the desired frequency range [158]. Power consistency between homogenized and full scale model is ensured by a generalized Hill-Mandel condition. The essential result of the homogenization is to average out the fast-scale contributions returning a corresponding slow-scale function in real space. Next, the problem is localized to a single unit cell by assuming a higher-order spatial-temporal gradient expansion to express the full scale displacements in terms of macroscale displacements. This provides a series of recursive unit cell problems giving appropriate micromechanical corrections to the homogenized macroscale governing equations, which are finally derived as a set of high-order partial differential equations with constant coefficients, solvable by a finite-element approach. The proposed multiscale framework was validated against numerical Bloch analysis of dispersion spectra in two-dimensional unit cell designs proving capable of capturing multiple high-order branches generated by local resonance and/or Bragg scattering, using a fourth-order spatial and second-order temporal expansion and a relatively small homogenization basis.

5 Concluding Remarks

Phenomenological models, high-order homogenization methods and averaging techniques are most typical nonlocal approaches to capture wave propagation phenomena in metamaterials. The brief overview of this Chapter cannot be exhaustive but is meant to describe relevant achievements and stimulate further investigations in this emerging field. Indeed, it is believed that nonlocal models may play a crucial role in promoting metamaterials to the engineering community, providing valuable insight into fundamental mechanics at a relatively-low computational costs.

References

1. Pennec Y, Vasseur JO, Djafari-Rouhani B, Dobrzyński L, Deymier PA (2010) Two-dimensional phononic crystals: examples and applications. Surf Sci Rep 65:229–291
2. Hussein MI, Leamy MJ, Ruzzene M (2014) Dynamics of phononic materials and structures: historical origins, recent progress, and future outlook. Appl Mech Rev 66:040802
3. Sridhar A, Liu L, Kouznetsova VG, Geers MGD (2018) Homogenized enriched continuum analysis of acoustic metamaterials with negative stiffness and double negative effects. J Mech Phys Solids 119:104–117
4. Liu Z, Zhang X, Mao Y, Zhu YY, Yang Z, Chan CT, Sheng P (2000) Locally resonant sonic materials. Science 289:1734–1736
5. Wen J, Zhao H, Lv L, Yuan B, Wang G, Wen X (2011) Effects of locally resonant modes on underwater sound absorption in viscoelastic materials. J Acoust Soc Am 130:1201–1208
6. Sheng P, Mei J, Liu Z, Wen W (2007) Dynamic mass density and acoustic metamaterials. Phys B 394:256–261

7. Ding Y, Liu Z, Qiu C, Shi J (2007) Metamaterial with simultaneously negative bulk modulus and mass density. Phys Rev Lett 99:093904
8. Bigoni D, Guenneau S, Movchan AB, Brun M (2013) Elastic metamaterials with inertial locally resonant structures: Application to lensing and localization. Phys Rev B 87:174303
9. Zhu R, Liu XN, Hu GK, Sun CT, Huang GL (2014) Negative refraction of elastic waves at the deep-subwavelength scale in a single-phase metamaterial. Nat Commun 5:5510
10. Lai Y, Wu Y, Sheng P, Zhang Z (2011) Hybrid elastic solids. Nat Mater 10:620–624
11. Mitchell SJ, Pandolfi A, Ortiz M (2014) Metaconcrete: designed aggregates to enhance dynamic performance. J Mech Phys Solids 65:69–81
12. Pendry JB (2000) Negative refraction makes a perfect lens. Phys Rev Lett 85:3966–3969
13. Engheta N (2002) Advances in electromagnetics of complex media and metamaterials. Springer, Dordrecht
14. Monti A, Alù A, Toscano A, Bilotti F (2015) Optical scattering cancellation through arrays of plasmonic nanoparticles: a review. Photonics 2:540–552
15. Guenneau S, Ramakrishna SA (2009) Negative refractive index, perfect lenses and checkerboards: trapping and imaging effects in folded optical spaces. C R Phys 10:352–378
16. Narimanov EE, Kildishev AV (2009) Optical black hole: Broadband omnidirectional light absorber. Appl Phys Lett 95:041106
17. Andrianov IV, Bolshakov VI, Danishevs'kyy VV, Weichert D (2008) Higher order asymptotic homogenization and wave propagation in periodic composite materials. Proc R Soc A 464:1181–1201
18. Deymier PA (2013) Acoustic metamaterials and phononic crystals. Springer, New York
19. Mnasri K, Khrabustovskyi A, Plum M, Rockstuhl C (2019) Retrieving effective material parameters of metamaterials characterized by nonlocal constitutive relations. Phys Rev B 99:035442
20. Hu R, Oskay C (2019) Multiscale nonlocal effective medium model for in-plane elastic wave dispersion and attenuation in periodic composites. J Mech Phys Solids 124:220–243
21. Del Vescovo D, Giorgio I (2014) Dynamic problems for metamaterials: review of existing models and ideas for further research. Int J Eng Sci 80:153–172
22. Srivastava A (2015) Elastic metamaterials and dynamic homogenization: a review. Int J Smart Nano Mater 6:41–60
23. Yu X, Zhou J, Liang H, Jiang Z, Wu L (2018) Mechanical metamaterials associated with stiffness, rigidity and compressibility: a brief review. Prog Mater Sci 94:114–173
24. Eringen AC (1964) Simple microfluids. Int J Eng Sci 2:205–217
25. Eringen AC (1964) Mechanics of micromorphic materials. In: Gortler H (Ed) Proceedings of the 11th international congress of applied mechanics. Springer
26. Wang X, Lee JD (2010) Micromorphic theory: a gateway to nano world. Int J Smart Nano Mater 1:115–135
27. Forest S (2009) Micromorphic approach for gradient elasticity, viscoplasticity, and damage. J Eng Mech 135:117–131
28. Forest S, Sievert R (2006) Nonlinear microstrain theories. Int J Solids Struct 43:7224–7245
29. Neff P, Ghiba ID, Madeo A, Placidi L, Rosi G (2014) A unifying perspective: the relaxed linear micromorphic continuum. Contin Mech Thermodyn 26:639–681
30. Madeo A, Neff P, Ghiba ID, Placidi L, Rosi G (2015) Wave propagation in relaxed micromorphic continua: modeling metamaterials with frequency band-gaps. Contin Mech Thermodyn 27:551–570
31. Madeo A, Barbagallo G, d'Agostino MV, Placidi L, Neff P (2016) First evidence of nonlocality in real band-gap metamaterials: determining parameters in the relaxed micromorphic model. Proc R Soc A 472:20160169
32. Madeo A, Neff P, Ghiba ID, Rosi G (2016) Reflection and transmission of elastic waves in non-local band-gap metamaterials: a comprehensive study via the relaxed micromorphic model. J Mech Phys Solids 95:441–479
33. Barbagallo G, Madeo A, D'Agostino MV, Abreu R, Ghiba ID, Neff P (2017) Transparent anisotropy for the relaxed micromorphic model: Macroscopic consistency conditions and long wave length asymptotics. Int J Solids Struct 120:7–30

34. Neff P, Eidel B, D'Agostino MV, Madeo A (2020) Identification of scale-independent material parameters in the relaxed micromorphic model through model-adapted first order homogenization. J Elast 139:269–298
35. D'Agostino MV, Barbagallo G, Ghiba ID, Eidel B, Neff P, Madeo A (2020) Effective description of anisotropic wave dispersion in mechanical band-gap metamaterials via the relaxed micromorphic model. J Elast 139:299–329
36. Shaat M (2018) A reduced micromorphic model for multiscale materials and its applications in wave propagation. Compos Struct 201:446–454
37. Reda H, Ganghoffer JF, Lakiss H (2017) Micropolar dissipative models for the analysis of 2D dispersive waves in periodic lattices. J Sound Vib 392:325–345
38. Chen Y, Frenzel T, Guenneau S, Kadic M, Wegener M (2020) Mapping acoustical activity in 3D chiral mechanical metamaterials onto micropolar continuum elasticity. J Mech Phys Solids 137:103877
39. Madeo A, Neff P, d'Agostino MV, Barbagallo G (2016) Complete band gaps including non-local effects occur only in the relaxed micromorphic model. C R Mécanique 344:784–796
40. Madeo A, Barbagallo G, Collet M, d'Agostino MV, Miniaci M, Neff P (2018) Relaxed micromorphic modeling of the interface between a homogeneous solid and a band-gap metamaterial: new perspectives towards metastructural design. Math Mech Solids 23:1485–1506
41. Barbagallo G, Tallarico D, d'Agostino MV, Aivaliotis A, Neff P, Madeo A (2019) Relaxed micromorphic model of transient wave propagation in anisotropic band-gap metastructures. Int J Solids Struct 162:148–163
42. Shaat M, El Dhaba AR (2019) On the equivalent shear modulus of composite metamaterials. Compos. Part B 172:506–515
43. Shaat M, Ghavanloo E, Emam S (2020) A micromorphic beam theory for beams with elongated microstructures. Sci Rep 10:7984
44. El Dhaba AR (2020) Reduced micromorphic model in orthogonal curvilinear coordinates and its application to a metamaterial hemisphere. Sci Rep 10:2846
45. Lurie SA, Kalamkarov AL, Solyaev YO, Ustenko AD, Volkov AV (2018) Continuum micro-dilatation modeling of auxetic metamaterials. Int J Solids Struct 132:188–200
46. Solyaev Y, Lurie SA, Ustenko SA (2019) Numerical modeling of a composite auxetic metamaterials using micro-dilatation theory. Contin Mech Thermodyn 31:1099–1107
47. Bigoni D, Drugan WJ (2007) Analytical derivation of Cosserat moduli via homogenization of heterogeneous elastic materials. J Appl Mech 74:741–753
48. Shaat M (2020) Topological mechanics of micromorphic metamaterials, arXiv:2006.01100
49. Mindlin RD (1964) Micro-structure in linear elasticity. Arch Ration Mech Anal 16:51–78
50. Papargyri-Beskou S, Polyzos D, Beskos DE (2009) Wave dispersion in gradient elastic solids and structures: A unified treatment. Int J Solids Struct 46:3751–3759
51. Berezovski A, Engelbrecht J, Salupere A, Tamm K, Peets T, Berezovski M (2013) Dispersive waves in microstructured solids. Int J Solids Struct 50:1981–1990
52. Mindlin RD (1965) Second gradient of strain and surface-tension in linear elasticity. Int J Solids Struct 1:417–438
53. Mindlin RD, Eshel N (1968) On first strain-gradient theories in linear elasticity. Int J Solids Struct 4:109–124
54. Altan SB, Aifantis EC (1992) On the structure of the mode III crack-tip in gradient elasticity. Scr Metall Mater 26:319–324
55. Lam DCC, Yang F, Chong ACM, Wang J, Tong P (2003) Experiments and theory in strain gradient elasticity. J Mech Phys Solids 51:1477–1508
56. Münch I, Neff P, Madeo A, Ghiba ID (2017) The modified indeterminate couple stress model: Why Yang et al.'s arguments motivating a symmetric couple stress tensor contain a gap and why the couple stress tensor may be chosen symmetric nevertheless. J Appl Math Mech 97:1524–1554
57. Askes H, Metrikine AV, Pichugin AV, Bennett T (2008) Four simplified gradient elasticity models for the simulation of dispersive wave propagation Phil. Mag 88:3415–3443
58. Metrikine AV (2006) On causality of the gradient elasticity models. J Sound Vib 297:727–742

59. Pichugin AV, Askes H, Tyas A (2008) Asymptotic equivalence of homogenisation procedures and fine-tuning of continuum theories. J Sound Vib 313:858–874
60. Lombardo M, Askes H (2012) Higher-order gradient continuum modelling of periodic lattice materials. Comput Mater Sci 52:204–208
61. Dontsov EV, Tokmashev RD, Guzina BB (2013) A physical perspective of the length scales in gradient elasticity through the prism of wave dispersion. Int J Solids Struct 50:3674–3684
62. Rosi G, Auffray N (2016) Anisotropic and dispersive wave propagation within strain-gradient framework. Wave Motion 63:120–134
63. Rosi G, Placidi L, Auffray N (2018) On the validity range of strain-gradient elasticity: a mixed static-dynamic identification procedure. Eur J Mech-A/Solids 69:179–191
64. Khakalo S, Niiranen J (2018) Form II of Mindlin's second strain gradient theory of elasticity with a simplification: For materials and structures from nano-to macro-scales. Eur J Mech-A/Solids 71:292–319
65. Khakalo S, Balobanov V, Niiranen J (2018) Modelling size-dependent bending, buckling and vibrations of 2D triangular lattices by strain gradient elasticity models: applications to sandwich beams and auxetics. Int J Eng Sci 127:33–52
66. Bacigalupo A, Paggi M, Dal Corso F, Bigoni D (2018) Identification of higher-order continua equivalent to a Cauchy elastic composite. Mech Res Commun 93:11–22
67. Askes H, Aifantis EC (2011) Gradient elasticity in statics and dynamics: an overview of formulations, length scale identification procedures, finite element implementations and new results. Int J Solids Struct 48:1962–1990
68. Huang HH, Sun CT, Huang GL (2009) On the negative effective mass density in acoustic metamaterials. Int J Eng Sci 47:610–617
69. Huang HH, Sun CT (2012) Continuum modeling of a composite material with internal resonators. Mech Mater 46:1–10
70. Liu AP, Zhu R, Liu XN, Hu GK, Huang GL (2012) Multi-displacement microstructure continuum modeling of anisotropic elastic metamaterials. Wave Motion 49:411–426
71. Zhou Y, Wei P, Tang Q (2016) Continuum model of a one-dimensional lattice of metamaterials. Acta Mech 227:2361–2376
72. Zhou Y, Wei P, Li Y, Tang Q (2017) Continuum model of acoustic metamaterials with diatomic crystal lattice. Mech Adv Mater Struct 24:1059–1073
73. Zhou Y, Wei P, Li Y, Li L (2019) Continuum model of two-dimensional crystal lattice of metamaterials. Mech Adv Mater Struct 26:224–237
74. Ghavanloo E, Fazelzadeh SA (2019) Wave propagation in one-dimensional infinite acoustic metamaterials with long-range interactions. Acta Mech 230:4453–4461
75. Mnasri K, Khrabustovskyi A, Stohrer C, Plum M, Rockstuhl C (2018) Beyond local effective material properties for metamaterials. Phys Rev B 97:075439
76. Bedford A, Drumheller DS, Sutherland HJ (1976) On modeling the dynamics of composite materials. Mech Today 3:1–54
77. Erofeyev VI (2003) Wave Processes in Solids with Microstructure. World Scientific, Singapore
78. Nassar H, He QC, Auffray N (2016) On asymptotic elastodynamic homogenization approaches for periodic media. J Mech Phys Solids 88:274–290 As references [79] and [153] are the same, we have deleted the duplicate reference and renumbered accordingly. Please check and confirm
79. Zhikov VV (2000) On an extension of the method of two-scale convergence and its applications. Sbor Math 191:973–1014
80. Ávila A, Griso G, Miara B (2005) Bandes phoniques interdites enélasticitélinéarisée. CR Math 340:933–938
81. Babych NO, Kamotski IV, Smyshlyaev VP (2008) Homogenization of spectral problems in bounded domains with doubly high contrasts. Netw Heterog Media 3:413–436
82. Boutin C (1996) Microstructural effects in elastic composites. Int J Solids Struct 33:1023–1051
83. Triantafyllidis N, Bardenhagen S (1996) The influence of scale size on the stability of periodic solids and the role of associated higher order gradient continuum models. J Mech Phys Solids 44:1891–1928

84. Smyshlyaev VP, Cherednichenko KD (2000) On rigorous derivation of strain gradient effects in the overall behaviour of periodic heterogeneous media. J Mech Phys Solids 48:1325–1357
85. Toupin RA (1962) Elastic materials with couple stresses. Arch Ration Mech Anal 11:385–414
86. Fleck NA, Hutchinson JW (1997) Strain gradient plasticity. In: Hutchinson JW, Wu TY (Eds) Advances in applied mechanics. Academic Press, New York
87. Bacigalupo A, Gambarotta L (2014) Second-gradient homogenized model for wave propagation in heterogeneous periodic media. Int J Solids Struct 51:1052–1065
88. Auriault JL, Bonnet G (1985) Dynamique des composites élastiques périodiques. Arch Mech 37:269–284
89. Auriault JL, Boutin C (2012) Long wavelength inner-resonance cut-off frequencies in elastic composite materials. Int J Solids Struct 49:3269–3281
90. Vasseur JO, Deymier PA, Prantziskonis G, Hong G (1998) Experimental evidence for the existence of absolute acoustic band gaps in two-dimensional periodic composite media. J Phys: Condens Mater 10:6051–6064
91. Liu Z, Chan CT, Sheng P (2005) Analytic model of phononic crystals with local resonance. Phys Rev B 71:014103
92. Sheng P, Zhang XX, Liu Z, Chan CT (2003) Locally resonant sonic materials. Phys B 338:201–205
93. Smyshlyaev VP (2009) Propagation and localization of elastic waves in highly anisotropic periodic composites via two-scale homogenization. Mech Mater 41:434–447
94. Cherednichenko KD, Smyshlyaev VP, Zhikov VV (2006) Non-local homogenized limits for composite media with highly anisotropic periodic fibres. Proc R Soc Edinb A 136:87–114
95. Sandrakov GV (1999) Homogenization of elasticity equations with contrasting coefficients. Sbor Math 190:1749–1806
96. Bellieud M, Gruais I (2005) Homogenization of an elastic material reinforced by very stiff or heavy fibers. Non-local effects. Memory effects. J Math Pure Appl 84:55–96
97. Briane M (2002) Homogenization of non-uniformly bounded operators: Critical barrier for nonlocal effects. Arch Ration Mech Anal 164:73–101
98. Camar-Eddine M, Seppecher P (2003) Determination of the closure of the set of elasticity functionals. Arch Ration Mech Anal 170:211–245
99. Camar-Eddine M, Milton GW (2005) Non-local interactions in the homogenization closure of thermoelectric functionals. Asymptot Anal 41:259–276
100. Cherednichenko KD (2006) Two-scale asymptotics for non-local effects in composites with highly anisotropic fibres. Asymptot Anal 49:39–59
101. Fish J, Chen W, Nagai G (2000) Nonlocal dispersive model for wave propagation in heterogeneous media: one dimensional case. Int J Numer Methods Eng 54:331–346
102. Fish J, Chen W, Nagai G (2000) Nonlocal dispersive model for wave propagation in heterogeneous media: multidimensional case. Int J Numer Methods Eng 54:347–363
103. Nagai G, Fish J, Watanabe K (2004) Stabilized nonlocal model for wave propagation in heterogeneous media. Comput Mech 33:144–153
104. Fish J, Chen W (2004) Space-time multiscale model for wave propagation in heterogeneous media. Comput Method Appl Mech Eng 193:4837–4856
105. Hu R, Oskay C (2018) Spatial-temporal nonlocal homogenization model for transient antiplane shear wave propagation in periodic viscoelastic composites. Comput Method Appl Mech Eng 342:1–31
106. Boutin C, Auriault JL (1993) Rayleigh scattering in elastic composite materials. Int J Eng Sci 31:1669–1689
107. Hu R, Oskay C (2017) Nonlocal homogenization model for wave dispersion and attenuation in elastic and viscoelastic periodic layered media. J Appl Mech 84:031003
108. Hui T, Oskay C (2013) A nonlocal homogenization model for wave dispersion in dissipative composite materials. Int J Solids Struct 50:38–48
109. Hui T, Oskay C (2014) A high order homogenization model for transient dynamics of heterogeneous media including micro-inertia effects. Comput Method Appl Mech Eng 273:181–203

110. Craster RV, Kaplunov J, Pichugin AV (2010) High-frequency homogenization for periodic media. Proc R Soc A 466:2341–2362
111. Craster RV, Kaplunov J, Postnova J (2010) High-frequency asymptotics, homogenisation and localisation for lattices. Quart J Mech Appl Math 63:497–519
112. Nolde E, Craster RV, Kaplunov J (2011) High frequency homogenization for structural mechanics. J Mech Phys Solids 59:651–671
113. Boutin C, Rallu A, Hans S (2014) Large scale modulation of high frequency waves in periodic elastic composites. J Mech Phys Solids 70:362–381
114. Yvonnet J, Bonnet G (2014) A consistent nonlocal scheme based on filters for the homogenization of heterogeneous linear materials with non-separated scales. Int J Solids Struct 51:196–209
115. Yvonnet J, Bonnet G (2014) Nonlocal/coarse-graining homogenization of linear elastic media with non-separated scales using least-square polynomial filters. Int J Multiscale Com Eng 12:357–395
116. Tognevi A, Guerich M, Yvonnet J (2016) A multi-scale modeling method for heterogeneous structures without scale separation using a filter-based homogenization scheme. Int J Numer Meth Eng 108:3–25
117. Filonova V, Fafalis D, Fish J (2016) Dispersive computational continua. Comput Method Appl Mech Eng 298:58–79
118. Fish J, Filonova V, Fafalis D (2015) Computational continua revisited. Int J Numer Meth Eng 102:332–378
119. Sridhar A, Kouznetsova VG, Geers MGD (2016) Homogenization of locally resonant acoustic metamaterials towards an emergent enriched continuum. Comput Mech 57:423–435
120. Geers MGD, Kouznetsova VG, Brekelmans WAM (2008) Multi-scale computational homogenization: trends and challenges. J Comput Appl Math 234:2175–2182
121. Pham K, Kouznetsova VG, Geers MGD (2013) Transient computational homogenization for heterogeneous materials under dynamic excitation. J Mech Phys Solids 61:2125–2146
122. Craig PR, Bampton MCC (1968) Coupling of substructures for dynamic analyses. AIAA J 6:1313–1319
123. Sridhar A, Kouznetsova VG, Geers MGD (2017) A semi-analytical approach towards plane wave analysis of local resonance metamaterials using a multiscale enriched continuum description. Int J Mech Sci 133:188–198
124. Chen Y, Wang L (2014) Tunable band gaps in bio-inspired periodic composites with nacre-like microstructure. J Appl Phys 116:063506
125. Chen Y, Wang L (2016) Bio-inspired heterogeneous composites for broadband vibration mitigation. Sci Rep 5:17865
126. Yin J, Huang J, Zhang S, Zhang HW, Chen BS (2014) Ultrawide low frequency band gap of phononic crystal in nacreous composite material. Phys Lett A 378:2436–2442
127. Yin J, Peng HJ, Zhang S, Zhang HW, Chen BS (2015) Design of nacreous composite material for vibration isolation based on band gap manipulation. Comput Mater Sci 102:126–134
128. Suiker ASJ, Metrikine AV, de Borst R (2001) Comparison of wave propagation characteristics of the Cosserat continuum model and corresponding discrete lattice models. Int J Solids Struct 38:1563–1583
129. Suiker ASJ, de Borst R (2005) Enhanced continua and discrete lattices for modelling granular assemblies. Phil Trans R Soc A 363:2543–2580
130. Bacigalupo A, Gambarotta L (2017) Dispersive wave propagation in two-dimensional rigid periodic blocky materials with elastic interfaces. J Mech Phys Solids 102:165–186
131. Bazant ZP, Christensen M (1972) Analogy between micropolar continuum and grid frameworks under initial stress. Int J Solids Struct 8:327–346
132. Bacigalupo A, Gambarotta L (2017) Wave propagation in non-centrosymmetric beam-lattices with lumped masses: discrete and micropolar modeling. Int J Solids Struct 118:128–145
133. Bacigalupo A, Gambarotta L (2019) Generalized micropolar continualization of 1D beam lattices. Int J Mech Sci 155:554–570

134. Torrent D, Pennec Y, Djafari-Rouhani B (2015) Resonant and non-local properties of phononic metasolids. Phys Rev B 92:174110
135. Ponge MF, Poncelet O, Torrent D (2017) Dynamic homogenization theory for nonlocal acoustic metamaterials. Extreme Mech Lett 12:71–76
136. Méndez JF, Villanueva MS, Hernández-Rodríguez S, Mora JIR (2017) Dynamic homogenization in the Nonlocal and Local regimes for a phononic superlattice: Resonant elastic metamaterial. Res Phys 7:1376–1378
137. Méndez JF, Rodríguez FP (2013) Metasolid with anisotropic mass density. Europhys Lett 103:54001
138. Willis JR (2009) Exact effective relations for dynamics of a laminated body. Mech Mater 41:385–393
139. Willis JR (2011) Effective constitutive relations for waves in composites and metamaterials. Proc R Soc A 467:1865–1879
140. Willis JR (1981) Variational and related methods for the overall properties of composites. In: Yih CS (Ed) Advances in applied mechanics. Academic Press, New York
141. Willis JR (1981) Variational principles for dynamic problems for inhomogeneous elastic media. Wave Motion 3:1–11
142. Milton GW, Willis JR (2007) On modifications of Newton's second law and linear continuum elastodynamics. Proc R Soc A 463:855–880
143. Willis JR (2012) A comparison of two formulations for effective relations for waves in a composite. Mech Mater 47:51–60
144. Willis JR (2012) The construction of effective relations for waves in a composite. C R Mécanique 340:181–192
145. Nemat-Nasser S, Srivastava A (2011) Overall dynamic constitutive relations of layered elastic composites. J Mech Phys Solids 59:1953–1965
146. Nemat-Nasser S, Srivastava A (2013) Bounds on effective dynamic properties of elastic composites. J Mech Phys Solids 61:254–264
147. Nemat-Nasser S, Willis JR, Srivastava A, Amirkhizi AV (2011) Homogenization of periodic elastic composites and locally resonant sonic materials. Phys Rev B 83:104103
148. Srivastava A, Nemat-Nasser S (2011) Overall dynamic properties of three-dimensional periodic elastic composites. Proc R Soc A 468:269–287
149. Shuvalov AL, Kutsenko AA, Norris AN, Poncelet O (2011) Effective Willis constitutive equations for periodically stratified anisotropic elastic media. Proc R Soc A 467:1749–1769
150. Norris AN, Shuvalov AL, Kutsenko AA (2012) Analytical formulation of three-dimensional dynamic homogenization for periodic elastic systems. Proc R Soc A 468:1629–1651
151. Nassar H, He QC, Auffray N (2015) Willis elastodynamic homogenization theory revisited for periodic media. J Mech Phys Solids 77:158–178
152. Meng S, Guzina B (2018) On the dynamic homogenization of periodic media: Willis' approach versus two-scale paradigm. Proc R Soc A 474:20170638
153. Nassar H, He QC, Auffray N (2016) A generalized theory of elastodynamic homogenization for periodic media. Int J Solids Struct 84:139–146
154. Sridhar A, Kouznetsova VG, Geers MGD (2018) A general multiscale framework for the emergent effective elastodynamics of metamaterials. J Mech Phys Solids 111:414–433
155. J. Gazalet, S. Dupont, J.C. Kastelik, Q. Rolland, B. Djafari-Rouhani (2013) A tutorial survey on waves propagating in periodic media: Electronic, photonic and phononic crystals. Perception of the Bloch theorem in both real and Fourier domains. Wave Motion 50:619–654
156. Bensoussan A, Lions JL, Papanicolaou G (1978) Asymptotic analysis for periodic structures: studies in mathematics and its applications. Elsevier
157. Farzbod F, Leamy MJ (2011) Analysis of Bloch's method and the propagation technique in periodic structures. J Vib Acoust 133:031010
158. Hussein MI (2009) Reduced Bloch mode expansion for periodic media band structure calculations. Proc R Soc A 465:2825–2848
159. Silveirinha MG, Alù A, Engheta N (2008) Infrared and optical invisibility cloak with plasmonic implants based on scattering cancellation. Phys Rev B 78:075107

Gradient Extension of Classical Material Models: From Nuclear & Condensed Matter Scales to Earth & Cosmological Scales

Elias C. Aifantis

Dedicated to the unforgettable memory of my mentor James Serrin and my mentee Hussein Zbib. And to the inspiring work of my classmate Constantinos Vayenas and my daughter Katerina.

Abstract The various mathematical models developed in the past to interpret the behavior of natural and manmade materials were based on observations and experiments made at that time. Classical laws (such as Newton's for gravity, Hooke's for elasticity, Navier-Stokes for fluidity, Fick's/Fourier's for diffusion/heat transfer, Coulomb's for electricity, as well as Maxwell's for electromagnetism and Einstein's for relativity) formed the basis for shaping our current technology and civilization. The discovery of new phenomena with the aid of recently developed experimental probes have led to various modifications of these laws across disciplines and scales: from subatomic and elementary particle physics to cosmology and from atomistic and nano/micro to macro/giga scales. The emergence of nanotechnology and the further advancement of space technology are ultimately connected with the design of novel tools for observation and measurements, as well as with the development of new methods and approaches for quantification and understanding. This chapter first reviews the author's previously developed weakly nonlocal or gradient models for elasticity, diffusion and plasticity within a unifying internal length gradient (ILG) framework. It then proposes a similar extension for fluids and Maxwell's equations of electromagnetism. Finally, it ventures a gradient modification of Newton's law of gravity and examines its implications to some problems of elementary particle

E. C. Aifantis (✉)
School of Engineering, Aristotle University of Thessaloniki,
54124 Thessaloniki, Greece
e-mail: mom@mom.gen.auth.gr

Michigan Technological University,
Houghton, MI 49931, USA

Mercator Fellow Friedrich-Alexander University, Erlangen-Nüremberg,
90762 Fürth, Germany

© The Author(s), under exclusive license to Springer Nature Switzerland AG 2021
E. Ghavanloo et al. (eds.), *Size-Dependent Continuum Mechanics Approaches*,
Springer Tracts in Mechanical Engineering,
https://doi.org/10.1007/978-3-030-63050-8_15

physics, also relevant to cosmology. Along similar lines, it suggests an analogous extension of London's quantum mechanical potential to include both an "attractive" and a "repulsive" branch. It concludes with some comments on a fractional generalization of the ILG framework.

1 Introduction

In a recent chapter in Advances of Applied Mechanics [1], a detailed account is presented of the author's internal length gradient (ILG) mechanics framework. It is based on the assignment of internal lengths (ILs) (associated with the local geometry/topology of material substructure) as scalar multipliers of extra Laplacian terms that are introduced to account for heterogeneity effects and weak nonlocality. Related background work for this framework can be found in the references quoted therein, as well as in earlier published articles by the author and his coworkers [2–10].

The motivation for the development of the initial continuum mechanics-based ILG framework was the need for describing deformation pattern-forming instabilities that emerge when an externally applied stress reaches a certain threshold. Beyond that threshold, the evolution equations governing the system's homogeneous response were becoming ill-posed and further analysis was not possible. The method proposed earlier by the author to overcome the difficulty for macroscopic deformation and fracture instabilities, was to introduce higher-order gradients (in the form of Laplacians) in the constitutive equations and corresponding ILs accounting for the heterogeneity of the underlying micro/nano structures. The resulting differential equations eliminate ill-posedness, estimate the width/spacings of deformation bands, dispense with the mesh-size dependence in finite element calculations, and remove stress/strain singularities at crack tips. A similar approach has been employed by the author for higher-order diffusion and heat conduction theories, as well as for phase transitions by revisiting van der Waals theory of liquid-vapor interfaces and Cahn-Hilliard theory of spinodal decomposition [11–16] through the introduction of chemical ILs. In these works, mechanical and chemical ILs were treated separately as phenomenological parameters, depending on the material local configuration and scale of observation. Their calibration and/or estimation was left to numerical and/or laboratory experiments. Moreover, statistical features emerging at sub-macroscopic scales were not considered. A preliminary effort to address these issues has been outlined in [1], and is further elaborated upon herein. In particular, the powerful multiscale technique proposed by Kevrekidis and co-workers [17, 18]—the equation-free method (EFM)—can be utilized for the hierarchical calibration of mechanical ILs. Their experimental estimation, usually inferred from "indirect" measurements of spatio-temporal features (width/spacing/velocity of deformation bands) and related size effects, can be based on "direct" measurements through novel nanoindentation (NI) tests by monitoring the local strain gradients at various indentation depths.

The enhancement of the above deterministic ILG considerations to include stochastic effects associated with internal stress fluctuations that manifest as stress drops/strain bursts in micro/nanopillar experiments and popins/popouts in nanoindentation tests, may be pursued along the lines also outlined in [1]. Corresponding gradient-stochastic models can be derived to capture intermittent plasticity and serrated stress-strain graphs, as well as to determine statistical features such as fractal dimensions (FDs) and probability density functions (PDFs). This task can be carried out by employing Tsallis q-statistics [19–21]—based on "nonextensive entropy" (as opposed to Boltzmann-Gibbs (B-G) "extensive thermodynamics")—resulting to q-dependent multifractal spectra and q-dependent PDFs, as well as q-generalization of B-G universal power laws. Novel NI tests may be conducted on multiple specimen sites and at different penetration depths for the determination of q-distributions by recording the observed popins/popouts and comparing them with corresponding determinations from micro/nanopillar serrated stress-strain curves. Such a statistical mechanics enhanced ILG framework may also be conveniently employed to consider the Portevin Le Chatelier (PLC) and persistent slip band (PSB) plastic instabilities, along with related size effects, as outlined below.

At very small scales, mechanical and chemical effects are often equipresent, and an extended chemomechanical ILG framework is necessary in order to consider higher-order IL couplings, as suggested in [1]. In view of the fact that mechanical and chemical ILs are introduced as scalar multipliers of corresponding Laplacian terms, it turns out that such coupled chemomechanical formulation is appealing and robust. Since in mathematical biology models cells are represented by scalar concentration fields (i.e. in the same way as chemical species), the formulation could be easily adapted for the description of higher-order couplings between mechanical and biochemical ILs. Such an extended ILG mechanics framework, including synergistic effects between mechanical and chemical or biological ILs, can be employed to consider chemomechanical instabilities in Lithium ion Battery (LiB) anodes and biomechanical instabilities in brain tumors, as also outlined below.

As mentioned above, we conclude this introductory section by summarizing main results of the ILG framework and its potential to be employed for considering a variety of problems of current or emerging interest as follows:

- *Plastic Instabilities and Size Effects*: Recent experiments at micro/nano scales [22] have revealed a strong dependence on specimen size. Ongoing work in several Labs has revealed, in particular, that PLC and PSB instabilities may be suppressed when the ratio of the specimen size over the internal length is reduced below a certain threshold. The previous deterministic ILG models earlier advanced by the author and collaborators for these instabilities at macro/meso scales can be revisited and evaluated for "small-volumes" and strain localization phenomena observed in nanocrystalline (NC) and ultrafine-grain (UFG) polycrystals. New combined gradient-stochastic ILG models for both PLC and PSB instabilities can be employed to capture spatio-temporal periodicity, fractality, and transition to chaos. FDs for the observed deformation bands and PDFs for the recorded serrations in stress-strain curves can be determined through Tsallis q-statistics.
- *Chemomechanical Instabilities & Lithiation Fronts in LiB Anodes*: A deterministic version of our coupled chemomechanical ILG framework can be employed to

address chemostress damage instabilities in nanostructured LiB anodes leading to cracking and capacity fade during Li insertion/de-insertion under electrochemical cycling. This is due to the huge local volume expansions (up to 400%) and associated internal stress generation occurring in Si active particles during lithiation [23–26]. A related issue is to understand the size/stress dependence of lithiation, as well as the propagation of stress-assisted lithiated fronts which controls battery efficiency. The interplay between higher-order mechanical and chemical ILs has not yet been sufficiently considered to address these chemomechanical instabilities in LiB, despite of their wide use in microelectronics, laptops and electric car technologies.

- *Biomechanical Instabilities & Cancer Growth/Metastasis in Human Brain*: A striking analogy[1] exists between the Walgraef-Aifantis (W-A) model [27–29] of dislocation patterning in PSBs and the Go or Grow (GoG) model for glioblastoma cancer cells [30, 31]. Both processes are described by similar reaction-diffusion (R-D) type equations for mobile-immobile dislocations in the W-A model (under the action of applied stress) and the motile-immotile cancer cells in the GoG model (under the action of internally generated stress). Such internal stress effects have not been explicitly accounted for in the GoG model, despite of the fact that Murray [32, 33]—the father of modern mathematical biology—had already introduced cell-tractions and corresponding strain gradients (in the form of Laplacians, as in author's work; see, for example, the related discussions in [1]) to revisit Turing's seminal R-D work of morphogenesis. The interplay between higher-order mechanical and biochemical ILs in the GoG model can be studied, and the role of internal stress can thus be evaluated. The results can provide new insight on brain cancer progression and potential therapeutic procedures.

2 State-of-the Art: Previous Literature & Current State of Affairs

An extensive bibliography on gradient theories has already been mentioned that can be found in [1]. Specific aspects pertaining to the present review and related work on continuum mechanics models at micro/meso/macro scales are discussed in this section. For the convenience of the reader, we present this section in two parts. In the first part, we provide background on relatively recent phenomenological strain gradient models that have been developed to capture mechanically-induced plastic instabilities and size effects under the action of applied loads. In the second part, we provide a brief account of earlier, more fundamental work on phase transitions which was a direct motivation for the author's initial ILG deterministic models, as well as

[1] An elaboration of this analogy is given in a forthcoming article by H. Hatzikirou and E.C. Aifantis: On the similarities between the W-A model for dislocations and the GoG model for cancer cells (in preparation).

for the development of coupled chemomechanical/ biomechanical ILG models to be used for addressing instabilities in LiBs and brain tumors.

2.1 Plastic Instabilities & Size Effects

The terms “material instabilities” and “dislocation patterning” were quoted by the author in the mid 1980s [2–5, 34–37] to denote the self-organization of localized strain bands and dislocations in deforming solids. Various gradient dislocation dynamics and gradient plasticity models were generated to deal with dislocation pattern formation and shear band thickness/spacing evolution, as well as for interpreting size effects [38–44]. Soon afterwards, in the beginning of the 1990s and later on [2–5, 7, 9, 34–36, 45–47], the author incorporated the Laplacian of the Hookean stress into the standard constitutive equation of linear elasticity, to remove the singularities from dislocation lines and crack tips. Some of the aforementioned author’s early work on gradient theory is reviewed in a number of specific book chapters [38–44] by leading authors in the field.

Subsequently, or in parallel to the above developments, other types of gradient models have been developed such as the Fleck-Hutchinson and the Gao-Nix-Huang strain gradient theories, as well as improved gradient theories taking into account surface effects. As an outgrowth of the initial W-A phenomenological model for dislocation patterning, a substantial effort has been devoted to discrete dislocation dynamics (DDD) modeling. Due to computational limitations of DDD for obtaining dislocation patterns and motivated by the initial W-A model, alternative dislocation density based methods or continuum dislocation dynamics (CDD) have been pursued. Related references connected to the above named authors and works can be found in the bibliography listed in [38–43, 48–53]. In this connection, it is pointed out that our gradient elasticity model has recently been successfully utilized by Ghoniem’s group in UCLA to dispense with near-core singularities causing code-malfunctioning in 3D discrete dislocation dynamics simulations [54]. Moreover, our related non-singular strain/stress crack tip solutions have been successfully used by Isaksson’s group in Uppsala to interpret experimental measurements on crack-tip profiles in micro-heterogeneous materials such as solid foams and bone tissue [55, 56].

With the exception of author’s preliminary efforts described in [1], all the above works on gradient models for addressing plastic instabilities and size effects do not account for internal stress/structural defect fluctuations and synergistic gradient-stochastic effects. There are no attempts for a hierarchical IL calibration through EFM multiscale simulations and novel NI tests. The same holds for the use of Tsallis q-statistics to determine FDs and PDFs. All these are open inter-dependent issues that need to be addressed. It is also pointed out that none of the above gradient elasticity/plasticity models incorporate diffusion and chemical reaction effects. An exception can again be found in [1] where higher-order chemomechanical IL couplings are discussed. This issue needs further be addressed to consider chemomechanical and biomechanical instabilities as described below.

2.2 Chemomechanical Instabilities in LiBs & Biomechanical Instabilities in Brain

The main reason that the author's Laplacian-based ILG models can easily be extended to include chemical and biochemical ILs is due to the fact that their motivation stems from his earlier treatment (with Serrin [11, 12]) of van der Waals theory [13] of fluid interfaces, which was also the predecessor of Ginzburg-Landau theory [14] of phase transitions and the Cahn-Hilliard theory [15, 16] of spinodal decomposition. This is in contrast to the aforementioned Fleck-Hutchinson and related strain gradient models which were motivated by Cosserat-type generalized continuum mechanics theories that do not contain explicitly the Laplacian and, thus, they do not exhibit the corresponding mathematical and physical properties that this operator implies. On the other hand, chemical reactions and phase transformations have traditionally been treated with R-D equations involving the Laplace operator. The fact that both mechanical and chemical or biochemical ILs can be treated on the same footing, through the introduction of the corresponding Laplacians, allows for a robust formulation of a chemomechanical and biomechanical ILG framework that can be used to consider corresponding instability phenomena in LiB anodes and brain glioblastomas, respectively.

There is a large number of recent articles on LiB capacity fade due to colossal volume changes in anodes (up to 400% for Li-Si based anodes) during lithiation/delithiation [23–26]. While in some of these works diffusion and coupled deformation-diffusion effects have been accounted for, higher-order strain gradients and corresponding mechanical ILs have not been considered. An exception is the recent article by the author and coworkers [57] employing strain gradients and mechanical ILs to model size effects in LiB anodes, as well as in [58] employing both mechanical and chemical ILs to model the propagation of lithation fronts. This work can be used as a guide to develop criteria for the most optimum nanocomposite configuration (size/spacing of active Si-nanoparticles) for LiB anodes to prevent cracking and/or accelerate lithtiation/delithiation.

Similarly to the case of LiBs, there is an abundance of mathematical models for brain cancer. However, related ILG models accounting for internal stress gradient effects due to tumor growth and cancer cell migration/proliferation are missing. This is also true for the aforementioned GoG [30, 31] phenotypic plasticity model of cancer cell migration and its impact on tumor progression. It was found that low-grade tumor micro-ecology potentially exhibits an emergent Allee effect, i.e. a critical tumor cell density implying tumor growth or control. The precise quantification of this critical tumor cell density could be a relevant prognostic criterion for the tumor fate through biopsy measurements. It was shown that the GoG mechanism explains the fast tumor recurrence time of high-grade brain tumors after resection. These findings can be re-evaluated by transferring the stability analysis results earlier derived for the W-A model (of mobile-immobile dislocations) to the GoG model (of motile-immotile cells), by also considering stochastic heterogeneity and internal stress gradient effects.

3 ILG Formulation Through Continuum & Statistical Mechanics

In this section we briefly review the formulation of ILG framework by using ingredients of generalized continuum and statistical mechanics.

- *ILs in Elasticity/Plasticity & Diffusion/Gradient Dislocation Dynamics*: For elastic deformations, the term $\ell_\varepsilon^2 \nabla^2 [\lambda \varepsilon_{mm} \delta_{ij} + 2G\varepsilon_{ij}]$—where ℓ_ε denotes elastic IL, ε_{ij} is the elastic strain ($\varepsilon_{ij} = 1/2[u_{i,j} + u_{j,i}]$; u_i designates displacement), and (λ, G) are the Lamé constants—is incorporated into classical Hooke's law. Previous results of the author and his coworkers (see [1] and references quoted therein) show that the resulting ILG model can eliminate stress/strain singularities from dislocation/disclination lines and crack tips, as well as interpret elastic size effects. Similarly, the term $\ell_p^2 \nabla^2 \gamma^p$—where ℓ_p denotes plastic IL and $\gamma^p = \int \dot{\gamma}^p dt$ ($\dot{\gamma}^p = \sqrt{2\dot{\varepsilon}_{ij}^p \dot{\varepsilon}_{ij}^p}$) is the equivalent plastic strain with ε_{ij}^p denoting the plastic strain tensor—is introduced in the classical von-Mises yield condition or the flow rule to derive differential equations that remain well-posed in the unstable flow regime. Previous results of the author and his coworkers (e.g. [1] and references therein) show that the resulting ILG model can determine shear band widths and spacings, as well as interpret plasticity induced size effects in micro-torsion/bending and micro/nano indentation experiments. For elastic deformations at the atomic scale (near dislocation lines in crystals), ℓ_ε relates to the subatomic configuration and electronic state (e.g. through DFT calculations), while at the microscale ℓ_ε relates to particle size/spacing (e.g. through MD simulations). For plastic deformations at micro/meso scales (deformation bands, dislocation cells), ℓ_p relates to dislocation source distance/pileup length/grain size (e.g. through DDD simulations). This suggests that our earlier practice of treating the ILs as "fitting" constants needs to be revised and consider them as evolving parameters in the course of deformation. This point of view can be adopted for exploring the IL-dependence on the current state of deformation and underlying micro/nanostructural configuration, also in relation to the size of the volume considered.
  For diffusion problems, the ILs enter through the additional term $\ell_d^2 \nabla^2 j_i$ which generalizes the classical Fick's law (ℓ_d is a diffusional internal length and j_i denotes the diffusion flux) in a manner similar to the Cahn-Hilliard theory for spinodal decomposition. For collective dislocation phenomena, the IL enters through the extra Laplacian term $D\nabla^2 \rho$, where ρ denotes dislocation density and D is an "effective" diffusion-like transport coefficient. Unlike in random diffusion processes however, D here is a strain rate driven parameter. Since the strain rate depends (through Orowan's equation) linearly to the average dislocation velocity (which, in turn, relates to the local stress), the coefficient D is treated as a stress-dependent parameter that relates to individual dislocation interactions. It is noted, in this connection, that the original W-A model for dislocation patterning, which is based on such type of $D\nabla^2 \rho$ terms for the mobile and immobile dislocation densities, was initially criticized for the phenomenological nature of these Laplacian

terms. However, recent work on continuum dislocation density based dynamics (CDD)—in contrast to discrete dislocation based dynamics (DDD) simulations which were unable to produce dislocation patterns—generate such type of Laplacian terms which are necessary for dislocation pattern formation interpretations.

- *Stochasticity and Tsallis q-Statistics*: The enhancement of the above discussed ILG deterministic models through the incorporation of stochastic terms is necessary in order to account for the heterogeneity and fluctuations of internal stresses, as well as deformation-induced random micro/nanostructures. The resulting combined gradient-stochastic models can capture the observed behavior at micro/nano scales, including size dependent serrated stress-strain graphs and intermittent plasticity phenomena. Some initial results along this direction have recently been reported by the author and his coworkers [59, 60] by resorting to empirical Weibull distribution functions, as also reviewed in [1]. This approach can be adopted to describe existing experimental data on stress drops/strain jumps routinely observed in micro tension/compression and nanoindentation laboratory tests. An additional issue that can be explored here is to employ time-dependent probability distributions guided by our earlier [61, 62] and most recent [63, 64] works based on the formalism of stochastic differential equations.

  A convenient way to consider the competition between deterministic gradient and random effects is to introduce (in analogy to Wiener processes in statistical mechanics) an additive stochastic term of the form $h(\gamma)g(x)$; $\langle g(x)g(x')\rangle = l_{corr}\delta(x - x')$—with l_{corr} denoting a correlation length, and δ being the usual Dirac delta function—into the gradient expression of the flow stress. This is not an arbitrary assumption but emerges generically if one aims at a description above the scale of the discrete substructure which defines the correlation length—i.e. within a continuum model. The delta function then simply emerges because the individual volume elements of the continuum theory are effectively uncorrelated. The function $h(\gamma)$ also covers the limiting case where only the material parameters fluctuate while the evolution is deterministic (e.g. in the case of flow stress fluctuations due to fluctuating grain orientation or in the presence of a chemical environment).

  Standard deterministic ILG models cannot provide any information on measured statistical aspects of plastic deformation, such as fractal dimensions for deformation patterns; power-law exponents for dislocation avalanches [65, 66]; and strain bursts recorded during nanoindentation [67] or micro/nanopillar compression tests [22, 59, 60]. When differential equations cannot be invented to interpret experimental data and simulations, system characterization is left to statistical analyses for investigating, among other things, fractality and universal power-laws. In many cases, however, the usual power-laws based on Boltzmann-Gibbs statistics exclude the regime of low intensity-high probability events. Tsallis q-statistics [19–21] based on nonextensive entropy thermodynamics remove this difficulty and can be employed here to analyze intermittent plasticity and deformation patterned images obtained experimentally. This information can also allow the construction of appropriate PDFs to be used in the aforementioned combined gradient-stochastic models. Tsallis nonextensive (non-additive) q-entropy reads $S_q = k\left(1 - \sum_i p_i^q\right)/(q-1)$ and by letting $q \to 1$ recovers the familiar Boltzmann-Gibbs extensive entropy. Corresponding q-distribution functions (q-Gaussian, q-exponential, q-Weibull) can thus be obtained, which for $q \to 1$ reduce to their standard counterparts.

4 ILG Applications: Mechanics, ChemoMechanics, and BioChemoMechanics

In this section, we discuss applications of the ILG framework to describe deformation instabilities and intermittent plasticity phenomena, as well as chemomechanical instabilities in lithium-ion battery anodes and tumor glioblastomas.

4.1 Mechanical Deformation Instabilities & Intermittent Plasticity

In this subsection we briefly discuss earlier developed ILG deformation models that were used to capture two types of propagating and stationary instabilities in metallic specimens under monotonic or cyclic applied loads. As the specimen size decreases these instabilities may be suppressed or manifest in a more complex non-deterministic manner when stochastic effects appear on equal footing with deterministic ones. This is the case for micro/nano pillar deformation where intermittent plasticity prevails and combined gradient-stochastic models are needed for interpreting size-dependent serrated stress-strain curves.

- *Propagating Portevin Le Chatelier Bands/PLC*: In order to provide insight on the applicability of ILG framework to capture propagating plastic deformation bands routinely observed in Al-Mg alloy specimens under tension, we list below an initial strain gradient model equation used by the author and coworkers for that purpose. It reads

$$\sigma = h\varepsilon + f(\dot{\varepsilon}) + c\varepsilon_{,xx} \tag{1}$$

where σ denotes stress, ε denotes strain, h is a hardening modulus, $f(\dot{\varepsilon})$ is a non-monotone function with a branch of negative slope modeling strain rate softening, and the gradient coefficient c (units $[\mathrm{m}]^2\times[\mathrm{sec}]$) is a phenomenological parameter. For constant stress rate tests ($\sigma = \dot{\sigma}_0 = h\,\dot{\varepsilon}_s$) and travelling wave solutions $\dot{\varepsilon} = z(x - Vt)$—where x denotes the spatial coordinate, t time, and V the deformation band propagation velocity—we obtain the following Lienard type nonlinear equation

$$Z_{\eta\eta} - \mu f'(Z)Z_\eta + (Z - Z_s) = 0 \tag{2}$$

where $\eta = \sqrt{h/c}\,(x - Vt)$ and $\mu = V/\sqrt{ch}$. This equation exhibits periodic solutions for propagating strain rate bands traveling through the specimen with constant velocity. It also leads to the staircase stress-strain graphs [3]. This model, which may be considered as a predecessor of later developed more elaborate PLC models, can be revisited for a strain-dependent gradient coefficient c to account for the observed increase of the strain jumps in the course of deformation. In addition, it can be used for applied constant strain-rate conditions to interpret serrated stress-strain curves exhibiting stress drops (instead of strain jumps). Internal stress fluctuations can be accounted for by introducing a stochastic term in Eq. (1) for the constitutive expression of the gradient-dependent stress. The resulting combined gradient-stochastic model can be evaluated according to the method discussed below to interpret non-deterministic serrations and intermittent plasticity phenomena in micro/nanopillar tests. Statistical characteristics for the serrations and corresponding PDFs can be obtained through Tsallis nonextensive q-entropy procedures. Additional typical experimental results for PLC bands and serrations in NC and UFG polycrystals can be analyzed in a similar way, as in the recent work of the author and coworkers [68, 69].

- *Stationary Persistent Slip Bands/PSBs*: Next, we briefly discuss the model equations describing the periodic ladder structure of stationary PSBs. The initial W-A model for the densities of immobile (ρ_i) and mobile (ρ_m) dislocations reads

$$\dot{\rho}_i = g(\rho_i) + D_i \nabla^2_{xx} \rho_i - h(\rho_i, \rho_m) \tag{3a}$$

$$\dot{\rho}_m = D_m \nabla^2_{xx} \rho_m + h(\rho_i, \rho_m) \tag{3b}$$

where (D_i, D_m) denote transport stress-dependent gradient coefficients for the two dislocation populations; $h(\rho_i, \rho_m)$ is an exchange term modeling dislocation reactions of the form $h(\rho_i, \rho_m) = \beta\rho_i - \gamma\rho_m\rho_i^2$; and $g(\rho_i)$ is a generation term for immobile dislocations. The coefficients (β, γ) depend on stress with β measuring the rate of production of mobile dislocations at the expense of immobile, and γ measuring the rate of immobilization of mobile dislocations by immobile dipoles. Since the stress remains constant during PSB formation, all these model coefficients may be assumed as constants. Then, linear stability analysis of Eq. (3) around an equilibrium homogeneous state $\left(\rho_i^0, \rho_m^0\right)$ results to a Turing instability for a critical value of the bifurcation parameter $\beta = \beta_c = \left[\sqrt{\alpha} + \rho_i^0\sqrt{\gamma\,(D_i/D_m\,)}\right]^2$, where $\alpha = -g'(\rho_i^0)$. The critical wave number q_c is given by the expression $q_c = \left[\alpha\gamma\rho_i^2/D_i D_m\,\right]^{1/4}$ and the corresponding critical wavelength $\lambda_c = 2\pi/q_c$ turns out to be of the same order of magnitude as in the experiments (see, for example, [3] and references quoted therein).

The above linear stability results were obtained for infinite domains, i.e. for specimen sizes much larger than the internal length. For finite size specimens, corresponding linear and nonlinear stability results were obtained recently by the author

and coworkers [29]. The periodic ladder structure of PSBs is revealed again but it now turns out that below a critical specimen size comparable to the internal length, the PSB instability is suppressed. This size-dependence is consistent with recently obtained experimental results [70, 71]. Stochasticity can also be introduced in Eq. (3) and the implication of a corresponding gradient-stochastic W-A model can readily be investigated.
An additional issue that can be considered is the coupling of the W-A model with an equation for the local stress τ related to the macroscopic stress σ through the gradient expression

$$\tau - l_\tau^2 \nabla^2 \tau = \sigma \tag{4}$$

where now the model parameters in Eq. (3) depend on τ rather than σ. This could offer an alternative simpler way to arrive at the result of the plateau stress during PSB formation than the method followed earlier by the author and coworkers [61, 62].

- *Intermittent Plasticity*: We conclude this section with some preliminary results on intermittent plastic instabilities by elaborating on a one dimensional combined gradient-stochastic model and Tsallis q-statistics, as an illustrative example. The combined gradient-stochastic expression for the flow stress σ reads

$$\sigma = \sigma^{ys} + h\varepsilon - \ell_p^2 \left(\partial^2\varepsilon/\partial x^2\right) \tag{5}$$

where the yield stress σ^{ys} contains both an average and a fluctuating part given by $\sigma^{ys} = (1+\delta)\,\bar{\sigma}^{ys}$– where $\bar{\sigma}^{ys}$ denotes mean value and δ follows a Weibull distribution fitted to experimental data. The rest of the quantities have their usual meanings; ε is the strain; h is a linear hardening modulus, and ℓ_p is a deterministic internal length.
When this model is incorporated into a cellular automaton (CA) grid, it results to serrated stress-strain curves and power-law interpretations of the corresponding statistical events. Appropriate expressions for the stochastic component of the flow stress can be more fundamentally deduced by employing the formalism of random processes and stochastic differential equations [61–64]. Another possibility is to resort to a class of Tsallis q-distributions that are used in many non-equilibrium physics problems where the usual power-laws based on Boltzmann-Gibbs statistics fail to predict observed behavior. An expression used for interpreting intermittent deformation behavior of Mo micropillar compression is Tsallis q-exponential PDF of the form $P(s) = A[1+(q-1)Bs]^{1/(1-q)}$: (A, B) are constants and the q-index is a measure of the system fractality, whereas s denotes the burst size. A power-law relationship between the internal length ℓ_p and the entropic index q seems to hold, but this needs to be examined further (this issue is also reviewed in [1]).
Further elaboration along the above lines on combined gradient-stochastic models for the interpretation of size dependent serrated stress-strain graphs by employ-

ing Tsallis q-statistics and the relationship of these findings to corresponding image observations on deformation patterns, is a challenging task that needs to be addressed in the future. In this connection, it is pointed out that the needed experimental information on appropriate PDF forms for the stochastic component of the flow stress can be deduced from multiple nanoindentation/NI tests and associated measurements of strain burst events. The PDF of the strain bursts would be related to a corresponding PDF for the flow stress on the assumption that a strain burst of a certain magnitude is the outcome of a number of material points yielding simultaneously. NI measurements at different locations and penetration depths can be conducted to deduce the statistical properties (mean, variance) for the local hardness which, in turn, can be used to extract information on the form of the stochastic component of the flow stress. From these multiple NI measurements one can extract direct information for both the deterministic ILs and the form of the stochastic contribution to the gradient-dependent flow stress.

4.2 Chemomechanical Instabilities in LiB Anodes

In this subsection, we provide elements of the ILG formulation that can be used to address chemomechanical instabilities in LiB anodes. In particular, we briefly present the basics of the stress-assisted diffusion and coupled ILG chemoelasticity theory that can be employed to consider local volume expansion in lithiated anodes and propagation of lithiation fronts.

- *Size Dependent Stress-Assisted Diffusion*: The standard equations that are usually employed to model coupled elasto-diffusion processes are of the form

$$\sigma_{ij} = \lambda \varepsilon_{mm} \delta_{ij} + 2G\mu \varepsilon_{ij} - \alpha \rho \delta_{ij} \tag{6a}$$

$$\mathbf{j} = -D\nabla\rho + M\rho\nabla\sigma_{ii} \tag{6b}$$

for the chemostress σ_{ij} and the diffusive flux $\mathbf{j}$ where the coefficients (α, M) denote chemomechanical coupling constants and D is the diffusivity. The fields (ρ, ε_{ij}) denote concentration of the diffusing chemical agent and mechanical strain, while (λ, G) are the Lamé constants. Since these constitutive equations do not contain higher-order ILs, related chemomechanical size effects may not be captured.
Within our Laplacian-based ILG formulation, it turns out that the above constitutive equations are generalized by replacing σ_{ij} with $\sigma_{ij} - \ell_\sigma^2 \nabla^2 \sigma_{ij}$; ε_{ij} with $\varepsilon_{ij} - \ell_\varepsilon^2 \nabla^2 \varepsilon_{ij}$; and ρ with $\rho - \ell_\rho^2 \nabla^2 \rho$, with $(\ell_\sigma, \ell_\varepsilon, \ell_\rho)$ denoting stress, strain and diffusional ILs. Under suitable assumptions, it is possible to uncouple the deformation and chemical fields by first computing a "ground" hydrostatic stress component $\sigma_h^0 = \sigma_{ii}^0$ (or $\sigma_{ii}^0/3$) from a conventional or gradient elasticity theory, and then derive the concentration ρ from a stress-assisted diffusion equation of the form

$$\partial\rho/\partial t = \left(D + N\sigma_h^0\right)\nabla^2\left[\rho - \ell_\rho^2\nabla^2\rho\right] - M\nabla\sigma_h^0\cdot\nabla\left[\rho - \ell_\rho^2\nabla^2\rho\right] \quad (7)$$

where N is a new phenomenological constant accounting for the effect of hydrostatic stress on diffusivity. This model with $\ell_\rho = 0$ has been used extensively to model hydrogen embrittlement and stress corrosion cracking in metals [72]. It can be adapted here, to consider chemomechanical damage and failure in LiB anodes.

- *Size-Dependent Lithiation Fronts*: To consider the propagation of lithiation fronts one may start with an expression for the free energy density ψ of the form

$$\psi\left(\boldsymbol{\varepsilon}, \nabla e, \rho, \nabla\rho\right) = f(\rho) + \frac{1}{2}\kappa\,\nabla\rho\cdot\nabla\rho + \frac{1}{2}\boldsymbol{\varepsilon}\cdot\boldsymbol{C}\boldsymbol{\varepsilon} + \frac{1}{2}c\nabla e\cdot\nabla e \quad (8)$$

where (κ, c) are respectively chemical and mechanical gradient coefficients, $\boldsymbol{C}$ denotes elasticity tensor, $\boldsymbol{\varepsilon}$ is the strain tensor and e its hydrostatic part, while ρ denotes concentration as before. In this case, both chemical ILs (through κ) and mechanical ILs (through c) enter into the formulation. Minimization of a corresponding energy functional yields field equations (and associated boundary conditions) for the local stress/strain and concentration of Li species, including the synergistic effect or interplay between higher-order mechanical and chemical ILs. The resulting governing coupled chemoelasticity equations for the stress and chemical potential read

$$\boldsymbol{\sigma} = 2G\boldsymbol{\varepsilon} + \lambda(\mathrm{tr}\,\boldsymbol{\varepsilon})\mathbf{1} - \ell_\varepsilon^2\nabla^2[2G\boldsymbol{\varepsilon} + \lambda(\mathrm{tr}\,\boldsymbol{\varepsilon})\mathbf{1}] - (2G + 3\lambda)M\rho\,\mathbf{1} \quad (9)$$

$$\mu = \mu^0 + RT\left[\ln\left(\frac{\rho}{1-\rho}\right) + \alpha(1 - 2\rho)\right] - \kappa\nabla^2\rho - \Omega_{Li}\sigma_h \quad (10)$$

where R is the universal gas constant, T is the absolute temperature, μ_0 a reference value of the chemical potential, $\sigma_h = tr\boldsymbol{\sigma}/3$ the hydrostatic stress, and $\Omega_{Li} = 3M/\rho_{\max}$ is the partial molar volume of the diffusing species. [It should be noted that the coefficients (α, M) in Eq. (6) have different meaning than those in Eqs. (9) and (10).]

4.3 *Glioblastoma Instabilities in Brain*

In this final subsection, we present some details on the GoG model, along with its mathematical similarities to the W-A model, and outline the potential new results to be expected from this comparison. Recent evidence in glioblastoma shows that one-size-fits-all vaso-modulatory interventions usually fail because control of glioma invasion characteristics, such as tumor front speed and infiltration width, vary widely and may require more personalized therapeutic interventions, in contrast to existing

GoG models, which assume that all glioma cells have an identical GoG mechanism. In reality, each cell may have an idiosyncratic migration and proliferation regulation due to internal stress dependence and associated intrinsic heterogeneity. The relevant question is "how can we model and analyze the impact of such internal stress dependence and intrinsic heterogeneity of a tumor cell population, where migration and proliferation are regulated by the GoG mechanism".

This question can be addressed by incorporating internal stress effects in the D terms of the GoG model in analogy to the W-A model discussed earlier for structural defects. The GoG model as formulated by Hatzikirou and coworkers [30, 31] reads

$$\partial \rho_m/\partial t = D_m \Delta \rho_m + E(\rho_m, \rho_i) \tag{11a}$$

$$\partial \rho_i/\partial t = D_i \Delta \rho_i - E(\rho_m, \rho_i) + g(\rho_i) \tag{11b}$$

where (ρ_m, ρ_i) denote respectively the motile and immotile glioma cell densities, with (D_m, D_i) being the corresponding diffusion coefficients. The term $E(\rho_m, \rho_i)$ signifies the switching between the two different phenotypes. Finally, the function $g(\rho_i)$ denotes the cell proliferation of the immotile population. The phenomenological resemblance of the GoG model for motile-immotile cells to the W-A model for mobile-immobile dislocations is striking. The results obtained from the earlier study of the W-A model to consider heterogeneity, stochasticity and local stress dependence can be used to improve predictions of the GoG model. It is expected that these predictive results can enable to understand the impact of intratumoral heterogeneity in glioma progression: in particular, the persistence and size dependence of the Allee effect under different heterogeneity and internal stress distributions, as well as the role of the pertinent spatio-temporal instabilities on potential therapeutic failures.

5 ILG and Rheology: Newtonian and Complex Fluids

In this section, we suggest possibilities for a gradient enhancement of constitutive equations used in fluid mechanics and rheology. In this connection, it is pointed out that following the author's work on gradient theory, a number of such generalizations have been proposed in these communities. For Newtonian fluids, such generalizations have been proposed by Silber and coworkers [73], as well as in more rigor and detail by Fried and Gurtin [74]. For complex fluids, such generalizations can be found in the pioneering articles by Olmsted and coworkers (e.g. [75] and references quoted therein), as well as in the enlightening review by Cates and Fielding [76]; see also an earlier one by Dhont and Briels [77].

In the spirit of the ILG formulation such type of generalizations can be readily deduced by replacing the local fields for the fluid density ρ, stretching tensor $\mathbf{D} = {}^1\!/_2\,[grad\,\mathbf{v} + (grad\,\mathbf{v})^T]$, and vorticity tensor $\mathbf{W} = {}^1\!/_2\,[grad\,\mathbf{v} - (grad\,\mathbf{v})^T]$ with their gradient-dependent counterparts $\rho - \ell_\rho^2 \nabla^2 \rho$, $\mathbf{D} - \ell_\mathbf{D}^2 \nabla^2 \mathbf{D}$, $\mathbf{W} - \ell_\mathbf{W}^2 \nabla^2 \mathbf{W}$.

Another possibility is to include the Laplacian of the viscoelastic stress $\mathbf{\Sigma}$ as proposed in the diffusive Johnson-Segalman (DJS) model employed to study shear banding flows of wormlike micelles or polymer solutions. In such wormlike micellar systems, it is assumed [75] that the total stress $\mathbf{T}$ is separated into contributions from the Newtonian solvent and a viscoelastic stress $\mathbf{\Sigma}$ from the micelles, so that for creeping flow we have

$$\mathbf{T} = -p\mathbf{1} + 2\mu\mathbf{D} + \mathbf{\Sigma} \tag{12a}$$

$$div\mathbf{T} = 0 \tag{12b}$$

with p denoting the pressure, μ being the solvent's shear viscosity, and the second equation standing for quasi-static equilibrium. It is further assumed that the viscoelastic stress $\mathbf{\Sigma}$ obeys the following evolution equation

$$\overset{o}{\mathbf{\Sigma}} + \frac{1}{\tau}\mathbf{\Sigma} = \frac{2\mu^*}{\tau}\mathbf{D} + D\nabla^2\mathbf{\Sigma} \tag{13}$$

where τ denotes relaxation time, μ^* is the micelle polymer-like viscosity and D is a diffusion-like coefficient. The corotational time derivative $\overset{o}{\mathbf{\Sigma}}$ may assumed to take various forms depending on the local micelle microstructural configuration. The above model and variants of it have been used extensively to address shear banding in complex fluids. The introduction of the Laplacian is needed to deal with ill- posedness in the negative slope regime of the shear stress—shear strain rate graph, i.e. the nonmonotonicity of the flow curve that also requires the introduction of Laplacians in the author's gradient plasticity theory used to address shear banding in the deformation softening regime [3].

On returning to the topic of an appropriate generalization of the Navier-Stokes (N-S) equations for incompressible fluids, i.e. of the constitutive equation $\mathbf{T} = -p\mathbf{1} + 2\mu\mathbf{D}$, we can propose, in analogy to the author's gradient elasticity theory [7], the following gradient model

$$\mathbf{T} - \ell_{\mathbf{T}}^2\nabla^2\mathbf{T} = -p\mathbf{1} + 2\mu(\mathbf{D} - \ell_{\mathbf{D}}^2\nabla^2\mathbf{D}) \tag{14}$$

where $\ell_{\mathbf{T}}$ and $\ell_{\mathbf{D}}$ denote internal lengths associated with stress and strain rate inhomogeneities. On assuming that $\ell_{\mathbf{T}}$ can be neglected and introducing Eq. (14) in the equation of momentum balance $\rho\dot{\mathbf{v}} = div\mathbf{T}$ (ρ is now the constant fluid density and $\dot{\mathbf{v}}$ its acceleration), we obtain the following gradient generalization of the N-S equations

$$\rho\dot{\mathbf{v}} = -\nabla p + \mu(\Delta\mathbf{v} - \ell_{\mathbf{D}}^2\Delta^2\mathbf{v}) \tag{15}$$

where $\Delta = \nabla^2$ and $\Delta^2 = \nabla^4$ denote the Laplacian and biharmonic or bi-Laplacian operators respectively. It is noted that Eq. (15) is identical to the equation used by Fried and Gurtin [74] to discuss plane Poiseuille liquid flow at small-length scales. A slightly generalized model was also used by the same authors to consider turbulence. The governing differential equations for this model read (in their notation)

$$\rho\dot{\mathbf{v}} = -\nabla p + \mu(1-\alpha^2\Delta)\Delta\mathbf{v} + 2\rho\alpha^2 div\overset{\nabla}{\mathbf{D}} \tag{16}$$

where the parameter α denotes a statistical correlation length and $\overset{\nabla}{\mathbf{D}} = \overset{\bullet}{\mathbf{D}} + \mathbf{DW} - \mathbf{WD}$ denotes the usual corotational Jaumann rate.

Steady-state solutions of Eq. (16) with $\alpha = 0$, may be determined by employing the operator split method (or the use of Ru-Aifantis theorem [78]) utilized to eliminate singularities from dislocation lines and crack tips in the theory of gradient elasticity (see also [1]). This same procedure leads to the cancelation of singularities in typical fluid flow calculations involving immersed objects. It turns out, for example, that the resulting gradient Oseen tensor $\mathcal{O}_{ij}^G$, which generalizes its classical counterpart $\mathcal{O}_{ij}$

$$\mathcal{O}_{ij} = \frac{1}{8\pi\mu r}\left(\delta_{ij} + \frac{r_i r_j}{r^2}\right) \tag{17}$$

where r_i denotes the position vector and r its magnitude, reads

$$\begin{aligned}\mathcal{O}_{ij}^G &= \frac{1}{8\pi\mu r}\left[1 - 2e^{-r/\ell} - \frac{2\ell}{r}e^{-r/\ell} + \frac{2\ell^2}{r^2}\left(1-e^{-r/\ell}\right)\right]\delta_{ij} \\ &+ \frac{1}{8\pi\mu r}\left[1 + 2e^{-r/\ell} + \frac{6\ell}{r}e^{-r/\ell} - \frac{6\ell^2}{r^2}\left(1-e^{-r/\ell}\right)\right]\frac{r_i r_j}{r^2}\end{aligned} \tag{18}$$

which resembles the exponential regularization of Green's tensor in gradient elasticity, resulting to nonsingular gradient expressions for the stresses and strains in dislocation lines and crack tips. More details can be found in [79] where the authors seemed to be unaware of analogous developments in gradient elasticity.

6 ILG in Other Disciplines & Scales

In this section we summarize the applicability of the ILG framework to other disciplines and scales ranging from earth scales to quantum scales.

- *ILG in Geology*: Some initial work on introducing internal lengths and Laplacians of strain has been published by the author and coworkers to model shear banding and related instability phenomena in geomaterials including granular materials, soils, rocks and snow/ice (see, for example, [80–91]). Various types of gradient-dependent constitutive equations for such classes of geomaterials have also been proposed and elaborated upon in detail by many other authors. This was mainly due to the fact that the Laplacian was regularizing unstable behavior in the geomaterial's softening regime and allowed for the determination of shear band thickness and convergence of corresponding finite element calculations. The popularization of the approach in the geomechanics community is mainly due to the follow-up works by Vardoulakis and collaborators for soils, as well as de Borst and collaborators for concrete. These are too many to mention and can be found in the web.

In connection with the above, it is worth noting that the W-A model for dislocation patterning has recently been used by Ord and Hobbs [92] to interpret fracture patterns in frictional, cohesive, granular materials. Their article was one contribution of seventeen to a Theme Issue "Patterns in our planet: applications of multi-scale non-equilibrium thermodynamics to Earth-system science".

- *ILG in Electrodynamics*: The inclusion of higher-order gradients in deforming materials under the action of electromagnetic fields has also become very popular in recent years due to emerging applications and design of piezoelectric (induction of electricity due to applied stress) and flexoelectric (induction of electricity due to strain gradients) components. The large number of published articles on these topics makes it prohibitive to mention them here and we only refer to few recent ones by the author and coworkers [93–95], as well as the bibliography listed there for related literature on size effects in electromechanical components.

  In relation to the issue of eliminating singularities and introducing screening effects (e.g. Debye screening) in the electric and magnetic fields, the following gradient modification of Coulomb's law of electrostatics has been proposed (see, for example, [96] where a fractional generalization of Debye screening is also discussed)

$$\Delta\Phi(\mathbf{r}) - \frac{1}{r_D^2}\Phi(\mathbf{r}) = -\frac{1}{\varepsilon_0}\rho(\mathbf{r}) \tag{19}$$

  where Φ is the electrostatic potential [$\mathbf{E}(\mathbf{r}) = -\nabla\Phi(\mathbf{r})$; $\mathbf{E}(\mathbf{r})$ is the electric field], $\rho(\mathbf{r})$ denotes now the charge density, ε_0 is the vacuum permittivity, r_D is the Debye screening distance, and $\mathbf{r}$ denotes as usual the position vector. The classical Coulomb's potential for spherical symmetry at a point charge of strength Q has the form $\Phi(\mathbf{r}) = Q/4\pi\varepsilon_0 r$, while its Debye screened counterpart obtained from Eq. (19) (which is identical in form to the reduced Ru-Aifantis equation for gradient elasticity [78]) reads

$$\Phi(\mathbf{r}) = \frac{1}{4\pi\varepsilon_0}\frac{Q}{r}e^{-r/r_D} \tag{20}$$

  In concluding this discussion on gradient electrodynamics, reference is made to an author's unpublished work where MacCullagh's 1850 proposal for an interesting formal analogy between elasticity and electromagnetism [97] is extended to include rotational gradients of the elastic aether. On assuming that the aether behaves as an elastic medium with its stress $\mathbf{T}$ depending linearly on rotations (instead of strains), we have

$$\mathbf{T} = 2k\boldsymbol{\omega}, \quad \boldsymbol{\omega} = 1/2\,[\nabla\mathbf{u} - (\nabla\mathbf{u})^T], \quad div\mathbf{T} = \rho\ddot{\mathbf{u}} \tag{21}$$

where $\mathbf{u}$ denotes displacement, ρ is density and k is an elastic constant. These lead to the equation $k\,curl\,curl\,\mathbf{u} + \rho\ddot{\mathbf{u}} = 0$ and by setting the terms $k\,curl\,\mathbf{u}$ and $\rho\dot{\mathbf{u}}$ to be proportional to the electric ($\mathbf{E}$) and magnetic ($\mathbf{B}$) fields respectively, we arrive at Maxwell's equations

$$\frac{\partial\mathbf{B}}{\partial t} + curl\,\mathbf{E} = 0, \quad div\mathbf{B} = 0, \quad \frac{\partial\mathbf{E}}{\partial t} - \frac{1}{\mu_0\varepsilon_0}curl\,\mathbf{B} = 0, \quad div\mathbf{E} = 0 \tag{22}$$

where the identities $div\,curl\,\mathbf{u} = 0$ and $curl\,(\partial\mathbf{u}/\partial t) - \partial(curl\,\mathbf{u})/\partial t = 0$, along with the following identification of the various coefficients ($\beta = k\varepsilon_0$, $\mu_0 = \rho/\beta$; with β being an arbitrary constant), were used. By adopting the above procedure, but replacing Eq. (21) for aether's elastic stress with its gradient counterpart

$$\mathbf{T} = 2k(\boldsymbol{\omega} - \ell^2\nabla^2\boldsymbol{\omega}) \tag{23}$$

we arrive at the following generalization of Maxwell's equations

$$\begin{aligned} &\frac{\partial\mathbf{B}}{\partial t} + curl\,\left[(1 - \ell^2\nabla^2)\mathbf{E}\right] = 0, \quad div\mathbf{B} = 0 \\ &\frac{\partial\mathbf{E}}{\partial t} - \frac{1}{\mu_0\varepsilon_0}curl\,\mathbf{B} = 0, \quad div\mathbf{E} = 0 \end{aligned} \tag{24}$$

It is noted that for electrostatics under the assumption that the electric field $\mathbf{E}$ is proportional to a potential gradient $\nabla\Phi$, Podolsky's non-quantum electromagnetics equation $\Delta\left[(1 - \ell^2\Delta)\right]\Phi = 0$ is obtained.[2]

- *ILG in Atomistics and Quantum Mechanics*: We conclude this section on applicability of the ILG framework to various disciplines and scales by focusing on two specific topics: A possible gradient generalization of the microscopic or molecular dynamics (MD) stress, and an analogous generalization of the quantum mechanical (QM) stress. In this connection, it is noted that the following expressions were proposed for these stresses [98, 99]:

$$\langle\sigma\rangle = \frac{1}{V}\left[\left\langle\frac{1}{2}\sum_i \mathbf{f}_{ij}\otimes(\mathbf{r}_i - \mathbf{r}_j)\right\rangle - \left\langle\sum_i m_i\boldsymbol{v}_i\otimes\boldsymbol{v}_i\right\rangle\right] \tag{25}$$

and

[2]Podolsky [B. Podolsky, A generalized electrodynamics Part I" Non-quantum. Phys. Rev. **62**, 68–71 (1942); B. Podolsky, P. Schwed, Review of a generalized electrodynamics. Rev. Mod. Phys. **20**, 40–50(1948)] has derived a generalization of Maxwell's equations through a variational principle, leading to the appearance of $\nabla^2\mathbf{B}$ in addition to $\nabla^2\mathbf{E}$. This is also possible through the aforementioned analogy by replacing $\mathbf{u}$ with $\mathbf{u} - \ell^2\nabla^2\mathbf{u}$.

$$\sigma_{\alpha\beta} = -\frac{1}{V}\sum_i \left\langle \frac{p_{i\alpha}p_{i\beta}}{m_i} \right\rangle - \frac{1}{2V}\sum_{\substack{i,j \\ (j\neq i)}} \left\langle \frac{(\mathbf{r}_i - \mathbf{r}_j)_\alpha (\mathbf{r}_i - \mathbf{r}_j)_\beta}{|\mathbf{r}_i - \mathbf{r}_j|} U'_{ij}\left(|\mathbf{r}_i - \mathbf{r}_j|\right) \right\rangle \tag{26}$$

where the various symbols have their usual meaning [98, 99]. The striking formal similarity between these two expressions and their resemblance with the virial stress and other statistical stress measures is noted. However, the problem to connect such discrete "microscopic" stress measures with the continuum "macroscopic" measure of Cauchy stress in a "seamless" way is a challenging issue. A gradient generalization of the force fields $\mathbf{f}_{ij}$ in Eq. (25) and the interaction potential U'_{ij} in Eq. (26) may be appropriate which, among other things, could naturally introduce screening distances and eliminate associated singularities.

The effect of strain $\boldsymbol{\varepsilon}$ on the electronic structure has been described [100] through the equations

$$\begin{aligned} (E_c - \frac{\hbar^2}{2m^*}\nabla^2)\,\psi(\mathbf{r}) + a_c tr(\boldsymbol{\varepsilon})\,\psi(\mathbf{r}) &= E\psi(\mathbf{r}) \\ \boldsymbol{\varepsilon} = \boldsymbol{\varepsilon}^0; \quad \boldsymbol{\sigma} = [\mathbf{C}]\,\boldsymbol{\varepsilon}\,; \quad div\,\boldsymbol{\sigma} &= \mathbf{0} \end{aligned} \tag{27}$$

where $\psi(\mathbf{r})$ denotes the wavefunction, $\mathbf{C}$ is the Hookean elasticity matrix, a_c the so-called deformation potential constant, and the rest of the symbols have their usual quantum mechanical meaning [100]. This is an uncoupled framework where strain can affect the electronic state but not vice-versa. A generalization to also account for the inverse effect on strain due to changes in the quantum field through the wavefunction $\psi(\mathbf{r})$, has already proposed as follows [101]:

$$\begin{aligned} (E_c - \frac{\hbar^2}{2m^*}\nabla^2)\,\psi(\mathbf{r}) + a_c tr(\boldsymbol{\varepsilon})\,\psi(\mathbf{r}) &= E\psi(\mathbf{r}) \\ \boldsymbol{\varepsilon} = \boldsymbol{\varepsilon}^0 - \frac{a_c}{3K}|\psi(\mathbf{r})|^2\mathbf{1}; \quad \boldsymbol{\sigma} = [\mathbf{C}]\,\boldsymbol{\varepsilon}\,; \quad div\,\boldsymbol{\sigma} &= \mathbf{0} \end{aligned} \tag{28}$$

where K is the isotropic bulk elastic modulus. A possible gradient modification is then to replace $\boldsymbol{\varepsilon}$ with its gradient counterpart $\boldsymbol{\varepsilon} - \ell_{\varepsilon}^2\nabla^2\boldsymbol{\varepsilon}$, and this formal generalization may be of interest to further explore.

7 ILG Modification of Newton's Gravitational Law

In this section, we venture a gradient generalization of Newton's Law which allows for the corresponding gravitational force to attain values larger than the electromagnetic force and even reach the levels of the nuclear and strong force which keeps matter together. The proposed modification is analogous to that earlier adopted by the author for gradient elasticity through the introduction of a Laplacian and a corresponding internal length.[3]

We begin with the following integral generalization of the gravitational force **f** in its component form (f_i):

$$f_i(\mathbf{r}) = \int G_{ij}(\mathbf{r} - \mathbf{r}')\, F_j(\mathbf{r}')\, d^3\mathbf{r}' \tag{29}$$

where $G_{ij}(\mathbf{r} - \mathbf{r}')$ is a nonlocal interaction kernel and F_j is the classical Newton's force. By assuming spherical symmetry/isotropy, Fourier transforming Eq. (29), Taylor series expanding up to the second order term, and inverting, we arrive at the following differential equation

$$\left(1 - \ell^2\nabla^2\right)\mathbf{f} = \mathbf{F} \tag{30a}$$

$$\ell^2\delta_{ij} = \frac{1}{2}\left|\frac{d^2\tilde{G}_{ij}(0)}{dk^2}\right| \tag{30b}$$

where $k = |\mathbf{k}|$ denotes wave vector, $\tilde{G}_{ij}$ is the Fourier transform of G_{ij} and ℓ is an internal length, with δ_{ij} appearing due to the assumed isotropy/spherical symmetry. In general, the sign in front of the Laplacian term of Eq. (30a) may be positive or negative depending on the sign of $d^2\tilde{G}_{ij}(0)/dk^2$ of the second order term in the Taylor expansion. In other words, for $G_{ij}(0) = \delta_{ij}G(0)$ and $\ell^2 = |\,l\,| = \left|d^2\tilde{G}\,(k)\,/dk^2\right|_{k=0}$, the term in the parenthesis of Eq. (30a) becomes $\left(1 - \ell^2\nabla^2\right)$ for $l > 0$ and $\left(1 - \mathrm{sgn}[l]\,\ell^2\nabla^2\right)$ for $l < 0$. Stability and related arguments may be employed to decide on the sign of l in a particular application.

Such a formal derivation can also be established by considering the two point masses M_0 and M in the classical Newton's Law, as being distributed and bounded by spheres of finite radii. By considering, for example, the mass M_0 ($M_0 = \sum_i m_i$) being distributed within a sphere of radius R_0, summing up the interactions of each point mass m_i (located at distance r_i from the center of the sphere where $r_i = 0$) with the point mass M, and expanding in Taylor series the density $\rho(\mathbf{r}_i)$ around $\rho_0 = \rho(0)$ keeping terms up to the second order we obtain the following relationship

$$\mathbf{f} = -\frac{GMV}{R^2}(\rho_0 + \ell^2\nabla^2\rho_0)\mathbf{e}_R \quad \text{with} \quad \ell^2 = \frac{R_0^2}{10} \tag{31}$$

---

[3] In fact, the question of exploring the consequences of such generalization to gravitation emerged during initial discussions with my daughter K.E. Aifantis during my visit in February 2019 to the University of to Florida at Gainesville and follow-up discussions with my former classmate C. Vayenas of the Academy of Athens during his visit in June 2019 to Thessaloniki. The initial numerical calculations reported herein started with the help of KEA's students in Gainesville and completed with the assistance of my Ph.D. student K. Parisis in Thessaloniki. The same holds for the results on gradient interatomic potentials listed in Sect. 8.

where $\mathbf{e}_R = \mathbf{R}/R$ denotes the unit vector along the line connecting the center of M_0 with the point mass M. On setting $\int_V \rho_0 dV = M_0$, we then have $\mathbf{f} = (1 + \ell^2\nabla^2)\mathbf{F}$ which by inversion leads to $(1 - \ell^2\nabla^2)\mathbf{f} = \mathbf{F}$. This simplified and rather intuitive calculation is similar to that earlier adopted by the author and coworkers (e.g. [10, 82]) to produce a corresponding gradient-dependent plastic strain.

On assuming a radial dependence of $\mathbf{f}$ and $\mathbf{F}$ $\{\mathbf{f} = -f(r)\mathbf{e}_r, \mathbf{F} = -F\mathbf{e}_r, F = A/r^2$ with $A(= GMM_0)$ and G denoting now Newton's classical gravitational constant, where we have also adopted the notation $\mathbf{e}_R \equiv \mathbf{e}_r\}$, we can readily solve the radial scalar counterpart of $(1 - \ell^2\nabla^2)\mathbf{f} = \mathbf{F}$, i.e.

$$f - \ell^2\left(\frac{\partial^2 f}{\partial r^2} + \frac{2}{r}\frac{\partial f}{\partial r} - \frac{2f}{r^2}\right) = \frac{A}{r^2} \tag{32}$$

by also requiring that $f \to 0$ as $r \to \infty$. The result is

$$f = \frac{A}{r^2}[1 + Be^{-r/\ell}(1 + \frac{r}{\ell})] \tag{33}$$

where B is a new parameter to evaluate in connection with experiments. It is noted that the above expression of Eq. (33) reduces to Newton's classical force $F_N = A/r^2$ as $r \to \infty$ and to the expression $F_{SF} = AB/r^2$ as $r \to 0$. By adjusting the value of the new parameter B ($B >> 1$) we can attain values of the nuclear and strong force.

The internal length parameter can be identified with the de Broglie relativistic length, the Compton length, the Planck length or the Schwarzschild distance, according to the configuration at hand, i.e.

- De Broglie: $\ell = \hbar/\gamma m_0 c$ .... 6.309×10^{-16} m
- Compton: $\ell = \hbar/m_p c$ .... 2.10×10^{-16} m
- Planck: $\ell = \sqrt{\hbar G/c^3}$ .... 1.616×10^{-35} m
- Schwarzschild: $\ell = 2Gm_{BH}/c^2$ ...... $10^{10} - 10^{13}$ m

where $\hbar$ denotes the Planck constant, c is the speed of light; and (m_0, m_p, m_{BH}) denote rest masses for neutrino, proton, and black hole, respectively; whereas G in the above, as in Eq. (31), denotes the classical Newton's gravitational constant (not to be confused with the same symbol earlier used for the shear modulus), and γ is the Lorentz factor ($\gamma = 1/\sqrt{1 - (v/c)^2}$; with v denoting particle speed), not to be confused with a similar symbol used in earlier sections for the strain.

On adopting the Vayenas and coworkers Rotating Neutrino model (RNM) for the nucleus [102, 103] we now utilize the above expression for the gravitational force given by Eq. (33), in conjunction with the centrifugal force $F_C = \gamma m_0 c^2/r$, where r denotes the radius of the nucleus modeled by the three rotating neutrinos whose total relativistic mass is $m_N = 3\gamma m_0$. An estimate of γ can be obtained by equating the proton energy $m_p c^2$ with the relativistic neutrino mass. This gives the value of $\gamma = m_p/m_N$ which, according to experimental measurements for m_p and m_0 turns out to be equal to 7.818×10^9.

Having such a value of γ available, we can make effective use of the aforementioned equality between gravitational and centrifugal forces in Vayenas' RNM to deduce the relationship

$$F_R = \frac{A}{\sqrt{3}r^2}[1 + Be^{-\sqrt{3}r/\ell}(1 + \frac{\sqrt{3}r}{\ell})] = F_C = \frac{\gamma m_0 c^2}{r} \tag{34}$$

where the factor $\sqrt{3}$ rises by considering the resultant gravitational force $F_R = f/\sqrt{3}$ due to the interaction of the 3 symmetrically placed (at angles 120°) rotated neutrinos. One possibility for the constant A is to set it equal to $A = Gm_0^2\gamma^2$, to account for relativistic effects during the interaction of each pair of neutrinos in the assumed RNM configuration. The above relationship (for ℓ identified with de Broglie's relativistic length $\ell = \hbar/\gamma m_0 c = 6.31 \times 10^{-16}$ m) gives the following value for the coefficient B

$$B = \frac{\sqrt{3}e^{\sqrt{3}}\hbar c}{\left(1 + \sqrt{3}\right) Gm_0^2\gamma^2} = 5.47 \times 10^{39} \tag{35}$$

and a corresponding value of F_R

$$F_R = 7.92 \times 10^4\,\text{N} \tag{36}$$

i.e. the value of the strong force obtained for the RNM configuration [102, 103] by using an entirely different approach. In that approach Eq. (34) with $B = 0$ was used with $A = Gm_0^2\gamma^6$ giving a value for $\gamma = 3^{1/12}m_{pl}^{1/3}m_0^{-1/3} = 7.167 \times 10^9$, where m_{pl} is the Planck mass ($m_{pl} = \sqrt{\hbar c/G}$), and the value of m_0 was taken as $m_0 = 0.0436\ eV/c^2$. And since $\gamma = m_p/3m_0$, this gives $m_p = 9.38 \times 10^8\ eV/c^2$, i.e. the same value as the one used in the previous paragraph by properly adjusting the parameters (A, B), as well as by identifying the internal length parameter ℓ with de Broglie relativistic length. Other choices of (A, B, ℓ) are possible not only for the RNM configuration at hand, but also other more complex geometric models for elementary particles represented by several neutrinos where the aforementioned gradient enhanced gravitational potential can be used.

8 Gradient Interatomic Potentials

Motivated by the above extension of Newton's gravitational potential, we consider in this section a similar gradient generalization of London's quantum mechanical potential. Based on exact quantum mechanical calculations London [104, 105] has arrived at the following forms of the interatomic force $F\,(= -dw/dr)$ and interatomic potential w

$$w = w(r) = \begin{cases} -\frac{3\alpha_0^2 h\nu}{4(4\pi\varepsilon_0)^2}\frac{1}{r^6} = -\frac{C}{r^6};\ r \geq \sigma \\ \infty;\ r < \sigma \end{cases} \tag{37}$$

where $C = 3\alpha_0^2 h\nu/4(4\pi\varepsilon_0)^2$, α_0 is the atomic polarizability and ε_0 the vacuum dielectric permittivity. The quantities (h, ν) denote respectively the Planck constant $(h = 2\pi\hbar)$ and the electron orbital frequency. It is noted that the above form provides an explicit expression for the attractive interaction until a critical distance σ below which the model breaks down as the interaction becomes repulsive going to infinity as $r \to 0$. To describe quantitatively "repulsive" interactions for distances $r < \sigma$, Lennard-Jones [106] suggested the following modification of London's potential

$$w_{L-J}(r) = -\frac{A}{r^6} + \frac{B}{r^{12}} \tag{38}$$

where A and B are determined by fitting them to obtain through atomistic simulations the measured experimental values of macroscopic properties. Other type of interaction potentials can be found in [107].

The gradient modification of London's quantum mechanical potential, denoted as w_L^G, is obtained in terms of its classical counterpart w through the inhomogeneous Helmholtz equation

$$\left(1 - \ell^2\nabla^2\right) w_L^G = w(r) = -\frac{C}{r^6} \tag{39}$$

The solution of Eq. (39) for $(w_L^G \to 0,\ r \to \infty)$ is given by the expression

$$\begin{aligned} w_L^G(r) &= A\ell\frac{e^{-r/\ell}}{r} \\ &+ \frac{C}{48\ell^6}\left\{\frac{4\ell^4}{r^4} + \frac{2\ell^2}{r^2} + \frac{\ell}{r}\left[e^{r/\ell}\,\mathrm{Ei}(-\frac{r}{\ell}) - e^{-r/\ell}\,\mathrm{Ei}(\frac{r}{\ell})\right]\right\} \end{aligned} \tag{40}$$

where A is a new integration constant, ℓ is an internal length parameter, and Ei denotes the exponential integral $\mathrm{Ei}(x) = -\int_{-x}^{\infty}(e^{-t}/t\,)dt$. Near the origin $(r \to 0)$, it turns out that $w_L^G(r) \to (C/12\ell^2 r^4)$, while at large distances $(r \to \infty)$ it approaches the classical London's potential, i.e. $w_L^G(r) \to -(C/r^6)$ for $r >> \ell$.

As an example application of the newly derived gradient potential, we consider the case of Argon (Ar). It has been shown that the Lennard-Jones potential is able to describe accurately the simulated liquid argon properties in agreement with the experiment. Numerical/experimental values can be utilized by the data provided in [108] (see also Table 6.1 of [107]). Among these data, of particular interest are the minimum of the potential function, designated by ε (in units of Joules or eV), as well as its location r_m (in Å). Their estimated values are $\varepsilon = 1.95 \times 10^{-21}$ J and

$r_m = 0.37$ nm respectively. The Lennard-Jones potential can be uniquely determined from these parameters. For this purpose, Eq. (38) is written in the form $w_{L-J}(r) = \varepsilon((r_m/r)^{12} - 2(r_m/r)^6)$, where it is evident that the minimum occurs at r_m with $w_{L-J}(r_m) = -\varepsilon$ and $dw_{L-J}(r_m)/dr = 0$. This point determines the transition from "attractive" to "repulsive" branch for distances $r < r_m$. Additionally, the Lennard-Jones potential is zero at $r = \sigma = 2^{-1/6} r_m = 0.89\, r_m = 0.32$ nm. The parameters (ε, r_m) are related with (A, B) of Eq. (38) through the relationships $A = \varepsilon r_m^{12} = 4\varepsilon\sigma^{12}$ and $B = 2\varepsilon r_m^6 = 4\varepsilon\sigma^6$. The fitted London's constant is $C = 50 \times 10^{-79}$ J$\cdot$ m^6, which was determined such as the classical London's potential passes through the experimentally determined potential minimum exactly at r_m.

Next, and in order to demonstrate the ability of the gradient modification of London's potential to recover the behavior of the Lennard-Jones potential for the Ar-Ar interaction case, we adjust the gradient parameters (A, ℓ, C), such as the position of the minimum of the potential occurs at r_m, i.e. $w_L^G(r_m) = -\varepsilon$, and the corresponding potential curves are as close as possible by minimizing their mean square error. The obtained parameter values are $A = 1.392 \times 10^{-17}$ J, $\ell = 0.57$ Å, and $C = 90 \times 10^{-79}$ J$\cdot$ m^6, with the estimated value for the internal length being consistent with the atomistic simulations. As shown in Fig. 1a, the gradient modification of London's potential curve fits nicely the Lennard-Jones curve, with both having their minima at distance r_m. It is noted that the gradient potential has the same asymptotic behavior $O\left(r^{-6}\right)$ at distances $r > r_m$, in agreement with both Lennard-Jones and London's potential. Finally, as expected, the gradient modified London's potential becomes "repulsive" for $r < r_m$, where the change of slope occurs, in contrast to London's original $1/r^6$ monotonic potential.

Another indicative example of the applicability of the newly derived gradient potential is the Stillinger-Weber potential, which is broadly used to model the interatomic interactions of materials with diamond structure, such as crystalline semiconductors (Si, Ge). The analytical expression of the two-body Stillinger-Weber potential reads [109]

$$w_{S-W}(r) = \begin{cases} \varepsilon A \left[B \frac{\sigma^4}{r^4} - 1 \right] \exp\left(\frac{1}{r/\sigma - a} \right); & r < a\sigma \\ 0; & r \geq a\sigma \end{cases} \tag{41}$$

in a most simplified form, excluding anisotropy effects.

The appropriate fitted values for the Stillinger-Weber potential when applied to Si semiconductor read $A = 7.04955627$, $B = 0.602224558$, $\varepsilon = 50$kcal $\times$mol^{-1} = 3.4723×10^{-19}J, $\sigma = 2.0951$ Å, and $a = 1.8$ [109]. The Stillinger-Weber potential has a cutoff at distance $r = a\sigma$, confining the interatomic interaction within that range, while for short distances it has a "repulsive" branch with asymptotic behavior $w_{S-W}(r) \to \varepsilon e^{-1/a} AB\sigma^4/r^4$.

The estimated values for the gradient potential are $A = 5.702 \times 10^{-17}$ J, $\ell = 0.974$ Å, and $C = 1.333 \times 10^{-78}$J $\cdot$ m^6 respectively. They are adjusted such as the fitted minimum of w_L^G coincides with the corresponding one of the Stillinger-Weber potential, which satisfies $w_{S-W}(r_m) = -\varepsilon$ at $r_m = 1.118\sigma = 2.34$ Å. In Fig. 1b, it

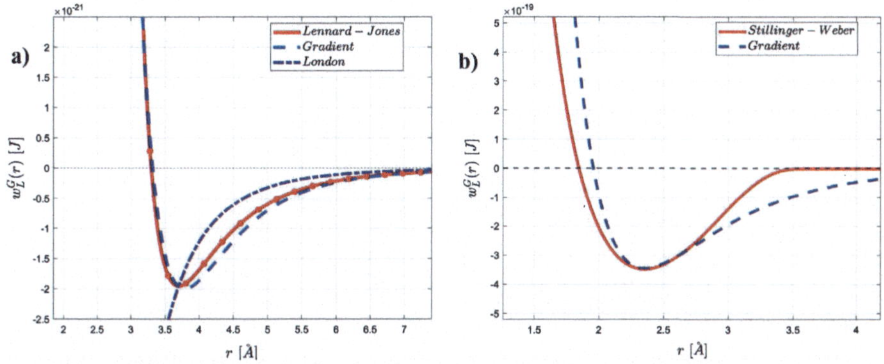

Fig. 1 Quantitative plots of the gradient London potential fitting to **a** Lennard-Jones (Ar-Ar) and **b** Stillinger-Weber (Si) potential

is illustrated how a suitable choice of the parameters (A, ℓ, C) can describe with desired accuracy the behavior of w_{S-W} for small distances. This is due to the fact that $w_L^G(r) \to (C/12\ell^2 r^4)$ as $r \to 0$, which is in agreement with the asymptotic behavior of w_{S-W} at the origin.

Another possible generalization of the gradient approach for constructing new interatomic potentials is through the introduction of an additional bi-Laplacian term in London's potential. Motivated by the 4th order GradEla extension used earlier by the author and co-workers [110–112], we can further generalize Eq. (39) to read

$$\left(1 - \tilde{\ell}_1^2 \nabla^2 + \tilde{\ell}_2^4 \nabla^4\right) w_L^G = w_L(r) \tag{42}$$

where $\left(\tilde{\ell}_1, \tilde{\ell}_2\right)$ now denote two internal lengths. In passing, it is noted that such a fourth-order equation for the elastic fields, derived within a second strain gradient elasticity/GradEla theory [46, 110–112], leads to the elimination of singularities of the dislocation density tensor, which remains singular in first strain GradEla. [The signs in front of the higher order terms in Eq. (42) may alter according to dynamic stability and related requirements–a subject partially addressed in [110] and further to be discussed elsewhere.]

Equation (42) can be factored as the product of two Helmholtz operators as $\left(1 - \ell_1^2 \nabla^2\right)\left(1 - \ell_2^2 \nabla^2\right) w_L^G = w_L(r)$ where the internal lengths $\left(\ell_1, \ell_2\right)$ are given by the expression $\ell_{1,2}^2 = (\tilde{\ell}_1^2/2)(1 \pm \sqrt{1 - 4(\tilde{\ell}_2^4/\tilde{\ell}_1^4)})$. The solution of Eq. (42) is obtained by applying the operator split approach of Ru-Aifantis theorem, arriving at the equation $\left(1 - \ell_2^2 \nabla^2\right) w_L^G = w_L^{G,1}\left(r; \ell_1\right)$ where $w_L^{G,1}\left(r; \ell_1\right)$ is the gradient London's potential of Eq. (40) with internal length ℓ_1. It is given by the expression

$$
\begin{aligned}
w_L^G(r) = & \frac{B\,\ell_2^3}{\ell_2^2-\ell_1^2}\frac{e^{-r/\ell_2}}{r} + \frac{A\,\ell_1^3}{\ell_1^2-\ell_2^2}\frac{e^{-r/\ell_1}}{r} \\
& - \frac{C}{48\ell_1^3\ell_2^3}\frac{\ell_2^3}{\left(\ell_1^2-\ell_2^2\right)r}\left[e^{r/\ell_1}\,\mathrm{Ei}(-\frac{r}{\ell_1}) - e^{-r/\ell_1}\,\mathrm{Ei}(\frac{r}{\ell_1})\right] \\
& - \frac{C}{48\ell_1^3\ell_2^3}\left[\frac{2\ell_1^3\ell_2}{\left(\ell_1^2-\ell_2^2\right)r^2} + \frac{2\ell_2^3\ell_1}{\left(\ell_2^2-\ell_1^2\right)r^2}\right] \\
& - \frac{C}{48\ell_1^3\ell_2^3}\frac{\ell_1^3}{\left(\ell_2^2-\ell_1^2\right)r}\left[e^{r/\ell_2}\,\mathrm{Ei}(-\frac{r}{\ell_2}) - e^{-r/\ell_2}\,\mathrm{Ei}(\frac{r}{\ell_2})\right]
\end{aligned}
\tag{43}
$$

where B is a new integration constant and (ℓ_1, ℓ_2) have been defined above, while C is London's constant and A the integration constant of Eq. (40). It is noted that the first two exponential terms in Eq. (43), which are related to the homogeneous part of the corresponding Helmholtz equation, are formally similar to expressions derived earlier for nuclear potentials in quantum electrodynamics based on an extension of Yukawa-type interactions. A quantitative elaboration for specific material types will be a subject of a future study.[4]

9 Fractional Considerations

In this final section, an extension of the ILG framework to incorporate fractional derivatives is presented. A fractional generalization of GradEla can be established by replacing the standard (integer) Laplacian $\Delta \equiv \nabla^2$ with a fractional one of the Riesz form $(-\Delta)^{\alpha/2}$ (or the Caputo form ${}^C\Delta_W^\alpha$) in the constitutive expression $\sigma_{ij} = \lambda\varepsilon_{kk}\delta_{ij} + 2G\varepsilon_{ij} - \ell^2\nabla^2[\,\lambda\varepsilon_{kk}\delta_{ij} + 2G\varepsilon_{ij}\,]$. An example of such a fractional generalization reads [113–117]

$$
\sigma_{ij} = (\lambda\varepsilon_{kk}\delta_{ij} + 2G\varepsilon_{ij}) + \ell^\alpha(-\Delta)^{\alpha/2}[\lambda\varepsilon_{kk}\delta_{ij} + 2G\varepsilon_{ij}] \tag{44}
$$

where $(-\Delta)^{\alpha/2}$ is the fractional generalization of the Laplacian in the Riesz form, defined in terms of the Fourier transform $\mathcal{F}$ by

$$
\left((-\Delta)^{\alpha/2}\varepsilon_{ij}\right)(\mathbf{r}) = \mathcal{F}^{-1}\left(|\mathbf{k}|^\alpha\tilde{\varepsilon}_{ij}(\mathbf{k})\right)(\mathbf{r}) \tag{45}
$$

where $\mathbf{k}$ denotes the wave vector, and $\tilde{\varepsilon}_{ij}(\mathbf{k}) = \mathcal{F}\left(\varepsilon_{ij}(\mathbf{r})\right)(\mathbf{k})$ is the Fourier transform of $\varepsilon_{ij}(\mathbf{r})$. On introducing the fractional GradEla constitutive relation given by Eq. (44) into the equilibrium relation $div\boldsymbol{\sigma} = 0$, we obtain

[4]For completeness, however, we may refer to the paper by Reid (R.V. Reid, Local phenomenological nucleon–nucleon potentials, Annal. Phys. **50**, 411–448 (1968)), where the following expression, among others, is proposed $V_C = h\left[e^{-x/3} - 13.8\,e^{-3x} + 138\,e^{-6x}\right]/x$, with $h = 10.463$ MeV and $x = \mu r$, $\mu = mc/\hbar \approx 0.7\,\mathrm{fm}^{-1}$.

$$(1 + \ell^{\alpha}(-\Delta)^{\alpha/2})(\lambda \nabla tr\, \boldsymbol{\varepsilon} + 2G\, div\, \boldsymbol{\varepsilon}) = 0 \tag{46}$$

where, as in Eq.(44), the notation ℓ^{α} is adopted for the corresponding fractional internal length. Noting the fact that the spatial operators commute and that the second bracket in Eq. (46) is also zero by replacing $\boldsymbol{\varepsilon}$ with $\boldsymbol{\varepsilon}_0$, where $\boldsymbol{\varepsilon}_0$ denotes the solution of the corresponding equation for classical elasticity, (i.e. $\lambda \nabla tr\, \boldsymbol{\varepsilon}_0 + 2G\, div\, \boldsymbol{\varepsilon}_0 = 0$), we can easily deduce that a solution of Eq. (46) satisfies the reduced fractional partial differential equation[5]

$$(1 + \ell^{\alpha}(-\Delta)^{\alpha/2})\boldsymbol{\varepsilon} = \boldsymbol{\varepsilon}_0 \tag{47}$$

which for the case $\alpha = 2$ reduces to the inhomogeneous Helmholtz equation. It is then useful to derive fundamental solutions for Eq. (47); i.e. for the equation

$$(1 + \ell^{\alpha}(-\Delta)^{\alpha/2})G_{\alpha}(\mathbf{r}) = \delta(\mathbf{r}) \tag{48}$$

where $G_{\alpha}(\mathbf{r})$ denotes the fundamental solution, $\delta(\mathbf{r})$ denotes the delta function and $\mathbf{r}$ is the radial coordinate in a 3D space. To obtain the fundamental solution of Eq. (48) with the natural boundary condition $G_{\alpha}(\mathbf{r}) \to 0$ as $\mathbf{r} \to \infty$, we employ the method of Fourier transforms. Using the properties of the Fourier transform of the Riesz fractional Laplacian as defined by Eq. (45), along with the well-known transform of the delta function $\mathcal{F}\{\delta(\mathbf{r})\}(\mathbf{k}) = 1$, we obtain the algebraic equation $(1 + \ell^{\alpha}|\mathbf{k}|^{\alpha})\, G_{\alpha}(\mathbf{k}) = 1$ which gives for the fundamental solution in Fourier space $G_{\alpha}(\mathbf{k}) = (1 + \ell^{\alpha}|\mathbf{k}|^{\alpha})^{-1}$. Consequently, the fundamental solution of Eq. (48) in the physical space is obtained through inversion as

$$G_{\alpha}(\mathbf{r}) = \frac{1}{(2\pi)^3} \int \frac{1}{1 + \ell^{\alpha}|\mathbf{k}|^{\alpha}} e^{i\mathbf{k}\cdot\mathbf{r}} d^3\mathbf{k} \tag{49}$$

The inversion of Eq. (49) is performed through application of the convolution property of the Mellin transform, along with a corresponding Mellin-Barnes integral representation, which yields the following corresponding Fox-H function expression [115]

$$G_{\alpha}(\mathbf{r}) = \frac{1}{2\alpha\, \pi^{3/2} \ell |\mathbf{r}|^2} H^{2,1}_{1,3} \left[\frac{|\mathbf{r}|}{2\ell};\ \begin{matrix} (1 - \frac{1}{\alpha}, \frac{1}{\alpha}) \\ (1 - \frac{1}{\alpha}, \frac{1}{\alpha})\ (1, \frac{1}{2})\ (\frac{1}{2}, \frac{1}{2}) \end{matrix} \right] \tag{50}$$

For more details concerning the definition, properties and applications of the Fox-H function in fractional analysis, the reader can consult [118–120]. A corresponding series expansion of Eq. (50) is also provided in [115]. It is noted that as

[5] The fact that solutions of Eq. (46) can be obtained in terms of solutions of Eq. (47) was first observed in [78] for the non-fractional GradEla and was extended later for more general fractional/fractal elastic materials in [114]

$\alpha \to 2$, we obtain the Green's function of the integer order Helmholtz equation, i.e. $G_\alpha(\mathbf{r})|_{\alpha\to 2} = (1/4\pi\ell^2 r)\, e^{-r/\ell}$.

Motivated by the above analysis of the fractional Helmholtz equation, as well as by noticing that Eq. (49) reduces to a Yukawa-like potential in the classical limit $\alpha \to 2$, a fractional treatment of Eq. (29), listed below again for convenience,

$$f_i(\mathbf{r}) = \int K_{ij}(\mathbf{r} - \mathbf{r}')\, F_j(\mathbf{r}')\, d^3\mathbf{r}' \tag{51}$$

is undertaken, where now K_{ij} denotes a fractional interaction kernel, and the forces f_i and F_j have been defined in Sect. 7. Such type of integral expressions have been previously introduced to model nonlocal constitutive relations in electrodynamics leading to fractional Debye screening effects [96]. Similar arguments have been recently applied to model fractional nonlocality in GradEla [116]. Through a Taylor series expansion (up to second order) of the fractional kernel K_{ij} in Fourier space involving non-integer powers of the wave vector $|\mathbf{k}|^\alpha$, and subsequent inversion through Eq. (45), the corresponding fractional counterpart of Eq. (30) is obtained as

$$\left(1 + \ell^\alpha(-\Delta)^{\alpha/2}\right)\mathbf{f} = \mathbf{F}; \quad \ell^\alpha \delta_{ij} = \frac{1}{\Gamma(\alpha+1)} \left|\left({}_0^C D_k^\alpha \tilde{K}_{ij}\right)(0)\right| \tag{52}$$

where ${}_0^C D_k^\alpha \tilde{K}_{ij}$ is the Caputo fractional derivative of order α with respect to k [118–120]. In the limit $\alpha \to 2$ the solution of Eq. (52) coincides with the one obtained from Eq. (30), since the fractional Laplacian and corresponding derivatives reduce to their classical counterparts. The solution of Eq. (52) can be obtained through convolution of the corresponding Green's function of Eq. (49) with the classical field $\mathbf{F}$, i.e. $f_i = G_\alpha * F_i$, resulting to the expression

$$f = \frac{A}{r^2}[1 + B\, K_\alpha(r/\ell)] \tag{53}$$

where A and B are the constants, defined in Sect. 7, and K_α is the fractional generalization of the modified (fractional-like) Bessel function $K_\alpha(r) \equiv (1/2\pi^2 r) \int_0^\infty [(k^2 \cos(k\,r))/(k^{2-\alpha}(1 + \ell^\alpha k^\alpha))]\, dk$. An analogous result can be obtained by further generalizing Eqs. (30) and (52) to include bi-Laplacian terms of the type appearing in Eq. (42) for both the integer and non-integer case. This will be a subject of future publication. However, for the completeness of this review, we list the corresponding formulas below:

$$f = \frac{A}{r^2}[1 + B_1 e^{-r/\ell_1}(1 + \frac{r}{\ell_1}) + B_2 e^{-r/\ell_2}(1 + \frac{r}{\ell_2})]; \text{ Integer case} \tag{54}$$

$$f = \frac{A}{r^2}[1 + B_1\, K_\alpha(r/\ell_1) + B_2\, K_\alpha(r/\ell_2)]; \text{ Non-integer case} \tag{55}$$

where B_1 and B_2 are dimensionless constants, (ℓ_1, ℓ_2) are two internal lengths, while the function $K_\alpha(r)$ has been defined above.

10 Conclusions

A concise review of gradient models (across scales, materials, and processes) was provided based on the author's ILG approach. As a result, earlier references on generalized continuum mechanics and recent contributions on gradient and non-local theories were not discussed due to space limitation. For solids, one should single out the contributions of Eringen [121], Fleck and Hutchinson [122, 123] Gurtin and Anand [124], Gao et al. [125], Nix and Gao [126], de Borst et al. [127, 128], Geers et al. [129], Peerlings et al. [130], Willis [131], Aifantis and Willis [132], and Polizzotto [133, 134]. Many more are included in a most recent and detailed article by Voyiadjis and Song [135] focusing on gradient plasticity. Gradients in fluid and granular flows were considered most recently by Goddard [136, 137]. For additional recent developments on granular flow, one may also consult references [138–142], while for internal length interpretations based on kinetic theory, one may consult [143]. However, the intention of the article was not to elaborate on and review the various important classical-like gradient models for solids and fluids, as well as for rheology and electrodynamics. Its main purpose was to explore the applicability of gradient theory for scales and processes not considered before, and point out its potential usefulness for atomistic simulations and elementary particles, as well as for earth and planetary processes. In this connection, it is noted that while completing this article, it came to the attention of the author that an expression similar to that derived herein and given by Eq. (33) was also proposed on rather intuitive grounds by Fischbach et al. [144] in an effort to re-interpret existing measurements on earth's gravity (see also [145]). The values of their constants were entirely different than ours, as they used it for a reanalysis of the Eötvös experiment on Earth's gravitational field. There has been a vast literature on this expression, subsequently referred to as the "fifth force," which we will discuss in a forthcoming publication, as this is beyond the scope of the present review. In concluding, the reader is referred to another review-like article [146] where nonlocal and gradient models with applications to biophysics are discussed. Forthcoming work on elementary particle physics is in progress by using the gradient Newton's gravitational force instead of the classical one and adjusting the new phenomenological parameter B to describe a variety of internuclear potentials. The same holds for gradient interatomic potentials by using both classical and fractional/fractal Laplacians[6].

---

[6]In fact, two preprints (by C. G. Vayenas, D. Tsousis, D. Grigoriou, K. Parisis and E. C. Aifantis) on Kaons and Deuteron are available and scheduled for arXiv and journal publication. Two more articels on gradient London's potential (by K. Parisis, F. Shuang, B. Wang, P. Hu, A. Glannakoudakis and A. Konstantinidis: J. Appl. Math. Phys.) and its fractional extension (by K. Parisis and E. C. Aifantis: TMS Proc. 2021) are forthcoming

Note added in proof: A few words are due on "why" the dedication under the title of this chapter. *Jim Serrin* – formerly a Regents Professor of Mathematics at the University of Minnesota and member of the US Academy of Sciences – was a strong advocate of gradient theory and my work with him on revisiting van der Waals theory on vapor-liquid transitions and relocating Maxwell's equal area rule within a purely mechanical framework without assuming the existence of a free energy function, was the predecessor of gradient plasticity. *Hussein Zbib* – my graduate student at Michigan Tech and later Chair at the Washington State University – was the first, in his PhD Thesis, to illustrate the need of gradient plasticity theory for deriving breakthrough results on shear band widths and spacings. The recent work (based on Newton's classical gravitational law and Einstein's special relativity) of *Costas Vayenas* – my undergraduate classmate at the National University of Athens, later a faculty member at MIT and the University of Patras and currently a member of the Academy of Athens and the US Academy of Engineering – was an additional motivation for extending gradient theory to the field of elementary particles and revisiting Newton's gravitational law, thus stimulating a challenging area of research after my retirement. Finally, my daughter *Katerina* – born the same year of writing first publication on gradient theory – was the first to enhance gradient plasticity with surface effects and guide me through recent advances on intermolecular potentials, lithium-ion batteries, and biomedical research, thus opening-up new paths for scientific and personal endeavors for the rest of my life.

Acknowledgements The work was greatly benefited from the RISE/FRAMED project no. 734485 (https://cordis.europa.eu/project/rcn/207050/factsheet/en) for which Aristotle University of Thessaloniki (AUTh) acts as coordinator. In this connection, thanks are extended to all beneficiary nodes and international partners of FRAMED. The gradient fluids section is a topic of the RISE/ATM2BT project no. 824022 (https://cordis.europa.eu/project/rcn/219192/factsheet/en) for which AUTh is a beneficiary. This section was included in anticipation of follow-up joint work between AUTh, Akita Univeristy/Japan and Aston University/UK (which acts as project's ATM2BT coordinator). The remaining of the sections were benefited from discussions with my former classmate C. Vayenas, my daughter K.E. Aifantis and my Ph.D. student K. Parisis, as well as by the continuous support of my former Ph.D. students A. Konstantinidis (successor of my Lab in Thessaloniki) and I. Tsagrakis (current collaborator in Crete). Finally, the author is also gratefully acknowledging the support of Deutsche Forschungsgemeinschaft/DFG grant No. 377472739/GRK 2423/1-2019 at Friedrich-Alexander University (FAU) where he has recently been appointed as a Mercator Fellow.

References

1. Aifantis EC (2016) Internal length gradient (ILG) material mechanics across scales and disciplines. Adv Appl Mech 49:1–110
2. Aifantis EC (1984) On the microstructural origin of certain inelastic models. J Eng Mater Technol 106:326–330
3. Aifantis EC (1987) The physics of plastic deformation. Int J Plast 3:211–247
4. Aifantis EC (1992) On the role of gradients in the localization of deformation and fracture. Int J Eng Sci 30:1279–1299

5. Aifantis EC (1995) Pattern formation in plasticity. Int J Eng Sci 33:2161–2178
6. Aifantis EC (2009) On scale invariance in anisotropic plasticity, gradient plasticity and gradient elasticity. Int J Eng Sci 47:1089–1099
7. Aifantis EC (2011) On the gradient approach - relation to Eringen's nonlocal theory. Int J Eng Sci 49:1367–1377
8. Aifantis EC (2011) Gradient nanomechanics: applications to deformation, fracture, and diffusion in manopolycrystals. Metall Mater Trans A 42:2985–2998
9. Askes H, Aifantis EC (2011) Gradient elasticity in statics and dynamics: an overview of formulations, length scale identification procedures, finite element implementations and new results. Int J Solids Struct 48:1962–1990
10. Aifantis EC (2014) Gradient material mechanics: perspectives and prospects. Acta Mech 225:999–1012
11. Aifantis EC, Serrin JB (1983) The mechanical theory of fluid interfaces and Maxwell's rule. J Coll Inter Sci 96:517–529
12. Aifantis EC, Serrin JB (1983) Equilibrium solutions in the mechanical theory of fluid microstructures. J Coll Inter Sci 96:530–547
13. Van der Waals JD (1895) Théorie thermodynamique de la capillarité, dans l'hypothèse d'une variation continue de densité. Arch Neerl Sci Exactes Nat 28:121–209
14. Ter Haar D (Ed) (1965) Collected papers of L.D. Landau. Pergamon, London
15. Cahn JW, Hilliard JE (1958) Free energy of a nonuniform system. I. Interfacial free energy. J Chem Phys 28:258–267
16. Cahn JW (1959) Free energy of a nonuniform system. II. Thermodynamic basis. J Chem Phys 30:1121–1124
17. Kevrekidis IG, Gear CW, Hyman JM, Kevrekidis PJ, Runborg O, Theodoropoulos C (2003) Equation-free, coarse-grained multiscale computation: enabling macroscopic simulators to perform system-level analysis. Comm Math Sci 1:715–762
18. Kevrekidis IG, Samaey G (2009) Equation-free multiscale computation: algorithms and applications. Annu Rev Phys Chem 60:321–344
19. Tsallis C (1988) Possible generalization of Boltzmann-Gibbs statistics. J Stat Phys 52:479–487
20. Tsallis C (2009) Entropy. In: Meyers RA (Ed) Encyclopedia of complexity and systems science. Springer, New York
21. Tsallis C (2009) Introduction to nonextensive statistical mechanics: approaching a complex world. Springer, Berlin
22. Greer JR, de Hosson JThM (2011) Plasticity in small-sized metallic systems: intrinsic versus extrinsic size effect. Prog Mat Sci 56:654–724
23. Aifantis KE, Hackney SA (eds) (2010) High energy density lithium batteries: materials. Engineering, Applications (Wiley-VCH
24. Ryu I, Choi JW, Cui Y, Nix Y (2011) Size-dependent fracture of Si nanowire battery anodes. J Mech Phys Solids 59:1717–1730
25. Cui Z, Gao F, Qu J (2013) Interface-reaction controlled diffusion in binary solids with applications to lithiation of silicon in lithium-ion batteries. J Mech Phys Solids 61:293–310
26. Cheng YT, Verbrugge MW, Desphande R (2013) Understanding diffusion-induced stresses in lithium ion battery electrodes, In: Kocks A, Wang J (Eds) IUTAM symposium on surface effects in the mechanics of nanomaterials and heterostrucures. Springer, Dordrecht
27. Walgraef D, Aifantis EC (1985) Dislocation patterning in fatigued metals as a result of dynamical instabilities. J Appl Phys 58:688–691
28. Pontes J, Walgraef D, Aifantis EC (2006) On dislocation patterning: multiple slip effects in the rate equation approach. Int J Plasticity 22:1486–1505
29. Spiliotis KG, Russo L, Siettos C, Aifantis EC (2018) Analytical and numerical bifurcation analysis of dislocation pattern formation of the Walgraef-Aifantis model. Int J Non-Linear Mech 102:41–52
30. Hatzikirou H, Basanta D, Simon M, Schaller K, Deutsch A (2010) 'Go or Grow': the key to the emergence of invasion in tumour progression? Math Med Biol 29:49–65

31. Boettger K, Hatzikirou H, Voss-Böhme A, Cavalcanti-Adam EA, Herrero MA, Deutsch A (2015) An emerging Allee effect is critical for tumor initiation and persistence. PLoS Comp Biol 11:E1004366
32. Murray JD (2002) Mathematical biology I: an introduction. Springer, New York
33. Murray JD (2003) Mathematical Biology II: spatial models and biomedical applications. Springer, New York
34. Aifantis EC, Hirth JP (eds) (1985) The mechanics of dislocations. ASM, Metals Park
35. Aifantis EC, Walgraef D, Zbib HM (Eds) Material instabilities. Special Issue of Res Mechanica 23:97–305
36. Estrin Y, Kubin LP, Aifantis EC (1993) Introductory remarks to the viewpoint set in propagative plastic instabilities. Scripta Met Mater 29:1147–1150
37. Aifantis EC (2003) Update on a class of gradient theories. Mech Mater 35:259–280
38. Kubin LP (1993) Dislocation patterning. In: Mughrabi H (Ed) Plastic deformation and fracture of materials. WILEY-VCH
39. Kubin LP, Fressengeas C, Ananthakrishna G (2002) Collective behaviour of dislocations in plasticity. In: Nabarro FRN and Duesbery MS (Eds) Dislocations in solids. Elsevier
40. Ananthakrishna G (2007) Current theoretical approaches to collective behavior of dislocations. Phys Rep 440:113–259
41. Sauzay M, Kubin LP (2011) Scaling laws for dislocation microstructures in monotonic and cyclic deformation of fcc metals. Prog Mater Sci 56:725–784
42. Carpinteri A (ed) (1996) Size-scale effects in the failure mechanisms of materials and structures. CRC Press
43. Muhlhaus HB (ed) (1995) Continuum models for materials with microstructure. Wiley, Chichester
44. de Borst R, van der Giessen E (eds) (1998) Material instabilities in solids. Wiley, Chichester
45. Gutkin MY, Aifantis EC (1999) Dislocations and disclinations in gradient elasticity. Phys Stat Sol B 214:245–284
46. Lazar M, Maugin GA, Aifantis EC (2006) Dislocations in second strain gradient elasticity. Int J Sol Struct 43:1787–1817
47. Aifantis EC (2014) On non-singular GRADELA crack fields. Theor App Mech Lett 4:051005
48. Aifanti EC, Gittus J (eds) (1986) Phase transformations. Elsevier, New York
49. Suresh S (1991) Fatigue of materials. Cambridge University Press, Cambridge
50. Walgraef D (1997) Spatio-temporal pattern formation. Springer, New York
51. Gutkin MY, Ovid'ko IA (2004) Plastic deformation in nanocrystalline materials. Springer, Berlin
52. Ghoniem N, Walgraef D (2008) Instabilities and self-organization in materials. Oxford Science Publications, Oxford
53. Gurtin ME, Fried E, Anand L (2010) The mechanics and thermodynamics of continua. Cambridge University Press, New York
54. Po G, Lazar M, Seif D, Ghoniem N (2014) Singularity-free dislocation dynamics with strain gradient elasticity. J Mech Phys Solids 68:161–178
55. Isaksson P, Dumont PJJ, du Roscoat SR (2012) Crack growth in planar elastic fiber materials. Int J Solids Struct 49:1900–1907
56. Isaksson P, Hägglund R (2013) Crack-tip fields in gradient enhanced elasticity. Eng Fract Mech 97:186–192
57. Bagni C, Askes H, Aifantis EC (2017) Gradient-enriched finite element methodology for axisymmetric problems. Acta Mech 228:1423–1444
58. Tsagrakis I, Aifantis EC (2018) Gradient elasticity effects on the two-phase lithiation of LIB anodes. In: Altenbach H, Pouget J, Rousseau M, Collet B, Michelitsch T (Eds) Generalized models and non-classical approaches in complex materials 2. Springer
59. Konstantinidis AA, Aifantis KE, de Hosson JThM (2014) Capturing the stochastic mechanical behavior of micro and nanopillars. Mater Sci Eng, A 597:89–94
60. Konstantinidis AA, Zhang X, Aifantis EC (2015) On the combined gradient-stochastic plasticity model: application to Mo-micropillar compression. AIP Conf Proc 1646:3–9

61. Zaiser M, Avlonitis M, Aifantis EC (1998) Stochastic and deterministic aspects of strain localization during cyclic plastic deformation. Acta Mater 48:4143–4151
62. Avlonitis M, Zaiser M, Aifantis EC (2000) Some exactly solvable models for the statistical evolution of internal variables during plastic deformation. Prob Eng Mech 15:131–138
63. Chattopadhyay AK, Aifantis EC (2016) Stochastically forced dislocation density distribution in plastic deformation. Phys Rev E 94:022139
64. Chattopadhyay AK, Aifantis EC (2017) Double diffusivity model under stochastic forcing. Phys Rev E 95:052134
65. Zaiser M, Aifantis EC (2003) Avalanches and slip patterning in plastic deformation. J Mech Behav Mater 14:255–270
66. Zaiser M, Aifantis EC (2006) Randomness and slip avalanches in gradient plasticity. Int J Plasticity 22:1432–1455
67. Li H, Ngan AHW, Wang MG (2005) Continuous strain bursts in crystalline and amorphous metals during plastic deformation by nanoindentation. J Mater Res 20:3072–3081
68. Iliopoulos AC, Nikolaidis NS, Aifantis EC (2015) Analysis of serrations and shear bands fractality in UFGs. J Mech Behav Mater 24:1–9
69. Iliopoulos AC, Aifantis EC (2018) Tsallis q-triplet, intermittent turbulence and Portevin-Le Chatelier effect. Phys A 498:17–32
70. Kawazoe H, Yoshida M, Basinski ZS, Niewczas M (1999) Dislocation microstructures in fine-grained Cu polycrystals fatigued at low amplitude. Scripta Mater 40:639–644
71. Wang D, Volkert CA, Kraft O (2008) Effect of length scale on fatigue life and damage formation in thin Cu films. Mat Sci Eng A 493:267–273
72. Unger DJ, Gerberich WW, Aifantis EC (1982) Further remarks on the implications of steady state stress assisted diffusion on environmental cracking. Scripta Metall 16:1059–1064
73. Silber G, Trostel R, Alizadeh M, Benderoth G (1998) A continuum mechanical gradient theory with applications to fluid mechanics. J de Phy 4(8):365–373
74. Fried E, Gurtin ME (2006) Tractions, balances, and boundary conditions for nonsimple materials with application to liquid flow at small-length scales. Arch Rat Mech Anal 182:513–554
75. Adams JM, Fielding SM, Olmsted PD (2008) The interplay between boundary conditions and flow geometries in shear banding: hysteresis, band configurations, and surface transitions. J Nonnewton Fluid Mech 151:101–118
76. Cates ME, Fielding SM (2006) Rheology of giant micelles. Adv Phys 55:799–879
77. Dhont JKG, Briels WJ (2008) Gradient and vorticity banding. Rheol Acta 47:257–281
78. Ru CQ, Aifantis EC (1993) A simple approach to solve boundary-value problems in gradient elasticity. Acta Mech 101:59–68
79. Giusteri GG, Fried E (2014) Slender-body theory for viscous flow via dimensional reduction and hyperviscous regularization. Meccanica 49:2153–2167
80. Vardoulakis I, Aifantis EC (1989) Gradient dependent dilatancy and its implications in shear banding and liquefaction. Ingenieur-Archiv 59:197–208
81. Vardoulakis I, Muhlhaus HB, Aifantis EC (1991) Continuum models for localized deformations in pressure sensitive materials. In: Beer G, Booker JR, Carter J (Eds) Computer methods and advances in geomechanics. Balkema Publishers, Rotterdam
82. Vardoulakis I, Aifantis EC (1991) A gradient flow theory of plasticity for granular materials. Acta Mech 87:197–217
83. Vardoulakis I, Aifantis EC (1994) On the role of microstructure in the behavior of soils: Effects of higher order gradients and internal inertia. Mech Mat 18:151–158
84. Oka F, Yashima A, Sawada K, Aifantis EC (2000) Instability of gradient-dependent elasto-viscoplastic model for clay and strain localization. Comp Method Appl Mech Eng 183:67–86
85. di Prisco C, Imposimato S, Aifantis EC (2002) A visco-plastic constitutive model for granular soils modified according to non-local and gradient approaches. Int J Num Anal Meth Geomech 26:121–138
86. Fyffe B, Schwerdtfeger J, Blackford JR, Zaiser M, Konstantinidis A, Aifantis EC (2007) Fracture toughness of snow: the influence of layered microstructure. J Mech Behav Mater 18:195–215

87. Konstantinidis A, Cornetti P, Pugno N, Aifantis EC (2009) Application of gradient theory and quantized fracture mechanics in snow avalanches. J Mech Behav Mater 19:39–48
88. Haoxiang C, Qi C, Peng L, Kairui L, Aifantis EC (2015) Modeling the zonal disintegration of rocks near deep level tunnels by gradient internal variable continuous phase transition theory. J Mech Behav Mater 24:161–171
89. Qi C, Wei X, Hongsen W, Aifantis EC (2015) On temporal-structural dynamic failure criteria for rocks. J Mech Behav Mater 24:173–181
90. Efremidis G, Avlonitis M, Konstantinidis A, Aifantis EC (2017) A statistical study of precursor activity in earthquake-induced landslides. Comput Geotechn 81:137–142
91. Chen H, Qi C, Efremidis G, Dorogov M, Aifantis EC (2018) Gradient elasticity and size effect for the borehole problem. Acta Mech 229:3305–3318
92. Ord A, Hobbs BE (2010) Fracture pattern formation in frictional, cohesive, granular material. Philos Trans R Soc A 368:95–118
93. Yue YM, Xu KY, Aifantis EC (2014) Microscale size effects on the electromechanical coupling in piezoelectric material for anti-plane problem. Smart Mater Struct 23:125043
94. Yue YM, Xu KY, Chen T, Aifantis EC (2015) Size effects on magnetoelectric response of multiferroic composite with inhomogeneities. Phys B 478:36–42
95. Yue YM, Xu KY, Aifantis EC (2015) Strain gradient and electric field gradient effects in piezoelectric cantilever beams. J Mech Behav Mater 24:121–127
96. Tarasov VE, Trujillo JJ (2013) Fractional power-law spatial dispersion in electrodynamics. Annal Phys 334:1–23
97. Truesdell C, Toupin R (1960) The classical field theories. In: Flügge S (Ed) Principles of classical mechanics and field theory/Prinzipien der Klassischen Mechanik und Feldtheorie. Springer, Berlin
98. Zimmerman JA, Webb EB, Hoyt JJ, Jones RE, Klein PA, Bammann DJ (2004) Calculation of stress in atomistic simulation. Model Simul Mater Sci Eng 12:S319–S332
99. Maranganti R, Sharma P (2010) Revisiting quantum notions of stress. Proc Royal Soc A 466:2097–2116
100. Davies H (2000) The physics of low-dimensional semiconductors. Cambridge University Press, Cambridge
101. Zhang X, Gharbi M, Sharma P, Johnson HT (2009) Quantum field induced strains in nanostructures and prospects for optical actuation. Int J Solids Struct 46:3810–3824
102. Vayenas CG, Souentie S (2012) Gravity, Special Relativity, and the Strong Force. Springer, Boston
103. Vayenas CG, Souentie S, Fokas A (2014) A Bohr-type model of a composite particle using gravity as the attractive force. Phys A 405:360–379
104. London F (1930) Zur Theorie und Systematik der Molekularkräfte. Z Physik 63:245–279
105. London F (1937) The general theory of molecular forces. Trans Faraday Soc 33:8–26
106. Jones JE (1924) On the determination of molecular fields I. From the variation of the viscosity of a gas with temperature. Phil Trans A 106:441–462
107. Israelachvili JN (2011) Intermolecular and surface forces. Academic Press
108. Parson JM, Siska PE, Lee YT (1972) Intermolecular potentials from crossed-beam differential elastic scattering measurements. IV. Ar+Ar. J Chem Phys 56:1511–1516
109. Stillinger FH, Weber TA (1985) Computer simulation of local order in condensed phases of silicon. Phys Rev B 31:5262–5271
110. Lazar M, Maugin GA, Aifantis EC (2006) On the theory of nonlocal elasticity of bi- Helmholtz type and some applications. Int J Solids Struct 43:1404–1421
111. Kioseoglou J, Dimitrakopulos GP, Komninou Ph, Karakostas T, Aifantis EC (2008) Dislocation core investigation by geometric phase analysis and the dislocation density tensor. J Phys D 41:035408
112. Aifantis EC (2009) Non-singular dislocation fields. IOP Conf. Series 3:0712026
113. Tarasov VE, Aifantis EC (2014) Toward fractional gradient elasticity. J Mech Behav Mater 23:41–46

114. Tarasov VE, Aifantis EC (2015) Non-standard extensions of gradient elasticity: fractional non-locality, memory and fractality. Commun Nonlinear Sci Numer Simulat 22:197–227
115. Aifantis EC (2019) Fractional generalizations of gradient mechanics, In: Tarasov VE (Ed) Handbook of fractional calculus with applications. De Gruyter, Berlin
116. Tarasov VE, Aifantis EC (2019) On fractional and fractal formulations of gradient linear and nonlinear elasticity. Acta Mech 230:2043–2070
117. Parisis K, Konstantopoulos I, Aifantis EC (2018) Nonsingular solutions of GradEla models for dislocations: an extension to fractional GradEla. J Micromech Mol Phys 3:1840013
118. Samko S, Kilbas A, Marichev O (1987) Integrals and derivatives of fractional order and applications. Nauka i Tehnika, Minsk
119. Kilbas A, Srivastava M, Trujillo J (2006) Theory and applications of fractional differential equations. Elsevier
120. Mathai A, Saxena RK, Haubold HJ (2010) The H-function: theory and applications. Springer, New York
121. Eringen AC (1999) Microcontinuum field theories I: foundations and solids. Springer, New York
122. Fleck NA, Hutchinson JW (1997) Strain gradient plasticity. Adv Appl Mech 33:295–361
123. Fleck NA, Hutchinson JW (2001) A reformulation of strain gradient plasticity. J Mech Phys Solids 49:2245–2271
124. Gurtin ME, Anand L (2009) Thermodynamics applied to gradient theories involving the accumulated plastic strain: the theories of Aifantis and Fleck and Hutchinson and their generalization. J Mech Phys Solids 57:405–421
125. Gao HJ, Huang Y, Nix WD, Hutchinson JW (1999) Mechanism-based strain gradient plasticity - I. Theory J Mech Phys Solids 47:1239–1263
126. Nix WD, Gao HJ (1998) Indentation size effects in crystalline materials: a law for strain gradient plasticity. J Mech Phys Solids 46:411–425
127. de Borst R, Muhlhaus HB (1992) Gradient-dependent plasticity - formulation and algorithmic aspects. Int J Numer Method Eng 35:521–539
128. de Borst R, Pamin J, Sluys LJ (1995) Computational issues in gradient plasticity, In: Mühlhaus HB (Ed) Continuum models for materials with microstructure. Wiley, pp 159–200
129. Geers MGD, Peerlings RHJ, Brekelmans WAM, de Borst R (2000) Phenomenological nonlocal approaches based on implicit gradient-enhanced damage. Acta Mech 144:1–15
130. Peerlings RHJ, Poh LH, Geers MGD (2012) An implicit gradient plasticity-damage theory for predicting size effects in hardening and softening. Eng Fract Mech 95:2–12
131. Willis JR (2019) Some forms and properties of models of strain-gradient plasticity. J Mech Phys Solids 123:348–356
132. Aifantis KE, Willis JR (2005) The role of interfaces in enhancing the yield strength of composites and polycrystals. J Mech Phys Solids 53:1047–1070
133. Polizzotto C (2003) Unified thermodynamic framework-for nonlocal/gradient continuum theories. Eur J Mech A Solid 22:651–668
134. Polizzotto C (2009) Interfacial energy effects within the framework of strain gradient plasticity. Int J Solids Struct 46:1685–1694
135. Voyiadjis GZ, Song Y (2019) Strain gradient continuum plasticity theories: Theoretical, numerical and experimental investigations. Int J Plasticity 121:21–75
136. Goddard JD (2018) On linear non-local thermo-viscoelastic waves in fluids. Mat Mech Compl Sys 6:321–338
137. Goddard JD (2017) On the stability of the $\mu(I)$ rheology for granular flow. J Fluid Mech 833:302–331
138. Kamrin K, Koval G (2012) Nonlocal constitutive relation for steady granular flow. Phys Rev Lett 108:178301
139. Henann DL, Kamrin K (2013) A predictive, size-dependent continuum model for dense granular flows. Proc Natl Acad Sci USA 110:6730–6735
140. Forterre Y, Pouliquen O (2008) Flows of dense granular media. Annu Rev Fluid Mech 40:1–24
141. Fenistein D, van Hecke M (2003) Wide shear zones in granular bulk flow. Nature 425:256

142. Dijksman JA, Wortel GH, van Dellen LTH, Dauchot O, van Hecke M (2011) Jamming, yielding, and rheology ofweakly vibrated granular media. Phys Rev Lett 107:108303
143. Bocquet L, Colin A, Ajdari A (2009) Kinetic theory of plastic flow in soft gassy materials. Phys Rev Lett 103:036001
144. Fischbach E, Sudarsky D, Szafer A, Talmadge C, Aronson SH (1986) Reanalysis of the Eötös experiment. Phys Rev Lett 56:3–6
145. Fischbach E (2015) The fifth force: a personal history. Eur Phys J H 40:385–467
146. Bardhan JP (2013) Gradient models in molecular biophysics: progress, challenges, opportunities. J Mech Behav Mater 22:169–184

The manufacturer's authorised representative in the EU is Springer Nature Customer Service Centre GmbH, Europaplatz 3, 69115 Heidelberg, Germany. If you have any concerns regarding our products, please contact ProductSafety@springernature.com

Printed and bound by CPI Group (UK) Ltd, Croydon, CR0 4YY
30/09/2026
02225695-0016